Elsevier
Materials Selector
Vol 2

Elsevier
Materials Selector

Vol 2

Edited by

Norman A. Waterman, B.Sc., Ph.D.
Chief Executive, Quo-Tec Ltd,
Amersham, Buckinghamshire, UK

and

Michael F. Ashby, F.R.S.
Professor of Engineering,
University of Cambridge, UK

Elsevier Applied Science
London

ELSEVIER SCIENCE PUBLISHERS LTD
Crown House, Linton Road, Barking, Essex, IG11 8JU, England, UK

Published in the USA and Canada as *CRC–Elsevier Materials Selector* by CRC Press, Inc.,
2000 Corporate Blvd., N.W., Boca Raton, Florida 33431, USA

WITH 777 TABLES AND 441 ILLUSTRATIONS

British Library Cataloguing in Publication Data
  Elsevier materials selector.
  1. Materials. Selection
  I. Waterman, N. A. (Norman Allan) *1941*– II. Ashby,
  Michael F.
  620.11
  ISBN 1 85166 605 2

ISBN 1 85166 606 0 (Vol. 1)
ISBN 1 85166 607 9 (Vol. 2)
ISBN 1 85166 608 7 (Vol. 3)
ISBN 1 85166 605 2 (Set)

Library of Congress Cataloging-in-Publication Data
CRC–Elsevier materials selector / edited by Norman A. Waterman and
  Michael F. Ashby.
      p.   cm.
    ISBN 0–8493–7790–0 (set)
      1. Materials.   I. Waterman, Norman A.   II. Ashby, M. F.
  TA403.C733   1991   91–9391
  620.1'1—dc20   CIP

ISBN 0 8493 7791 9 (Vol. 1)
ISBN 0 8493 7792 7 (Vol. 2)
ISBN 0 8493 7793 5 (Vol. 3)
ISBN 0 8493 7790 0 (Set)

Indexed by Paul Nash MSc BTech

Typeset in Great Britain by Variorum Publishing Limited, Northampton

Printed in Great Britain at the Alden Press Ltd, Oxford

Preface

The main aim of these three volumes is to provide a system and the necessary information for the selection and specification of engineering materials and related component manufacturing processes.

These volumes are intended for use by designers, materials engineers and production engineers who are seeking to identify the most suitable materials and manufacturing methods for a specific application.

Throughout the volumes, extensive use has been made of tabular and graphical information to facilitate the comparison of candidate materials for a specific application. Every material and process is described in terms of what it will do for the component designer and product maker and no detailed knowledge of metallurgy, polymer chemistry or materials science is assumed or necessary.

The information is arranged in order of increasing detail with the aim that no section need be read unless it is likely to be of direct relevance to the reader and his quest; whether this is the selection of material for a new product or the search for a substitute material.

Disclaimer

Whilst every effort has been made to check the accuracy of the information contained in these volumes, no material should ever be selected and specified for a component or product on a paper exercise alone. The purpose of these volumes is to provide enough information for a short list of candidates for testing and to reduce the number of fruitless tests. No liability can be accepted for loss or damage resulting from the use of information contained herein.

How to use the Elsevier Materials Selector

There will be two main reasons for using this information system.

(a) To select and specify materials and manufacturing routes for a new product.

(b) To evaluate alternative materials or manufacturing routes for an existing product.

The method of using the *Elsevier Materials Selector* for each of these purposes is as follows:

(a) Selection and Specification of Materials and Manufacturing Routes for a New Product

1. Define the function of the product and translate into materials requirements of strength, stiffness, corrosion and wear resistance, etc. (see Volume 1, Chapters 1.1–1.7).

2. Define the production requirements in terms of number required, tolerances, surface finish, etc.

3. Search for possible combinations of materials and production routes using Volume 1 and compile a short list according to performance/cost relationship.

4. Investigate candidate materials in more detail using Volume 2 for metallic materials and ceramics, and Volume 3 for plastics, elastomers and composites.

5. Specify optimum materials and processing routes.

(b) Evaluation of Alternative Materials and Manufacturing Routes for an Existing Product

1. Characterise currently used materials in terms of performance (see Volume 2 for metallic materials and ceramics, and Volume 3 for plastics, elastomers and composites), manufacturing requirements and cost (from in-house data).

2. Evaluate which characteristics are necessary for product function (see Volume 1, Chapter 1.1).

3. Search for alternative materials and, if permissible, alternative manufacturing routes (using Volume 1).

4. Compile short list of materials and manufacturing routes and estimate costs.

5. Compare existing materials and production routes with alternatives.

Acknowledgements and history of the publication

The origins of the *Elsevier Materials Selector* may be traced back to the early 1970s. At that time I was employed as a materials engineer in industry and perceived the need for a selection system which would compare the performance and cost of materials in terms which could be understood by designers and production engineers without the benefit of degrees in metallurgy, polymer chemistry, materials science, etc. I am grateful to my employers at that time, Danfoss A/S, for providing a stimulating working environment in which the idea was born.

The Fulmer Research Institute, my next employer, provided the financial and technical resources to convert the idea into reality; the *Fulmer Materials Optimizer* which was published in 1976. I am particularly grateful for the support and encouragement of Dr W. E. Duckworth, Managing Director and Mr M. A. P. Dewey (then Assistant Director) of Fulmer. As the first editor of the *Optimizer*, I am acutely aware that the original production and successful launch would have been impossible without the insight and hard work of the contributors of individual sections, in particular:

Dr T. J. Baker	— Steels
Mr G. B. Brook	— Aluminium alloys
Mr J. N. Cheetham	— Polyurethanes
Mr D. G. S. Davies	— Ceramics and unit conversion tables
Dr H. Deighton	— Mechanical properties
Mr D. W. Mason	— Copper alloys
Mr V. Micuksi	— Surface coatings
Mr M. J. Neale	— Wear
Mr R. Newnham	— Ceramics
Mr G. Sanderson	— Corrosion
Mr J. A. Shelton	— Nylons & Polyacetals
Mr W. Titov	— PVC

who produced work of exceptional quality.

It is also a pleasure to acknowledge the painstaking and thorough work of Mr A.M. Pye who undertook, in 1979, the first major review and updating of the *Optimizer*.

Between 1979 and 1987, my only relationship with the *Optimizer* was that of user of the system in support of my activities as a consultant on materials selection and specification. In the intervening period, the *Optimizer* was edited by Dr M.A. Moore, Dr U. Lenel and Mr L. Wyatt at Fulmer. A particularly valuable section on adhesive bonding was added by Mr W. A. Lees at this time.

In 1988, Elsevier purchased the rights of the *Fulmer Materials Optimizer* and invited me to undertake the task of converting the information in the 1987 edition of the *Optimizer* into a new three-volume materials selection to be known as the *Elsevier Materials Selector*.

I am very grateful to the publishers for this opportunity and also wish to thank the following sub-editors for their efforts in updating and checking individual sections.

Volume 2, Chapter 2.1 Wrought steels—Dr T.J. Baker (*Imperial College*)
Volume 3—Dr James Maxwell (*Formerly ICI Advanced Materials*)
     Dr David Wright (*Technical Director, RAPRA*)
Also for Volumes 1–3   Amanda White,
              Mathew Poole,
              Martin Smith,
              and Michael Weston
              (*all of Quo-Tec*)

It is a special pleasure to acknowledge the help and inspiration of my associate editor, Professor M. F. Ashby, whose work in recent years has created the ideal introduction to the *Elsevier Materials Selector*.

Last, but not least, I wish to thank my wife, Margaret, for patience, sacrifice and support, without which I could not have started, let alone finished.

N. A. Waterman

Contents

Vol 2

Wrought steels

Contents

List of tables

List of figures

2.1.1 Introduction

The range of strengths which can be obtained from steels exceeds that exhibited by any other material class, and the specific strength is exceeded only by titanium alloys and advanced composites. The modulus is high and specific stiffness is exceeded only by molybdenum, titanium, nickel and cobalt alloys, some cast irons and advanced composites. The toughness of steel is also superior to most other materials given correct selection of composition and treatment.

Steel components may be manufactured by a variety of techniques which include casting, forging, hot and cold rolling, cold forming and powder forging. Joining is possible by a wide variety of techniques.

Steels are subdivided into types based upon method of manufacture, hardening or characteristic properties. For each type the approximate strength range, the advantages and limitations, are compared in:

Table 2.1.1 *Comparison of available steel classes*

Approximately equivalent international specifications are listed in:

Table 2.1.2 *Equivalent international specifications for wrought steels*

Physical properties of steels are summarised in:

Table 2.1.3 *Physical properties of steels*

TABLE 2.1.1 Comparison of available steel classes

Steel type	Strength range MN/m²	Characteristic properties	
		Advantages	Disadvantages
Hot rolled steel	230–550	Low cost, good formability weldability and machinability. Availability.	Low strength, low toughness.
Cold finished steel	250–1000 (2000 MN/m² patented wire)	Good surface finish, dimensional tolerance, good machinability, high strength at low cost. Availability.	Low toughness, directionality in mechanical properties. High impact transition temperature.
High strength low alloy steel	300–650	High strength, high toughness, good formability, good weldability.	Mostly only available as plate and sheet.
Hardened and tempered steel	400–1850	Wide range of strengths available. High toughness and low impact transition temperature. High temperature strength. Wear resistance.	High cost. Poor weldability and machinability with increasing alloy content.
Case-hardened steel	350–1400 (Surface hardness 750–1100 HV)	Wear resistance combined with high core toughness. Fatigue resistance.	
Stainless steel	200–1400	Austenitic—High toughness, no impact transition temperature, corrosion resistance, creep resistance, non-magnetic.	High cost. Low strength.
		Ferritic— Corrosion resistance. Lower cost than austenitics.	Tendency to grain in coarsening.
		Duplex—Better mechanical properties than austenitics. Corrosion resistance.	
		Martensitic—High strength with moderate corrosion resistance.	
		Precipitation Hardening— Very high strength with corrosion resistance.	Complex heat treatment required.
Tool steel	Hardness 370–990 HV	High strength. Wear resistance.	

Table 2.1.1

TABLE 2.1.1 Comparison of available steel classes—*continued*

Steel type	Strength range MN/m²	Characteristic properties	
		Advantages	Disadvantages
Spring steel	620–2250	High yield strength produced by cold work or heat treatment.	
Intermetallic strengthened steel	700–2400	High strength, high toughness. May be machined in soft condition and subsequently hardened with minimum distortion or oxidation.	High cost.
Valve steel	465–1150 (RT) 170–670 at 600°C	High temperaure strength and low creep with resistance to oxidation and high temperature corrosion.	High cost. Specialist products.

Table 2.1.1—*continued*

TABLE 2.1.2   Equivalent international specifications for wrought steels

Part 1—Carbon and carbon–manganese steels

United States AISI	United Kingdom BS970 Pt1 1983	Federal Republic of Germany DIN Werkstoff Nummer	Federal Republic of Germany DIN Kurzname	France AFNOR	Sweden SIS	Japan JISG4051–65	USSR GOST.1050.60	EN.No.	Classification and Treatment
1005		1.0312	D5–1		14.11.60				
1006 S		1.0312	D8–2						
1008 S	040 A04	1.0204	UQ 5t36	FD		SPHTI S6DI SBC	05KP	2A	Carbon steels usually supplied 'As Rolled'
1009 S		—							
1010 S	040 A10	1.1121	Ck10	XC10	14J1/12/13	S10C, 59CK	08KP	2A/2B	
—		1.0301	C10	C10	1256		1050,08,10	2B	
1012	040 A12					S12C	4231		
1015		1.1141	Ck15	C12, C620		15KPP,15,15G	1050.15		
1016	080M15ab,0,80A15	1.1141	Ck15	XC15 XC18	St5C, SISCK 1370	S17C		3A/3C	
1017									
1018	055 M15a	—							
1019	070 M20b	—							
1020 S	080 A20	1.0402	C22	C20			20	3A/3C	
1021 S		1.1133	21Mn4	XC185 XC25		S20C,S22C,S200K	23570, 18Gsp		
1023 S		1.1151	Ck22				1050, 20		
1025 S		1.1158	Ck25			S25C, S28C		5A	
1026 S	070 M26b	—		C28					
1029		—				S33C, 535C		5	
1030 S	080 M30b	1.172	Cq35						
—									
C 1033		1.1181	C35	XC32,XC35,XC38					
1035 S		1.0501	C35	C35	1550	S38C	1050.35	8A	Carbon steels usually supplied 'Normalised'
1037 S	080 M36 H+								
1038 S		1.1176	Ck38						
1038 H		1.1176	Ck38						
1039 S		1.1157	40Mn4						
1040 S	080 M40b	1.1186	Ck40			S40C		8	
1042 S	080 H41	1.1191	CS–Ck45	XC42,XC45,XC48	1672	S45C, S48C	1050.45		
1043 S		1.0503	C+5	C45	1650				
1044		1.5011	C40						
1045 S		1.1191	Ck45						
1045 H	080 M46 H+	1.1191	Ck45					43B	
1046 S		—						43A	
1049 S	080 M50b	1.1202	Cm45		1600	S50C		43A	
1050 S	080 M50b	1.1210	Cm45			553C, 555C			

Table 2.1.2

TABLE 2.1.2 Equivalent international specifications for wrought steels—*continued*

Part 1—Carbon and carbon–manganese steels—*continued*

AISI	BS 970	Werkstoff No.	DIN	NF	NF No.	JIS	SS	ISO	Remarks
1053	070M55 [b]080M55	1.1273	C55	C55			1050.60	43E	Carbon steels usually supplied 'Softened'
1055 S		1.1209	Cm55						
1059	—	—							
1060 S	060 A62	1.0601	C60	XC60,XC55,XC58	1665, 1678				
C1062		1.1260	66Mn4						
1064 S	060 A67	1.1221	Ck604						
1065 S	080 A67	1.1230	Feder-Stahl-draght FD						
—									
1070 S	060 A72	1.1231	Ck67		1778		1090.75	42	
1074 S		1.0605	C75						
1075 S		—							
1078 S	060 A78	1.1248	Ck75	XC75	1774				
1080 S	060 A78	1.0647	85Mn3						
1084 S	060 A81	1.1273	90Mn4						
1085 S		1.1269	Ck854						
1086 S		1.1273	90Mn4						
1090 S		1.1274	Ck101	XC90					
1095 S									
1108		1.0721	10S20	10F1		SUM32			Carbon and carbon manganese free cutting steels supplied 'As Rolled' 'Normalised' or 'Hardened and Tempered'
1110		1.0702							
1117		1.0722	10SPb20	10PbF2	1922			8M	
1118		1.0723	15S20		1922				
1137		—							
1139	216 M36[b]	—							
1140		—							
1141		1.0726	35S20		1957				
1144	226 M44[b]	—							
1146		1.0727	45S20	45MF4	1973				
1151									
1211									

Table 2.1.2—*continued*

TABLE 2.1.2 Equivalent international specifications for wrought steels—continued

Part 1—Carbon and carbon–manganese steels—continued

United States AISI	United Kingdom BS970 Pt1 1983	Federal Republic of Germany DIN Werkstoff Nummer	Federal Republic of Germany DIN Kurzname	France AFNOR	Sweden SIS	Japan JISG4051–65	USSR GOST.1050.60	EN.No.	Classification and Treatment
1212	220 M07b	1.0711	9S20	S250	1912	SUM21		1A	
1213	230 M07b	1.0715	95Mn28S250						
1215		1.0736	95Mn36		1914				
12L13		1.0718	95MnPb28		1926				
12L14		1.0737	95MnPb36						
1513		—	GS20Mn5	120Mn19		SMnC420			Carbon Manganese steels supplied 'Normalised' or 'Hardened and Tempered'
1522	120M19	1.1133	22Mn6						
1524 S		1.1160	26Mn5						
1526		1.1161	26Mn5						
1527 S	120M28	1.1161	34Mn5	40M5	2120	SMn433	972 35GL	15B	
1536 S	120M36b	1.1166	36Mn5			SMn438			
1541 S		1.1167	52Mn5						
1548 S		1.1226							
1551		—	52Mn5						
1552 S		1.1226							
1561		—	66Mn4						
1566		1.1260							
1330	150M19b	1.1165	GS30Mn5	40M5	2120	SMn433H SCMn2	977 30GSL	14A 14B	
1335	150M28	1.1167	36Mn5			SMn438H SCMn2		15	
1340	150M36b	1.5069	36Mn7						
1345	150M40	1.0912	46Mn7						
	L80M01								Micro alloyed carbon manganese steel
	170H20 170H36 170H41 170H40								Boron steels

a Supplied normalised.
b Also supplied as Bright Bar.
S Supplied as Sections or Flat Products as well as Billet & Bar.
H+ Also supplied to Hardenability Requirements.

Table 2.1.2—continued

TABLE 2.1.2 Equivalent international specifications for wrought steels—continued

Part 2—Through hardening alloy steels

United States AISI	United Kingdom BS970[a] Pt1 1973	Federal Republic of Germany DIN Werkstoff Nummer	Federal Republic of Germany DIN Kurzname	France AFNOR	Sweden SIS	Japan JISG4051–65	USSR GOST.1050.60	EN.No.	Classification
4023	—	—	—	—	—	—	—	16	Molybdenum steels
4024	—	—	—	—	—	—	—		
4027	—	—	—	—	—	—	—		
4028	—	—	—	—	—	—	—		
4037	605 H36	—	—	—	—	—	—		
4047	605 A32, H	—	—	—	—	—	—		
—	605 A37, H	—	—	—	—	—	—		
—	606 M36	—	—	—	—	—	—	16 M	Free cutting
4118 S	708 A30	1.7218	GS25CrMo4	25CD4	2225	SCM420/430	30 Ch M		Chromium Molybdenum steels
4130 S	708 A37, H	1.7220	34CrMo4	35CD4	2234	SCM432	AS38Ch GM		
4137 S						SCCrM3	35ChM		
4140 S	708 M40	1.7225	42CrMo4	45CD4	2244	SCM440(H)	40Ch4	19A	
4142 S	708 A42, H	1.7223	41CrMo4	42CD4TS	2244	SCM440		19C	
4145 S	708 A45, H	—							
4147	708 A47	1.7228	50CrMo4			SCM445 (H)	50Ch4		
4150		1.7228	50CrMo4			SCM445(H)	50Ch4		
4161		—							
—	708 A25								
—	709 A37								
—	709 A40,M	1.7225	42CrMo4	45CD4	2244	SCM440(H)	40Ch4	19	
—	709 A42								
—	720 M22	1.7361	32CrMo12	30CD12	2246		—		
—	722 M24								
—	897 M39	1.8523	39CrMoV13.9	—	—	—	—		Chromium Molybdenum Vanadium steel
4320	817 M40	1.6565	40NiCrMo6			SNCM 439		24	1¾ Nickel Chromium Molybdenum steels
4340 S	817 A37								
	817 A32								

Table 2.1.2—continued

TABLE 2.1.2 Equivalent international specifications for wrought steels—*continued*

Part 2—Through hardening alloy steels—*continued*

United States AISI	United Kingdom BS970ª Pt1 1973	Federal Republic of Germany DIN Werkstoff Nummer	Federal Republic of Germany DIN Kurzname	France AFNOR	Sweden SIS	Japan JISG4051–65	USSR GOST.1050.60	EN.No.	Classification
	826 M31	1.6743	32NiCrMo10 4					25	
	826 M40	1.6745	40NiCrMo10 4					26	
	835 M30	1.6747	30NiCrMo16 6	30ND16				30B	
4615 S	—	—	—		—	—	—		Nickel Molybdenum Steels
4620 S	—	—	—		—	—	—		
4626	—	—	—		—	—	—		
4718	945 M38	1.6755	22NiMoCr47					100	1% Nickel Chromium Molybdenum Steel
4720		—	—						
4815	—	—							3½% Nickel Molybdenum Steel
4817	—	—							
4820		—							
50 B44									½% Chromium Boron Steel
50 B46		1.7138	52 MnCrB3						
50 B50									
50 B60									
5117		1.7016							
5120		1.7147							
5130		1.7030				S Cr 430(H)	30Ch		
5132		1.7033		32C4		S Cr 435(H)	35Ch		Chromium Steels
5135		1.7034		38C4		S Cr 440(H)		18	
5140		1.7035		42C4			40Ch		
5147		—							
5150		—							
5155		1.7176		55C3		SUP 9(A)	—		
5160 S		—							
51B60		—							
E51100	535 A99	1.3505		100C6	2258	SUS 2	S Ch Ch 15	31	1% Carbon Chrome
E52100									
6118		1.7511							Nickel Chromium Molybdenum Vanadian Steel
6150 S									

Table 2.1.2—*continued*

TABLE 2.1.2 Equivalent international specifications for wrought steels—*continued*

Part 2—Through hardening alloy steels—*continued*

SAE/AISI							Type of steel
8115	—						0.3% Nickel Chromium Molybdenum Steel
81B45	—						
8615 S	1.6523	21 NiCrMo 2	20 NCD 4	2506	SNCM 220(H)		½% NiCrMo ½% Nickel Chromium Molybdenum Steel
8617 S	1.6523	40 NiCrMo 2	—	—	—		
8620 S	1.6543	21 NiCrMo 22	—	—	—		
8622 S	—						
8625 S	1.6545	30 NiCrMo 22					
8630 S	—						
8637 S	—						
8640 S	1.6546	40 NiCrMo 22			SNCM 240		
8642	—						
8645 S	—						
8655 S	—						
8720	1.6543	21 NiCrMo 22					Nickel Chromium 0.25 Molybdenum Steel
8740	1.6546	40 NiCrMo 22					
8822	1.6543	21 NiCrMo 22					Nickel Chromium 0.35 Molybdenum Steel
9260	1.0909	60 Si 7	60 S 7			6052	Silico-Mar Steel
94B17	—	—					Nickel Chromium Molybdenum Boron Steel
94B30	—	—					

aBS steels may be supplied to close limits of chemical composition (4th character 'A') to hardenability requirements (4th character H) or to mechanical property requirements (4th character M). Where, for a specific steel, more than one version may be supplied one full specification number is included but alternative versions are indicated by additional letters following.

Table 2.1.2—*continued*

TABLE 2.1.2 Equivalent international specifications for wrought steels—*continued*

Part 3—Carbon and carbon–manganese steels for case hardening

United States AISI	United Kingdom BS970[a] Pt1 1983	Federal Republic of Germany		France AFNOR	Sweden SIS	Japan JISG4051–65	USSR GOST.1050.60	EN.No.	Classification
		Werkstoffe Nummer	Kurzname						
1009, 1010, 1012	045M10 A	1.0305 1.0345	St. 35.8 HI	XC10F C10D C12D	14 13 31 14 13 32		10	32A	.10 Carbon
1015, 1016	080M15 A	1.0419	R St. 44-2	XC12F	14 13 70 14 21 01	S 15C	14G 15G	32C	.15 Carbon
	210M15 A	1.0723	15S20	12MF4	14 19 22			32M	.15 Carbon, free cutting
1513	130M15 A				14 14 31 14 14 32 14 21 01		14G 15G	201	.15 Carbon 1.3Mn
1118	214M15			20MF4	14 19 22			202	.15 Carbon 1.4Mn free cutting
	170H15 173H16 174H20 175H23								Boron steels

Table 2.1.2—*continued*

TABLE 2.1.2 Equivalent international specifications for wrought steels—*continued*

Part 4—Alloy steels for case hardening

United States AISI/SAE	United Kingdom BS970[a] Pt1 1983	Federal Republic of Germany		France AFNOR	Sweden SIS	Japan JISG4051–65	USSR GOST.1050.60	EN.No.	Subclassification
		Werkstoffe Nummer	Kurzname						
	523 M15, H	1.7015		18C3		G4104 Scr22		206	½% Chromium
5120	527 A17, H	1.7147	20MnCr5	18C3		Scr22	20KH	207	¾% Chromium
	590 A15 / 590 M17								1% Chromium
	635 M14, A, H						20KHN	351	¾% Nickel–Chromium
	637 A16						20KHN	352	1% Nickel–Chromium
	637 M17, H	1.5752		16NCD5					
	655 M13, H	1.5752	14NiCr14	10NC12 / 12NC15		G4120	12KHN3A	36A	3¼% Nickel–Chromium
	655 M17	1.5860	14NiCr18	12NC15		SNC22	20KH2N4A	39A	4% Nickel–Chromium
4617	655 M17			15NDB / 20NDB	14 25 20	SNCM9	15NM	34	1¾% Nickel–Chromium
							15NMA		
4620	655 M20, H			20NDB			20NM	35A	
	665 M23, H			20NDB			20NM	35	
	708 M20						20NM	35B	½% Chromium Molybdenum

Table 2.1.2—*continued*

TABLE 2.1.2 Equivalent international specifications for wrought steels—*continued*

Part 4—Alloy steels for case hardening—*continued*

United States AISI/SAE	United Kingdom BS970[a] Pt1 1983	Federal Republic of Germany		France AFNOR	Sweden SIS	Japan JISG4051–65	USSR GOST.1050.60	EN.No.	Subclassification
		Werkstoffe Nummer	Kurzname						
8615 8617 8620 8622 8625	805 M17, H 805 M20 A, H 805 M22 A, H 808 M17 808 H17	1.6523, 1.6543	21NiCrMo	20NCD2		SNCM21		361 362 363	½% Nickel–Cr–Mo
	815 M17, H			16NCDS	14 25 11	(SNCM9)	20KHN	353	1½% Nickel–Cr–Mo
	820 M17, H	1.6587		18NCD6				354	1¾% Nickel–Cr–Mo
	822 M17, H							355	2% Nickel–Cr–Mo
AMS6260E	832 M13, H	1.6657	14NiCrMo 13.4			(SNCM5)		36C	3½% Nickel–Cr–Mo
	835 M15, H	1.6723	14NiCrMo 16.5			SNCM25		39B	4% Nickel–Cr–Mo

Table 2.1.2—*continued*

TABLE 2.1.2 Equivalent international specifications for wrought steels—continued

Part 5—Carbon and alloy spring steels

United States AISI/SAE	United Kingdom BS970 Pt1 1983	Federal Republic of Germany		France AFNOR	Sweden SIS	Japan JISG4451–65	USSR GOST.1050.60	EN.No.	Subclassification
		Werkstoffe Nummer	Kurzname						
1050, 1055	080 A52			XC55	14 16 06		50, 55	43	.52 Carbon
1065	080 A67	1.0603	C67	XC65			65	43E	.67 Carbon
1074, 1080	070 A72			XC80F			80	42	.72 Carbon
1074, 1080	070 A78			XC80F				42	.78 Carbon
1095	060 A96							44	.96 Carbon
9260	250 A53	1.0970	38Si7	5557	14 20 90	SUP6	55S2, 55S9	45	Silicon—
9260	250 A58	1.0971	60SiCr7			SUP7	60S2, 60S9	45A	
9260	250 A61	1.0971				SUP7	60S2A	45A	Manganese
5147	527 A60 527 H60			50C4		SUP9	50CH	48	¾% Chromium
6150	735 A50	1.8159	50CrV4	50CV4	14 22 30	SUP10	50KHCA	47	1% Chromium-V
	805 A60 805 H60							— —	½% Nickel–Cr–Mo
	925 A60							—	Silicon–Mn–Cr–Mo

Table 2.1.2—continued

TABLE 2.1.2 Equivalent international specifications for wrought steels—continued

Part 6—Tool steels

United States AISI/SAE	United Kingdom BS	Federal Republic of Germany Werkstoffe Nummer	Federal Republic of Germany Kurzname	France AFNOR	Sweden SIS	Japan JISG	USSR GOST.	Type
Water hardening steels								
W108	BW1A							
W109	BW1B			Y90			U8A	
W110	BW1C	1.1525	C80 WI			U		
W112		1.1545	C105 WI	Y105	1550			
W209	BW2	1.1663	C125 W	Y120		SK2	U10A	
W210								
W310								
Shock resisting steels								
S1	BS1							
S2	BS2	1.2542	45WCrVT					
S5	BS5							
Cold work tool steels								
01	B01	1.2510	100MnCrW4					Oil hardening type
02	B02							
06								Med
A8								
D2	BD2							High Carbon
D3	BD3							High Chrome
D5								
D7								
Hot work tool steels								
H11	BH11	1.2343	X35CrMoV51	Z38CDV55		SKD6	4Ch5NFS	Chromium Base
H12	BH12							
H13	BH13	1.2349	X40CrMoV51	Z100CDV5	2242	SKD61	4Ch5NFIS	
H21	BH21	1.2581	X30WCrV9.3	Z30WCV9		SKD5	3Ch2W8F	Tungsten Base
High speed steels								
T1	BT1	1.3355	S18-0-1	Z80WCV 18-04-01		SKH2	R18	Tungsten Base
T2	BT2							
T4	BT4	1.3255	S18-1-2-5	Z80WKCV18-08-04-01		SKH3		
T5	BT5	1.3265	S18-1-2-10			SKH4A		
T8								
M1	BM1	1.2346	S2-9-1	Z85DCWV08-04-02-01	2723		M42	
M2	BM2			Z85DCWV08-04-05-02				
M3				Z120WDCV-06-05-04-03			H41	
M4	BM4	1.3344	S653					Molybdenum Base
Special purpose tool steels								
L6								Low alloy Nickel
L7								Chromium

[a] The description 'base' indicates that the metal named is the most significant alloying constituent AFTER iron.

Table 2.1.2—*continued*

TABLE 2.1.2 Equivalent international specifications for wrought steels—*continued*

Part 7—Stainless and heat resisting steels

Austenitic Stainless Steels

United States AISI/SAE	United Kingdom BS	Federal Republic of Germany Werkstoffe Nummer	Kurzname	France AFNOR	Sweden SIS	Japan JISG	USSR GOST
301	—	1.4310	X 12 CrNi 17 7	Z 12 CN 17.07	2331	SUS 301	—
303	303 S 21	1.4305	X 12 CrNi 18 8	Z 10 CNF 18.09	2346	SUS 303	—
304	304 S 15	1.4301	X 5 CrNi 18 9	Z 6 CN 18.09	2332	SUS 304	08Ch18N10
304 L	304 S 12 / 304 C 12	1.4306	X 2 CrNi 18 9	Z 2 CN 18.10 / Z 3 CN 19.10	2352; 2333	SCS 19 / SUS 304 L	03Ch18N11
—	304 C 15	1.4308	G-X 6 CrNi 18 9	Z 6 18.10 M	—	SCS 13	07Ch18N9L
304 LN	304 S 62	1.4311	X 2 CrNi N 18 10	Z 2 CN 18.10	2371	SUS 304 LN	—
309	309 S 24	1.4828	X 15 CrNiSi 20 12	Z 15 CNS 20.12	—	SUH 309	20Ch20N14S2
309 S	—	1.4833	X 7 CrNi 23 14	Z 15 Cn 24.13	—	SUS 309 S	—
—	309 C 30	1.4837	G-X 40 CrNiSi 25 12	—	—	SCH 17 / SCS 17	40Ch24N12SL
310 S	310 S 24	1.4845	X 12 CrNi 25 21	Z 12 CN 25.20	2361	SUH 310	—
—	310 C 40	1.4848	G-X 40 CrNiSi 25 20	—	—	SCH 21; SCH 22	—
314	—	1.4841	X 15 CrNiSi 25 20	Z 12 CNS 25.20	2347	—	20Ch25N20S2
316	316 S 16	1.4401	X 5 CrNiMo 18 10	Z 6 CND 17.11	2348	SUS 316	—
316 L	316 S 12	1.4404	X 2 CrNiMo 18 10	Z 2 CND 17.12 / Z 3 CND 19.10 M	—	SUS 316 L	—
316 LN	316 S 61	1.4406	X 2 CrNiMoN 18 12	—	—	SUS 316 LN	—
—	316 C 16	1.4408	G-X 6 CrNiMo 18 10	Z 2 CND 17.13	—	SCS 14	—
316 LN	—	1.4429	X 2 CrNiMoN 18 13	Z 2 CND 17.13	2375	SUS 316 LN	—
316 L	316 S 12	1.4435	X 2 CrNiMo 18 12	Z 2 CND 17.13	2353	SCS 16 / SUS 316 L	03Ch17N14M2
316	316 S 16	1.4436	X 5 CrNiMo 18 12	Z 6 CND 17.12	2343	SUS 316	10Ch17N13M3T / 08Ch17N13M2T / 08Ch16N13M2B
316 Ti	320 S 33	1.4573	X 10 CrNiMoTi 18 12	Z 8 CNDT 17.13 B	—	—	—
316 Ch	—	1.4580	X 10 CrNiMoNb 18 10	Z 6 CNDNb 17.12	2367	—	—
317 L	317 S 12	1.4438	X 2 CrNiMo 18 16	Z 2 CND 19.15	—	SUS 317 L	—
317	317 S 16	1.4449	X 5 CrNiMo 17 13	—	—	SUS 317	—
318	318 C 17	1.4581	G-X 5 CrNiMoNb 18 10	Z 4 CNDNb 18.12 M	—	SCS 22	—
—	—	1.4583	X 10 CrNiMoNb 18 12	Z 6 CNDNb 17.13 B	—	—	—
321	321 S 12 / 321 S 20	1.4878	X 12 CrNiTi 18 9	Z 6 CNT 18.12 B	2337	SUS 321	12Ch18N10T
329	—	1.4460	X 8 CrNiMo 27 5	—	2324	SUS 329 J1 / SCH 11; SCS 11	—
330	—	1.4864	X 12 NiCrSi 36 16	Z 12 NCS 35.16	—	SUH 330	—
—	330 C 11	1.4865	G-X 40 NiCrSi 38 18	—	—	SCH 15; SCH 16	—
347	347 S 17	1.4550	X 10 CrNiNb 18 9	Z 6 CNNb 18.10	2338	SUS 347	08Ch18N12B
—	347 C 17	1.4552	G-X 5 CrNiNb 18 9	Z 6 CNNb 18.10 M	—	SCS 21	—

Table 2.1.2—*continued*

TABLE 2.1.2 Equivalent international specifications for wrought steels—*continued*

Part 7—Stainless and heat resisting steels—*continued*

United States AISI/SAE	United Kingdom BS	Federal Republic of Germany		France AFNOR	Sweden SIS	Japan JISG	USSR GOST
		Werkstoffe Nummer	Kurzname				
Austenitic Stainless Steels—continued							
348	—	1.4546	X 5 CrNiNb 18 10	—	—	—	—
EV 8	349 S 54	1.4871	X 53 CrMnNiN 21 9	Z 52 CMN 21.09	—	SUH 35; SUH 36	55Ch20G9AN4 5Ch20N4A69
Ferritic and Austenitic Steels							
HNV 3	401 S 45	1.4718	X 45 CrSi 9 3	Z 45 CS 9	—	SUH 1	40Ch9S2
403	403 S 17	1.4000 1.4001	X 7 Cr 13 X 7 Cr 14	Z 6 C 13	2301	SUS 403	—
405	405 S 17	1.4002	X 7 CrAl 13	Z 6 CA 13	—	SUS 405	—
405	403 S 17	1.4724	X 10 CrAl 13	Z 10 C 13	—	SUS 405	—
—	—	1.4731	X 40 CrSiMo 10 2	Z 40 CSD 10	—	SUH 3	40Ch10S2M
409	409 S 19	1.4512	X 5 CrTi 12	—	—	SUH 409	—
430 Ti	—	1.4510	X 8 CrTi 17	—	—	—	08Ch17T
430	430 S 15	1.4742	X 10 CrAl 18	Z 10 CAS 18	—	SUS 430	—
HNV 6	443 S 65	1.4747	X 80 CrNiSi 20	Z 80 CSN 20.02	—	SUH 4	—
446	—	1.4762	X 10 CrAl 24	Z 10 CAS 24	2322	SUH 446	—

All British stainless steels have as a fourth character the letter S.
Austenitic steels in Britain and US standards are listed 300–399.
There are 200 stainless steels in British Standards.
Ferritic and martensitic stainless steels are listed 400–499.

Table 2.1.2—*continued*

TABLE 2.1.3 Physical properties of steels

Type Specification Condition	Temperature (°C)	Specific Gravity (g/cm³)	Specific heat (J/kgK)	Coefficient of expansion (10^{-6}/K)	Thermal conductivity (W/mK)	Electrical resistivity (μΩ cm)
CARBON STEELS						
C0.06 Mn 0.4 040A04–En2 Annealed	RT 100 600	7.87	— 482 754	— 12.6 14.7	65.3 60.3 36.4	12.0 17.8 72.5
C0.2 Mn 0.6 060A22–En3 Annealed	RT 100 600	7.86	— 486 749	— 12.2 14.4	51.9 51.1 35.6	15.9 21.9 75.8
C0.4 Mn 0.6 060A42–En8 Annealed	RT 100 600	7.85	— 486 708	— 11.2 14.6	51.9 50.7 33.9	16.0 22.1 76.6
C0.8 Mn 0.3 SAE 1078 Annealed	RT 100 600	7.85	— 490 712	— 11.1 14.2	47.8 48.2 32.7	17.0 23.2 77.2
C0.2 Mn 1.5 150M 19–En 14	RT 100 600	7.85	— 477 741	— 11.9 14.7	46.1 46.1 34.3	19.7 25.9 78.6
LOW ALLOY STEELS						
1% Ni 503M40 En 12 850°C OQ T600°C OQ	RT 100 600	7.85	— 486 586	— 11.9 14.5	— 49.4 34.8	21.9 26.4 77.5
Mn–Mo 605A37 En 16 845°C OQ T600°C OQ	RT 100 600	7.85	— 456 599	— 12.5 14.8	— 48.2 33.9	25.4 30.6 88.5
1% Cr 530A32–En 18B Annealed	RT 100 600 800	7.84	— 494 741 934	— 12.2 14.5 12.1	48.6 46.5 31.8 26.0	20.0 25.9 77.8 110.6
1% Cr-Mo 708A42–En 19 850°C OQ T600°C OQ	RT 100 600	7.83	— 477 561	— 12.3 14.5	— 42.7 33.1	22.2 26.3 64.6
3% Ni En 21 Annealed	RT 100 600 800	7.85	— 482 749 604	— 11.2 13.9 11.1	36.4 37.7 32.7 25.1	25.9 32.0 81.4 112.2
3% Ni-Cr En 23 Hardened and Tempered	RT 100 600	7.85	— 494 775	— 11.4 13.7	34.3 36.0 31.8	25.6 31.7 81.7
1½% Ni-Cr–Mo 317M40–En 24 830°C OQ T630°C OQ	RT 100 600	7.84	— — —	— — 14.3	— — —	24.8 29.8 79.7

Table 2.1.3

TABLE 2.1.3   Physical properties of steels—*continued*

Type Specification Condition	Temperature (°C)	Specific Gravity (g/cm³)	Specific heat (J/kgK)	Coefficient of expansion (10^{-6}/K)	Thermal conductivity (W/mK)	Electrical resistivity (μΩ cm)
HIGH ALLOY STEELS						
MARTENSIC STAINLESS						
13%Cr 420S29–En 56B Annealed	RT 100 600	7.74	— 473 779	— 10.1 12.1	26.8 27.6 26.4	48.6 58.4 102.1
FERRITIC STAINLESS						
17% Cr 430S15–En 60 Annealed	RT 100	7.7	— —	— 10.0	21.8 —	62.0 —
30% Cr–Ni Ferritic Hardened and Tempered	RT 100	7.9	— —	— 10.0	12.6 —	80.0 —
AUSTENITIC STAINLESS						
18% Cr–8% Ni 302S25–En 58A Softened	RT 100 600 800	7.92	— 511 642 641	— 14.8 18.4 19.0	15.9 16.3 23.9 26.8	69.4 77.6 102.2 114.1
25% Cr–20% Ni BS1648F Cast–Normalised	RT 100 800	7.90	544 — —	— 16.5 19.2	15.9 — —	90.0 — —

Table 2.1.3—*continued*

2.1.2   Carbon steels—hot rolled

2.1.2.1   General characteristics

Hot rolled carbon steels still cover the majority of common engineering applications, and their properties, as obtained by normalising, are compared to the properties obtained in the cold drawn and hardened and tempered conditions in:

Table 2.1.4   *Properties of carbon steel*

The advantages and limitations of hot rolled carbon steels are:

Advantages	Disadvantages
Low cost Good formability Good weldability } in lower carbon grades Good machinability Availability—plate, rod, hollows.	Modest strength Modest toughness

AVAILABLE TYPES

Steels used in the hot rolled condition are almost invariably carbon steels, and because toughness, machinability and weldability decrease with increasing carbon content, this is usually less than 0.25%. High strength low alloy steels are used in the hot rolled condition (see Section 2.1.5) but generally the potential of alloy steels is only realised by subsequent heat treatment.

Carbon steels may be subdivided into three major classes on the basis of their deoxidation practice. The characteristic features of each class are summarised in:

Table 2.1.5   *Hot rolled steel. Classification and characteristics*

2.1.2.2   Properties

TENSILE PROPERTIES

Both yield and tensile strength increase slowly but progressively with increasing carbon content, and depend on section thickness, smaller sections having higher strengths due to faster rate of cooling. Typical values for 25 mm bar are shown in:

Fig. 2.1.1   *Influence of carbon content on the strength of 25 mm bars of carbon steel*

Strengths and toughness of hot rolled carbon steels are given in:

Table 2.1.6   *Available hot rolled carbon steels*

FATIGUE STRENGTH

Fatigue strength of machined samples is approximately 40% of the tensile strength. In hot worked condition fatigue strength is lowered due to surface roughness, decarburisation and surface defects.

TOUGHNESS

The impact transition temperature of hot rolled carbon steel containing ~0.2%C is typically 0–+20°C. This increases markedly with increasing carbon content (see Section 2.1.12). Fully killed steels have lower transition temperatures due to the grain refining influence of aluminium and the removal of soluble nitrogen. Manganese (up to 1.5%) improves low temperature properties by refining grain size.

The fibrous fracture (or upper shelf) energy is controlled by the volume fraction of pearlite and non-metallic inclusions. Thus the toughness decreases with increasing carbon content and increases with increased cleanliness (i.e. reduced sulphur and oxygen content). Also these steels display a pronounced anisotropy in toughness due to deformation of non-metallic inclusions during working (see also Section 2.1.12).

MACHINABILITY

Machinability of low carbon hot rolled steels is generally good.

For enhanced machinability resulphurised steels may be employed (containing up to 0.3% sulphur) but this reduces toughness.

FORMABILITY

Formability of hot rolled steels decreases with increasing carbon content. With less than 0.2% C formability is good and these steels are suitable for bending, riveting, pressing, etc. For optimum formability, e.g. deep drawing, the carbon content should be reduced to less than 0.15%.

2.1.2.3   Influence of alloying additions and processing on properties

Hot rolling carbon steel is the most economical way of making steel plate. Properties may be improved with increased costs by alloying additions and further processing operations (see Tables 2.1.7, 2.1.8 and 2.1.9):

Table 2.1.7   *Influence of alloy additions on hot rolled carbon steels*
Table 2.1.8   *Influence of alloying additions on normalised steels*
Table 2.1.9   *Influence of alloying elements on properties of controlled rolled steels*

TABLE 2.1.4 Properties of carbon steel

Type	Low carbon free cutting	20 Carbon	30 Carbon	40 Carbon	55 Carbon	90 Carbon
Specification BS970	230 M07 (En1)	070 M20 (En3)	080 M30 (En5)	080 M40 (En8)	080 M50 (En9)	060 A96 (EN44)[b]
Carbon content	Less than 0.15	0.16–0.24	0.26–0.34	0.36–0.44	0.5–0.6	0.9–1.2
Normalised — Yield MN/m²	216	200–216	232–247	247–278	309–355	395–555
Normalised — UTS MN/m²	355	400–432	463–494	510–540	602–695	917–1034
Normalised — Elongation %	20–22	21	19–20	16–17	12–13	11
Cold drawn — Section (mm)	10 / 40 / 75	10 / 40 / 75	10 / 40 / 75	10 / 40 / 75	10 / 40 / 75	wire
Cold drawn — UTS (MN/m²)	463 / 432 / 355	525 / 463 / 432	602 / 556 / 525	650 / 602 / 570	788 / 741 / 710	1930–1961
Cold drawn — Elongation (%)	7 / 8 / 10	12 / 13 / 14	10 / 12 / 12	8 / 10 / 10	7 / 8 / 9	2/5
Hardened and tempered to 600°C — Section (mm)		18 / 62	18 / 62	18 / 62	18 / 62 / 100	25 / 50 / 100
Hardened and tempered to 600°C — UTS (MN/m²)		540–700	618–772 / 540–695	695–850 / 618–772	849–1003 / 772–926 / 695–849	752–1290 / 927–1157 / 896–1138
Hardened and tempered to 600°C — Yield (MN/m²) min.		355	417 / 340	463 / 386	571 / 479 / 417	669 / 605 / 547
Hardened and tempered to 600°C — Elongation (%) min.		20	16 / 18	16 / 16	12 / 14 / 14	10–26 / 12–18 / 12–17
Hardened and tempered to 600°C — Impact J min.		23[a]	22[a] / 22[a]	22[a] / 22[a]	– / – / –	– / – / –
Formability	—	Good	Limited	Limited	Nil	Nil
Weldability	Not recommended	Good	Pre- and post-heat treatment	Pre- and post-heat treatment	Not recommended	Not recommended
Available forms	Bar	Sheet, plate bar, tube	Sheet, plate bar, tube	Plate, bar, tube, forgings	Sheet, plate, bar, forgings, wire	Sheet, strip bar, forgings, wire
Machinery Index — Normalised	100	100	70	60	43–50	43
Machinery Index — Hardened and tempered	—	60	60	50	30–40	30

[a] The impact properties apply only to the fine-grain aluminium treated steels. BS970 gives the impacts in terms of the Izod Test. The figures have been converted to Charpy-V which is now the more common test, but cannot be regarded as definitive.
[b] 060A96 can be ordered only on chemical analysis. The mechanical properties quoted are based on actual test results but are not guaranteed.

Table 2.1.4

TABLE 2.1.5 Hot rolled steel. Classification and characteristics

Steel class	Description	Advantages	Disadvantages
Rimmed steel	Steel containing less than 0.25% carbon which has not been deoxidised. Considerable evolution of CO occurs during casting and eliminates pipe formation.	Possesses a 'rim' of pure low carbon steel having a clean, smooth surface. Ideal for applications where good surface finish is required.	Low strength (low carbon). Pronounced segregations in interior of steel giving non-uniform properties. Contains internal voids, defects and general unsoundness.
Killed steel	Prior to casting the steel is deoxidised with aluminium and silicon.	Best quality, uniform mechanical properties, minimum segregation. Inclusion number and size reduced. Improved impact transition temperature.	Most expensive practice due to extensive pipe formation.
Semi-killed or balanced steel	Intermediate degree of deoxidation. Some rimming action occurs: reduces pipe formation.	Most economical steel. Improved surface finish compared with killed steel and less internal segregation and unsoundness than rimmed steel.	Inferior toughness compared with killed steel.

TABLE 2.1.6 Available hot rolled carbon steels

Steel type	Composition (%)				Mechanical properties			Comments
	C	Si	Mn	S	Yield strength (MN/m^2)	Tensile strength (MN/m^2)	Notch toughness (J)	
Free cutting steel	0.07 0.15	0.10 max.	0.80 1.4	0.2 0.6	230	400	35	Poor transverse toughness
Cold forming mild steel	0.10 0.15	0.10 0.35	0.5	0.5	230	400	35	Especially suitable for cold forming, e.g. bending, riveting deep drawing.
Mild steel	0.25 max.	0.05 0.35	0.6 1.0	0.05	300	450	30	Limited cold formability.
'30' Carbon steel	0.25 0.35	0.05 0.35	0.6 1.0	0.05	300	550	30	Poor toughness and high transition temperature.

For high strength low alloy steels see Section 2.1.5.

Table 2.1.5 and Table 2.1.6

TABLE 2.1.7 Influence of alloy additions on hot rolled carbon steels

Element	Strength	Fibrous fracture energy	Impact transition temperature	Remarks
Carbon 0.15 – 0.25	←	→	↑↑	For effect of carbon on strength (see Fig. 2.1.1). For effect on toughness (see Section 2.1.12).
Manganese 0.6 – 1.8			→	Refines grain size.
Aluminium (up to 0.2%)	←	←	→	Refines grain size, eliminates free nitrogen. Removes silicate inclusions (See Section 2.1.12).
Niobium	←		→	Grain refinement and some precipitation hardening.
Nitrogen	←		↑↑	Very harmful effect on transition temperature unless Al and V present.
Sulphur		→		Lowers toughness, gives rise to directionality (see Section 2.1.12).

High carbon concentrations increase strength but reduce weldability. Also impact transition temperature is high unless Al and Nb are added for grain refinement.

Table 2.1.7

TABLE 2.1.8 Influence of alloying additions on normalised steels

Element	Strength	Fibrous fracture energy	Impact transition temperature	Remarks
Carbon 0.1 – 0.25	↑↑	↓	↑↑	For effect of carbon strength (see Fig. 2.1.1).
Manganese 0.6 – 1.8	—	—	↓	For effect on toughness see Section 2.1.12.
Niobium	↓	—	↓	Refines grain size. Improvement in strength due to grain size. No precipitation hardening.
Vanadium	↑↑	↓	↓	Refines grain size. Increases strength by carbide precipitation. May increase fibrous energy by reducing pearlite.
Aluminium	—	↓	↓	Refines grain size, eliminates free nitrogen. Removes silicate inclusions (see Section 2.1.12).
Nitrogen	↓	—	↑↑	Very harmful effect on transition temperature unless Al and V present.
Aluminium + Nitrogen	↓	—	↓	Enhanced grain size refinement.
Aluminium + Nitrogen + Vanadium	↑↑	↓	↓	Grain size refinement and precipitation hardening.
Sulphur	—	↓	—	See selection for toughness Section 2.1.12

Table 2.1.8

TABLE 2.1.9 Influence of alloying elements on properties of controlled rolled steels

Element	Strength	Fibrous fracture energy	Impact transition temperature	Remarks
Carbon	↑	↓	↑	In controlled rolled steels carbon is usually restricted to 0.20%. Optimum toughness is obtained at 0.1%.
Manganese	↑	—	↓	Refines grain size and enables lower finishing temperatures to be employed.
Niobium	↑	↑	↓	Refines grain size. Precipitation harden and reduce volume fraction of pearlite.
Vanadium	↑	↑	↓	
Aluminium		↑	↓	Grain refinement and removes soluble nitrogen.
Sulphur		↓		Sulphur is especially harmful in controlled rolled steels due to low rolling temperature. Treatment with calcium or rare earth is recommended. See selection for toughness Section 2.1.12.

Controlled rolling involves subjecting the steel to a final rolling reduction of about 50% in the temperature range 950–800°C. This produces an extremely fine grain size with associated high strength and low impact transition temperature. In this way properties equal to or superior to those in the normalised condition can be obtained in the hot rolled steel.
The presence of niobium and/or vanadium is essential in order to prevent grain growth during hot working.
The effect of alloying additions on strength and toughness are summarised in the following table.

Table 2.1.9

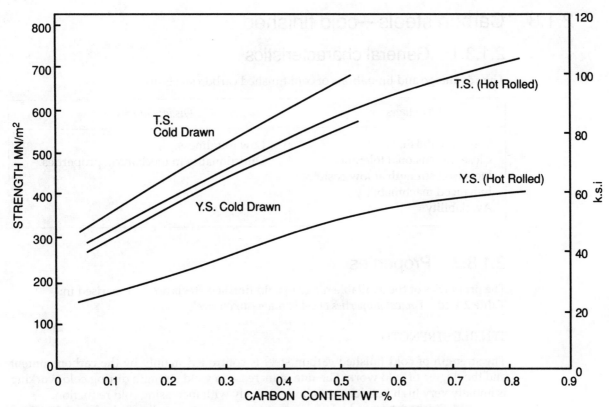

FIG 2.1.1 Influence of carbon content on the strength of 25 mm bars of carbon steel

Fig 2.1.1

2.1.3   Carbon steels—cold finished

2.1.3.1   General characteristics

The advantage and limitations of cold finished carbon steels are:

Advantages	*Disadvantages*
Scale-free finish. Close dimensional tolerance. Increased strength at low cost. Enhanced machinability. Availability	Low toughness. Directionality in mechanical properties.

2.1.3.2   Properties

The properties of the available range of cold finished steels are summarised in:
Table 2.1.10   *Typical properties of cold drawn carbon steels*

TENSILE STRENGTH

The strength of cold finished carbon steel is controlled mainly by the carbon content and the degree of cold work. The rate of increase in yield strength during cold working is initially very high but decreases progressively with increasing cold reduction.

In cold finished bar and plate, carbon contents are usually limited to 0.5% since higher carbon grades must be annealed prior to cold working and the resultant mechanical properties are frequently inferior to those obtained from lower carbon grades which are cold worked after hot rolling. Cold working reductions are normally limited to less than 10% but increases in yield strength of up to 50% can be achieved.

In cold rolled steel reductions of between 30 and 70% are employed and yield strengths of up to 550 MN/m^2 are developed. However, such steels may be supplied in a range of intermediate tempers or in an annealed condition for forming applications. For increased strength with good toughness and formability, fine-grained, low carbon high strength low alloy (HSLA) steels are also available in the cold rolled condition, see Section 2.1.5.

In steel wire a range of strengths between 300 and 3000 MN/m^2 may be obtained by control of composition and cold working reduction. For maximum strength combined with good ductility, high carbon steel (0.65–0.85%C) is used in the patented condition. This involves heating the steel above the austenite transformation temperature and then cooling in air or molten lead to produce a very fine pearlite structure which is subsequently cold drawn to reductions of up to 90%.

The temperature of the lead bath must be adjusted to suit the carbon content and the section size. Zinc phosphate and lime treatments are used prior to drawing.

TOUGHNESS

The toughness of cold finished steels decreases and the impact transition temperature increases with increasing cold working reduction. Also with increasing carbon content and strength the toughness decreases progressively:
Fig 2.1.2   *Influence of cold working reduction on the strength and toughness of cold finished steel*

With carbon contents in excess of about 0.3% impact transition temperatures are in excess of room temperature.

Toughness anisotropy develops during cold working due to deformation of non-metallic inclusions. Minimum properties are obtained normal to the direction of working and this can give rise to cracking during forming. Techniques for alleviating the problems associated with inclusions are described in Selection for Toughness, Section 2.1.12.

FATIGUE

The fatigue strength of cold finished steels increases in proportion to the tensile strength. The fatigue limit of machined specimens is about 40% of the tensile strength, but at high cold working reductions the improvement may be offset by residual stresses developed during working.

MACHINABILITY

The machinability of low and medium carbon steels is enhanced by cold working.

FORMABILITY

Cold rolled steel sheet for forming applications usually contains less than 0.15% carbon. For moderate draws, commercial quality sheet produced from rimming steel is generally suitable and most economical. Such steels are susceptible to 'Luders bands' or 'stretcher strain' formation but this can be alleviated by temper rolling immediately prior to forming. For severe forming and deep drawing applications, aluminium-killed steels containing low sulphur contents are recommended. Alternatively HSLA steels containing niobium and vanadium may be employed. Such steels have more uniform properties than rimming steels, they exhibit little tendency for strain ageing (Luders bands formation) and resist local thinning during drawing. Also due to their finer grain size (and possibly precipitation hardening) they have higher strength than rimmed steel.

TABLE 2.1.10 Typical properties of cold drawn carbon steels

Steel designation	BS 970	Composition				Mechanical properties			Notch-tough-ness (J)	Comments
		C	Si	Mn	S	Yield strength (MN/m²)	Tensile strength (MN/m²)	Elon-gation (%)		
Cold Forming Steel	040A10 (En2)	0.08 0.13	0.05 0.35	0.03 0.50	0.05 max	250 400	350 500		50	Good formability.
Free Cutting Steel	220M07 (En1A)	0.15 max	0.10	0.9 1.3	0.2 0.3	270 400	335	7 10	10 30	Good machinability. Low toughness.
Mild Steel '20' Carbon Steel	070M20 (En3)	0.16 0.24	0.05 0.35	0.5 0.9	0.05 max	340 390	432 525	12 14	10 40	High impact transition temperature.
'26' Carbon Steel	070M26 (En4)	0.22 0.30	0.05 0.35	0.5 0.9	0.05	370 432	495 570	11 13	10 40	High impact transition temperature.
'30' Carbon Steel	080M30 (En6)	0.26 0.34	0.05 0.35	0.6 1.00	0.05	390 450	525 600	10 12	15 30	Strength dependent on section size.
'40' Carbon Steel	080M40 (En8)	0.36 0.44	0.05 0.35	0.6 1.0	0.5 (0.12–0.20)	430 510	570 650	8 10	25	Sulphur may be added for machinability.
'55' Carbon Steel	070M55 (En9)	0.50 0.60	0.05 0.35	0.5 0.9	0.05	570 618	710 790	7 9	—	Very low toughness.
Steels for hard drawn wire	BS1408 (En49)	0.40 0.85	0.05 0.35	1.0 max	0.05	—	1000 2300 (patented 2470)	— —	—	Strength increases with carbon content. Very low toughness.

For composition and properties of high strength low alloy steels see Section 2.1.5.

Table 2.1.10

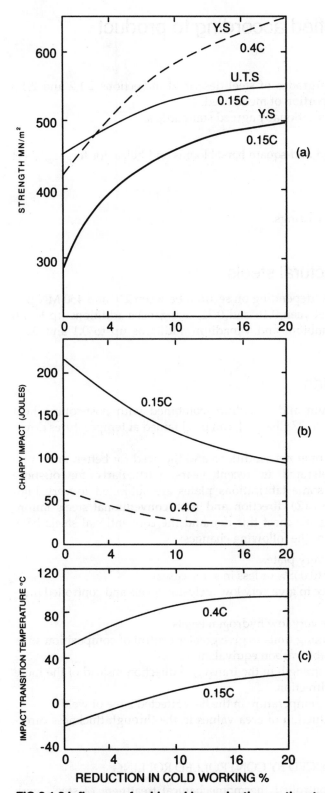

FIG 2.1.2 Influence of cold working reduction on the strength and toughness of cold finished steel

Fig 2.1.2

2.1.4   Carbon steels classified according to product

2.1.4.1   Introduction

Rolled or hollow sections of the grades of steel described in Sections 2.1.2 and 2.1.3 constitute by far the highest proportion of metal used.

These products are sold to internationally agreed standards as:

Weldable structural steels;
plates, flats, sections, round and square bars, blooms and billets for forging, sheet and strip and steel tubes.
Reinforcing bar.
Pressure vessel steels;
plates, tubes forgings and castings.
Materials for cold forming.

2.1.4.2   Weldable structural steels

These steels have yield strengths (depending on section) between 210 and 450 MN m^{-2} achieved by carbon additions between 0.16 and 0.22, manganese additions up to 1.6 max., and for some qualities niobium and vanadium additions up to 0.1 and 0.2% respectively.

2.1.4.3   Structural plates

By selective additions of niobium and vanadium combined with post-rolling heat treatments, impact properties up to 27 Charpy J, can be obtained at temperatures down to –60°C.

The tendency for use in more onerous conditions and the need for better, and more consistent properties has accelerated in recent years, particularly for offshore structures. As an example, on some fabrications plates are subjected to very high stresses in the through thickness (Z) direction and with conventional steels failure would occur by lamellar tearing. To meet this demand the conventional steels have been modified with one or more of the following changes:

(a)   More rigorous testing in every plate.
(b)   Lower maximum sulphur (0.008% or less in some cases).
(c)   Special deoxidation practice to give very low inclusion levels and controlled morphology.
(d)   Vacuum degassing to give very low hydrogen levels.
(e)   Use of secondary steel-making units to give greater control of composition so as to achieve the lowest possible carbon equivalent values
(f)   Guarantee of high impact strength in the transverse direction instead of the more usual (and easier) length direction.
(g)   High impact values at low temperature in the heat affected zone of welds.
(h)   Guaranteed minimum reduction of area values in the through thickness direction.

SUPERIOR PROPERTIES PRODUCED BY CONTROLLED ROLLING

This process depends on the controlled thermomechanical treatment of steels with minor vanadium, niobium or titanium additions. A very fine grain size can be achieved together with considerable extra strength derived from the precipitation of alloy carbonitrides. The carbon and manganese content of the steel can be reduced to give a low carbon equivalent and thus a much more weldable steel.

2.1.4.4   Weldable sections

The range of qualities available for sections which include flats (up to 610mm in width), round and square bars and hollow and solid sections, is slightly reduced compared with plates and the available properties show minor variations.

However, a whole range of beams, girders and columns can be fabricated from plate steels by automatic welding.

Hollow sections are increasingly used in engineering construction because they take up less space than angles or 'I' sections, allow increased natural lighting and/or decreased wind resistance and are cheaper to paint. With reasonable design no internal protective coating is required. Cold formed sections show higher strength and improved surface finish. Tubes are available with yield strengths between 170 and 355 MPa.

Generally speaking the cheapest steels and the minimum testing (one test every 40 tons weight) are adequate for most engineering requirements. More demanding specifications and an increased number of tests are specified for more onerous service conditions.

2.1.4.5   Sheet steels

Sheet (or strip steel) can be up to about 12 mm in thickness but is generally much thinner.

There are two principal qualities: materials having specific requirements based on formability and materials having specific requirements based on mechanical strength. Steel to be used in forming processes must be suitable in tensile strength and ductility, grain size, cleanness and surface finish. Sheet is usually supplied on a 'fitness for purpose' basis.

A high proportion of sheet steel is supplied in the form of tin plate. The only mechanical properties specified are hardness and number of bends withstood. The main requirements are freedom from coating defects, and tin coating weight which may range from approximately 40 g/m (heaviest hot dipped) to approximately 3 g/m lightest dectrolytic. The tin facilitates welding and lubricates during drawing. Lacquered 'blackplate' suffices for many applications.

Besides the standard forming quality steels a number of higher strength qualities have been developed, mainly for vehicle manufacture. One of these, 'Hypress' supplied by British Steel, is controlled rolled to give an ultra-fine grain structure, combined with precipitation of titanium or niobium carbide. The steel is normally hot rolled but can, to produce thinner gauges, improved tolerances or bright surface finish, be cold rolled, bright annealed and temper rolled. Hypress is readily welded, with negligible reduction in properties for grades up to 29. Above this level there are local reductions in strength in the heat-affected zone for certain types of welding. 'Dual phase steels' with properties comparable to Hypress have been developed in other countries.

The group of steels with requirements based on specified minimum strengths are listed in:

Table 2.1.11   *Summary of material grades, chemical compositions, mechanical properties and types of carbon and carbon-manganese sheet and strip steels (from BS1449 Part 1, 1972).*

Other important applications of sheet steel are:

Steels for gas bottle applications.
Hot dip zinc coated sheet steel.
Hot dip zinc coated corrugated steel sheet.

2.1.4.6   Reinforcing bars for concrete

Hot rolled bars are manufactured to a yield strength of 250 MN/m^2.

Cold rolled bars show a proof stress rather than a yield point. A typical product is 'Torbar' (manufactured by Reinforcement Steel Services, which ranges from 6 to 50mm diameter and has a patented rib pattern). It is readily bent on site, is easily welded and has good fatigue properties.

2.1.4.7   Tubes for general engineering

Tubes for automobile, mechanical and general engineering purposes are specified in terms of:

(a) Material

1. Carbon steel, ranging in carbon content from 0.2 to 0.55%
2. Case hardening steel
3. Manganese, molybdenum and chrome molybdenum steels.
4. Stainless steel (austenitic steels with low and medium carbon, stabilised and unstabilised).

(b) Method of manufacture

1. Seamless (austenitic steel tubes are not specified).
2. Welded (electric resistance, induction and submerged arc).
3. 'Longitudinally welded' (stainless by inert gas shielded arc).

(c) Finish

1. Hot finished.
2. Cold finished.

(d) Heat treatment

1. No heat treatment after cold finishing or welding and sizing.
2. Heat treated and given a light finishing pass.
3. Annealed or normalised.
   (a) In a controlled atmosphere.
   (b) In air and descaled.
4. Softened
   (a) In a controlled atmosphere.
   (b) In air and descaled.

2.1.4.8   Pressure vessel steels

Steels for fired and unfired pressure vessels include plates, tubes, forgings and castings. The specifications cover carbon steels, carbon manganese steels and alloy and austenitic steels. This section is confined to carbon and carbon manganese steels.

PLATE STEELS

Pressure vessel plate steels are similar to the structural steels, but there are some important differences.

These are:

1. Pressure vessel plates supplied to positive dimensional tolerances, the specified thickness being also the minimum, whereas for structural steel plates it is the mean.

2. Whereas only one tensile test may be specified for every 40 tons of structural steel, every pressure vessel plate is tested (and each end of every plate for plates weighing more than 2½ t).
3. Some pressure vessel plates have the content of soluble aluminium specified.
4. All pressure vessel plates are supplied in the normalised condition or hardened and tempered.
5. All pressure vessel plates have a specified nitrogen content.
6. Elevated temperature proof stress values are specified for pressure vessel plates. It should be noted that the elevated temperature proof stresses for some British steels (and also for certain overseas steels) are higher in value than for those specified in the international standard. This is achieved by supplying the steels to a narrower tensile range than the international standard.

TUBE STEELS FOR PRESSURE VESSELS

Tube steels for pressure purposes are processed rather more drastically than plates, particularly cold bending as for boiler and superheater tubes. Therefore, though essentially the same kind of steel is used as for plates, some of the grades are softer.

There are three different tube types, seamless, electric (resistance or induction welded) and submerged arc welded. For lower temperature service, steels with up to 9% nickel are used.

DESIGN FOR HIGH TEMPERATURE SERVICE WITH PRESSURE VESSEL STEELS

Three measured high temperature stress values have been used for pressure vessel design, the lower yield stress (or 0.2% proof stress) the 100 000-h rupture strength and the 1% creep strain at 100 000 h.

The high temperature proof stress is the ruling parameter for carbon manganese steels at temperatures up to the region of 400°C where its design factor corrected value is below that of the rupture strength.

The measured value of the high temperature yield is very dependent on the testing procedure, in particular the rate of strain, and the agreed procedure must be rigidly adhered to. The code values which incorporate a very large number of tests carried out world-wide are the lower 95% confidence limit plus 30 MN/m^2 and are applicable to steels from all countries.

The 100 000-h rupture stress is the ruling parameter at high temperatures. The values for this are average stress rupture values devised by parametric extrapolation from a large number of tests. Several points must be noted:

1. The permitted analyses of pressure vessel steels may contain up to 0.1% Mo. This quantity of molybdenum can greatly increase the rupture strength of a steel. Therefore all casts containing more than 0.02% Mo have been excluded from consideration.
2. For steels containing less than 0.8% Mn active nitrogen increases rupture strength. De-oxidation with more than 0.015% aluminium combines much of the nitrogen as aluminium nitride which is ineffective. There is therefore, a considerable difference in rupture strength between semi and silicon-killed, and silicon- and aluminium-killed carbon steels.

   The standardised rupture strengths of semi- and silicon-killed carbon steels are listed in:
   Table 2.1.12 *Creep rupture strengths of carbon steels (semi- and silicon-killed)*
   The standardised rupture strengths of aluminium-killed carbon steel are given in:
   Table 2.1.13 *Creep rupture strengths of carbon steels (silicon- and aluminium-killed)*
   The effect of aluminium on the rupture strength of steels containing over 0.8% Mn is small. The standardised rupture strengths of these steels is shown in:

Table 2.1.14 *Creep rupture strengths of carbon–manganese steels (semi-killed or fully-killed carbon–manganese steels including Nb-treated steels)*

3. Long-term high temperature heat treatment precipitates active nitrogen. Therefore, any steel component that has been tempered, stress relieved or post-weld heat treated should be considered to have a rupture strength 10% less than those in Table 2.1.12, manganese contents below 0.8% and 10% less than those in Table 2.1.14 for manganese contents above this figure.

4. The Larson–Miller type of extrapolation procedure used generates a curve with a steadily increasing downward slope on a log time/log stress plot. Many very long-term stress rupture determinations show an inflection with a reducing downward slope at very long times. The vast majority of the tests which were taken into consideration terminated at 30 000 h or less and very few extended for 80 000 h.

Design procedures based on the ISO agreed properties usually incorporating a design stress of 60% of the rupture stress provide an excellent guarantee against failure at service times up to 100 000 h.

In the past, design stresses have been based on the stress required to produce a deformation of 1% in 100 000 h, estimated from the secondary creep rate. This procedure is of value for plant of complex design or rotating plant where the maintenance of dimensions is essential to prevent interference in operation. The primary and secondary creep rates depend very much on structure and pretreatment, which makes the 1% deformation technique unsatisfactory for pressure vessels.

The rupture ductility of carbon manganese pressure vessel steels is usually in excess of 40% and therefore, provided reasonable fabrication procedures are employed, low ductility failures will not occur.

TABLE 2.1.11 Summary of material grades, chemical compositions, mechanical properties and types of carbon and carbon–manganese sheet and strip steels (from BS1449 Part 1, 1972)

Section	Material grade	Condition	C % min	C % max	Si % min	Si % max	Mn % min	Mn % max	S % min	P % max	Types	Tensile strength, P_m min. (MN/m²)	Lower yield stress R_{eL} min. (MN/m²)	Elongation, A, min, L_o=200mm (%)	Elongation, A, min, L_o=50mm (%)	Bend test (180°)
2						Materials having specific requirements based on formability										
	1	HR,HS	—	0.08	—	—	—	0.45	0.030	0.025	Extra deep drawing aluminium-killed steel					
	1	CR,CS	—	0.08	—	—	—	0.45	0.030	0.025	Extra deep drawing stabilised steel					
	2	HR,HS CR,CS	—	0.08	—	—	—	0.45	0.035	0.030	Extra deep drawing					
	3	HR,HS CR,CS	—	0.10	—	—	—	0.50	0.040	0.040	Deep drawing					
	4	HR,HS CR,CS	—	0.12	—	—	—	0.60	0.050	0.050	Drawing or forming					
	14	HR,HS	—	0.15	—	—	—	0.60	0.050	0.050	Flanging					
	15	HR,HS	—	0.20	—	—	—	0.90	0.060	0.060	Commercial					
3						Materials having specific requirements based on minimum strengths										
	34/20	HR,HS CR	—	0.15	—	—	—	0.70	0.050	0.050	Supplied as either rimmed (R), balanced (B), or killed (K) steels	340	200	21	29	R = 1t (see note 4)
	37/23	HR,HS CR	—	0.20	—	—	—	0.80	0.050	0.050		370	230	20	28	R = 1t (see note 4)
	43/25	HR,HS	—	0.25	—	—	—	0.90	0.050	0.050		430	250	16	25	R = 1½t (see note 2)
	43/28	HS	—	0.25	—	—	—	0.90	0.050	0.050		430	280	18	26	R = 1½t
	40/30	HR,HS	—	0.15	—	—	—	1.20	0.040	0.040	Niobium treated fully killed steels having high yield strength and good formability	400	300	20	28	R = 1t
	43/35	HR,HS	—	0.15	—	—	—	1.20	0.040	0.040		430	350	18	25	R = 1t
	46/40	HR,HS	—	0.15	—	—	—	1.20	0.040	0.040		460	400	15	22	R = 1t
	50/45	HR,HS	—	0.20	—	—	—	1.50	0.040	0.040		500	450	14	22	R = 1½t
	50/35	HR,HS	—	0.20	—	—	—	1.50	0.050	0.050	Grain refined balanced or killed steel	500	350	15	20	R = 1½t
	54/35	HR,HS	—	0.23	—	—	—	1.70	0.050	0.050	Balanced or killed plain carbon manganese steel possibly with niobium additions	540	350	13	18	R = 1½t (see note 2)

R = internal radius of bend
t = thickness of material

NOTE 1: The properties of HS materials are only applicable up to 8mm in thickness. For materials thicker than 8mm, the properties are to be agreed between the manufacturer and the purchaser. Grade HS 43/28 is available in thicknesses of up to and including 4 mm only.

NOTE 2: The properties indicated in the table are applicable to all grades in the as-rolled condition. The properties of Grades 34/20, 37/23, 43/25, 50/35 and 54/35 are applicable to the as-rolled and normalised condition. For special applications Grades 43/25 and 54/35 may be ordered having a bend test value of R = 1t.

NOTE 3: For a particular grade, higher strength values may be available for thinner material. The purchaser shall establish these values with the manufacturer.

NOTE 4: In the case of Grades 34/20 and 37/23 for material of 3mm and over rolled on wide mills the bend test (180°C) reqirement shall be R = 1½t when a balanced (B) or killed (K) steel is supplied.

NOTE 5: The bend radii quoted in this table are for specially prepared test pieces and conditions during fabrication may be more severe and not simulate those during laboratory testing.

Table 2.1.11

TABLE 2.1.12 Creep rupture strengths of carbon steels (semi- and silicon-killed)

Conditions of steel to which the properties apply

	Details of materials actually tested		Range for which data are expected to apply agreed by TC17/SC10/ETP	
			min.	max.
Chemical composition (wt %)	C	0.07–0.24	—	0.30
	Si	0.005–0.330	—	0.60
	Mn	0.32–0.80	→ 0.40	—
	P	0.003–0.048	—	0.050
	S	0.003–0.050	—	0.050
Heat treatment	1. 899–950°C AC		1. Normalised	
	2. 850–920°C AC		2. Normalised and tempered	
	+		3. Hot finished	
	T500–690°C		4. Hot finished and tempered	
Products	Form	Size (mm)		
	Plates	18–75 thk	All wrought product forms	
	Tubes	6.5–970 thk x 25–191 dia.		
	Bars	178–305 thk 16–25 dia.		
	Forgings	25–146 thk x 575–1194 dia.		

Quantity and duration of data upon which the properties are based

Temperature (°C)	Test Duration					
	< 10 000	10 000–20 000	20 000–30 000	30 000–50 000	50 000–70 000	> 70 000
	Number of test points available					
400	292	41	18	19	6	2
450	461	55	16	7	3	1
500	463	33	16	9	4	3

Average rupture stresses (N/mm²)

Temperature (°C)	10 000 h	30 000 h	50 000 h	100 000 h	150 000 h	200 000 h	250 000 h
380	277	251	238	219	207	199[a]	192[a]
390	255	228	215	196	184	175[a]	167[a]
400	233	206	193	173	160	151[a]	143[a]
410	213	185	171	151	137	128[a]	121[a]
420	193	164	150	129	116	107[a]	101[a]
430	173	144	129	109	98[a]	90[a]	84[a]
440	154	124	110	92	82[a]	76[a]	71[a]
450	136	107	94	78	70[a]	64[a]	60[a]
460	118	91	80	67	60[a]	55[a]	50[a]
470	102	79	69	57	50[a]	44[a]	
480	89	68	60	48	39[a]		
490	77	59	51				
500	68	51	41				
510	60	41	36				
520	52	36					

[a] Values which have involved extended time extrapolation.
→ This minimum Mn level was selected since lower levels are known to reduce stress rupture properties.

Table 2.1.12

TABLE 2.1.13 Creep rupture strengths of carbon steels (silicon- and aluminium-killed)

Conditions of steel to which the properties apply

	Details of materials actually tested		Range for which data are expected to apply agreed by TC17/SC10/ETP	
			min.	max.
Chemical composition (wt %)	C	0.10–0.185	—.	0.30
	Si	0.01–0.32	—	0.50
	Mn	0.36–0.79	→ 0.40	—
	P	0.007–0.029	—	0.050
	S	0.011–0.028	—	0.050
	Al (sol)	0.016–0.102	0.015	—
Heat treatment	1. 899–950°C AC 2. 850–925°C AC + T 600°C		1. Normalised 2. Normalised and tempered 3. Hot finished 4. Hot finished and tempered	
Products	**Form**	**Size (mm)**		
	Tubes	4–28 thk x 38–194 dia.	All wrought product forms	
	Plates	15–50 thk		
	Bars	20 dia.		
	Forgings	25 thk		

Quantity and duration of data upon which the properties are based

Temperature (°C)	Test Duration (h)				
	< 10,000	10 000–20 000	20 000–30 000	30 000–50 000	> 70 000
	Number of test points available				
400	30	7	6	2	—
450	61	6	2	3	—
500	24	3	2	2	—

Average rupture stresses (N/mm²)

Temperature (°C)	10 000 h	30 000 h	50 000 h	100 000 h	150 000 h	200 000 h	250 000 h
380	213	192	183	171[a]	164[a]	159[a]	155[a]
390	197	176	167	155[a]	149[a]	144[a]	140[a]
400	181	161	152	141[a]	134[a]	130[a]	126[a]
410	166	147	138	127[a]	121[a]	116[a]	113[a]
420	151	133	125	114[a]	108[a]	104[a]	101[a]
430	138	120	112	102[a]	96[a]	92[a]	89[a]
440	125	107	100	90[a]	84[a]	80[a]	77[a]
450	112	95	88	78[a]	73[a]	69[a]	66[a]
460	100	84	77	67[a]	62[a]	58[a]	55[a]
470	89	73	66	57[a]	52[a]	48[a]	45[a]
480	78	63	56[a]	47[a]	41[a]	37[a]	34[a]
490	67	52	46[a]	36[a]	29[a]	23[a]	
500	57	42	35[a]				
510	47	31					
520	37						

[a] Values which have involved extended time extrapolation.
→ This minimum Mn level was selected since lower levels are known to reduce stress rupture properties.

Table 2.1.13

TABLE 2.1.14 Creep rupture strengths of carbon–manganese steels (semi-killed or fully-killed carbon–manganese steels including Nb-treated steels)

Conditions of steel to which the properties apply

	Details of materials actually tested		Range for which data are expected to apply agreed by TC17/SC10/ETP	
			min.	max.
Chemical composition (wt %)	C	0.09–0.29	—	0.30
	Si	0.006–0.49	—	0.50
	Mn	0.80–1.64	0.80	—
	P	0.008–0.048	—	0.050
	S	0.001–0.103	—	0.050
	Nb	0.001–0.077	—	0.10
Heat treatment	1. 860–960°C AC 2. 840–960°C AC + T550–720°C		1. Normalised 2. Normalised and tempered 3. Hot finished 4. Hot finished and tempered	
Products	**Form**	**Size (mm)**		
	Plates	1–133 thk	All wrought product forms	
	Tubes	6–666 thk x 6–273 dia.		
	Forgings	25–150 thk x 25–1425 dia.		

Quantity and duration of data upon which the properties are based

Temperature (°C)	Test Duration					
	< 10 000	10 000–20 000	20 000–30 000	30 000–50 000	50 000–70 000	> 70 000
	Number of test points available					
400	454	107	51	75	9	4
450	639	84	36	49	16	3
500	596	88	24	38	21	3

Average rupture stresses (N/mm²)

Temperature (°C)	10 000 h	30 000 h	50 000 h	100 000 h	150 000 h	200 000 h	250 000 h
380	291	262	248	227	215	206[a]	199[a]
390	266	237	223	203	190	181[a]	174[a]
400	243	214	200	179	167	157[a]	150[a]
410	221	192	177	157	144	135[a]	128[a]
420	200	171	156	136	124	115[a]	108[a]
430	180	151	136	117	105	97[a]	91[a]
440	161	132	118	100	89	82[a]	77[a]
450	143	115	102	85	76	70[a]	66[a]
460	126	99	87	73	65	60[a]	56[a]
470	110	86	75	63	56	52[a]	(48)[a]
480	96	74	65	55	(49)	(44)[a]	(41)[a]
490	84	65	57	(47)	(42)	(37)[a]	(32)[a]
500	74	57	50	(41)	(34)		
510	65	50	(44)	(32)			
520	58	(44)	(37)				

[a] Values which have involved extended time extrapolation.
( ) Values which have involved extended stress extrapolation.

Table 2.1.14

2.1.5   High strength low alloy (HSLA) structural steels

2.1.5.1   General characteristics

This class of steels is used primarily for structural applications, the major functional requirements being:

1. high strength (up to 700 MN/m^2)
2. high toughness
3. low impact transition temperature
4. good formability
5. good weldability

The traditional way of achieving increased strength in steels is to increase the carbon content. However, this is incompatible with the other properties required in structural steels. Thus carbon has a pronounced detrimental effect on toughness and impact transition temperature and it impairs formability. Also in welded structures a major problem is heat-affected zone cracking which derives from the formation of low transformation products such as martensite and bainite. The susceptibility to this form of cracking is controlled by the carbon equivalent value (CE) as determined by the following formula.

$$CE = C + \frac{Mn}{6} + = \frac{Cr + Mo + V}{5} + \frac{Ni + Cr}{15}$$

To avoid cracking the CE value should be as low as possible and hence high carbon contents must be avoided. CE <0.14 = excellent weldability; 0.14–0.45 = requires low hydrogen electrodes; >0.45 = very difficult to weld.

An alternative way of increasing strength is by refinement of grain size and compared to all the strengthening mechanisms this has the outstanding advantage that it simultaneously lowers the impact transition temperature. Provided that a sufficiently fine grain size can be attained, further strengthening may be obtained by carbide (or nitride) precipitation whilst maintaining low impact transition temperatures. Such fine grained, low carbon steels form the basis of all modern high strength low alloy steels.

Another major factor controlling the properties of structural steels is the dispersion of non-metallic inclusions, these being silicates, manganese sulphides and alumina. If present as high volume fraction or in highly elongated condition these give rise to low transverse toughness, cracking during forming operations and short transverse cracking (lamellar tearing) in restrained welds. Techniques for alleviating the problems associated with non-metallic inclusions are discussed in Section 2.1.12.

2.1.5.2   Processing techniques for HSLA steels

These steels may be processed in a number of ways to achieve a range of properties which are summarised in:

Table 2.1.15 *Comparison of HSLA steels*, and

Fig 2.1.3 *Comparison of strength and toughness of various types of HSLA steels*

In normalised steels aluminium, niobium and vanadium are all effective in refining grain size. Conventional normalising temperatures are insufficient to achieve re-solution of niobium carbide and hence the improvement in the strength and toughness of Nb steels derives solely from grain size refinement. Similarly no precipitation hardening is obtained with aluminium grain refined steels but the impact properties are further improved by the removal of soluble nitrogen. However, a disadvantage of aluminium nitride grain refinement is that it can only be employed with fully killed steels with an inevitable loss of ingot yield. With vanadium-containing steels both grain refinement and precipitation hardening can be obtained due to the solubility of the carbide at the normalising temperature. Also by adding nitrogen or aluminium the finest possible grain size can be attained whilst accentuating the hardening response. For effect of composition on mechanical properties of normalised HSLA steels see:

Fig 2.1.4   *Effect of alloying additions on the properties of hardened and tempered steels*

2.1.5.3   Bainitic steels

This class of steel represents an intermediate stage between the normalised ferrite/pearlite steels and the quenched and tempered martensitic steels. The ferrite transformation is suppressed by the addition of Mo, Cr and B, thus enabling a bainitic structure to be developed in the hot rolled or normalised condition. Due to the fine effective grain size and carbide precipitation the strength is higher than the normalised steels but the impact transition temperature also tends to be higher.

TABLE 2.1.15 Comparison of HSLA steels

Steel type	Processing technique	Yield strength (MN/m²)	Tensile strength (MN/m²)	Impact toughness (J)	Transition temperature (°C)	Remarks
Hot rolled	Rolled to finishing temperature 0f ~1 000°C and air cooled.	300 450	450 550	55	up to +50	Low cost, low strength, low toughness. Addition of Al lowers impact temperature, and improves toughness. Nb and V increase strength.
Normalised	Hot rolled, reheated to ~ 900°C and air cooled.	300 550	450 650	70	0 to –60	Heat treatment required after rolling.
Control rolled	Final rolling at 800–950°C to give grain refinement by precipitation of Nb and V carbides.	350 550	500 600	70	–20 to –100	High strength and toughness obtained in as-rolled steel.
Bainitic	Alloying additions increase hardenability and ensure bainitic structure	450 500	600 650	70	20 to –20	High strength obtained in as-rolled steel. Impact transition temperature tends to be high.
Hardened and tempered	Hot rolled, reheated to 850–900°C, quenched and tempered at 500–650°C.	450 650	500 750	70	–40 to –100	Highest strength and lowest transition temperature. High cost.

Table 2.1.15

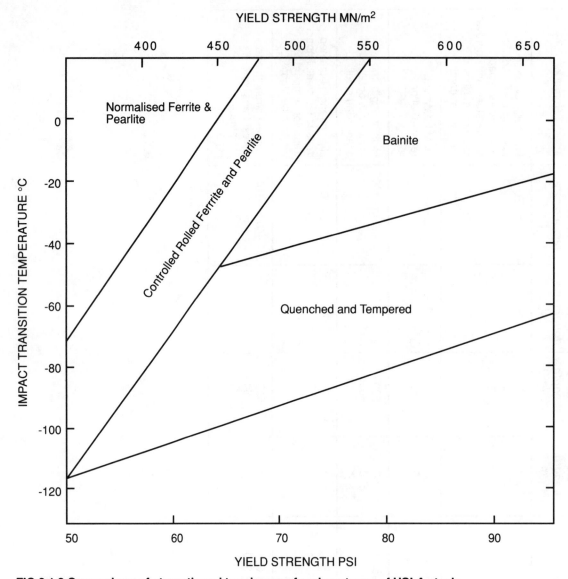

FIG 2.1.3 Comparison of strength and toughness of various types of HSLA steels

Fig 2.1.3

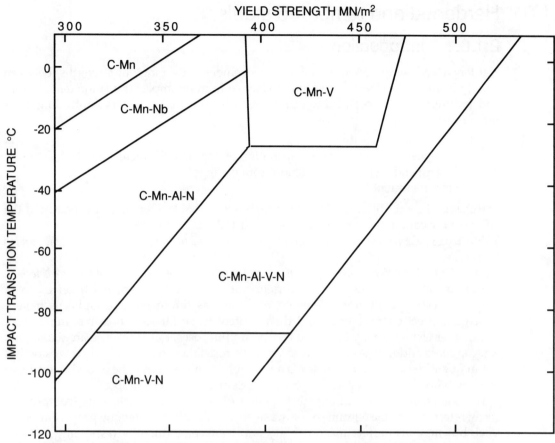

FIG 2.1.4 **Effect of alloying additions on the properties of hardened and tempered steels**

Fig 2.1.4

2.1.6   Hardened and tempered steels

2.1.6.1   Introduction

Hardened and tempered steels have an extremely wide range of strengths, between about 350 and 2000 MN/m^2, and tend to have minimum impact transition temperature and optimum toughness at a specific strength. The major factors controlling their properties are:

1. Chemical composition.
2. Composition, content and distribution of non-metallic inclusions.
3. Degree and method of reduction from the ingot.
4. Heat treatment.

The chemical composition controls the properties available after heat treatment, and the effects of the common alloying elements are listed in:

Table 2.1.16   *General summary of effects of alloying elements on properties of hardened and tempered steels*

Table 2.1.17   *The effects of alloying additions on the properties of hardened and tempered steels*

Most of these elements are added deliberately, but other elements (referred to as 'impurity' elements, 'tramp' elements, etc.) such as tin, arsenic, phosphorus have a damaging effect on properties and their content is, so far as is possible, minimised. Some other elements may be useful for certain purposes, but otherwise are damaging, e.g. copper is added to some steels to improve resistance to atmospheric corrosion but not in quenched and tempered steels and sulphur and lead may be added to improve machinability, but in other steels these three elements are undesirable.

The composition, content and distribution of non-metallic inclusions (comprehensively referred to as the 'cleanness' of the steel) mainly affect the fatigue properties of the steel, but may also affect machinability and transverse ductility (especially in large sizes). Non-metallic inclusion content is controlled or influenced by steel-making and casting processes. Certain special processes (vacuum-arc remelting, electro-slag remelting, vacuum steel-making) produce exceptionally clean steels.

The degree of reduction from the ingot (mostly produced by rolling, except that larger-section billets, above say 300mm diameter, are reduced by forging the ingot) influences the degree of breakdown of the original cast structure or, more specifically, the properties of the material in the longitudinal direction as compared with those in a transverse direction. For most engineering steels the degree of reduction in area is in practice at least about 10:1, and usually much more; variations beyond this ratio have little further effect on the steel. Some forgings may however, have reductions as low as 2:1; variations between 2 and 10 can have significant effects and must be taken into account when considering the properties attainable.

2.1.6.2   Heat treatment

All steels in this class are used in the hardened and tempered condition, but intermediate processing such as rough-machining may be done in the 'softened' or annealed condition, i.e. slow-cooled after rolling (provided the finish-rolling temperature is above the critical temperature Ac_1) or reheated into the austenitic phase and cooled slowly. Indeed, most of these steels are delivered to the customer in the softened condition.

HARDENING

To harden, the steel is reheated into the austenitic condition and cooled quickly enough to prevent transformation to either pearlite or bainite, so that after cooling the structure is 100% martensite, or as near this as possible. During this operation the outer layers of the piece will cool more rapidly than the inner ones, so that the outer layers may transform fully to martensite but the inner layers fail to do so.

HARDENABILITY

Hardenability is a measure of the depth to which the steel will harden fully, and this is usually considered in terms of a round bar of considerable (theoretically 'infinite') length—in practice, of lengths greater than about 3 times the diameter, so that the cooling of the ends does not augment the radial cooling of the middle portions.

Thus, with a C–Mn steel with Mn $\leq$1%, very high cooling rates are required to obtain complete transformation to martensite. Consequently, through-hardening can be obtained only in small diameters (up to about 5mm) water-quenched. With increasing diameter, still water-quenched, the cooling rate decreases so that the extent of martensite transformation falls and there are increasing amounts of bainite and ferrite/ pearlite transformation products. This results in a decrease of both strength and toughness. These steels have *low* hardenability. The effect of diameter on the as-quenched hardness is shown in:

Fig. 2.1.5. *Hardness traverse curves on quenched bars showing effect of alloying on hardenability*

Most of the alloying elements in steel increase hardenability, their effectiveness in decreasing order being roughly carbon, manganese, molybdenum, chromium and nickel, although this order may be affected by such variables as grain size, hardening temperature, etc. If the optimum hardening temperature (about 20–30°C above Ac_1) is used for each steel and other conditions are more-or-less constant, the effect of each element is about proportional to its concentration, and when used in combination, the total effect on hardenability is roughly the sum of the individual contributions. This principle (in association with others relating to the tempering operation—see below) enables nomograms to be constructed to determine the approximate hardness of steel—see:

Fig 2.1.6 *Hardness and strength of plain carbon steels after austenitising, quenching and given various tempering treatments*

Fig 2.1.7 *Hardness increments to be added to the tempered hardness of plain carbon steel because of alloying elements*

As well as increasing the diameter at which full hardening may be obtained, increased hardenability may allow slower cooling rates (e.g. cooling in oil or air) to be used. This reduces residual stresses, distortion, and susceptibility to quench-cracking (see below). Thus, a low manganese steel (0.3–0.6% Mn) in a diameter that requires water-quenching for complete hardening becomes oil-hardening with 0.6–0.9% Mn.

Equivalent diameter

The diameter of the bar (in association with the mode of cooling and the tempering operation), is the main parameter which determines the final properties. It is therefore usual to specify the properties of the steel in terms of the maximum diameter at which a steel-maker would guarantee that those properties could be obtained. This procedure is adopted in most national standards (although steels may alternatively be specified in terms of composition alone or in combination with a hardenability test, see below).

The 'diameter' referred to here is that of the 'infinite' bar, but in practice bars of square or oblong section, plates and forgings that may have quite complex shapes must also be dealt with. The hardening of these other pieces are referred to the 'equivalent diameter' of the piece, which is defined as the diameter of an infinite bar, the centre of which would cool at the same rate as the slowest cooling point in the piece. The value of the equivalent diameter will change slightly according to the mode of cooling, in air, oil or water. For most shapes in general use, i.e. non-circular section bars, plates, irregular shapes etc., the equivalent diameter can be estimated from the rules given in the national standards.

Note that the term *'section'* is often used to mean *'diameter'*, and *'ruling section'* to refer rather vaguely to the section of most importance in determining the properties of the steel. However, the terms *'diameter'* and *'equivalent diameter'* are more specific and are to be preferred. The term *'equivalent section'* is synonymous with *'equivalent diameter'*.

Specification of hardenability

For specification purposes, hardenability is assessed by means of the Jominy End quench test, whereby a bar of the steel ¾" (1.9cm) dia. × 4" (10cm) long is heated to hardening temperature and then placed in a jig in such a way that a stream of water under standard pressure is directed against one end-face, while the other faces cool in air, thus giving a gradation of cooling rate along the bar. A flat is then ground down one side of the bar and VPN, or Rockwell C, hardness measurements taken along this flat, starting at the quenched end. Typical 'end-quench curves' are shown for certain low alloy steels with various carbon contents:

Fig 2.1.8 *Effect of carbon content on end-quench hardenability of low alloy steels containing 0.5% Ni, 0.5% Cr, 0.2% Mo*

There is a rough correlation between distance along the bar and equivalent diameter, but this is seldom used in practice. Instead, the test is used to check that the hardenability of a consignment of steel is up to an appropriate standard. It is usually known from experience that steel of a particular analysis will harden satisfactorily at a certain diameter, so that subsequent casts of the same quality of steel should give an end-quench curve falling within a band derived from such satisfactory casts. This may be reduced to a requirement that the hardness at a specific distance along the bar must lie between two specified values.

Residual stress, quench-cracking and dimensional instability

During the hardening operation, different parts of the piece will cool at different rates (e.g. the outside of the bar will cool more rapidly than the central parts) and will consequently transform at different times and temperatures. Hence the various parts will contract (with the cooling) or expand (with the transformation) at different times, thus setting up considerable internal stress. In certain cases this stress can cause cracking on the surface known as 'quench-cracking'. Indeed high-hardenability steels may continue to transform at the inner regions (with expansion) long after the whole body has reached final temperature and so-called 'delayed cracking' can occur. Thus it is generally good practice to start reheating for tempering as soon as the inner temperature is judged to have fallen to a safe low temperature.

Residual stresses can also cause distortion, although in ordinary bar treatment a small variation of diameter is not usually detrimental. However, if the bars are to be kept straight or if the pieces have been machined almost to size, appropriate jigs may be needed to minimise distortion.

Straightness can often be maintained by careful uniform heating in the hardening furnace and ensuring an end-on entry into the quenching medium.

TEMPERING

After hardening, the piece is reheated for tempering. As tempering temperature increases the as-quenched structure (preferably martensite) gradually alters to other structures depending on the alloy content, and the mechanical properties change according to a fairly standard pattern as exemplified in:

Fig 2.1.9 *Effect of tempering on mechanical properties of 0.4%C low alloy steel at room temperature*

As-quenched martensite contains a high degree of internal stress ('tessellated stresses') which cause the elastic limit (or proof stress) to be relatively low while the tensile strength is very high. As the tempering temperature rises, these stresses are gradually relieved causing the observed increase in proof stress; in practice, temperatures below about 250°C are rarely used.

With further increase of tempering temperature, the proof stress reaches a maximum and then begins to fall, while the tensile strength drops continually. The drop in both these properties is very pronounced in straight carbon and carbon manganese steels. Other alloying elements help to maintain the strength to varying extents.

Tensile ductility (reduction of area % and elongation %) increase progressively with tempering temperature.

The impact toughness of the as-quenched structure is usually very low but increases slightly with rise of tempering temperature. However, with further rise in temperature, the impact values fall again to give low values between about 250 and 550°C (the 'blue brittleness' range), the exact position of the trough varying considerably from one steel to another. Tempering in this range is therefore generally avoided. However, in certain applications where strength is the primary consideration and toughness of less importance (e.g. Springs, see Section 2.1.7 some Tools, see Section 2.1.8), low tempering temperatures may be used and the damaging effect of the blue-brittleness trough can be to some extent offset by increasing the silicon content; thus silico-manganese spring steels with 1.8% Si are usually tempered at 450–550°C.

The normal tempering temperatures for most steels fall within the range 550–650°C, the exact value being chosen to give the optimum combination of strength and toughness. After heating to tempering temperature, and allowing the temperature to equalise over the whole section, the pieces are usually cooled in air, unless the surface must be protected from excessive oxidation, in which case both heating and cooling may be done in an atmosphere-controlled furnace. However, certain steels when cooled relatively slowly through the 250–500°C range become very brittle ('temper-brittleness'). This can be avoided by cooling more rapidly, e.g. quenching from tempering, but if the piece is large this may cause internal stresses due to the unhomogeneous cooling rate, and these stresses can induce distortion when the pieces are machined or lead to failure if the final component is highly stressed. The other alternative is to use a steel with up to 0.5% Mo which generally gives freedom from temper brittleness, unless the cooling is exceptionally slow or service conditions involve very long periods in the embrittling temperature range. It has also been shown that temper brittleness is promoted by phosphorus and certain tramp elements such as arsenic, antimony and tin.

2.1.6.3 Weldability

Low-alloy steels with up to 0.4%C are considered nominally 'weldable', but pre-heating and post-heat treatment (stress relieving) may be required.

All further additions of alloying elements decrease weldability, and all the other standard alloy steels are not considered suitable for *fabrication* by welding, because the deposit is seldom of comparable strength to the parent metal, the properties of the parent metal in the heat-affected zone are completely destroyed, and the heat-affected zone is very prone to cracking.

However, minor welds such as sealing welds or those subject to negligible stress can usually be made provided appropriate precautions are taken, e.g.

— very high pre-heat—for the higher alloy steels, this may have to be almost up to the tempering temperature;
— slow-cooling after welding; and post-heat treatment—if to allow operator access, the pre-heat temperature has been relatively low, the post-treatment may be as

high as the original tempering temperature in order to temper the very brittle heat-affected zone;
— careful avoidance of any restraint on the weld.

2.1.6.4   Machinability

It is generally possible with modern machine tools to machine alloy steels of tensile strengths up to about 1200 MN/m², and such steels may indeed be supplied to the customer in the hardened and tempered condition. Even steels up to about 2000 MN/m² tensile strength can be machined by special techniques, but generally for strengths over about 1400 N/mm² it is preferable to rough-machine in the softened condition before hardening and tempering, and grind to size afterwards; in such cases it may be preferable to use an air-hardening steel to minimise distortion during hardening.

Sulphur and/or lead may be added to certain steels to produce 'free-machining' versions. The improvement in machinability is usually quite dramatic as indicated in:

Fig 2.1.10   *Approximate assessment of machinability of a free-machining 0.25-0.20%C steel as compared with that of the equivalent steel (both normalised)*

In very general terms, the machinability is controlled by the tensile strength of the material, and a broad indication of how the machinability assessment varies with strength is given in Fig 2.1.10. However, it must be emphasised that this diagram can be used only in rough qualitative terms.

TABLE 2.1.16 General summary of effects of alloying elements on properties of hardened and tempered steels

Element	Advantages	Drawbacks
All	Increases hardenability	Decreases weldability.
All except sulphur and lead	—	Decreases machinability.
Carbon	Considerably increases strength and hardenability.	Decreases ductility
Manganese	Increases hardenability.	Increases tendency to quench cracking.
Nickel	Improves impact toughness. Lowers impact transition temperature. $3\frac{1}{2}$ + 9% steels are standard for low temperature service.	
Chromium	Improves wear resistance. Improves corrosion resistance. Improves creep strength.	
Molybdenum	Alleviates temper embrittlement. Increases resistance to tempering. Improves creep strength.	
Vanadium	Increases resistance to tempering. Improves high temperature strength. Improves creep strength.	
Silicon	Inexpensive. Increases hardenability. Improves resistance to scaling at high temperatures. May be used for applications such as springs where toughness is not important.	Large additions reduce toughness. Promotes decarbonisation.
Boron	Increases hardenability considerably when added in small amounts.	
Sulphur	Improves machinability. For highest quality steels must be below 0.02%. For general purpose steels must be below 0.045%. Where machinability is the major consideration may be added up to 0.60%.	Reduces ductility. Reduces cleanness.
Lead	Where machinability is a major consideration may be added up to 0.15%.	Reduces ductility.

Table 2.1.16

TABLE 2.1.17 The effect of alloying additions on the properties of hardened and tempered steels

Element	Strength	Toughness	Impact transition temperature	Remarks
Carbon	←	→	←	Decreases formability and impairs weldability.
Manganese	←		→	Increases hardenability.
Chromium	←		→	Increases hardenability.
Molybdenum	←		→	Increases hardenability and resistance to tempering. Increases high temperature strength.
Vanadium	←		—	Gives increased strength at high tempering temperatures and improved high temperature strength.
Nickel	—		→	Increases hardenability.
Boron	—	—	→	Very effective in lowering input transition temperature. Increases hardenability.

Table 2.1.17

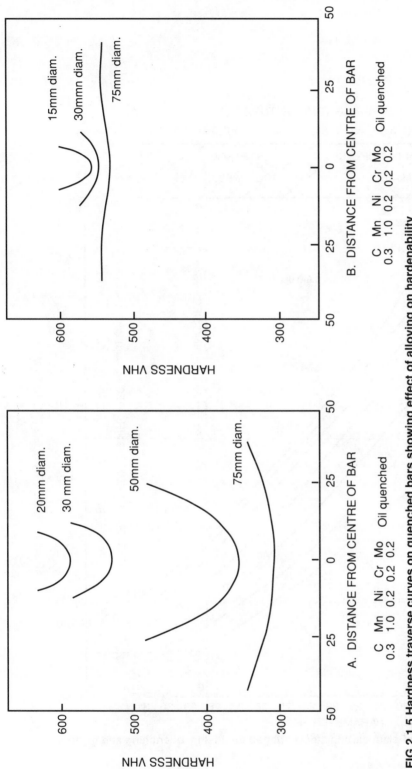

FIG 2.1.5 Hardness traverse curves on quenched bars showing effect of alloying on hardenability

Fig 2.1.5

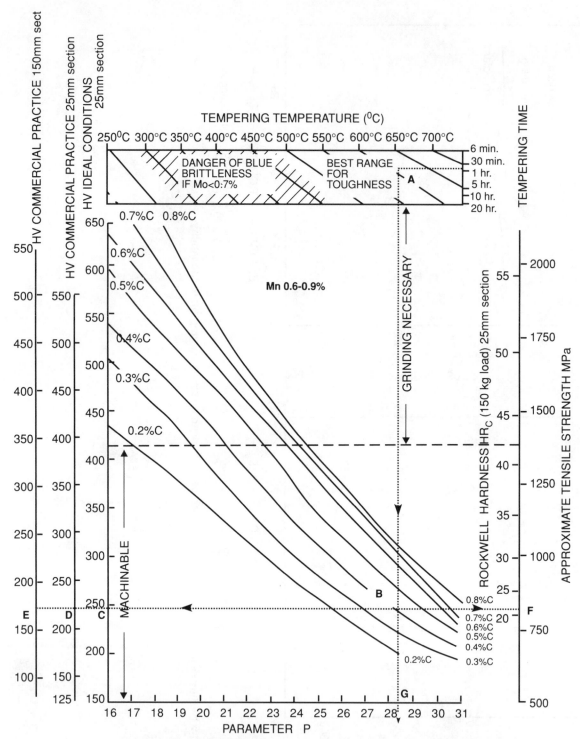

FIG 2.1.6 Hardness and strength of plain carbon steels after austenitising, quenching and given various tempering treatments

Fig 2.1.6

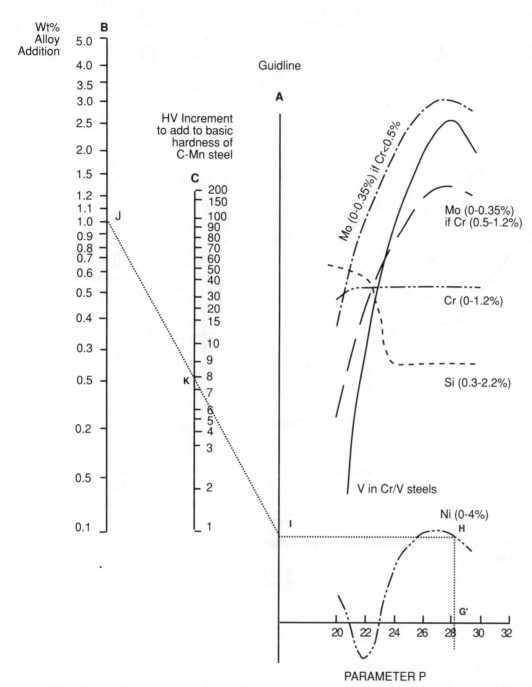

FIG 2.1.7 Hardness increments to be added to the tempered hardness of plain carbon steel because of alloying elements

Fig 2.1.7

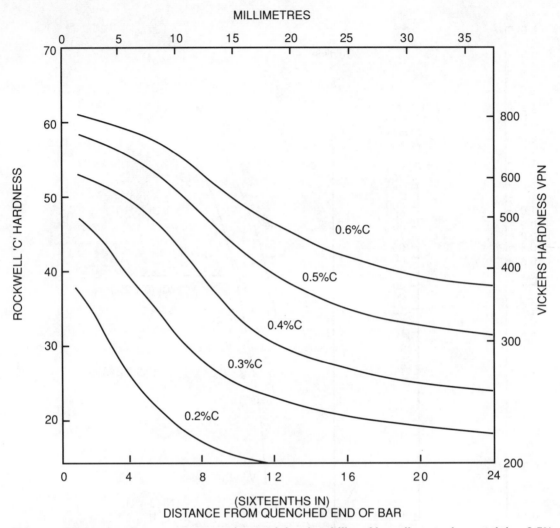

MILLIMETRES

ROCKWELL 'C' HARDNESS

VICKERS HARDNESS VPN

0.6%C

0.5%C

0.4%C

0.3%C

0.2%C

(SIXTEENTHS IN)
DISTANCE FROM QUENCHED END OF BAR

FIG 2.1.8 Effect of carbon content on end-quench hardenability of low alloy steels containing 0.5% Ni, 0.5% Cr, 0.2% Mo

Fig 2.1.8

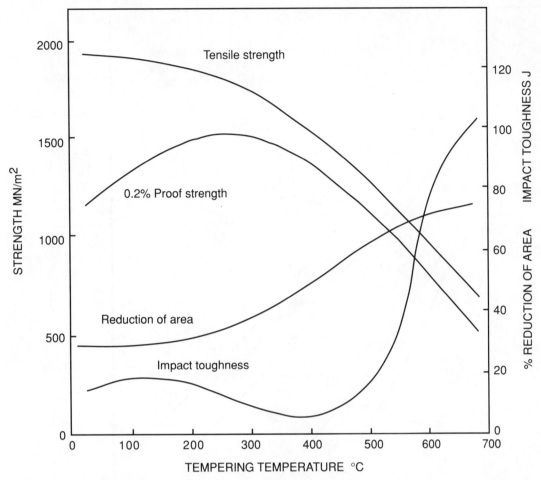

FIG 2.1.9 Effect of tempering on mechanical properies of 0.4%C low alloy steel at room temperature

Fig 2.1.9

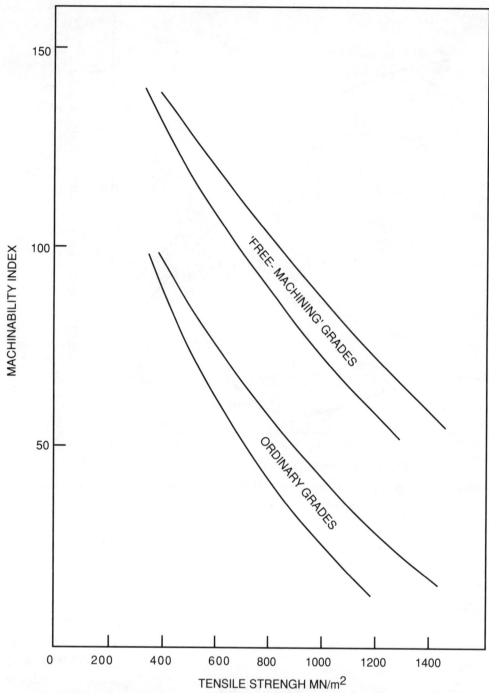

FIG 2.1.10 Approximate assessment of machinability of a free-machining 0.25–0.20%C steel as compared with that of the equivalent steel (both normalised)

Fig 2.1.10

2.1.7   Steels for springs

2.1.7.1   Introduction

The important property required for any spring steel is fairly high surface hardness, and hence high elastic limit. Toughness is not, of itself, a particularly vital property as springs nearly always work under fatigue or relatively static loading and fatigue initiation is not related to toughness (but see Section 2.1.7.3). However, very brittle materials are prone to fracture from small imperfections, damage marks, etc., in the surface.

The highest stresses on all types of spring when under load occur at the surface so that surface condition and freedom from decarburisation (i.e. loss of carbon from the surface layers) which is one of the main causes of failure in alloy steel springs, is particularly important.

Silico-manganese steel decarburises very readily at rolling, coiling and hardening temperatures, and one of the advantages of the other alloy steels is their reduced proneness to this condition. Because of its importance, the bars for all alloy steel springs usually undergo a decarburisation test before acceptance by the spring maker.

It is usual practice for all but the smallest springs to condition the surface by shot peening, which besides cleaning the surface (removing scale and smoothing out other defects) induces a thin layer of compressive stress. Also, most springs are scragged (i.e. for coil-springs, pressed to closure; for leaf-springs, pressed flat), which in effect overloads the spring in the direction it will be used in service. This induces plastic flow in the surface and therefore on release produces residual stresses which oppose the applied service stresses in the surface layers. However, a badly made spring with deep decarburisation cannot be converted into a good one merely by peening or scragging. If the decarburisation is very deep, peening may not affect the full depth, and a layer of weak, decarburised material may be present just below the surface where the stresses will still be nearly equal to the full surface value, and hence early failure may occur.

After peening the steel is prone to rust quickly, so there should be no delay in coating the spring with oil, enamel, or a zinc-rich paint, which is currently the most common coating. It is not considered necessary to phosphate the spring before applying the paint.

The importance attached to surface condition and protection is emphasised by the fact that for a high proportion of springs design is limited by fatigue.

2.1.7.2   Carbon spring steels

Small coil springs are usually made from straight high-carbon steel that has been 'patented' and cold-drawn into wire. 'Patenting' consists of heating the billet before drawing to a high austenitising temperature (usually about 1000°C) which develops a coarse grain-size, so that after slow cooling (often with an initial quench into a molten lead bath) the steel has a coarse pearlite/bainite structure which is much easier to draw into wire than the conventional softened structure.

The steels used, and the hardness available after drawing are summarised in:
Table 2.1.18   *Patented cold-drawn carbon steel wire for springs BS5216*

Springs are cold-coiled from the wire, and no further treatment is required. The table gives only the hardest and the softest wires available in each grade; between these are many ranges (e.g. 35 for Grades 2 and 3) of tensile strength available according to the reduction by drawing.

2.1.7.3   Alloy spring steels

Larger springs, both coil and laminated, are made from carbon and alloy steels, hardened and tempered.

Coil springs are usually made from hot rolled and ground bar that is already at the diameter required for the final spring. After cutting to the length required to make the spring, the bar is heated to coiling temperature within the hardening temperature range, coiled on a mandrel, slipped off the mandrel, quenched in water or oil and tempered.

2.1.7.4 Corrosion-resistant spring steels

Steels are also available for springs that must resist rusting or other corrosive environments, or moderately high temperatures. These are of the ferritic, martensitic and austenitic types of stainless steel, as described in Volume 1, Chapter 1.4.

The ferritic and martensitic steels are used in either the softened condition or lightly cold-drawn to gain a slight improvement of strength. The austenitic steels are too soft for use as softened, so they are usually supplied in the cold-drawn condition, when very high strengths can be obtained.

Precipitation-hardening spring steel is available. This may be coiled in the softened condition and then precipitation-hardened by heating to about 980°C (some springs made from this steel are coiled at this temperature) followed by ageing (soaking) at various times up to almost 16 h at 720°C to develop the required hardness. Alternatively, the steel may be supplied as bar already precipitation-hardened, or hardened by cold-drawing, and the springs made by cold-drawing with no further treatment.

Most of the above springs are made in relatively small diameters, but for larger heavy-duty springs, a high-carbon 12%Cr steel may be used. The tensile strength after hardening and tempering is specified as 95 tons/in^2, minimum, with no diameter given. However, this steel has a very high hardenability, and diameters up to 4in (100mm) can easily be fully hardened, even by air cooling from the hardening temperature.

For the corrosion resistance of all these steels, see Volume 1, Chapter 1.4.

TABLE 2.1.18 Patented cold-drawn carbon steel wire for springs BS5216

| Type | Analysis | | | Grade | Range of tensile strengths and diameters | | | | |
| | C | Si | Mn | | Hardest | | Softest | | |
					Tensile strength (MN/m²)	Diameter (mm)	Tensile strength (MN/m²)	Diameter (mm)	
0.45/0.85C	0.45–0.85	0.35 max.	0.4–1.0	1	1370–1570	2.0	940–1140	9.0	For spring applications under relatively static load, both normal and high duty.
				2	2340–2640	0.2	1060–1260	12.5	
				3	2640–2940	0.2	1260–1460	12.5	
0.55/0.85C	0.55–0.85	0.35 max.	0.3–1.0	2	2340–2640	0.2	1060–1260	12.5	For springs under dynamic load, normal and high duty.
				3	2640–2940	0.2	1260–1460	12.5	
0.7/1.0	0.7–1.0	0.35 max.	0.25–0.75	4	3020–3400	0.1	1770–1920	4.0	For heaviest duty springs, (commercially known as 'music' or 'piano' wire).
				5	3400–3780	0.1	2020–2170	2.8	

Table 2.1.18

2.1.8   Tool steels

2.1.8.1   General characteristics

The name 'tool steels' covers a wide variety of steels that are used for forming or cutting materials ranging from soft woods and plastics to high tensile steels and superalloys. These are usually grouped into five different classes:

1. Carbon tool steels, mainly straight high carbon steels and used for many less arduous purposes.
2. High speed steels, intended for cutting hard materials at a rapid rate.
3. Steels for hot work, used for forming by hot pressing, extrusion and die-casting.
4. Steels for cold work, used for cold-forging, cold extrusion, etc.
5. Shock-resistant steels.

Essential properties are:
High hardness.
Resistance to wear and abrasion.
Reasonable toughness.

Additional requirements of importance to particular classes are:
Red hardness (i.e. the ability to maintain a high hardness even when the tool has become heated into the visible red-heat).
Hardenability.
Machinability in the softened condition, resistance to cracking during heating.
Resistance to decarburisation.
Ability to undergo the full hardening and tempering treatment with little distortion.

METALLURGY

Since hardness and wear resistance are so important, all these steels contain much more carbon than the ordinary engineering steels described in Sections 2.1.2–2.1.6, and to exploit this high carbon as well as to endow red hardness and the various other properties described above, there are usually high contents of alloying elements. The structures generally consist of carbides (especially of W, V and Cr, but often very complex in constitution) in a matrix of martensite which may have been tempered to some extent depending on the proposed use of the material. Because of the high alloy contents, the austenite formed on heating to hardening temperature is very stable and often retained on quenching. Tempering then causes transformation to martensite, so that a second tempering may be needed to continue the transformation process and temper the martensite formed previously. In addition, many of the steels exhibit a 'secondary hardening' in which more carbides are precipitated during tempering, giving an increase in hardness for the upper tempering ranges. Some of the high speed steels, after quenching, may be cooled to a low temperature (e.g. –80°C), which causes considerable transformation of the austenite, before the tempering operation.

Most tools are machined to size, except for a grinding allowance, before heat treatment. It is good practice to stress-relieve them before commencing treatment.

Carbide networks

Because of the high alloy contents, most of the steels in the as-cast condition have a grossly 'cored' structure, with heavy networks of carbides and heavy segregation of elements in and around the network.

It is the function of the forging operation (apart from bringing the ingot to a size appropriate for machining) to break up this network and distribute the carbides more uniformly throughout the material. It is essential, therefore, that forging should be done throughout the temperature ranges quoted below, not completed at the highest workable temperatures as for ordinary steels, i.e. the quoted temperature range is for starting *and finishing* the forging operation.

The carbide problem can be avoided by producing the steel in powder form, through atomisation of the liquid steel and allowing the droplets to fall through a large argon-filled chamber where they each solidify. The powder is then isostatically compacted, first cold and then at about 1000/1020°C to produce a billet which is then processed in the usual way and has a very fine carbide structure.

Decarburisation

Because the high hardness of tool steel derives from the high carbon content, and as most of them are used for processes (e.g. cutting, abrasion, etc.) that operate primarily on the surface of the tool, any loss of carbon from the surface may cause considerable deterioration in performance. The steels vary considerably in their proneness to decarburisation but, in general, good modern tool steel practice involves the use of atmosphere-controlled furnaces for all operations.

2.1.8.2 Carbon tool steels

These are relatively cheap steels used for a large number of hand tools and other applications where high levels of toughness, wear resistance, etc., are not critical, and where some distortion in treatment, and negligible hot hardness can be tolerated. The hardenability of these steels is low, but they will often develop adequate hardness on the surface, even if the inside is much softer—indeed, this supplies a ductile support for the surface material.

Heat treatment is easy, except that quenching from the hardening temperature should always be done in water or brine, and the tool tempered as soon as its temperature has fallen to 'hand warm'.

2.1.8.3 High speed steels

These steels are used for the more onerous cutting conditions, the high contents of carbide forming elements (W, V and Cr) conferring high hardness and wear resistance, and the Co and W conferring resistance to softening at high temperatures. As may be expected from their complex metallurgy, considerable care is required in heat treatment.

All these steels are prone to decarburisation, so that heating for all operations should be carried out in atmosphere-controlled furnaces. Further, the hardening temperature is very critical, so that accurate instrumentation and careful observation is required.

Annealing should be followed by slow cooling in the furnace; indeed the cooling rate for the highest alloyed steels should not exceed 20°C/h. After machining the material should be stress-relieved by heating slowly to the indicated temperature and cooled slowly in the furnace.

Heating for hardening should be done slowly up to about 800/850°C, and then quickly up to the quoted temperature (although a short arrest at about 1100°C is advocated by some authorities before going to final temperature) without soaking. Cooling should be done in either an air-blast, oil or into a salt bath at about 500/550°C followed by air-cooling. As soon as the tools are hand-warm they should be heated for tempering. However, it is becoming more widely accepted that a refrigeration treatment (at –80°C or –196°C, liquid nitrogen temperature) helps to continue the austenite transformation (up to about 15% can be transformed in this way) thus making the subsequent tempering more effective. However, if such a treatment is to be used, the as-quenched pieces should be given a stress-relieving reheat to about 150°C before refrigeration.

All these steels are secondary hardening. As soon as they are hand-warm they should be heated for secondary hardening (i.e. for 'tempering'), soaked for 1–3 h according to section size and then air-cooled. This should be followed by a second tempering, soaking for at least 1 h and air-cooling. Some steels are improved by a third tempering reheat. The steels should be allowed to become hand-warm between each reheating.

Tempering graphs on these steels can be very misleading, as the hardness is dependent on the actual hardening temperature used, the cooling conditions, the amount of retempering etc.

2.1.8.4   Hot work steels

These steels are intended for use in forming (not cutting) hot materials, as in hot pressing or extrusion processes, die-casting of molten metals, etc. They must, therefore, not soften as their temperature rises, and must show good wear-resistance. Also, because the surface is being continually heated and cooled (often by water jets), they must be able to resist the consequent thermal fatigue processes that may cause the surface to crack ('hot-checking', 'craze-cracking').

The basic metallurgy is similar to that of the high speed steels, and all the remarks on heat treatment in Section 2.1.8.3 above apply to them also. However, the pre-heat temperature during hardening for the chromium grades should be 760/820°C instead of the 800/850°C recommended for the high speed steels.

2.1.8.5   Cold work steels

These steels are intended for forming cold materials, as in cold-forging of steels, cold extrusion, etc., and resistance to abrasive wear is of major importance. Also some tools need to be very accurate in final size, or are of intricate shape, or carry some pattern that is to be impressed onto the final product. The necessary machining, which may be very complicated and time-consuming, must be done in the annealed condition, and it is essential that the subsequent hardening and tempering operations do not alter the dimensions significantly.

2.1.8.6   Shock-resisting steels

In certain applications a tool may be subject to heavy vibration or hammering, making it prone to failure by fatigue. It must be hard, as it is often required to cut or penetrate into various hard materials (e.g. stone cement) but a reasonable toughness is required in order to avoid early failure by brittle fracture before the fatigue process has started.

2.1.9   Case hardening of steels

2.1.9.1   Processes

In case hardening the surface of the steel is hardened but the hardness, maximised at the surface, grades into a softer, more ductile, core. There are two distinct processes. In one, the surface composition is changed by a diffusion process so that there are, in effect, two steels; that of the case composition, grading into the core composition. Heat treatment provides dissimilar properties between case and core. The case is hard and wear-resistant whilst the core is strong and tough. The elements diffused into the surface are either carbon or nitrogen or both. The respective treatments are carburising, nitriding and carbonitriding.

In the other process, a hard case is formed by heat treating the surface by induction or flame hardening. No change takes place in steel composition.

2.1.9.2   Diffusion case hardening (carburising, carbonitriding and nitriding)

The several methods of diffusion case hardening are listed, classified, described and compared in:
Table 2.1.19   *Diffusion processes for case hardening*

SELECTION OF CASE HARDENING PROCESS

Carburising: Thick or thin cases can be formed. Process choice depends on work type, and throughput. Box carburising is employed for one-off jobs or very small-scale work. Liquid processes are very versatile and have fairly low capital costs. Gas plants are very suitable for large quantities. Fatigue resistance is improved and a hard-wearing surface of about HB800 is formed.

Nitriding: Very hard but thin cases are produced. Case hardness HV1050–1100. Wear, corrosion and fatigue resistance components are machined and heat-treated before case hardening. Distortion is low because of low treatment temperature. Steel costs and equipment costs are rather high.

SELECTION OF CARBURISING STEELS

The choice of carburising steel depends on the requirement for:

> Strength of core.
> Toughness of core.
> Component size.
> Design, e.g. smooth uniform vs sharply varying section.
> Requirement for case, e.g. maximum hardness but thin or thick.
> Sensitivity to cost of raw material or machining.

The parameters UTS, comparative hardenability of core, hardness of case, severity of quench, toughness of core and comparative raw material cost (which depends mainly on the molybdenum, nickel and chromium content) may be determined from:
Table 2.1.20   *Carburising steels*

SELECTION OF NITRIDING STEELS

The same parameters determine the choice of steel for nitriding as for carburising, but the steels must be harder and therefore stronger than the majority of carburising steels. Hardenability of the nitriding steel is, because of the higher carbon and usually the higher alloy content, in general superior to that of the carburising steels. These parameters are listed in:

Table 2.1.21   *Nitriding steels*

HEAT TREATMENT OF CASE HARDENED STEELS

Four processes in the heat treatment of case hardened steels can be recognised. In individual schemes, these may be combined, or some may be omitted.

(a) Carburising or nitriding. Diffusion of carbon or nitrogen into the surface. Carburising: between 880 and 930°C. Nitriding: diffusion of nitrogen into the surface—usually by means of ammonia gas at about 500°C.

(b) Core refining. A heat treatment to reduce the grain size in the core by austenitising and quenching. Quenching may be water–oil–air–slow cool. Water is often unacceptable because of distortion.

(c) Hardening. A heat treatment to a lower temperature that will austenitise the high carbon case, followed by a quench.

(d) Tempering. A low-temperature heat treatment to give stress relief and reduce brittleness. Usually not above 200°C so that no case softening takes place.

For the sake of economy, a single quench is sometimes effected. After carburising, the temperature is allowed to fall to between that required for (b) and (c) and then quenched. Steels for which this is acceptable are shown in Table 2.1.20.

The fall-off of hardness from the surface varies with steel composition. Some typical curves for hardness against distance into the work for core and case compositions are shown in:

Fig 2.1.11   *Hardenability of typical case hardening steels*

These show hardness profiles for both core and case compositions. In the real piece, the hardness will fall from the case curve to the core curve as the case and core compositions merge. The location of this on the diagram will depend upon case depth.

2.1.9.3   Surface hardening (induction and flame hardening)

Surface hardening of steel is possible by austenitising and quenching the surface only. This is achieved by the application of high heat flux onto the surface by either electrical induction or by direct flame impingement. In both cases a quench follows the surface heating. A uniform bar may be surface hardened continuously by traversing it through the heating medium and then the quench.

INDUCTION HARDENING

The induced currents from an induction coil penetrate deeply at low frequency and are progressively restricted into thinner skins of the work piece as the frequency rises. Choice of frequency is therefore important for the depth of induction hardening required.

Quenching is often in water but oil and air blast are used depending on steel requirements.

FLAME HARDENING

Heat is applied to the surface by means of one or more high-temperature gas burners followed by a quench. The process is versatile and can be affected on work pieces that cannot readily be furnace treated or case hardened as described above.

Any steel of appropriate hardenability may be induction or flame surface hardened. The case is not so hard as can be achieved by a diffusion process but may be significantly thicker.

The results achieved by surface hardening typical steels are listed in:

Table 2.1.22   *Surface hardness of steels flame or induction hardened and variously quenched*

TABLE 2.1.19 Diffusion processes for case hardening

Process class	Process	Description of process	Process temperature	Case depth	Advantages	Limitations
Carburising (Carbon diffusion)	Pack	Parts packed within a heat-resistant box surrounded by a carburising powder consisting of alkali carbonates, charcoal or coke tar, and molasses with a barium carbonate energizer.	~925°C	1.25mm (5h) 1.8mm (10h) 2.5mm (20h)	Low capital cost. Simple. Low distortion.	Labour intensive. Heat wasted.
	Liquid	Parts suspended in molten salt bath containing sodium cyanide (≤ 23%), barium chloride, sodium chloride and accelerators. The salt bath can be heated externally with oil, gas or electricity (submerged electrodes).	Vary according to depth required.	0.075–3 mm	Simple controls can be automated. Bath heat reusable. Can combine carburising, refining and heat treatment.	Poisonous salts and vapours. Equipment maintenance necessary.
	Gas	A special muffle furnace allows carburising gas mixture to pass around the workpieces. The carbon source is usually a hydrocarbon, often natural gas.	925°C max. Higher temperatures shorten furnace life and cause core grain growth	0.5–5 mm	Good control. Suited to mass production. Can be combined with quenching.	High capital cost—not suited to jobbing work.
Carbonitriding	Cyaniding	Similar to liquid carburising but 30–40% sodium cyanide. To provide nitrogen, bath must react with air. A freshly made bath is therefore aged for a few hours at 700°C before use.	870°C	Thinner cases than liquid carburising. ≤0.75 mm	Thinner but harder and more temper-resistant case than liquid carburising. Hard wear resistant case. Longer equipment life.	Poisonous salts.
	Gas	Proportions of carbon and nitrogen may be varied. Ammonia is used to provide nitrogen.	Lower than with gas carburising. Lower temperatures increase nitrogen percentage.	Thinner cases than gas carburising. ≤0.75 mm	Hard and temper-resistant case.	Similar to gas carburising.
Nitriding	Liquid	Low-temperature cyanide bath, pre-aged to allow cyanate formation. With low temperatures and long times the case is mostly nitride.	550°C	Thin	Hard and wear-resistant. Improved fatigue properties. Machined and hardened before casing. No distortion or grinding necessary.	Thin case. Slow process. Not suitable for heavy coarse work.
	Gas	Fully machined and heat-treated parts are nitrided in a muffle in contact with ammonia gas.	500–565°C		Similar to liquid nitriding; used for crankshafts, camshafts, gear shift forks, etc.	Case brittle and can crack or spall if used with plain carbon steels hence special steels necessary.

Table 2.1.19

TABLE 2.1.20 Carburising steels

Part 1. Low hardenability (see footnotes)

Steel designation	BS970	En. No.	C	Mn	Other	Refine (°C)	Harden (°C)	Tensile strength (MN/m²)	Elongation (%)	Impact toughness (J)	Remarks
Carbon	045M10	32A	0–1	0–5		870–900 (1)	760–780 (W)	500	20	55	Properties sensitive to section size.
	080M15	32C	0–15	0–75		870–900 (1)	760–780 (W)	500	20	55	As above.
	210M15	32M	0–15	1.0	0–1–0.18S	870–900 (1)	760–780 (W)	500	20	55	Free-cutting.
Carbon manganese	130M15	201	0–15	1.3		870–900 (1)	770–790 (W)	620	20	55	Hardenability improves core strength
	214M15	202	0.15	1.5	0.1–0.18S	870–900 (1)	770–790 (W)	600	20	40	Free-cutting (as above).
Low chromium	—	207	0.17	0.7	0.6–0.8 Cr	—	—	600–800	—	55	Water quench for higher strength.

Part 2. Medium hardenability

Steel designation	BS970	En. No.	C	Mn	Other	Refine (°C)	Harden (°C)	Tensile strength (MN/m²)	Elongation (%)	Impact toughness (J)	Remarks
3% Nickel	—	33	0.12	0.5	3 Ni	850–880 (1)	760–780 (0)	700	18	55	Good shock resistance.
2% Nickel–Mo	665M17	34	0.17	0.5	2 Ni 0.25 Mo	850–880 (1)	760–780 (0)	700	18	55	Good shock resistance.
2% Nickel–Mo (higher C)	665M33	35	0.23	0.5	1.7 Ni 0.25 Mo	850–880 (1)	760–780 (0)	850	15	30	Moderate shock resistance.
'20' Carbon low alloy	805M20	362	0.20	0.8	0.5 Ni 0.5 Cr 0.2 Mo	850–880 (1)[a]	780–820 (0)[a]	850	15	20	Low shock resistance.
Low Nickel–Cr–Mo	—	325	0.22	0.5	1.7 No 0.5 Cr 0.25 Mo	850–880 (1)[a]	770–800 (0)[a]	850	15	40	Tough, medium strength.
³⁄₄ Nickel–Cr	635M15	351	0.15	0.8	0.75 Ni 0.5 Cr	850–880 (1)[a]	780–820 (0)[a]	700	18	40	Good shock resistance.
15' Carbon low alloy	805M17	361	0.17	0.8	0.5 Ni 0.5 Cr 0.2 Mo	850–880 (1)[a]	780–820 (0)[a]	700	18	35	Moderate shock resistance.
1% Nickel–Cr	635M17	352	0.17	0.8	1.0 Ni 0.75 Cr	850–880 (1)[a]	780–820 (0)[a]	850	15	27	Moderate shock resistance

Part 3. High hardenability

Steel designation	BS970	En. No.	C	Mn	Other	Refine (°C)	Harden (°C)	Tensile strength (MN/m²)	Elongation (%)	Impact toughness (J)	Remarks
3% Nickel–Cr	655M13	36A	0.15	0.5	3.5 Ni 0.75 Cr	850–880 (1)	760–780 (0)	850	15	47	Good shock resistance.
	828M13	36C	0.15	0.5	3.5 Ni 0.75 Cr	850–880 (1)	760–780 (0)	1000	13	40	As above.
5% Nickel	—	37	0.16	0.45	5 Ni 0.15 Mo	850–880 (1)	750–780 (0)	620	20	68	Max. toughness.
4% Nickel–Cr (Mo)	835M15	39B	0.15	0.4	4.0 Ni 1.2 Cr 0.2 Mo	850–880 (2)	760–780 (0)[b]	1310	12	34	Heavy duty components.
1¼ Nickel–Cr	815M17	353	0.17	0.75	1.25 Ni 1.0 Cr	850–880 (2)	780–820 (0)	1000	12	27	Moderate shock resistance.
1¾ Nickel–Cr–Mo	820M17	354	0.17	0.75	1.75 Ni 1.0 Cr 0.1 Mo	850–880 (2)[a]	780–830 (0)[a]	1150	12	27	As above.
2% Nickel–Cr	822M17	355	0.17	0.5	2 Ni 1.5 Cr 0.2 Mo	850–880 (2)[a]	780–820 (0)[a]	1310	12	34	Heavy duty gears, etc.

[a] Or single quench 810–830
[b] Temper 200°C max.

(0) = Carburise at 880–930°C.
(1) = A/O/W
(2) = W/O

A = air cool
O = oil quench
W = water quench

Table 2.1.20

TABLE 2.1.21 Nitriding steels

Steel designation	BS1970 Part 1 1983	Composition (%)								Typical mechanical properties				Remarks
		C	Si	Mn	Ni	Cr	Mo	V	Al	Section size (mm)	Yield strength (MN/m²)	Tensile strength (MN/m²)	Impact toughness (J)	
1% Chromium–molybdenum	708M40	0.36 0.44	—	0.7 1.0	—	0.9 1.20	0.15 0.25	—	—	150	480	625 775	16	Lowest strength steel, suitable for nitriding.
1% Chromium–molybdenum	709M40	0.36 0.44	—	0.7 1.0	—	0.9 1.20	0.25 0.35	—	—	150	585	775 925	50	
3% Chromium–molybdenum	722M26	0.2 0.28	0.1 0.35	0.45 0.7	0.4 max.	3.0 3.5	—	—	—	150	680	850 1,000	50	
3% Chromium–molybdenum–vanadium	897M29	0.35 0.43	0.1 0.35	0.45 0.7	0.4 max.	3.0 3.5	0.80 1.10	0.15 0.25	—	60	1,160	1,310	16	Higher strength. High temperature strength. Higher hardenability.
1½% Chromium–Aluminium–molybdenum	905M39	0.35 0.43	0.1 0.45	0.40 0.65	0.4 max.	1.4 1.8	0.15 0.25	—	0.9 1.3	60 150	680 525	850–1,000 700–850	42 50	High hardenability. Higher case hardness. High wear resistance. Corrosion resistance.
Nitralloy N	—	0.2 0.27	0.2 0.4	0.4 0.7	3.25 3.75	1.00 1.50	0.20 0.30	—	0.85 1.20	—	1,200	1,300	—	High strength.
5% Ni–2% Al	—	0.20 0.25	0.2 0.3	0.25 0.45	4.75 5.25	0.40 0.80	0.20 0.30	0.09 0.15	1.80 2.20	—	1,350	1,400	20	Very high core strength.

Table 2.1.21

TABLE 2.1.22 Surface hardness of steels flame or induction hardened and variously quenched

Steel type	Basic steel composition						Rockwell hardness (HRc)		
	C	Mn	Cr	Ni	Mo	S	Air quench[a]	Oil quench[b]	Water quench[b]
Plain carbon	0.25–0.35	0.6	—	—	—	—	—	—	35–50
	0.4–0.52	0.6–0.8	—	—	—	—	—	52–58	55–60
	0.55–0.75	0.6–0.8	—	—	—	—	50–60	58–62	60–63
	0.83–0.96	0.5–0.6	—	—	—	—	55–62	58–62	62–65
Free cutting	0.25–0.37	1.3–1.6	—	—	—	0.08–0.13	—	—	45–55
	0.38–0.44	1.3–1.6	—	—	—	0.08–0.13	45–55	52–57	55–62
	0.46–0.51	0.7–1.8	—	—	—	0.08–0.2	50–55	55–60	58–64
Alloy	0.38–0.48	1.6–1.9	—	—	—	—	45–55	52–57	55–62
	0.36–0.44	0.6–0.9	0.5–0.8	1.1–1.5	—	—	50–60	55–60	60–64
	0.28–0.38	0.7–0.9	0.3–1.1	—	0.15–0.25	—	—	50–55	55–60
	0.38–0.43	0.2–0.35	0.8–1.1	—	0.15–0.25	—	52–56	52–56	55–60
	0.48–0.53	0.7–1.0	0.8–1.1	—	0.15–0.25	—	58–62	58–62	62–65
	0.38–0.43	0.6–0.8	0.7–0.9	1.6–2.0	0.2–0.3	—	55–57	53–57	60–63
	0.28–0.33	0.7–0.9	0.4–0.6	0.4–0.7	0.15–0.25	—	48–53	52–57	58–62
	0.4–0.64	0.7–1.0	0.4–0.6	0.4–0.7	0.15–0.25	—	55–63	55–63	62–64

[a] Parts away from heated surface must be kept cool.
[b] Thin sections are susceptible to cracking when oil or water quenched.

Table 2.1.22

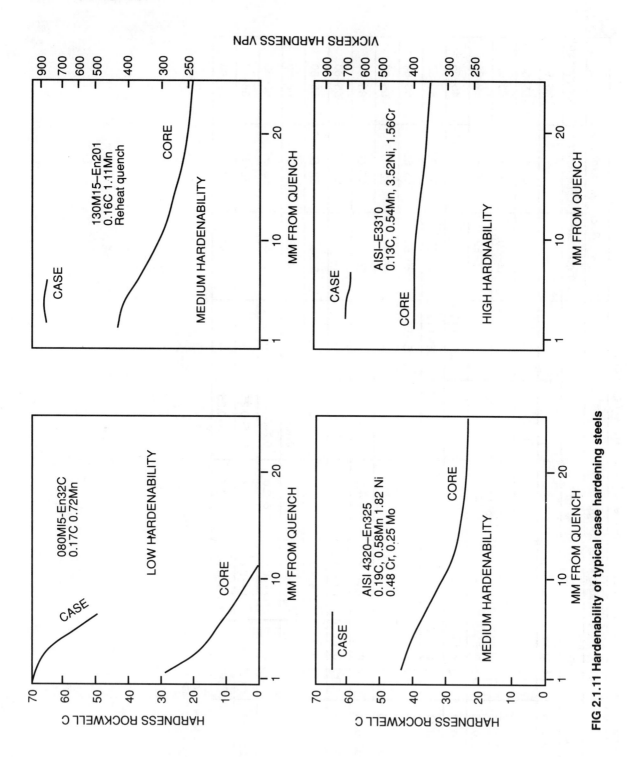

FIG 2.1.11 Hardenability of typical case hardening steels

Fig 2.1.11

2.1.10 Stainless steels

2.1.10.1 General characteristics

INTRODUCTION

The advantages and limitations of stainless steels are listed in:
Table 2.1.23 *Characteristics of stainless steels*

Stainless steels are characterised by their resistance to corrosion and oxidation deriving from the presence of at least 12% chromium which results in the formation of a continuous, adherent and self-healing oxide film.

Because their high alloy content makes these steels expensive they are used mainly in applications where corrosion resistance or another of their specific properties, oxidation resistance, creep resistance, toughness at low temperature or one of their magnetic or thermal characteristics is of paramount importance.

APPLICATIONS

A list of stainless steels and their current uses is given in:
Fig 2.1.12 *Applications of stainless steels*

2.1.10.2 Available types

Stainless steels may be subdivided into five separate types, as shown in:
Table 2.1.24 *Classification of stainless steels*

A general rating of properties influencing the choice between the various grades of stainless steel is shown in:
Table 2.1.25 *Factors influencing the choice of class of stainless steel*

The available forms of stainless steels are given below:

Austenitic stainless	— sheet, strip, plate, bar, wire, tubing.
Ferritic stainless	— bar, plate, sheet, strip.
Martensitic stainless	— plate, sheet, strip, bar, tubing, shapes, wire.
Duplex stainless	— Most wrought forms.
Precipitation hardened	— Most wrought forms.

AUSTENITIC STAINLESS STEELS

There are two main families of austenitic steels:
— The '300' series based on the initially developed 18/8 chromium nickel composition.
— The higher strength '200' series, in which some of the nickel required to ensure an austenitic structure has been replaced by nitrogen and manganese.

Intermediate between these are the 'high proof' austenitics, 18/8 type steels to which has been added roughly 0.2% of nitrogen. This produces alloys of significantly higher strength than, but similar in other respects (including cost) to, the '300 series' steels from which they are derived.

There are in addition a large number of 'non-standard steels' which have been developed for a variety of reasons, usually to provide exceptional resistance to attack by a specific environment but in some cases to provide a material of added strength, toughness or creep resistance.

The influence of alloying elements on the characteristics of austenitic stainless steels is illustrated in:
Table 2.1.26 *Influence of alloying elements on the austenitic stainless steels*

Austenitic steels can be hardened by cold working but not by heat treatment. All are non-magnetic when annealed although some may become slightly magnetic when cold worked. They have excellent resistance to general corrosion, good formability, excellent low temperature toughness and most are readily weldable.

They are, however, expensive compared to most other stainless steels. They suffer from stress corrosion in chloride environments and they may also be prone to pitting and crevice corrosion and weld decay unless composition, heat treatment and conditions of service are carefully controlled.

FERRITIC STAINLESS STEELS

Ferritic stainless steels, which share with the martensitic stainless steels a '400' series number, are generally straight chromium types containing from 11 to 30% chromium and a maximum of 0.1% carbon. They are magnetic, have good ductility (although inferior in this respect to the austenitics), are highly resistant to stress corrosion cracking and cost less than the corresponding austenitic steels. Most are weldable, but prone to grain growth with corresponding reduction in ductility. Type 430 with about 17 Cr is a general purpose grade used for automotive trim and similar applications where weldability is not important. Type 405 is the cheapest ferritic steel, used mainly for automobile exhausts.

Other ferritic stainless steels contain titanium for improved weldability, molybdenum for increased resistance to pitting and crevice corrosion and aluminium for high temperature scaling and sulphur resistance. The influence of alloying elements on the characteristics of ferritic stainless steels is illustrated in:

Table 2.1.27 *Influence of alloying elements on the ferritic stainless steels*

The low interstitial stainless steels contain a maximum of 0.03% of each of carbon and nitrogen. Their ductility, toughness and weldability is superior to the conventional ferritic stainless grades. They range from the general purpose type, 444, with pitting and crevice corrosion resistance equivalent to type 316, to highly alloyed grades such as A1 29–4C which is claimed to compete with titanium and nickel base alloys in many environments. These highly alloyed steels have been used in the USA for sea-water-cooled condensers and heat exchangers.

DUPLEX STAINLESS STEELS

The structure of the duplex stainless steels is a mixture of austenite and ferrite. Their resistance to chloride stress-corrosion-cracking is better than comparable austenitic steels, but may be slightly inferior to the best ferritics. Their high chromium and molybdenum contents confer good resistance to pitting corrosion. In this, they are probably superior to 316 and 317 but inferior to those austenitics specially designed to withstand pitting.

Because the duplexes have better toughness than do the ferritics they can be fabricated in plate thicknesses and may, for example, be used for tube sheets.

Besides their use as tube plates they are very suited to oil and gas production, particularly for sour gas welds and for acetic acid production. They make excellent consumables for welding dissimilar metals and other difficult ferrous materials. The addition of nitrogen produces duplex stainless steels which can be cold worked to strengths unobtainable by ferritic and most austenitic grades. They are highly weldable and exhibit better as-welded corrosion resistance than the conventional duplex grades.

MARTENSITIC STAINLESS STEELS

Martensitic stainless steels have an austenitic structure at elevated temperatures that can be transformed into martensite (i.e. hardened) by suitable cooling to room temperature. They generally contain 11–18% chromium. The lower limit is governed by corrosion resistance, and the upper limit by the requirement for the alloy to convert fully to austenite on heating.

Martensitic stainless steels are magnetic, resist corrosion in mild environments, have fairly good ductility and are (for stainless steels) relatively inexpensive. Some can be heat-treated to tensile strengths exceeding 1400 MPa. The most widely produced grade is the general purpose type 410. Martensitics are used for parts such as valves, tools, cutlery, turbine blades, coal mining equipment and surgical instruments. The influence of alloying elements on the characteristics of martensitic stainless steels is illustrated in:

Table 2.1.28   *Influence of alloying elements on the martensitic stainless steels*

PRECIPITATION-HARDENING STAINLESS STEELS

Precipitation-hardening stainless steels are martensitic or duplex chromium nickel types containing alloying elements such as copper or aluminium which form precipitates during processing. They can be hardened to high strengths by solution treatment and ageing, and combine very high corrosion resistance and mechanical properties. They are used for gears, fasteners, cutlery and aircraft and steam turbine parts.

2.1.10.3   Corrosion resistance of stainless steels

The corrosion resistance of stainless steels is dealt with in relation to other materials in Vol. 1, Chapter 4. In this section Table 2.1.29 lists stainless steels in order of increasing general resistance to corrosion. Table 2.1.30 gives approximate guidance for the main types of stainless steels under the influence of a wide range of corrodants. Since corrosion depends upon many variables including environment, concentration, availability of oxygen, time, temperature, and adjacent materials, a testing programme is always advised. Table 2.1.31 gives the oxidation resistance of stainless steels:

Table 2.1.29   *Stainless steels listed in order of increasing corrosion resistance*
Table 2.1.30   *Corrosion resistance of stainless steels*
Table 2.1.31   *Oxidation resistance of stainless steels*

In general high chromium and molybdenum contents lead to improve corrosion resistance. The dry corrosion or oxidation resistance in air at elevated temperatures is an important attribute of stainless steels. The maximum recommended service temperature for a range of wrought and cast stainless steels in continuous and intermittent use is given in Table 2.1.30 (creep resistance of stainless steels is covered in Section 2.1.14).

Austenitic steels possess excellent resistance to atmospheric corrosion and scaling at elevated temperatures. Molybdenum is added for enhanced resistance to reducing acids and chloride attack. Reduced carbon content or the addition of titanium or niobium alleviates susceptibility to intergranular attack in sensitised or welded conditions. Increased chromium and silicon gives increased scaling resistance up to 1150°C.

Ferritic steels give good resistance to atmospheric corrosion but are inferior to austenitic steels. Corrosion and oxidation resistance increase with increasing chromium content and 25% Cr confers excellent resistance to high-temperature scaling. The addition of molybdenum, niobium and tantalum increases corrosion resistance.

The corrosion resistance of *martensitic steels* is best in the fully hardened condition but only marginally inferior when tempered above 650°C. Resistance to corrosion and stress corrosion cracking decreases markedly when tempered in the range 350–650°C. Corrosion resistance increases with increasing chromium content.

The corrosion resistance of precipitation-hardened steels is equal to or superior to that of martensitic stainless steels.

Ferritic and duplex stainless steels and high nickel alloys resist corrosion in boiling concentrated NaCl but may be attacked by more aggressive environments.

If there is any risk of exposure to aqueous liquid solutions of halides at any temperature above ambient it is preferable not to specify austenitic or martensitic steels.

CORROSION FATIGUE

In corrosion fatigue the separate actions of corrosion and fatigue are mutually reinforced. There are two forms, corrosion-initiated and corrosion-propagated fatigue cracking, which may coexist.

2.1.10.4   Choice of steels for specific environments

ATMOSPHERIC CORROSION

All stainless steels will give good service in a dry, unpolluted atmosphere and all except 410 (which dulls and blackens) will normally retain a light polished appearance.

The ferritic steels such as 430 (and the less resistant austenitic types) have adequate corrosion resistance for domestic and transport hardware (which is usually maintained in a reasonably clean condition). In exceptional locations, or where cleaning is infrequent, 304 should be used. With increasing atmospheric pollution and designs with crevices, steels with higher corrosion ratings such as 316 must be used. Austenitic steels will withstand a normal industrial atmosphere, but the environment in the vicinity of chemical plants will stain and pit any (with the possible exception of the low interstitial steels) in a few decades.

MARINE CORROSION

The selection of a stainless steel for use in sea-water is complicated by the influence of composition changes, flow variations, biological action and the possibility of crevices.

304 is satisfactory for clean, aerated, continually flowing sea-water. 316 usually gives adequate service in clean water and a flow rate not less than 1.5m/s, particularly when the water is aerated. However, even short interruptions of flow, biofouling or the presence of crevices (even under plastic washers) will cause 316 or even 317 to pit.

In conditions marginally too severe for 317 the duplex steel SF22/5 may be used and the interstitial steels 26.1 and 29.4 are stated to give good service. Tests on Avesta 254SMO have shown it to resist pitting in sea-water at temperatures up to 60°C. Austenitic steels are very resistant to cavitation erosion and are therefore recommended for uses such as pump impellers and ship's propellers.

CHEMICAL PLANT

The choice of a material for use in a chemical environment may be influenced by general corrosion, stress corrosion cracking, pitting corrosion, crevice corrosion and galvanic corrosion. The requirements may vary between achieving an economic life for the plant to avoiding contamination of the product by material leached from the container. Material performance may be affected by a variety of factors including concentration, temperature gradients, flow variations, the effect of contaminants, design and fabrication procedures.

CHEMICAL ENVIRONMENT

Acetic acid   For the resistance of stainless steels to acetic acid see:
              Table 2.1.32   *Rate of attack of acetic acid on stainless steels* and
              Fig 2.1.13   *Resistance of nitrogen-containing and standard 304 and 316 steels to acetic acid*
              Nickel base alloys such as Hastelloy C are more resistant than stainless steels.

For castings, high chromium 316 S16 (CF8M) is comparable to 316.

Amino acids	Same corrosive characteristics as acetic acid.
Ammonia	All stainless steels with corrosion resistance equivalent to type 304 or better show good resistance to ammonia in all concentrations up to the boiling point. Pitting in ammonia plant may arise from chlorides in cooling water and may require the use of steels ranging from 430 to 29/4. A specific requirement is cast tubes for reformers which are usually HK40 castings (comparable with wrought 310 but with a higher creep resistance).
Ammonium sulphate	This chemical usually occurs in aqueous solutions containing free sulphuric acid. 316L gives satisfactory service but if 316 (or 316 C16 castings) are used, the component must be given a desensitising treatment after welding or heavy intergranular corrosion will occur adjacent to the weldment.
Bromoform	Type 304 is satisfactory at ambient temperatures. However, wet bromoform discolours in contact with stainless steel and if a white product is required, stainless steel may not be used.
Chlorinated solvents	Dry chlorinated solvents do not attack stainless steels but are hydrolysed in water to form hydrochloric acid. The hydrochloric acid environment may cause pitting, weld decay or stress corrosion cracking unless a steel with at least the pitting resistance of 316, low carbon content or desensitised is used and locked-in stresses are minimised.
Chlorosul- phonic acid	Liquid chlorosulphonic acid may be contained in carbon steel, but stainless steels better than 316 must be used to withstand vapours at and above the surface of the liquid. Stress corrosion and pitting may be caused by chloride impurities produced in sulphation and welds should be fully annealed.
Chromic acid	All stainless steels are satisfactory at temperatures below 24°C or concentrations below about 2%. Above a temperature of about 82°C combined with a concentration of 5% corrosion becomes excessive and alternative materials must be used.
Citric acid	At temperatures up to the boiling point and low concentrations all stainless steels resist citric acid. At concentrations of 10%, temperatures at boiling point and above and in the presence of chlorides, 430, 316 or better must be used. Pitting and stress corrosion can occur in the presence of chlorides.
Dyes	Dyes do not normally corrode stainless steels but chloride or hydrochloric acid impurities may cause pitting and stress corrosion.
Epichloro- hydrin	Epichlorohydrin will stress corrode austenitic stainless steels unless precautions are taken. A 29/4 type should be satisfactory.
Esters	Pure esters do not corrode stainless steel but, during the esterification process, sulphuric acid-resistant materials must be used.
Fatty acids	The acids of higher molecular weight (for example lauric, myristic, palmitic and stearic acids) are less corrosive than acetic acid.

Below 65°C:	stainless steels required only to maintain freedom from contamination and colour.
6–175°C:	all austenitic steels are satisfactory.
Above 175°C:	316 or better is required.
Above 300°C:	317 or better is required.

Contamination with chlorides causes stress corrosion and contamination with sulphuric acid markedly increases the corrosion rate.

Fertilisers	Dry fertilisers may be handled by 409, and liquid types by 304. Potash may contain chloride which may cause pitting.
Formic and Oxalic acids	Formic and oxalic acids are marginally more corrosive than acetic, but all the austenitic stainless steels are satisfactory at room temperature. The low interstitial steels 26/1 and 29/4, show a corrosion rate of 76 and 50 μm/year in boiling 45% formic and 60 μm/year in boiling 10% oxalic acids and are marginally acceptable.
Hydrochloric acid	Except for very dilute aerated environments where 316 or better may be used, stainless steels are not recommended for containing hydrochloric acid.
Hydrocyanic acid	Stainless steels are not necessary for pure hydrocyanic acid. Hydrocyanic acid stabilised with sulphur dioxide against polymerisation can be contained in any austenitic stainless steel (not straight chromium steels) at all concentrations and temperatures up to boiling. Molybdenum-containing steels show improved resistance and all unstabilised steels must be annealed.
Hydrofluoric acid	Stainless steels may only be used at very high or very low concentrations of HF at room temperature.
Lactic acid	Pure, below 40°C, use 304. Impure between 40 and 95°C, use 316 (preferably 316L). Above 95°C, stainless steels are unsuitable.
Monoethanolamine	The choice between carbon and stainless steel for monoethanolamine depends on the process used. For high rated stripping of carbon dioxide, 304 is adequate and it has been used for stripping H_2S. Often small process changes will allow carbon steel to be used.
Nitric acid	Stainless steels are satisfactory for containing nitric acid at moderate temperatures and pressures provided that intergranular attack in the heat-affected zone of welds is prevented, either by post fabrication heat treatment, by the addition of niobium, as in 347 (molybdenum does not improve corrosion resistance in nitric acid), or by the reduction of carbon to a minimum (304L).
	The corrosion rates are greatly aggravated by amounts in solution greater than 0.004% of chromium, which is derived from the corrosion reaction. This promotes intergranular corrosion, even in steels that are otherwise immune and must be borne in mind when corrosion products accumulate as in recycling nitric acid.
	Aluminium is commonly used for storing concentrated nitric acid. For hot concentrated nitric acid high-silicon iron may be used when its shape limitations and mechanical properties are acceptable.
	Agitation, velocity and aeration have little effect on corrosion of stainless steels by nitric acid. Halides however, increase general corrosion, and intergranular attack in susceptible steels. The presence of sulphuric acid does not increase corrosion in solutions in which the nitric acid level is high enough to maintain passivity.
Oxalic acid	See formic and oxalic acids.
Phenol	Phenol does not corrode stainless steels to a significant extent.
Phosphoric acid	The low interstitial steels 26/1 and 29/4 are appreciably more resistant than austenitic having corrosion rates of 30 and 120 μm/year respectively at the boiling points of 54 and 60% concentrated acids.
Silver nitrates	The corrosion of chemical plant producing silver nitrate may not exceed 25 μm/year. Type 310 can have the required corrosion resistance with 25% silver nitrate and 30% nitric acid at 95°C but composition, annealing and pickling are critical and material for each cast used must be tested.

Types 304 and 316 are satisfactory for pure silver nitrate at room temperature but not for elevated temperatures or acidified conditions.

Sodium hydroxide
All stainless steels resist corrosion by sodium hydroxide up to about 65°C. Above this temperature, corrosion may occur. Concentrated solutions at high temperatures stress corrode austenitic stainless steels. No problems have been reported on the behaviour of low interstitial stainless steels (particularly 26/1) in contact with concentrated (up to 50%) solutions of mixed sodium chloride/hydroxide at temperatures up to 150°C.

Sodium sulphide
Type 304 is satisfactory in contact with sulphide solutions in concentrations up to 50% and temperatures up to the boiling point. Specific rates of attack up to 230 μm/year have been reported but most results are in the region of 40 μm/year. No advantage appears to accrue from the use of 316.

It is essential when repairing plant which has been in service with sulphides to remove completely all absorbed sulphur, otherwise the welds will crack.

Stannic chloride
Type 304 will resist a solution containing 1% of stannic chloride at 95°C. The use of 316 extends this range to 10–15% at 20°C. Other materials must be used at higher temperatures and concentrations.

Stannous fluoride
There is evidence that 304 will resist stannous fluoride in aqueous solution at concentrations between 2 and 50% at 95°C. 316 shows a superior resistance.

Sulphuric acid
The resistance of steels to sulphuric acid is highly dependent on concentration.

Behaviour is strongly influenced by the presence of impurities. For example, the addition of 2% HNO_3, 1% CrO_3 and 0.5% $Na_2Cr_2O_7$ virtually eliminates the attack of 30% sulphuric acid at 93°C on 316 and reduces that on 304 to a very low level. Aeration will induce passivity in some regions and agitation will render suspended solids innocuous. On the other hand, hydrogen and carbonaceous deposits (e.g. in crevices) will accelerate corrosion and cause failure by pitting. Impurities may concentrate attack on welded zones and render even more important the need to use low carbon grades of steel.

Sulphonation processes
At room temperature both carbon and stainless steels are satisfactory for sulphuric acid concentrations between 80 and 100% and above 103%. Only stainless steels may be used between 100 and 103%.

Where accuracy of parts such as valves and controls are required, nickel base alloys must be used.

Care is required in sulphonation to ensure that dilute acids do not separate and pit austenitic stainless steels.

Sulphurous acid and sulphur dioxide
The molybdenum-containing stainless steels 316, 317 and 250 SMO will resist wet sulphur dioxide and sulphurous acid but care should be taken in design and operation to prevent:

Crevice-type pitting due to joints, corners and deposited solids.

Stress corrosion due to: stress, hardness above 215 Brinell and the presence of chlorides.

Failure of weld zones due to the use of high carbon steel.

Erosion corrosion which should be checked by rotating specimens in process slurry.

Fine chemicals and pharmaceuticals
The corrosion resistance requirements of materials for the fine chemical and pharmaceutical industries are exceptionally severe because of sanitation, colour and stability which may demand a minimum of metal ions. There are many specific requirements. For example:

Iron base alloys cannot be used for Vitamin B hydrochloride, because it forms a highly coloured complex.

Stainless steels which handle Vitamin C must be copper free because copper promotes decomposition of Vitamin C.

Chloroform can be handled in carbon steel but chloroform which contains hydrochloric acid must be carried in stainless steel 316.

Fine chemicals for which stainless steels are suitable include:

35% ammonium sulphate;

5% butanol at 20°C;

50% caustic soda at 20°C;

20% phenol at 20°C;

20% sodium sulphide at 25°C;

organic acids for protein extraction;

biological media;

metaphosphate chelating agents.

Foods   Stainless steels are used in the food industry for corrosion resistance and because they can easily be cleaned and made sanitary. They may have to withstand process temperatures as high as 150°C. They are also used for washing equipment where they avoid rust stains.

304 can be used for non-corrosive solutions and dry storage. Sources containing acetic acid and chlorides however, corrode 304, and in bad cases even 316. Where there is any doubt 316 should be tested and if pitting occurs 254 SMO or a similar type used in place. Low carbon stainless steels should be used to handle corrosive liquors.

2.1.10.5   Physical properties of stainless steels

The physical properties of typical stainless steels are listed in:

Table 2.1.33   *Physical properties of typical stainless steels*

2.1.10.6   Mechanical properties of stainless steels

The mechanical properties of stainless steels are summarised in:

Table 2.1.34   *Mechanical properties of typical non-hardening stainless steels*

Typical low temperature properties are given in:

Table 2.1.35   *Typical mechanical properties of austenitic stainless steels at low temperatures (annealed condition)*

More information on the strength and toughness of the main types of stainless steels is given below.

(A) STRENGTH

Austenitic steels cannot be hardened by heat treatment but considerable increase in strength can be obtained by cold working. See:

Fig 2.1.14   *Effect of cold working on the yield and tensile strengths of stainless steels*

For maximum work hardening, low chromium and nickel concentrations are recommended. When designing with austenitic steels, it is important to remember that the proof stress/ultimate tensile stress ratio may be only half as compared with three-quarters or more for other types of steel. These steels are suitable for use at very low temperatures since they increase in strength at low temperatures, but retain considerable ductility.

Ferritic steels are usually used in the annealed or cold worked condition and hence possess relatively low strength. For increased strength they may be hardened and tempered. The proof stress of the very low interstitial high-chromium molybdenum (superferritic) grades is higher in the soft condition than the corresponding austenitic steels.

Duplex steels have higher yield strengths than the austenitics and the addition of nitrogen gives them the capability of cold working to exceptionally high strengths.

They have been developed more recently than the austenitic stainless steels and appear to have very great potential but their use has so far been restricted to components such as tube plates, sour gas pipeline components and welding consumables.

Martensitic steels respond in the same way to hardening and tempering as carbon steels (except that the alloy content improves their quench-hardening characteristics). They are therefore capable of at least equivalent tensile strengths to alloy steels.

Strength is obtained on these grades by heat treatment. Austenite is produced by heating to a temperature around 1050°C and decarburisation must be avoided at this stage. Cooling to produce martensite is followed by tempering to produce the required condition.

For very high strength and wear resistance carbon contents up to 1.2% may be employed with tempering temperatures of 150–180°C. For medium strength and good toughness, 0.15–0.35%C is employed with tempering temperatures of 650–750°C. Due to their high chromium content these steels have high hardenability and hence may be hardened by cooling in air or oil from 950–1020°C.

The temperature at which martensite starts to form (Ms) is about 300°C for a basic 12% Cr, 0.14%C steel and the temperature at which transformation is completed (Mf) is about 150–200°C below this. Additions to this base steel depress the Ms temperature as follows:

$$- \quad \Delta \, Ms \; = \; 477 \, [C\%] \; + \; 33 \, [Mn\%] \; + \; 17 \, [Ni\%] \; + \; 17 \, [Cr\%] \; + \\ 21 \, [Mo\%] \; + \; 11 \, [W\%] \; + \; 11 \, [Si\%]$$

If the Ms is below 200°C refrigeration will be required for complete hardening since the Mf temperature will be below room temperature.

Precipitation-hardened stainless steels exhibit a wide range of high strengths which are derived from a combination of carbide and intermetallic strengthening. Aluminium, molybdenum, copper and titanium additions are made for intermetallic strengthening and titanium, molybdenum and vanadium for carbide strengthening. Most of the steels exhibit a peak strength condition but are often used in an under- or over-aged condition because of the increased toughness and corrosion resistance in this form. High-temperature strength is superior to austenitic or stainless steels. The treatment involves several stages and often requires refrigeration.

(B) TOUGHNESS

Austenitic steels have very high impact toughness and exhibit no impact transition temperature. They may be used for cryogenic applications. Impact values for typical austenitic steel (304) type are shown in:

Fig 2.1.15 *Effect of cryogenic temperatures on the properties of 18.8 stainless steel*

Ferritic steels show a marked ductile-brittle transition which is higher for slow cooling from austenitising than for water quenching. The transition temperature is very dependent upon the combined carbon and nitrogen content. If this is above 0.015% the transition temperature is above room temperature. See:

Fig 2.1.16 *Effect of carbon and nitrogen on the impact brittle-ductile transition temperature for ferritic steels*

The newly developed ferritic steels with very low interstitial content possess toughness at room temperature.

Duplex steels are tougher than the ferritics.

The toughness of the *martensitic steels* varies according to the heat treatment process but is in general lower than that of a carbon or low alloy steel at the same strength level.

The *precipitation hardening stainless steels* combine a high corrosion resistance with very high tensile strength levels. A 0.2% proof stress of 1500MPa can be achieved, higher even than martensitic steels (except the 440 types which are not normally subjected to stress other than that involved in the cutting actions). An important advantage is that they are strengthened by a relatively low temperature treatment (480–520°C) so that internal stress and warping are minimised. They may in many cases be finish-machined in the soft condition before ageing. Ductility is similar to that of martensitic grades at the same strength level. Optimum service performance is sometimes achieved by using them in the 'overaged' condition. Their main disadvantage is the complexity of the heat treatment procedures needed to produce their high strength properties consistently.

2.1.10.7   Working of stainless steels

Stainless steels are readily fabricated by conventional forming and joining processes but for certain manufacturing operations some steels are more suitable than others.

HOT FORMING

Stainless steels are readily formed by hot operations such as rolling, extrusion or forging. Forging capability can be extended by special operations such as drawing, piercing and coining. The hot forming characteristics of stainless steels differ on account of their different compositions and microstructures.

Austenitic grades are readily workable at elevated temperatures. They possess higher hot strengths than carbon, alloy, or even martensitic stainless steels and in fact work harden at temperatures within their recommended hot-working ranges, which are shown in:

Table 2.1.36   *Recommended hot-working temperatures for stainless steels*

Greater energy is required to forge austenitic stainless steels than carbon or alloy steels. Austenitic grades have lower thermal conductivities than carbon steels and consequently require longer soaking times. They also have higher coefficients of thermal expansion, for which allowance must be made in order to obtain accurate dimensions.

Problems may arise in hot working (particularly forging) ferritic stainless steels because the rapid grain growth which occurs at forging temperatures reduces the deformation capabilities. Ferritic steels may be forged, however, if the finish forging temperature is adjusted to provide grain refinement.

Martensitic stainless steels have similar forging characteristics to carbon and alloy steels. However, they have greater high-temperature strength and require greater forging loads. Annealing after forging is recommended.

COLD AND WARM WORKING

Because of their excellent mechanical properties stainless steels have good cold forming characteristics. In particular, the austenitic grades have excellent ductility and, because they work harden rapidly (having work hardening coefficient n values about 0.5), they are very amenable to operations which involve stretching, which increase the strength level of the product. Ferritic stainless steels have in general lower ductilities, and n values around 0.2. They are generally therefore less formable and operations which involve stretching require more generous bending ratios and intermediate anneals. Grades such as 409 with greatly improved formability are, however, now available. Formability usually decreases with increasing chromium content and the higher chromium grades can show pronounced brittle tendencies. These types may therefore require warm working. The extra low carbon low interstitial stainless steels have superior formability compared with the standard ferritic stainless steels.

Free-machining grades such as Type 430F are not recommended for cold-forming operations in view of the high volume fraction of inclusions, which significantly reduces ductility. When forming ferritic steels, care must be taken to avoid roping, which is a surface defect appearing as corrugations, that occurs when these steels are subjected to large strains during the forming process.

Martensitic grades have relatively moderate ductilities, which preclude their use in applications requiring a high degree of formability, although, in certain instances, multiple anneals and draws can be used. Forming is generally carried out in the annealed condition owing to the increased ductility that results when hardening is carried out after forming.

Bending

In bending processes all stainless steels within a group behave in the same manner. Most austenitics will withstand a free bend of 180° at a radius equal to one half the thickness.

The 400 series also bend easily, but at a higher radius than the austenitics. The lower chromium steels and the low interstitial steels are best. The others may require warming.

Deep Drawing

Certain austenitic grades which work harden at a relatively slower rate than the standard grades are specially suited for deep drawing. Specially tailored compositions of certain steels such as, for example, Type 304 DD sheet, is available with the most suitable composition and structure.

Ferritic grades require intermediate anneals between draws to prevent rapid thinning and fracture; low interstitial ferritics are better.

Spinning and Flow Forming

Most austenitic steels are suitable for spinning processes. The lower work hardening grades, 302, 304 and 305 can be flow formed to greater reductions.

Ferritic grades can be spun but their lower ductility restricts their formability.

Cold and Warm Heading

Many stainless steels are available in the form of cold heading wire or bar. Most generally used are the plain chromium steels, 430 or the higher strength heat treatable 410. For corrosion resistance or for multiple blow operations in which a number of forming steps are performed in rapid sequence, austenitic steel with good ductility and a low work hardening rate is required. This requires a low ratio of chromium to nickel, and three stainless steels, 305, 384 and UNS530430, work well. 305 and 384 are subject to carbide precipitation if heated or cooled slowly in the range 427–900°C. This can be corrected by annealing and water quenching from 1040°C.

A summary of this information is given in:

Table 2.1.37 *Relative suitability of stainless steels for forming operations*

MACHINING

Stainless steels may be machined by any of the customary methods but because of their toughness, the time taken, and therefore the cost, is higher than that required for carbon and alloy steel and much higher than that of easily machinable alloys such as, for example, brass.

Ferritic grades have good machining characteristics, and martensitics can, if necessary, be machined in the annealed condition.

Precipitation hardening steels may conveniently be machined in the soft condition because a subsequent ageing treatment induces little if any distortion.

The rapid work hardening of many austenitic steels necessitates heavier feeds and slower feeds, sharp tools and copious supplies of cutting fluid, water-soluble or chlorosulphonated oils. Even so, it is more difficult to provide a good machined finish. These problems may be overcome by using free-machining grades which contain sulphur, which minimises tool edge build-up or selenium which also improves surface finish. These additions, however, impair ductility, and may reduce both resistance to corrosion and weldability, and free-machining grades should be adopted only when these drawbacks are not important. A guide to machinability is given in:

Table 2.1.38   *Machinability guide for stainless steels*

HEAT TREATMENT OF STAINLESS STEELS

The heat treatment procedure to be employed depends on the type of stainless steel and the precise temperature and cooling rate varies between individual steels. There are, however, general principles that must be followed.

No stainless steel must enter into service in the 'sensitised' condition if there is any possibility that it may be exposed to an aggressive aqueous environment. Sensitisation arises from the grain boundary precipitation of chromium carbide and the local denudation of chromium from the matrix at temperatures between 500 and 900°C. Steels low in carbon or steel in which the carbon is stabilised by the addition of strong carbide formers, titanium or niobium, are less prone to sensitisation, but stabilised steels or steels with 0.03% carbon can be sensitised if heated for long enough within the critical range, shown in:

Fig 2.1.17   *Relationship between carbon content and sensitivity of stainless steels to intergranular corrosion after heating*

A corresponding problem with martensitic stainless steels is the severe embrittlement which results from tempering for a critical combination of time and temperature. In addition the martensitic steels are prone to stress corrosion in the hardened (untempered) condition.

Precipitation hardening stainless steels are prone to stress corrosion in the 'aged' condition.

To avoid these and other problems, austenitic steels should, wherever practicable, be put into service in the 'solution treated' condition. This implies heating to a temperature usually between 1050 and 1100°C and cooling in air. This treatment, besides eliminating sensitisation, usually ensures maximum ductility, minimum locked-in stress and minimum magnetic permeability.

The solution treatment temperature may be critical in specific cases, too high a temperature causing grain growth but too low a temperature failing to develop creep resistance in a steel such as 321.

The corresponding treatment for ferritic steels is annealing carried out at lower temperatures. In the case of both ferritic and austenitic steels, where advantage is taken of the enhancement of properties by cold or warm work, the solution or annealing treatment must be carried out before cold working. In these cases the final heat treatment may be a stress release at about 300°C. Martensitic steels are hardened and tempered like other ferritic alloy steels. Duplex steels may be solution-treated in approximately the same temperature range as austenitic steels but in general with more rapid cooling (air and water quenching). Duplex steels are in general very easy to handle because the duplex structure precludes problems associated with grain growth.

Obtaining the optimum properties in a precipitation hardening stainless steel may require very complex heat treatments consisting of a solution treatment and at least two ageing treatments. The second, or overageing treatment, is essential to prevent stress corrosion in aggressive environments. These steels have the advantage, however, that the later heat treatment processes introduce very little or no distortion, and the steel may be finish-machined in its soft condition before ageing and, where essential, overageing.

CLEANING OF STAINLESS STEEL

The processes used to manufacture stainless steels introduce surface contamination which must be removed before the component is put into service. Heat treatment thickens the oxide film and may form scale; welding introduces splatter, flux, scale and arc strikes; and grinding, fabrication and erection can attach or embed carbon steel particles and dirt. Heavy scale and welding contamination may be removed by grinding, brushing or shot blasting followed by pickling in a 15% nitric acid, 3% hydrofluoric acid solution. The acid solution should be removed after pickling by thorough rinsing in water. Fabrication and erection contamination may be removed by a passivation solution which consists of a 10–40% nitric acid solution at 50–60°C for as-fabricated austenitic, ferritic and duplex stainless steels, and 20 wt% nitric with a 5 wt% addition of sodium dichromate at 43–50°C for polished steel and martensitic stainless steels. In both cases, a thorough rinsing with water is essential.

An important point which must be borne in mind is that chromium may be oxidised preferably during working or heat treatment so that the surface of the metal beneath the scale becomes deficient in chromium. When the oxide film is removed, the surface metal will have reduced resistance to corrosion and oxidation. It is essential for critical applications to ensure that the whole of the material with reduced corrosion resistance is removed by machining or grinding.

Stainless steel should also be cleaned during service, both for reasons of appearance, and because corrosive environments may develop beneath layers of dirt. Normal industrial cleaning and washing with soap and detergents followed by brushing and, where necessary, polishing is usually satisfactory. Dirt and aggressive chemicals collect in crevices, and are difficult to remove. Careful design which eliminates such crevices and provides smooth, easily accessible surfaces can do much to ensure satisfactory performance for long periods.

2.1.10.8   Joining of stainless steels

Stainless steels may be joined by soldering, brazing, welding, adhesive bonding and fastening. Further details are given in Vol. 1, Chapter 7.

All austenitics can be soldered to any other solderable material. Soldering is, however, relatively weak and corrosion may be promoted by galvanic action. All flux must be removed after soldering. All stainless steels, except the free-machining grades, can be brazed, but to avoid intergranular corrosion the brazing action should be very rapid or restricted to stabilised or low carbon grades. Brazing is particularly useful in joining copper, bronze or nickel to stainless steel.

WELDING

Most fusion welding processes, including manual metal and submerged arc, tungsten and metal arc, inert gas, plasma and resistance, are suitable for almost all stainless steels. Carbon arc and gas welding in which carbon and oxygen could react with chromium and reduce corrosion resistance are unsuitable.

Certain precautions in welding practices must be observed. First, procedures should be followed to preserve corrosion resistance in the weld and in the area immediately adjacent to it, known as the heat affected zone. Second, optimum mechanical properties in the joint must be maintained. Third, certain steps are needed to minimise problems of heat distortion.

In welding, it is necessary to select a weld rod or wire of a weld-filler metal that has corrosion resistance properties nearly identical to—or better than—the base metal. This may well entail using a different specification number welding rod than that of the base material. For example, a 304 steel is welded with a 308 type rod and 400 series steels are often welded with a 300 series welding rod. Choice of the correct filler rod will also ensure optimum mechanical properties. The welding rod manufacturers will recommend the most suitable rod for the steel and conditions.

Welding Austenitic Stainless Steels

Almost all austenitic steels are freely weldable. Pre-heating is not required. Post-heating is necessary only to dissolve precipitated carbides and to stress-relieve components to be subjected to stress corrosion cracking environments. The high coefficient of expansion may cause weld bend cracking and distortion unless precautions are taken. It is necessary, especially when heat inputs and restraints are high, to adjust the composition of the electrode to ensure that the deposited metal always contains ferrite. For components in which a completely austenitic weld is essential, hot tearing must be avoided by attention to design and welding procedure.

A most important consideration in welding austenitic steels which will be subjected to corrosion environments is the prevention of weld decay caused by the precipitation of chromium carbides at or near grain boundaries in the heat-affected zone. This can be overcome by post-welding solution treatment, the choice of grades stabilised with niobium and titanium, or best of all, the choice of low carbon grades. Niobium-stabilised 347 may be susceptible to post-weld heat-affected zone cracking unless welded with a relatively soft filler metal.

The free-machining grades are difficult to weld because the sulphur or selenium additives may cause cracks to develop. Special welding consumables are available.

Welding Ferritic Stainless Steels

Ferritic stainless steels are more difficult to weld than austenitics because they are less ductile and subject to grain growth and sensitisation. These problems can be alleviated by post-weld heat treatment or by the use of the low interstitial grades. Filler metals can be identical to the base metal but austenitic or duplex filler metals have advantages in ductility and toughness.

Welding Martensitic Stainless Steel

Martensitic steels are liable to crack during welding and should be pre-heated in the range 200–315°C and post-heated to restore ductility. The filler metal can be identical to the base metal, or can be an austenitic stainless steel. Welding of the 440 grades is not recommended, but if this process cannot be avoided a 312S94-type electrode should be employed.

Welding Duplex Stainless Steels

Duplex stainless steels in general have excellent weldability.

Welding Precipitation Hardening Stainless Steels

Precipitation hardening stainless steels are suited to welding, a post-welding heat treatment being required only to restore or improve mechanical properties.

Consumables used for Welding Stainless Steel

Basic coated electrodes are used mainly for all positional welding and basic rutile-coated electrodes for high temperature creep resisting applications and for multipass welds in thicker sections. Powder-coated electrodes give deposition rates roughly 50% higher than other electrodes in down fillets and V-butt welds. Acid rutile coatings are used for all other applications. Only tensile strength and elongation are specified.

TABLE 2.1.23 Characteristics of stainless steels

Advantages	Limitations
Excellent general corrosion and oxidation resistance. Wide choice of mechanical properties. Austenitic steels have excellent ductility at ambient and at low temperatures. Steels are available for high temperature applications with high creep and creep rupture strengths. Attractive appearance with minimum maintenance.	Care needed in choice of steel; fabrication procedure and or operating parameters to avoid localised corrosion. Low yield strength of austenitic steels complicates design and choice of steel. High coefficient of thermal expansion (of austenitic steels) and low thermal conductivity amplifies problems of distortion and locked in stresses caused by thermal gradients. High hot strength requires high forging pressures. Great attention must be paid to steel composition, welding procedure, consumables and heat treatment to avoid localised defects, brittleness and corrosion in weldments. High material cost.

Table 2.1.23

TABLE 2.1.24 Classification of stainless steels

Type	Composition or alloy content	Microstructure	Mechanical properties	Physical properties	Advantages	Drawbacks	Applications
Austenitic	15–27% Cr, 8–35% Ni, 0–6% Mo, Cu, N (Mn and/or N may replace Ni)	Austenite	Tensile strength: 490–860 MPa Yield strength: 205–575 MPa Elongation in 50mm: 30–60%.	Non heat-treatable; non-magnetic.	Good ductility esp. at low temperature. Good general corrosion resistance. High creep strength. Good weldability.	Limited strength. Prone to localised corrosion. High thermal comparison. High cost.	Most widely used in general applications.
Ferritic	11–30% Cr, 0–4% Ni, 0–4% Mo	Ferrite	Tensile strength: 415–650 MPa Yield strength: 275–550 MPa Elongation in 50mm: 10–25%.	Non heat-treatable; magnetic.	Moderate cost. Resistant to chloride stress corrosion cracking.	Limited strength.	Parts requiring combination of good general corrosion resistance with good stress corrosion resistance, seawater applications.
Martensitic	11–18% Cr, 0–6% Ni, 0–2% Mo	Martensite	Tensile strength: 480–1000 MPa Yield strength: 272–860 MPa Elongation in 50mm: 14–30%	Hardenable by heat treatment.	Moderate cost. High hardness and strength.	Limited corrosion resistance. Limited weldability.	High-strength parts, pumps, valves and paper machinery.
Duplex	18–27% Cr, 4–7% Ni, 2–4% Mo, Cu, N	Austenite and ferrite	Tensile strength: 680–900 MPa Yield strength: 410–900 MPa Elongation in 50mm: 10–48%	Non heat-treatable.	High strength and corrosion resistance.		Shell-and-tube heat exchangers, wastewater treatment and cooling coils.
Precipitation hardening	12–28% Cr, 4–7% Ni, 1–5% Mo, Al, Ti, Co	Austenite and martensite	Tensile strength: 895–1100 MPa Yield strength: 276–1100 MPa Elongation in 50mm: 10–35%	Hardenable by heat treatment.	Very high strength and corrosion resistance.	Complicated heat treatments.	Parts requiring very high strength and corrosion resistance.

Table 2.1.24

TABLE 2.1.25 Factors influencing the choice of class of stainless steel

Type of steel	Austenitic			Duplex	Precipitation hardening	Martensitic	Ferritic		
Property	General	Low carbon	High nitrogen				High chromium	Low chromium	Low interstitial
Price	5	4	5	3	0	9	8	10	7
General corrosion	9	9	9	10	10	0	10	0	10
Intergranular corrosion	0	9	0	9	9	0	8	6	10
Strength up to 500°C	5	5	6	7	10	9	3	3	3
High temperature creep resistance	10	10	6	6	7	0	0	0	
Ductility	10	10	7	9	5	3	4	5	7
Low temperature ductility		10		4	3	0		3	5
Thermal expansion		0		5	10	10		10	
Thermal conductivity		0		3	1	7		10	
Weldability	5	9	4	10	10	0		6	
Machinability		0		8	7	5		10	

10=Most advantageous; 0=least preferred.

Table 2.1.25

TABLE 2.1.26 Influence of alloying elements on the austenitic stainless steels

General purpose 302–18 Cr 8Ni 0.15C

Element added (or reduced)	C	Ni	Cr	Si	Mo			Ti	Nb	S or Se	N	Ti & Al
Action (+ Add, ↑ Increase, ↓ Decrease)	→	→	←	←	+	←	←←	+	+	←	←←	+
Effect	Reduces weld decay	Increases austenite stability; Decreases stress corrosion	Increases oxidation resistance (requires augmentation of Ni)	Increases scaling resistance	Increases corrosion resistance; Increases creep resistance	Increases resistance to pitting corrosion	Greatly increases resistance to pitting corrosion	Eliminates weld decay	Eliminates weld decay; Increases creep resistance	Increases machinability (usually to the detriment of other properties)	Greatly increases strength which can be obtained by cold or warm working	Greatly increases creep resistance
Steel type	304	301 / Alloy 800	310	302B	316	317	318	321	347	303 & 303 Se	'Nitronic'[d] series	Superalloys
Additional element	N	N & Mn	Si		C	N	Cu		Mo + V			
Action	+	+	←		→	←	+		+			
Effect	Increases strength	Restores austenite stability at reduced cost. Increases strength	Highest scaling resistance		Improves weldability	Increases strength	Virtually eliminates pitting and crevice attack in seawater. Resist H_2SO_4		Increases creep resistance			
Steel type	304N / 304L / 304LN	201, 202, 205	314		316L	316N	254 SMO[a] / 20 Cb3[c]		Esshete[b] 1250			

a Product of Avesta Jernwerks (Sweden).
b Product of British Steel.
c Product of Carpenter.
d Product of Armco.

Table 2.1.26

TABLE 2.1.27 Influence of alloying elements on the ferritic stainless steels

Basis steel	General purpose 430–17 Cr								
Modified composition	11%Cr	13%Cr	15%Cr	17%Cr 1%Mo	17%Cr 0.6%Mo 0.15%S	21%Cr	25%Cr	26%Cr 1%Mo	29% Cr 4%Mo
Effect	Greatest reduction in cost	Reduces cost	Improves weldability	Improves corrosion resistance (for automobile trim)	Improves machinability (at expense of mechanical and corrosion properties)	Increases scaling resistance	Increases corrosion resistance		
Steel type	409		429	434	430 F	442	446		
Additional element	Ti	Al		Nb + Ta	Se (instead of S)	Mo	Al	C.N.	C.N.
Action + add ↑increase ↓decrease	+	+		+	+	+	+	↓↓↓	↓↓↓
Effect	Cheapest steel used for automobile exhausts	Cheapest steel that does not harden		Improves corrosion and scaling resistance	Improves surface finish of machined surfaces	Increases corrosion resistance	Increases scaling resistance	Improves resistance to crevice and pitting attack	Improves resistance to crevice and pitting attack
Steel type	409	405		436	430 F Se	444	Sicromal 12	26/1	29/4

Table 2.1.27

TABLE 2.1.28 Influence of alloying elements on the martensitic stainless steels

Basis steels   General purpose 410 (and 403) 12G 0.15C

Element added	Ni	C	Cr	Mo	P + S
+ Added ↑ Increased	+ +	↑	↑↑	↑	↑
Effect	Improves corrosion resistance	Improves mechanical properties			Improves machinability (often to detriment of other properties)
Steel type	414	420			416

Additional elements	Cr	Mo, V, W	P & S	C	C	C	Se
+ Added ↑ Increased	↑	+	↑	↑	↑↑	↑↑↑	+
Effect	Improves corrosion resistance and mechanical properties	Improves mechanical properties at high temperature	Improves machinability (often to detriment of other properties)	Highest hardness with toughness	Highest hardness with some toughness	Highest hardness	Improves machined surface finish
Steel type	413	422	420 F	440 A	440 B	440 C	416 Se

Table 2.1.28

TABLE 2.1.29 Stainless steels listed in order of increasing corrosion resistance

410	12 Cr Martensitic
430	17 Cr Ferritic
436	17 Cr 1 Mo Ferritic
304	18 Cr 9 Ni Austenitic
	also 304L, 304N, 304LN
316	17 Cr 12 Ni 2.5 Mo Austenitic
	also 316L, 316N, 316LN
317	18 Cr 15 Ni 3.5 Mo Austenitic
310	25 Cr 20 Ni Austenitic
SF22/5	22 Cr 5 Ni 3 Mo Low C Duplex
20 Cb3[a]	20 Cr 35 Ni 2.5 Mo 3.5 Cu (nickel alloy)
18/2	18 Cr 2 Mo Low interstitial ferritic
29/4	29 Cr 4 Mo Low interstitial ferritic
254 SMO	20 Cr 18 Ni 6 Mo 0.7 Cu Austenitic

[a] 20 Cb3 is a nickel alloy but has been included to indicate its relative position.

Table 2.1.29

TABLE 2.1.30 Corrosion resistance of stainless steels (see footnotes)

Corrodant	Concentration % (w/v)	Temperature °C	Austenitic		Ferritic	Martensitic
			AISI 304	AISI 316	AISI 430	AISI 420
Acetaldehyde	98	50	A	A	—	—
Acetic acid	0–100	0–118	C	B	C	—
Acetic anhydride	100	100	B	B	—	—
Acetone	—	30	A	A	—	—
Acetone	—	100	A	A	—	—
Acetyl chloride	100	25	B	A	—	—
Aluminium acetate	0–Sat.	20	B	—	—	—
Aluminium chloride	0–26	30	B	B	—	—
Alum. pot. sulphate	0–Sat.	25	—	A	—	—
Alum. pot. sulphate	0–Sat.	b.p.	—	B	—	—
Aluminium sulphate	Sat.	30	A	A	—	—
Ammonia	99	100	C	A	—	—
Ammonium carbonate	0–20	30	A	A	A	—
Ammonium chloride	0–Sat.	25–75	C	B	C	—
Ammonium hydroxide	0–30	30	—	A	—	—
Ammonium nitrate	0–Sat.	b.p.	—	A	—	—
Ammonium phosphate	5–10	0–50	A	A	—	—
Ammonium sulphate	20	25	C	A	C	—
Amyl alcohol	100	30	A	A	—	—
Amyl phenols	100	30–100	A	A	—	—
Aniline	100	30	A	A	—	—
Antimony trichloride	20	75	C	B	—	—
Arsenic acid	40	30	B	A	—	—
Atmos. marine[a]	—	UK	A	A	—	—
Atmos. rural[a]	—	UK	A	A	—	—
Atmos. industrial[a]	—	UK	B	A	—	—
Barium chloride	20	100	—	C	—	—
Beer	—	25	A	A	A	—
Beet sugar	0–40	30	A	A	—	—
Benzene	100	30	—	A	—	—
Benzoic acid	Sat.	25	A	A	A	—
Benzoyl chloride	100	20	B	A	—	—
Boric acid	0–20	100	B	B	—	—
Boron trichloride	100	12–15	—	A	—	—
Boron trifluoride	100	150	A	A	—	—
Bromine	0–Sat.	20	C	C	C	—
Butane	100	25	A	A	—	—
Butyl acetate	100	25	A	A	—	—
Butyl alcohol	100	25	A	A	—	—
Butyric acid	100	100	B	A	—	—
Calcium bromide	10–Sat.	25–100	A	—	—	—
Calcium chloride	0–25	25	—	—	C	—
Calcium hydroxide	0–Sat.	25–100	A	A	—	—
Calcium phosphate	Sat.	25	A	A	—	—
Carbon dioxide	100	60–70	A	A	—	—
Carbon dioxide	Sat.	25	A	—	—	—

Table 2.1.30

TABLE 2.1.30 Corrosion resistance of stainless steels (see footnotes)—*continued*

Corrodant	Concentration % (w/v)	Temperature °C	Stainless steel type			
			Austenitic		Ferritic	Martensitic
			AISI 304	AISI 316	AISI 430	AISI 420
Carbon disulphide	100	25	A	A	—	—
Carbon monoxide	100	200	A	—	—	—
Carbon tetrachloride	100	25	—	A	—	—
Chlorine (dry)	100	350	C	C	—	—
Chlorine (moist)	100	20	C	C	C	—
Chloroacetic acid	80	25–60	B	A	—	—
Chlorobenzine	100	25	B	B	—	—
Chlorophenols	100	55	B	B	—	—
Chlorosulphonic acid	100	40	—	C	—	—
Chromic acid	0–100	30	B	B	—	—
Citric acid	0–Sat.	20–100	B	A	B	C
Coffee	—	100	A	A	—	—
Cresols	100	40	A	—	—	—
Crotonaldehyde	100	25–100	A	A	—	—
Cumene	75	25–100	A	A	—	—
Cupric cyanide	Sat.	25	B	—	—	—
Cupric sulphate	0–10	25–100	B	A	B	C
Dibromoethane	100	25	A	A	—	—
Dichloroethane	100	30	A	A	—	—
Diethyl ether	100	20	A	A	—	—
Ethyl acetate	100	30	A	A	—	—
Ethyl alcohol	100	30	B	A	—	—
Ethyl chloride	100	25	—	A	—	—
Ethylene glycol	100	25	A	A	—	—
Fatty acids	—	100	C	C	—	—
Ferric chloride	0–60	30	B	B	B	B
Ferric sulphate	0–30	30	B	B	B	B
Ferrous chloride	0–Sat.	60	B	B	—	—
Ferrous sulphate	10	25	B	B	—	—
Fluoroboric acid	32	25	B	A	—	—
Fluorine (dry)	100	20–200	—	A	—	—
Formaldehyde	37	30	A	A	—	—
Formic acid	0–100	20	C	B	—	—
Freons (dry)	100	25	B	B	—	—
Fruit juices	—	25	A	A	A	—
Furfural	100	100	A	A	—	—
Gelatine	0–30	30	A	A	—	—
Glutamic acid	0–Sat.	30	A	A	—	—
Glycerol	100	30	A	—	—	—
Hexamine	25–80	25–80	A	A	—	—
Hydrocyanic acid	100	25	A	A	C	—
Hydrofluoric acid	0–60	20	C	C	C	—
Hydrogen chloride (dry)	100	25	A	A	—	—
Hydrogen peroxide	0–30	25	A	A	—	—
Hydrogen sulphide (dry)	100	25	A	A	—	—
Hydrogen sulphide (wet)	99	25	A	A	C	—

Table 2.1.30—*continued*

TABLE 2.1.30 Corrosion resistance of stainless steels (see footnotes)—*continued*

Corrodant	Concentration % (w/v)	Temperature °C	Austenitic		Ferritic	Martensitic
			AISI 304	AISI 316	AISI 430	AISI 420
Hydrogen sulphide (soln)	0–Sat.	0–b.p.	B	B	—	—
Hydroquinone	0–Sat.	35	A	A	—	—
Iodine	Sat.	25	A	A	—	—
Iodine (alcohol)	10	25	B	A	C	—
Lactic acid	0–100	25	B	A	—	—
Lactic acid	0–100	b.p.	C	C	—	—
Lead	100	327	A	A	—	—
Lead	100	600	C	C	—	—
Lead acetate	20	30	B	B	B	—
Lithium chloride	30	25	A	A	—	—
Magnesium carbonate	0–Sat.	30	A	—	—	B
Magnesium chloride	0–42	30	B	B	—	—
Magnesium hydroxide	0–Sat.	25	B	—	—	—
Magnesium sulphate	0–30	25	A	A	—	—
Maleic acid	0–Sat.	25	B	B	—	—
Malic acid	1–3	25	A	A	—	—
Manganese chloride	5–20	100	C	C	—	—
Manganese sulphate	25	25–80	A	A	—	—
Mercury	100	25	A	A	A	—
Mercury	100	300	—	—	—	A
Methyl alcohol	~100	30	A	A	—	—
Methyl chloride	100	30	A	A	—	—
Methylene dichloride	99	38	A	A	—	—
Methylene dichloride (wet)	22	120	A	A	—	—
Methylethylketone	100	100	A	A	—	—
Milk	—	30	A	A	A	—
Napthas	96	75	A	A	—	—
Naphthalene	100	210	A	A	—	—
Naphthamic acid	100	25	A	A	—	—
Naphthamic acid	100	227	C	A	—	—
Nickel chloride	0–Sat.	30	A	A	—	—
Nickel Nitrate	0–Sat.	30	B	—	—	A
Nickel sulphate	0–Sat.	30	A	A	—	—
Nitric acid	0–70	20	A	A	B	B
Nitric acid	98	20	B	B	B	—
Nitric acid	0–40	70	B	A	B	—
Nitric acid	40–70	70	B	B	C	—
Nitric acid	10–60	b.p.	B	B	—	—
Nitric acid	70	b.p.	C	C	—	—
Nitric acid	>98	30	A	A	A	A
Nitrobenzine	100	100	B	—	—	—
Nitrous oxide	100	25	B	—	—	—
Oxalic acid	0–Sat.	25	A	B	—	—
Oxygen	100	225–500	A	A	A	A
Oxygen	100	500–1000	B	B	B	—
Ozone	~1	20	A	A	—	B

Table 2.1.30—*continued*

TABLE 2.1.30 Corrosion resistance of stainless steels (see footnotes)—*continued*

Corrodant	Concentration % (w/v)	Temperature °C	Stainless steel type			
			Austenitic		Ferritic	Martensitic
			AISI 304	AISI 316	AISI 430	AISI 420
Paper liquors	—	100	A	A	—	—
Paper liquors	—	230	B	C	—	—
Perchloric Acid	100	25	B	B	B	—
Phenol	Sat.	25	A	A	—	—
Phosphoric acid	0–100	20–80	A	A	C	C
Phosphoric acid	0–40	b.p.	A	A	—	—
Phosphoric acid	40–80	b.p.	C	C	—	—
Phosphorus	100	60	A	A	—	—
Picric acid	0–Sat.	25	B	A	B	B
Potassium bromide	10–Sat.	25	B	B	—	—
Potassium carbonate	0–Sat.	30	A	A	—	—
Potassium chlorate	0–30	25	A	A	B	A
Potassium chloride	0–30	25	B	B	—	—
Potassium chromium sulphate	45	30	C	A	—	—
Potassium cyanide	0–30	30	B	B	—	—
Potassium dichromate	0–20	30	B	A	—	—
Potassium ferricyanide	10–30	25	B	—	—	—
Potassium hydroxide	10	b.p.	A	A	—	—
Potassium iodide	10–Sat.	25–100	B	—	—	—
Potassium nitrate	0–Sat.	30	B	—	—	—
Potassium permanganate	10–Sat.	25–100	B	—	—	B
Potassium persulphate	10	30	A	A	—	—
Potassium sulphate	10	25	B	—	—	—
Pyridine	10–100	25–60	A	A	—	—
Salicylic acid	0–Sat.	30	A	—	—	—
Sewage	—	25	A	A	—	—
Silver nitrate	0–50	25	B	A	B	B
Soap	—	30–100	A	A	—	—
Sodium	100	600	A	—	—	—
Sodium acetate	2–10	25	A	A	—	—
Sodium bicarbonate	0–Sat.	30	A	A	A	—
Sodium carbonate	10	20	B	B	—	—
Sodium chlorate	Sat.	30	A	A	—	—
Sodium chloride	10	25	—	B	—	—
Sodium chloride	10	b.p.	—	B	—	—
Sodium hydroxide	0–100	20	A	B	—	—
Sodium hydroxide	10	b.p.	B	B	—	—
Sodium hypochlorite	10kg/m³	20	C	B	C	—
Sodium nitrate	Sat.	25	B	A	B	—
Sodium triphosphate	0–Sat.	30	A	A	—	—
Sodium monophosphate	0–Sat.	30	B	—	—	—
Sodium diphosphate	0–Sat.	30	A	A	—	—
Sodium sulphate	0–10	30	A	A	—	—
Sodium sulphide	0–Sat.	25	B	B	—	—
Sodium sulphite	25	25	A	A	—	—

Table 2.1.30—*continued*

TABLE 2.1.30 Corrosion resistance of stainless steels (see footnotes)—*continued*

Corrodant	Concentration % (w/v)	Temperature °C	Stainless steel type			
			Austenitic		Ferritic	Martensitic
			AISI 304	AISI 316	AISI 430	AISI 420
Sodium tetraborate	0–Sat.	25	B	—	—	—
Stannic chloride	5–24	100	C	B	—	—
Steam	—	100–220	A	A	—	—
Sugar	0–Sat.	30	A	A	—	—
Sulphur	100	25	B	—	—	—
Sulphur	100	150	C	B	—	—
Sulphur	100	450–750	C	C	—	—
Sulphur dioxide (wet)	99	20	B	B	B	B
Sulphuric acid	0–25	20	C	B	C	—
Sulphuric acid	98	20	—	A	—	—
Sulphuric acid	0–5	80	—	B	—	—
Sulphuric acid	15%SO_3	25–80	—	C	—	—
Sulphurous acid	0–Sat.	20–b.p.	A	A	—	—
Tannic acid	10	25	B	—	—	—
Tar	—	150–320	B	A	—	—
Tartaric acid	0–50	50	A	A	—	—
Tin	100	500–700	C	C	—	—
Trichloroethylene	—	20	B	B	—	—
Trichloroethylene	100	100	A	A	—	—
Turpentine	100	30	A	A	—	—
Urea	10–50	30	B	B	—	—
Vinegar	—	30	A	A	A	—
Vinyl chloride	100	30	B	B	—	—
Water, distilled	—	25–30	A	A	—	—
Water, mine	—	15	B	A	—	—
Water, condensate	—	25	A	A	—	—
Water, sea	—	—	A	A	—	—
Water, town	—	25–100	A	A	—	—
Xylene	99	100	A	A	—	—
Zinc	100	425	C	C	—	—
Zinc chloride	0–Sat.	25	—	B	—	—
Zinc sulphate	20	30	A	A	C	—

A < 0.12mm/year.

0.12 < B < 0.50mm/year.

0.50 < C < 2.5mm/year.

[a] Classification as above, but μm/year (atmospheric exposure).

Table 2.1.30—*continued*

TABLE 2.1.31 Oxidation resistance of stainless steels

Type	AISI	Cr (%)	Ni (%)	C (%)	Other (%)	Maximum service temperature (°C)	
						Intermittent[a]	Continuous
Wrought alloys	201	18	4½	0.08	7Mn	815	845
	202	18	5½	0.07	8Mn	815	845
	302	18	9	0.12		870	925
	304	18	10	0.03–0.15		870	925
	310	25	20	0.15		1035	1150
	316	17	12	0.03–0.07	2½ Mo	870	925
	317	18	15	0.03–0.06	3½ Mo	870	925
	321	18	9	0.08–0.12	Ti	870	925
	347	18	9	0.08	Nb	870	925
	410	13	—	0.12		815	705
	416	13	—	0.12–0.17	Free machining	760	675
	420	13	—	0.17–0.32		735	620
	440	17	—	1.0	0.75Mo	815	760
	405	13	—	0.12	0.2Al	815	705
	430	17	—	0.1		870	815
	442	20	—	0.1		1035	980
	446	25	—	0.2		1175	1095
	BS1648						
Cast alloys	B1	27	4	0.5		—	1065
	B2	27	10	0.5		—	1065
	D	20	8	0.4		—	900
	E	25	12	0.5		—	1065
	F	25	20	0.5		—	1095
	G	20	25	0.5		Good	1095
	H1	17	35	0.5		V. good	1035
	H2	19	39	0.5		V. good	1095
	K	12	60	0.5		Excellent	1095

[a] In certain cases the intermittent maximum temperature is lower than the continuous maximum temperature, due to the effects of cyclic expansion on the integrity of the superficial protective oxide film.

TABLE 2.1.32 Rate of attack of acetic acid on stainless steels

Concentration (%)	Temperature (°C)	Steel	Rate (μm/year)	Notes
99	50	304	38	
50	75	310	38	Pure
99	BP	316	38	
90	110	316	180	3.5% Ionised halogens
10	106	316	50	5–10 ppm Ionised halogens
87	122	316	406	5–10 ppm Ionised halogens
99.5	130	316	76	5–10 ppm Ionised halogens
100	BP	26/1	13	Pure
100	BP	26/1	500	220 ppm Cl$^-$
100	100	316	84	4% Formic acid
25	104	316	38	1.25% Formic acid

Table 2.1.31 and Table 2.1.32

TABLE 2.1.33 Physical properties of typical stainless steels

Steel	Specific gravity at 20°C	Specific heat: mean coefficient 20–200°C (J/kg per K)	Thermal conductivity (W/m per K) 100°C	500°C	Thermal expansion (mean coefficient) microstrain K^{-1} 20–200°C	200–800°C	Electrical resistivity at 20°C (Ω m)	Temperature coefficient of electrical resistance 20–100°C	Magnetic properties B/H
Austenitic steels									
301	7.88	520	16.0	22.0	17.45	19.4	700	0.0012	1.62
302	7.91	520	16.0	22.0	18.0	19.9	700	0.0012	1.02
304 & 304L	7.905	520	16.0	22.0	17.1	19.4	700	0.0011	1.02
305	7.91	520	16.0	22.0	18.0	19.9	700	0.0012	1.02
309	7.80	520	15.7	21.5	16.0	18.2	780	0.0009	1.02
310	7.89	520	15.0	21.0	15.9	18.6	840	0.0006	1.00
315	7.92	520	16.0	22.0	17.4	20.2	730	0.0011	1.02
316 & 316L	7.97	520	16.0	22.0	17.2	19.6	760	0.0011	1.02
317	7.975	520	15.5	21.5	17.3	19.5	810	0.0008	1.02
320	7.96	520	15.5	21.5	17.2	19.6	770	0.0011	1.02
321	7.90	520	16.0	22.0	17.1	19.5	720	0.0012	1.02
347	7.915	520	16.0	22.0	17.2	19.6	710	0.0012	1.02
304N & 304LN } 316N & 316LN	7.9	500	15.1	21	16.8	19.4	750	0.001	1.09–1.02
321 S87ᵃ	7.88	430	15.4	17.6	17.4	19.2ᵇ	725	0.0008	1.8–1.2
Avesta 254 SMO	8.0	500	13.5		16.5		850	—	1.1
Nitronic 40	7.83		11.8	17.4	17.3	20	730	—	1.02
Ferritic steels									
403	7.715	490	25.0	27.0	10.9	12.8	540	0.0018	Magnetic
405	7.715	490	25.0	27.0	11.0	—	600	0.0014	Magnetic
409	7.715	490	25.0	27.0	10.9	12.7	600	0.0015	Magnetic
430	7.72	520	23.0	25.0	10.6	12.7	580	0.0016	600–1100
434	7.695	490	23.0	25.0	10.5	12.6	600	0.0015	600–1100
Low interstitial ferritic steels									
18.2	7.7	460	24.0	27.0	11.0	12.5	600	—	Magnetic
28.4 Z	7.7	500	19.5	—	10.5	—	—	—	Magnetic
Martensitic steels									
410	7.7	490	25.0	27.0	10.9	12.8	530	0.0019	700–1000
430	7.7	490	25.0	27.0	10.9	12.8	540	0.0015	700–1000
Precipitation hardening steels									
17.4 PH	7.8	460	18.3	23.0	11.2	—	770	—	95
17.7 PH	7.8	460	16.4	21.8	11.3	—	800	—	—
FV 520 B	7.83	460	21.0	26.5	11.7	14	850	0.0016	110
Duplex (ferritic-austenitic) steels									
329	7.7	460	21	25	12	13.5			
Ferralium 2553SF	7.81	475–532	16.3	—	11.5				

ᵃ Warm worked 321.
ᵇ 20–600°C.

Table 2.1.33

TABLE 2.1.34 Mechanical properties of typical non-hardening stainless steels

Steel	Tensile strength (MPa)	0.2% Proof stress (MPa)	Elongation on 50mm gauge length (%)			Hardness (max) sheet and strip (HV)	Hardness (max) plate (HB)	Room Temperature Charpy Impact (J)
			0.5–1.6 mm thick	1.6–3.0 mm thick	over 3.0 mm thick			
Austenitics								
201	655	275						
216	626	345						
301	540	215	30	35	40	245	245	
302	510	210	30	35	40	220	—	150
304	510	210	30	35	40	190	—	150
304L	490	195	30	35	40	190	192	150
304N & 304LN	587	293		35	40	190	192	150
305	460	170	30	35	35	185	—	150
309	540	215	30	35	40	—	207	150
310	540	215	30	35	40	205	207	
316	540	210	30	35	40	195	197	165
316L	490	200	25	35	40	195	197	165
316N & 316LN	618	316	25	35	35	195	205	165
317	540	210	30	30	35	—	197	165
317L	490	195	30	30	35	195	207	
320	540	210			40	205	202	
321	510	210			40	200		150
321 warm rolled	618	402			20			
347	510	210	30	35	40	202	200	150
347N & 347LN	648	342			35			
254 SMO[a]	650	300			35		210	
20 Cb3[a]	585	240			30			
Nitronic 5.0[a]	690	380			35			
Esshete 1250[a]	500	185			30			314
Ferritics								
403	420	245	17	20	22	175	192	120
405	420	245	17	20	22	175	192	25
409	420	230	17	20	22	180	179	
430	430	245	17	20	22	175	192	
434	430	245	17	20	22	185	—	25
444	414	241			20			30
Sicromal 10	490	295			12	217	217	
Duplex								
SF22/5[a]	620	450			30	290		120
Ferralium 255 3SF	760	550					297	

[a] Manufacturers' data.

Table 2.1.34

TABLE 2.1.35 Typical mechanical properties of austenitic stainless steels at low temperatures (annealed condition)

Steel	Temperature (°C)	Tensile strength (N/mm^2)	Yield stress (N/mm^2)	Elongation (%)	Charpy Impact value (J)
304	−18	870	260	57	150
	−73	1090	310	49	150
	−130	1310	345	44	145
	−185	1490	380	40	140
	−240	1675	450	32	140
304N	−196				26[a]
304L	−18	730	—	56	150
	−73	915	—	51	150
	−130	1100	315	46	145
	−185	1280	260	42	145
	−240	1460	240	41	140
304LN	−196				39[a]
316	−18	660	310	67	150
	−73	815	380	65	150
	−130	1000	470	62	140
	−185	1210	540	59	140
	−240	1485	615	56	140
316N	−196				26[a]
316LN	−196				39[a]
321	−18	745	305	57	150
	−73	925	345	54	150
	−130	1140	375	50	145
	−185	1380	400	45	145
	−240	1630	455	36	120
347	−18	760	295	58	150
	−73	925	330	53	145
	−130	1105	340	46	145
	−185	1310	350	41	140
	−240	1520	405		140
Nitronic 40	−79	924	600	59	197
	−196	1400	1034		88
	−253	1690	1352	15	62

[a] Guaranteed.

Table 2.1.35

TABLE 2.1.36 Recommended hot-working temperatures for stainless steels

Steel	Recommended hot working temperature (°C)	
	Start	Finish
301	1150–1200	900
302	1150–1200	900
304	1150–1200	900
304L	1150–1200	900
305	1150–1200	900
309	1150 max.	950
310	1200 max.	900
315	1150 max.	900
316L	1150 max.	900
317	1150 max.	900
317L	1150 max.	900
320	1150 max.	900
321	1100–1150	900
347	1100–1150	900
403	1150–1200	900
405	1150–1200	900
409	1050–1150	700
430	1050–1150	700
434	1050–1150	700
410	1150–1200	900
420	1150–1200	900
FV520	1180 max.	1000 min.
Nitronic 50	1232 max.	1177 min.

Table 2.1.36

TABLE 2.1.37 Relative suitability of stainless steels for forming operations

Stainless steel class	BS970	Forming operation							
		Blanking	Piercing	Press brake forming	Deep drawing	Spinning	Roll forming	Coining	Embossing
Austenitic	284S16	B	B	A	A	B/C	A	B	B
	301S21	B	C	B	A/B	C/D	B	B/C	B/C
	302S17	B	B	A	A	B/C	A	B	B
	303S41(FC)	B	B	D[a]	D	D	D	C/D	C
	304S15/16	B	B	A	A	B	A	B	B
	304S12	B	B	A	A	B	A	B	A/B
	305S19	B	B	A	B	A	A	A/B	B
	310S24	B	B	A[a]	B	B	A	B	B
	316S16	B	B	A[a]	B	B	A	B	B
	316S12	B	B	A[a]	B	B	A	B	B
	317S16	B	B	A[a]	B	B/C	B	B	B
	321S12/20	B	B	A	B	B/C	B	B	B
	347S17	B	B	A	B	B/C	B	B	B
Martensitic	410S21	A	A/B	A	A	A	A	A	A
	416S21(FC)	B	A/B	C[a]	D	D	D	D	D
	420S45	B	B/C	C[a]	C/D	D	C/D	C/D	C
	431S29	C/D	C/D	C[a]	C/D	D	C/D	C/D	C/D
Ferritic	403S17	A	A/B	A	A	A	A	A	A
	405S17	A	A/B	A[a]	A	A	A	A	A
	430S15	A	A/B	A[a]	A/B	A	A	A	A
	442S19	A	A/B	A[a]	B	B/C	A	B	B

The comparison is only valid WITHIN A CLASS, whereby the ratings are as follows:

A = Excellent.
B = Good.
C = Fair.
D = Not recommended.
[a] = Avoid sharp bends.

Table 2.1.37

TABLE 2.1.38 Machinability guide for stainless steels

Stainless steel class	BS970	Composition (%)				Machinability (surface feet per minute)
		Cr	Ni	C	Other	
Austenitic	284S16	18	5½	0.7	Mn 8 N	85
	301S21	17	7	0.15		85
	302S25	18	9	0.12		85
	303S21	18	9	0.12	S Free Mach.	150
	303S41	18	9	0.12	Se Free Mach.	150
	304S12	18	10	0.03		95
	305S15	18	10	0.06		95
	305S19	18	12	0.10		80
	309S24	25	14	0.15		70
	316S12	17	12	0.03	2½ Mo	100
	316S16	17	12	0.07	2½ Mo	100
	317S12	18	15	0.03	3½ Mo	70
	321S12	18	10	0.08	Ti	85
	347S17	18	10	0.08	Nb	80
Ferritic	403S17	13	—	0.08		95
	405S17	13	—	0.12		95
	430S15	17	—	0.1	(S, Se)	95 (150)
	442S19	20	—	0.1		95
	ASTM XM 34	18	—	0.05	0.15%S 1.5Mo	150
Martensitic	416S21	13	—	0.15	S	150
	416S29	13	—	0.20	S	160
	420S29	13	—	0.17		75
	420S37	13	—	0.24		75
	420S45	13	—	0.36		75
	431S29	17	2.5	0.2		75
	441S29	17	2.5	0.15	0.2S	150
Precipitation hardened	17.4PH	17	4	0.07	4 Cu, Nb, Ta	55
	17.7PH	17	7	0.09	1 Al	55

Table 2.1.38

Steel Type		Domestic hardware	Automobile trim	Exhaust components	Food and drink	Transport	Airframe	Architectural sections	Process plant	Cutting edges	Car bumpers	Superheater components	Aircraft engines	Blades and shafts	Furnace equipment	Sour gas piping	Marine	Sulphuric acid	Welding electrodes	Nuts and bolts
AUSTENITICS	301		X			X	X													
	302	X		X																
	304				X			X	X			X								
	308																			X
	309														X					
	310														X					
	316				X			X	X				X							
	317							X	X											
	320								X											
	321				X				X			X	X							
	347						X		X			X			X				X	
	254SMO																X			
	20CB3																	X		
	Nitronic 50					X			X								X			
	Esshete 1250											X								
FERRITICS	405								X					X						
	409				X									X						
	430	X	X																	
	434	X	X									X								
	Sichromel											X								
	Low Interstitial — 444									X							X			
	Low Interstitial — 29/4/4									X							X			
DUPLEX	SF22/5								X								X		X	
	Ferralium																X			
MARTENSITICS	403								X				X	X						X
	410												X							X
	420									X										
	431																	X		X
	440									X										

FIG 2.1.12 Applications of stainless steels

Fig 2.1.12

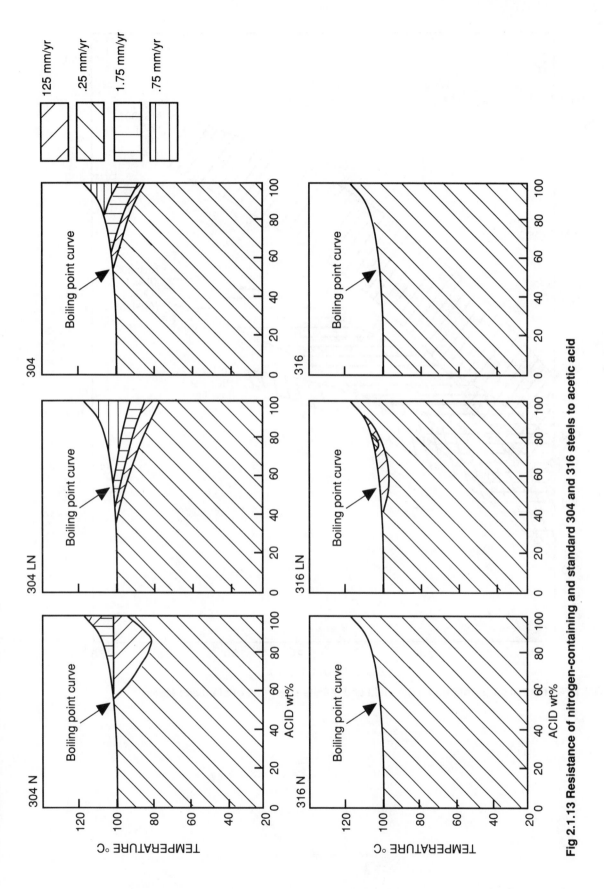

Fig 2.1.13 Resistance of nitrogen-containing and standard 304 and 316 steels to acetic acid

Fig 2.1.13

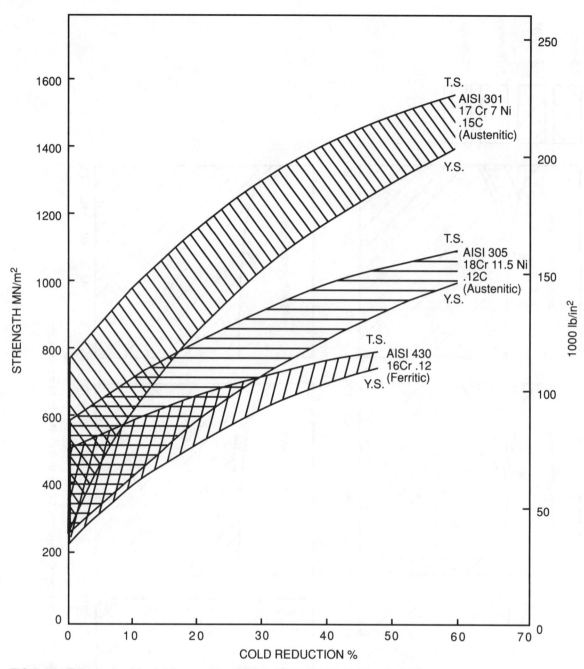

FIG 2.1.14 Effect of cold working on the yield and tensile strengths of stainless steels

Fig 2.1.14

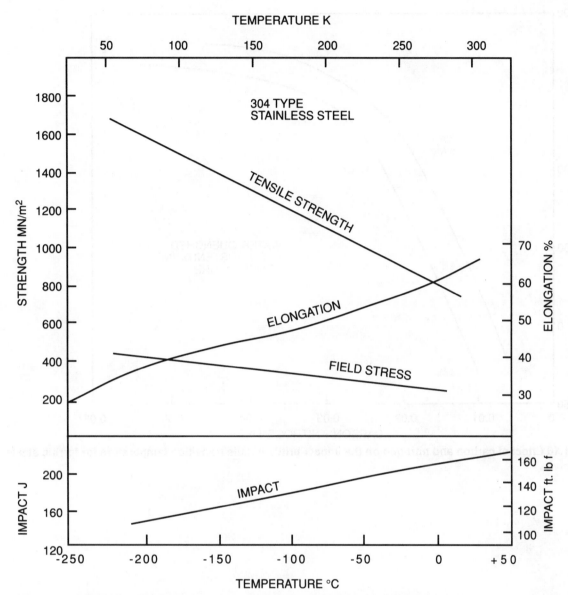

FIG 2.1.15 Effect of cryogenic temperatures on the properties of 18.8 stainless steel

Fig 2.1.15

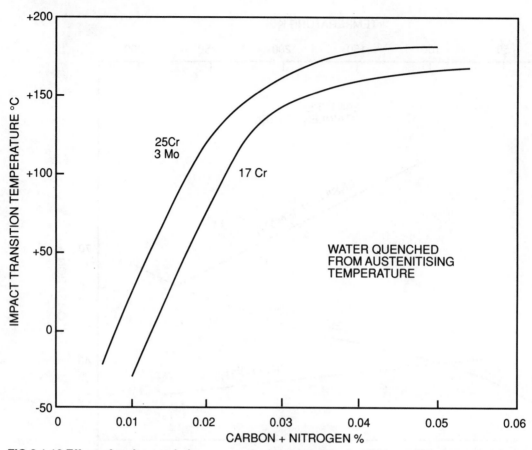

FIG 2.1.16 Effect of carbon and nitrogen on the impact brittle ductile transition temperature for ferritic steels

Fig 2.1.16

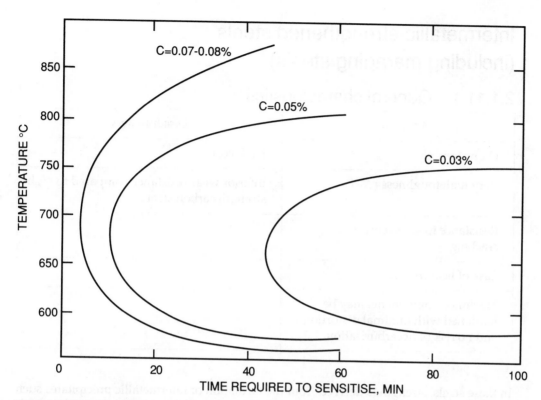

FIG 2.1.17 Relationship between carbon content and sensitivity of stainless steels to intergranular corrosion after heating

Fig 2.1.17

2.1.11   Intermetallic-strengthened steels (including maraging steels)

2.1.11.1   General characteristics

Advantages	Disadvantages
High strength	High cost
Very high toughness	Inferior wear resistance compared to high strength carbon steels
Resistance to stress corrosion cracking	
Ease of heat treatment	
Machined components may be hardened with minimal distortion and no risk of decarburisation	

In these steels, strength is derived from the formation of intermetallic precipitates such as Ni_3Al. Ni_3Ti, Ni_3Mo and Cu. In the annealed or as-rolled condition, the steels are relatively soft and may be readily machined or formed prior to being hardened by a low temperature ageing treatment. There are two main classes of intermetallic strengthened steels:

(a) Medium strength (nicuage)—
   Ni–Cu–Nb steels for structural applications having proof strengths of 540–690 MN/m^2.

(b) High strength (maraging)—
   12–18% nickel steels having proof strengths of 1350–2400 MN/m^2.

Their properties are listed in:
Table 2.1.39   *Composition and properties of intermetallic strengthened steels*

2.1.11.2   Medium strength steels (nicuage)

These steels may be used in the as-rolled and aged or normalised and aged condition. In the as-rolled condition they are strengthened by niobium carbide precipitation and further hardening is obtained on ageing by the precipitation of copper.

In the aged condition the proof strength is typically 650 MN/m^2. This is superior to the strengths exhibited by control rolled or normalised steels but is similar to that obtained from quenched and tempered low alloy steels, (see Section 2.1.6). The outstanding advantage of these steels is that they are readily cold formed in the soft as-rolled condition and may then be aged to peak strength at a temperature of about 560°C. Also because of the low carbon content they have a good weldability and are not susceptible to Heat Affected Zone (HAZ) cracking. Full strength is achieved after welding by the normal ageing treatment. Fibrous fracture energy (e.g. upper shelf energy) and impact transition temperature are comparable with quenched and tempered steels, i.e., very good.

2.1.11.3   High strength steels (marage)

A range of very high strength maraging steels is available with the peak strength being controlled predominantly by the concentrations of titanium and aluminium. In the as-rolled, forged or annealed condition the strength is relatively low and hardening is obtained by an ageing treatment at 500°C.

The advantages of these steels compared with quenched and tempered steels are as follows:

(i)   Ultra-high strengths may be obtained.

(ii)   Toughness is superior to all low alloy carbon steelsof similar strength. For optimum toughness double vacuum melted grades are recommended.

(iii)   Low temperature toughness is very good. No abruptductile to cleavage transition occurs and high toughness isobtained even at –196°C.

(iv)   Weldability, even in the fully hardened condition, isgood . This is a considerable advantage compared with lowalloy steels containing typically 0.4%C. Pre-heat isunnecessary and properties may be restored by normal ageingtreatment.

(v)   In the low strength grades the resistance to stresscorrosion cracking is superior to low alloy steels.

(vi)   Ease of heat treatment and machining.

In the low strength, solution-treated condition, machinability is good and components may be machined to finished dimensions. During the low temperature hardening treatment minimal distortion occurs and there is no risk of cracking or decarburisation. Also because of the low temperature, there is no need to employ controlled atmospheres or extensive surface finishing. As indicated by the production sequences compared below, considerable economies can be achieved in heat treatment and machining, and for complex-shaped, high strength components this may outweigh the higher material cost.

Production Sequence

Maraging steel	*Quenched and tempered steel*
1. As-rolled or as-forged stock (280–320 HV).	1. As-rolled or as-forged stock (260–650 HV).
2. Rough machine.	2. Soften, to 230–320 HV, by heating at 600–700°C; air cool.
3. Stress-relief anneal, if stock has been heavily or non-uniformly machined, by heating at 800–900°C; air cool.	3. Rough machine.
4. Machine to finished dimensions.	4. Stress-relieve if heavily or unevenly machined.
5. Harden, to 500–600HV, by heating at 450–500°C; air cool.	5. Machine to oversize dimensions to accommodate decarburisation and distortion.
6. Final surface finish.	6. Harden by heating to 800–1050°C and cooling at a controlled rate—usually water quench or oil quench.
	7. Temper, to 500–600HV, by heating at 200–600°C.
	8. Finish—machine to remove hardened oversize material (see 5 above).
	9. Final surface finish.

The main disadvantages of maraging steels apart from their high cost, is that their wear resistance and fatigue resistance are inferior to low alloy steels. On smooth test specimens, the fatigue ratio is typically 0.3–0.4 but the rate of fatigue crack growth is no different from low, alloy steels.

TABLE 2.1.39 Composition and properties of intermetallic strengthened steels

Specification ASTM A579	Steel type	Composition											Condition	Annealed hardness (HB)	Proof stress (MN/m²)	Tensile strength (MN/m²)	Elongation %	Reduction in Area (%)	Impact toughness (J)
		C max.	Ni	Mo	Ti	Co	Al	Cr	Cu	Nb min.	Mn	Si							
—	Nicuaging	0.6	0.7–1.0	0.15–0.25	—	—	—	—	1.0–1.3	0.02	0.40–0.65	0.20–0.35	As-rolled	—	490–590	540–640	—	—	65–140
													As-rolled and aged	—	590–690	640–740	—	—	40–120
74	Maraging 12% Nickel type	0.03	12	3	0.1	—	0.3	5	—	—	0.1 max.	0.1 max.		321	1105	1175	15	65	81
75		0.03	12	3	0.2	—	0.4	5	—	—	0.1 max.	0.1 max.		321	1240	1310	14	60	68
71	Maraging 18% Nickel type	0.03	18	3.25	0.2	8.5	0.1	—	—	—	0.1 max.	0.1 max.	Solution treated and aged to 480°C	321	1380	1450	12	55	48
72		0.03	18	5	0.4	8	0.1	—	—	—	0.1 max.	0.1 max.		321	1725	1760	10	45	27
73		0.03	18	5	0.6	9	0.1	—	—	—	0.1 max.	0.1 max.		321	1800	1930	9	40	20
—		0.03	18	4	1.7	12.5	0.1	—	—	—	0.1 max.	0.1 max.		—	2400	2450	—	—	8–15

Table 2.1.39

2.1.12   Selection of steels for toughness

2.1.12.1   Fracture in steels

Ferritic steels are characterised by an abrupt decrease in fracture energy with decreasing temperature. This is associated with a transition from a fibrous to a cleavage mode of failure. Fibrous fractures are usually tough and resist rapid crack propagation, whereas cleavage fractures are associated with a low energy of propagation and usually result in catastrophic brittle failure.

For resistance to brittle fracture (i.e. fracture before general yielding) it is necessary that *both* the transition temperature should be below the service temperature *and* the fibrous fracture should be associated with a high energy of crack propagation. These properties are usually assessed by measuring the energy absorbed in the fracture of notched test pieces (Izod or Charpy) under impact loading conditions and over a range of testing temperatures. The toughness is then characterised by the fibrous fracture energy (upper shelf energy), the impact energy at the service temperature and the transition temperature (this is usually taken as the temperature corresponding to either 50% of the upper shelf energy or 50% cleavage on the fracture surface).

These are valuable parameters for comparing the toughness of steels, particularly if they can be correlated with previous service experience. It is important to note that the transition temperature measured in a notched test specimen is only relevant to the testing conditions specified in the test. An increase in section size, strain rate or notch acuity will lead to an increase in transition temperature. Therefore for resistance to cleavage failure in service it is essential that the impact transition temperature should be below the service temperature. See:

Fig 2.1.18   *Effect of test variables on transition temperature*

The Fracture Mechanics method of assessing toughness (see Vol. 1, Chapter 2) is more reliable and provides a quantitative basis for design against brittle fracture, but valid fracture toughness data is only available for high strength steels.

2.1.12.2   Factors affecting toughness

In this section the various factors which influence the toughness of steel are summarised.

STEEL STRUCTURE

The effect of steel structure on toughness is summarised in the following table—

Austenitic steels	Austenitic steels (stainless) exhibit no impact transition temperature and possess high fibrous fracture energy. Recommended for cryogenic applications.
Ferrite–pearlite carbon steels	Impact transition temperature increases and upper shelf energy decreases with increased carbon content. See **Fig. 2.1.19** *Effect of carbon content on the impact toughness of normalised carbon steels.*
Hardened and tempered steel	Impact transition temperature decreases progressively as structure changes from ferrite/pearlite to bainite to martensite. See **Fig. 2.1.20** *Effect of microstructure on the notch-toughness of a 0.3% C steel.* Optimum toughness is obtained from fully martensitic structures. Thus for high toughness in large sections, high hardenability is required (see Section 2.1.6).

GRAIN SIZE

Impact transition temperatures decreases markedly with decreasing grain size. See:
Fig 2.1.21 *Reduction of impact transition temperature with reduction in grain size*
  Grain size has no effect on upper shelf energy.
  Techniques available for achieving a reduction in grain size are shown in the following table. For further information see Section 2.1.5.

Normalising	Effective in low carbon steels. Refinement enhanced by accelerated cooling and alloying additions to depress transformation temperature—principally manganese. In higher carbon steels (<0.4%) bainite forms on air cooling due to increased hardenability. This increases impact transition temperature markedly.
Grain refining additions	Principal alloy additions Al, Nb and V. Effective in hot rolled and normalised steels.
Control rolling	Very fine grain sizes achieved. Presence of grain refining additions (principally Nb) required.
Harden and temper	Finest grain size. Associated with minimum transition temperature commensurate with strength.

STRENGTH

Refinement of grain size increases strength and lowers transition temperature. All other strengthening mechanisms lead to an increase in transition temperature and a reduction in fibrous energy.

HEAT TREATMENT

The effect of heat treatment variables on toughness are summarised in:
Table 2.1.40 *Effect of heat treatment on toughness*

NON-METALLIC INCLUSIONS

Non-metallic inclusions have no effect on impact transition temperature but a major effect on fibrous fracture energy. Toughness is approximately inversely proportional to inclusion volume fraction and inclusion aspect ratio. Main problems are deformable silicates in semi-killed steels and alumina and manganese sulphides in killed steels. During hot working of steel, deformation of the inclusions leads to anisotropy in toughness. Minimum toughness is encountered in the short transverse direction and this can lead to lamellar tearing in welded fabrications. Transverse toughness is also lowered and can give rise to cracking in forming operations. For optimum toughness a low volume fraction of small, uniformly dispersed non-deformable inclusions is required. Techniques available for minimising effects of non-metallic inclusions are summarised below.

(a) Inclusion control during steel-making

Process	Comments
Basic oxygen	Al-killed steels have lower inclusion contents and silicate inclusions avoided. Sulphide inclusions are especially harmful in fully killed steels.
Basic electric	Al-killed steels have lower inclusion contents and silicate inclusions avoided. Sulphide inclusions are especially harmful in fully killed steels. Oxygen and sulphur contents may be lowered.
Vacuum degassing	Lowers oxygen content. Provides ideal opportunity for addition of shape control additives—see below.
Vacuum induction	Very low oxygen. High cost.
Consumable Electrode Vacuum Arc Remelting	Very low oxygen. Uniform dispersion of very fine inclusions. High cost.
Electro-Slag Remelted	Very low oxygen and sulphur. Uniform dispersion of fine inclusions. High cost.

(b) Inclusion control by alloy additions

Element	Comments
Sulphur	Reduced sulphur improves toughness but impairs machinability. For very high toughness, sulphur content should be less than 0.01%.
Selenium and tellurium	Maintains machinability but reduces anistropy.
Aluminium	Eliminates deformable silicates. Aggravates sulphide problem.
Cerium (Mischmetal) or cerium silicide	Reduces sulphur content and with residual Ce: S of 1.5:1 prevents sulphide deformation and minimises anistropy. Recovery erratic, preferably use with vacuum degassing.
Calcium	Added as alloy with Si, Al and Ba. Effect is similar to Ce at low sulphur contents, i.e. less than 0.03%. Not as effective at high sulphur contents. Recovery erratic, and vacuum degassing recommended.
Zirconium	Minimises sulphide deformation if present at minimum content of $1 \times \%S + 6 \times \%N$. Cannot be used in steels which utilise nitride precipitation e.g. vanadian containing steels. Excess Zr can lead to embrittlement.

EFFECT OF ALLOYING ELEMENTS ON TOUGHNESS

Carbon	In ferrite–pearlite steels, increasing carbon increases Impact Transition Temperature (ITT) and upper shelf energy dramatically. In hardened and tempered steels effect of increased carbon is to increase ITT and decrease upper shelf energy. Effect is less pronounced than in ferrite steels and is dominated by effect of carbon on strength.
Manganese	In ferrite–pearlite steels up to 1.5% lowers transition temperature at a rate of approximately 5°C per 0.1% Mn due to refinement of pearlite and reduction in grain size. In hardened and tempered steels manganese increases hardenability and hence increases toughness of large sections by ensuring a fully martensitic structure. Manganese increases susceptibility to temper embrittlement and quench cracking.
Chromium	Increases toughness of hardened and tempered steels in large sections due to increased hardenability.
Nickel	Increased nickel lowers transition temperature progressively. With greater than 9% nickel transition temperature is decreased to approx. −150°C.
Silicon	In low concentrations (0.15–0.30%) improves toughness by lowering oxygen content. Higher levels tend to raise ITT.
Molybdenum	Increases resistance to temper embrittlement in hardened and tempered steels. Increases hardenability.
Vanadium and Niobium	Decreases impact transition temperature in H.S.L.A. alloys due to grain refinement. Increases fibrous fracture energy by reducing volume fraction of pearlite.
Aluminium	Up to 0.1% lowers impact transition temperature due to grain refinement and removal of dissolved nitrogen. Toughness also improved by reduction in oxygen content and general improvement in steel cleanliness. High levels of aluminium increase ITT.
Sulphur	Toughness increases progressively with reduced sulphur.
Hydrogen	Decreases toughness dramatically under conditions of slow strain rate. If present must be removed by annealing treatment at 200°C.

2.1.12.3   Selection for toughness

(i) Select minimum strength commensurate with load bearing and fatigue requirements, etc.

(ii) For high strength (>550 MN/m^2) select hardened and tempered steel.

   a) consider carbon-manganese steel only if section size is very small.

   b) for large sections use alloy steel and ensure hardenability is sufficient to produce fully martensitic structure.

   c) avoid 300–500°C tempering range—if not possible ensure molybdenum is present.

   d) avoid rapid quenching if possible.

(iii) Where only high surface hardness is required use case hardened steel. (Section 2.1.9).

(iv) For medium strength structural applications (300–500 MN/m^2) yield strength consider either grain-refined normalised steels, controlled rolled steels or low carbon hardened and tempered steels. (Hardened and tempered steels have minimum impact transition temperature.)

(v) For low strength steels use grain-refined hot rolled or normalised steels.

(vi) For low temperature service use grain refined normalised steels, hardened and tempered steels containing up to 9% nickel or austenitic steels in order of decreasing service temperature.

(vii) Ensure inclusion volume fraction is low and if material is being stressed in the transverse or short-transverse direction avoid elongated inclusions. Select steel-making technique to satisfy toughness requirements.

TABLE 2.1.40 Effect of heat treatment on toughness

Process	Effect on toughness	Comment
Normalising	Lowers I.T.T. in low carbon steels.	Refines grain size. High normalising temperatures should be avoided due to grain growth.
Quenching	1. Rapid quenching may give rise to quench cracking.	Problem usually occurs in steels with low hardenability (where rapid quenching is required) or when carbon and manganese alone are employed for hardenability. Problem can be alleviated by substitution of nickel, chromium or molybdenum for increased hardenability and associated reduction in cooling rate. Alternatively employ martempering.
	2. Slack quenching increases I.T.T.	Insufficient cooling rate prevents complete transformation to martensite. Remedy—increase cooling rate or increase hardenability.
Tempering	Increase in tempering temperature generally improves toughness. Temper embrittlement (increased I.T.T.) may be encountered in 300–550°C temperature range (see Fig. 2.1.5).	Increased toughness associated with decreased strength. Major alloying elements responsible for temper embrittlement are P, As and Sb but effect is aggravated by high concentration of manganese. Problem can be alleviated by addition of molybdenum in concentrations up to 0.7%. Tempering range for optimum toughness 550–650°C.

Table 2.1.40

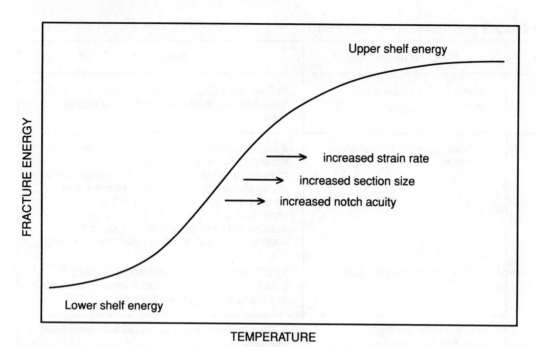

FIG 2.1.18 Effect of test variables on transition temperature

Fig 2.1.18

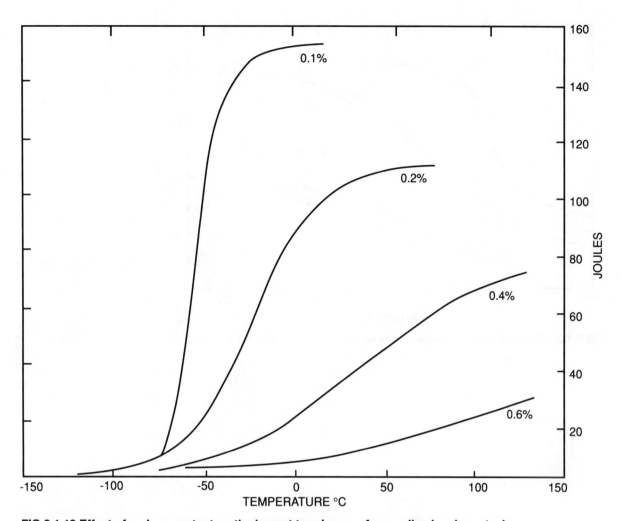

FIG 2.1.19 Effect of carbon content on the impact toughness of normalised carbon steels

Fig 2.1.19

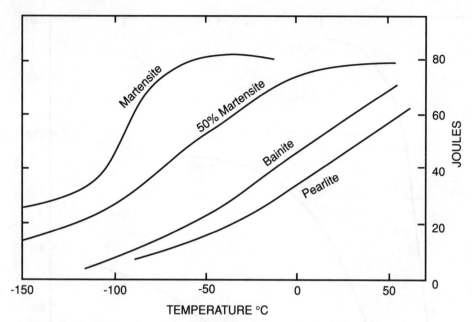

FIG 2.1.20 Effect of microstructure on the notch-toughness of a 0.3%C steel

Fig 2.1.20

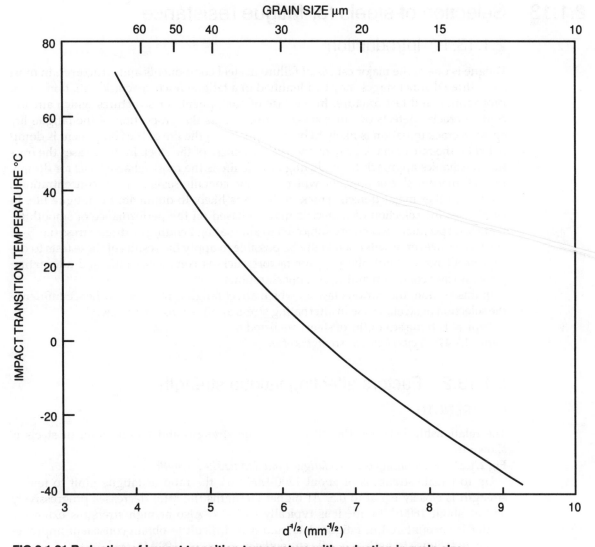

FIG 2.1.21 Reduction of impact transition temperature with reduction in grain size

Fig 2.1.21

2.1.13   Selection of steels for fatigue resistance

2.1.13.1   Introduction

Fatigue is one of the major causes of failure in steel components and structures. In most cases three distinct stages may be identified in a fatigue fracture, crack initiation, crack propagation and fast fracture. In the case of components or structures which already contain cracks, defects or other severe stress raisers the proportion of the fatigue life spent in crack initiation is likely to be very small and the life of the component is dominated by the rate of crack growth and the toughness of the steel. In these cases the fracture mechanics approach to predicting fatigue life is the most reliable basis for design. In well-machined components which do not contain severe stress concentrations, cracks or other major defects, crack initiation is likely to dominate the fatigue life. In such cases the selection of materials may be based on the performance of smooth fatigue test specimens which are subjected to alternating bending or direct stress (i.e. S–N curves). However, it will not usually be possible to apply the results of these tests to design problems without allowing for factors such as combined static and alternating loads, surface condition and stress concentrations.

In this section, the various factors which affect fatigue strength and hence influence the selection of steel for use in alternating stress applications are reviewed.

Typical Fatigue strengths of steels are listed in:

Table 2.1.41   *Typical fatigue strengths of steels*

2.1.13.2   Factors affecting fatigue strength

(A) STRENGTH

The relationship between the ultimate tensile strength and fatigue limit of steels is shown in:

Fig 2.1.22   *Relationship between fatigue limit and tensile strength*

Up to tensile strengths of about 1200 MN/m^2, the ratio of fatigue limit to tensile strength is closely equal to 0.5. At higher strengths the ratio decreases progressively and at about 1800 MN/m^2 it is typically 0.35–0.4. Also at high strengths extensive scatter is encountered in fatigue data, and it is difficult to obtain consistent improvements in fatigue strength at tensile strengths in excess of 1600 MN/m^2.

The above relationship between fatigue and tensile strength has been determined from tests on polished specimens. However, with rough-machined specimens or specimens containing notches very much lower fatigue strengths are obtained (Fig 2.1.22). The higher strength steels are especially sensitive to the presence of surface defects and in practical situations they may show no improvement over low strength steels unless extreme care is taken in surface finish. Also, once initiated the ratio of fatigue crack growth will generally be more rapid in high strength steels due to higher applied stresses being employed and the extent of fatigue cracking prior to fast fracture will be less due to the decreased toughness of the steel.

(B) STEEL COMPOSITION

Due to the relationship between strength and fatigue limit, any compositional modifications which result in increased tensile strength simultaneously increase the fatigue strength. At constant strength there is generally no significant difference between steels of different matrix composition. However, in quenched and tempered steels, increased fatigue strength may be obtained from higher carbon or alloy steels which are tempered at higher temperatures, this being due to the reduction of the residual stress originating from hardening.

(C) HEAT TREATMENT AND MICROSTRUCTURE

As with composition, heat treatment processes which increase tensile strengths generally also increase fatigue strength. However, in quenched or lightly tempered steels, the fatigue strength may be impaired by the presence of residual tensile stresses in the surface of components. The fatigue strength of homogenous structures is generally superior to that of mixed structures. For high strength steels, optimum fatigue properties are obtained from structures consisting wholly of tempered martensite. Even small amounts of bainite or ferrite and pearlite will produce a reduction of about 20% in fatigue limit.

Ferrite-pearlite steels have inferior fatigue strength to martensitic steels of the same strength. However, the fatigue strength of high carbon pearlite structures can be improved by spheroidisation.

(D) NON-METALLIC INCLUSIONS

Non-metallic inclusions have a very pronounced effect on fatigue strength, particularly in high strength steels. The most harmful inclusions are alumina and silicates and their effect on fatigue strength becomes increasingly detrimental with increased volume fraction and size. The reduction in fatigue strength is more pronounced in the transverse direction where the elongated inclusions are orientated perpendicular to the direction of applied stress. Sulphide inclusions do not appear to have detrimental effect on fatigue and when present at low volume fractions may ameliorate the effect of oxide inclusions.

Inclusion volume fraction and size may be reduced by vacuum melting, vacuum arc remelting and electro-slag remelting techniques and outstanding improvements in the fatigue strength steels may be obtained thereby.

(E) SURFACE CONDITION

(i) Surface finish

The fatigue properties of high strength steels are extremely sensitive to surface finish. As indicated by the following results optimum fatigue strength is obtained from polished surfaces.

Surface condition	Fatigue strength as % of maximum value
Polished	100
Ground	93
Smooth turned	74
Rough turned	65

The results were obtained from a Ni, Cr, Mo steel having a tensile strength of 900 MN/m^2. The effect of surface finish becomes more pronounced at higher strengths.

(ii) Decarburisation and oxidation

Decarburisation and oxidation have a very detrimental effect on fatigue strength as indicated in:

Fig 2.1.23 *Effect of surface condition on fatigue strength*

At strengths of about 900 MN/m^2 the fatigue strength of as-forged steels may be about half that of polished specimens, whereas with very high strength steels the fatigue strength may be lowered by 80%. Where decarburisation has occurred some improvement in fatigue strength may be obtained by recarburising but this is not effective on oxidised specimens.

(F) SURFACE TREATMENT

Shot peening, carburising, induction hardening and nitriding are all effective in increasing fatigue strength.

Shot peening has a relatively small effect on polished specimens but produces substantial improvements in the fatigue strength of oxidised, decarburised or roughly machined components (Fig 2.1.23). The improved fatigue performance derives predominantly from residual compressive stresses which are set up in the surface of the components as a result of local plastic deformation caused by shot impingement. Excessive peening should be avoided as fatigue strength may eventually deteriorate due to surface damage.

Both carburising and nitriding produces substantial improvements in fatigue strength under conditions of reverse bend and in notched specimens. The improvement derives from a combination of the higher strength of the case and compressive residual stresses which are induced in the surface during quenching.

Comparable improvements in fatigue strength are produced by both techniques and selection of case-hardening treatment depends on the type of steel, depth of case required and service temperature.

TABLE 2.1.41 Typical fatigue strengths of steels

Steel type	BS970	Treatment	Fatigue strength (10⁷ cycles, MNm⁻²)	UTS (MNm⁻²)	Yield Stress (MNm⁻²)	Fatigue Ratio
Carbon steels	070M20	N	193	400–430	200–230	0.45–0.48
	070M26	N	201	430–500	220–250	0.40–0.47
	080M30	N	232	460–500	230–250	0.46–0.50
	080M40	H + T	278	620–780	390–460	0.37–0.45
	070M55	H + T	293	700–1000	415–570	0.29–0.42
	150M19	N	278	510–540	290–325	0.51–0.54
Low alloy steels	En22	H + T	525	772–850	590–680	0.61–0.68
	722M24	H + T	293	850–1080	650–760	0.27–0.35
	653M31	H + T	432	770–1000	590–680	0.43–0.56
	976M33	H + T	486	900–1050	770–820	0.46–0.54
	853M30	H + T	525	1540	1235	0.34
Stainless steels	401S21	H + T	340	540–700	370	0.49–0.63
	420S37	H + T	402	770–930	590	0.43–0.52
	431S29	H + T	371	850–1000	680	0.37–0.44
	304S15	S	263	460	170	0.57
	302S17	S	278	510	210	0.55
	302S25	S	293	510	210	0.57
	321S20	S	270	510	210	0.53
	347S17	S	301	510	210	0.59
	315S16	S	270	460	170	0.59
	316S16	S	270	640	170	0.42
Steel castings	BS592–B	A	229	500	260	0.46
	BS1456–B1	H + T	334	620–770	370	0.43–0.54
	BS1458–A	H + T	372	700–850	500	0.44–0.53
	BS1458–G	H + T	534	1000–1160	700	0.46–0.53

N = normalised; H + T = hardened and tempered; S = softened; A = annealed.
Ranges due to variations with section size.

Table 2.1.41

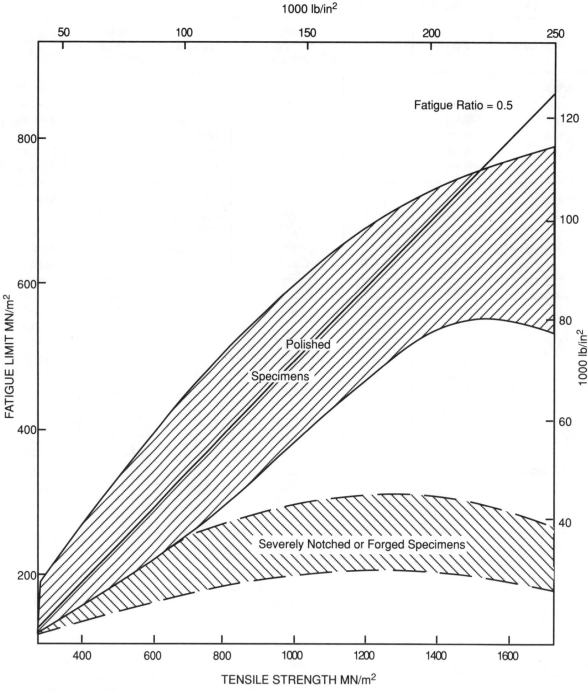

FIG 2.1.22 Relationship between fatigue limit and tensile strength

Fig 2.1.22

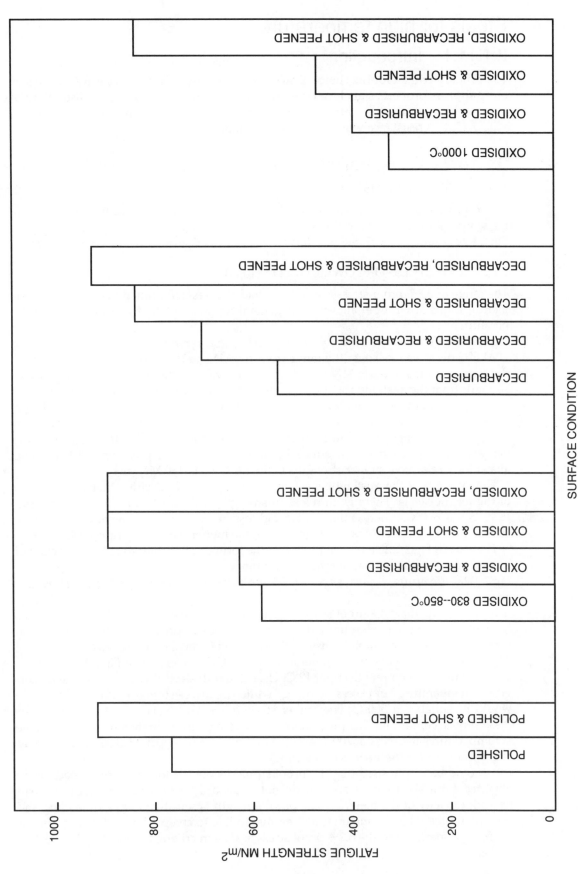

FIG 2.1.23 Effect of surface condition on fatigue strength

Fig 2.1.23

2.1.14   Steels for high temperature

2.1.14.1   Introduction

The primary criteria for the choice of steels for high temperature service are tensile and creep rupture properties at elevated temperature. Oxidation resistance is also essential but it so happens that the chromium additions which confer creep resistance also provide oxidation resistance which is usually adequate for the temperature range.

2.1.14.2   Comparative tensile properties

ELASTICITY MODULUS

The Young's moduli of both ferritic and austenitic steels decline steadily with increasing temperature as shown in:

Fig 2.1.24   *Variation of Young's modulus of austenitic and ferritic steels with temperature*

0.2% PROOF STRESS OR LOWER YIELD STRESS

The range of 0.2% proof stresses or lower yield stresses determined from various commercial forms of wrought alloys of the general types are plotted as a function of temperature in:

Fig 2.1.25   *Variation of 0.2% proof stress of steels with temperature*

At 0°C the strengths cover the range from 160 MN/m^2 to 1460 MN/m^2 and at 700°C the range is from 60 to 640 MN/m^2. At low temperatures there is a general increase in strength from the austenitic stainless steels, to the plain carbon steels, to super 12% Cr class I steels, to the precipitation hardened austenitic steels and to the highest strength super 12 class II and III steels; the Cr–Mo steels cover the range of the plain carbon steels up to the range of the low strength super 12 Cr class II and III steels. At 700°C the various types, apart from the super 12 class II and III and precipitation hardening austenitic steels, have similar strengths in the range 60 to 180 MN/m^2.

The strength of the austenitic steels is less temperature-sensitive than the strength of the ferritic steels. There is little loss in strength of the austenitic steels at temperatures below 600°C, whereas there is a rapid loss in strength of the ferritic steels above 300°C; the loss in strength is much greater for the higher strength ferritic steels.

The data in Fig 2.1.25 has been plotted in the form of a bar chart to aid the selection of a steel for a particular stress or temperature range in:

Fig 2.1.26   *Comparison of the ranges of 0.2% proof stresses of selected steel types at elevated temperatures*

In this figure, the 0.2% proof stress at temperatures of 0–800°C, in 100° intervals, has been plotted for the maximum and minimum values of proof stress for each general type of alloy. As an example, consider the selection of an alloy for service at a stress of 400 MN/m^2 and a maximum temperature of 500°C. By reference to Fig 2.1.26 it can be seen that the higher strength super 12 Cr class II and III steels would be suitable for service at temperatures in excess of 700°C, while the lower strength alloys of this type would be suitable at 500°C. The higher strength Cr–Mo steels would be suitable for these service conditions. The precipitation hardening austenitic steel would be suitable at temperatures in excess of 700°C and the higher strength super 12 class I steels would also be suitable for these service conditions.

It should be emphasised that the proof strengths quoted for the various steels were obtained from short-term tests and do not necessarily represent the long-term load bearing capacity of the material. The proof strength is sensitive to rate of loading and time under load, i.e. the material will be susceptible to creep. The relative importance of creep properties and the 0.2% proof stress as design criteria is considered in Section 2.1.14.3.

ULTIMATE TENSILE STRENGTH

The range of ultimate tensile strengths determined from various commercial forms of wrought alloy of the general types are plotted as a function of temperature in:

Fig 2.1.27 *Variation of ultimate tensile strength of steels with temperature*

At 0°C the strengths cover the range from 400 to 2000 MN/m^2 and at 700°C the range is from 150 to 850 MN/m^2. The relative order of the ultimate tensile strengths of the ferritic alloys is the same as the relative order of the proof strengths; there is about a two-fold increase in strength for the plain carbon and low alloy steels and about a 15% increase for the high alloy steels. The ultimate tensile strength of the austenitic steels, where the ultimate strength is about three times the proof strength, is raised relative to the ultimate strength of the ferritic steels. At 0°C, the austenitic stainless steels have superior strengths to the plain carbon steels and this superiority is retained to high temperature. The precipitation hardening austenitic steel[†] has superior ultimate tensile strengths to all but the creep resisting 12 Cr class II and III steels at low temperatures and becomes stronger than the latter at around 650°C.

The ultimate tensile strength of the austenitic steels, like the 0.2% proof stress, is less temperature-sensitive than the ultimate strength of the ferritic steels; there is little loss in strength below 550°C, whereas there is a rapid loss in strength of the ferritic steels above 300°C.

2.1.14.3   Time-dependent properties

For service at higher temperatures the time-dependent properties of a material have more design significance than the tensile properties. There are four important parameters:

— Creep rupture strengths at a specified time up to 10^5 h.
— Creep strength expressed as stress to produce a specified elongation ε
  (usually 1% in a specified time up to 10^5 h).
— Rupture elongation (measured extension ε at rupture).
— *m* Value (exponent of the stress–strain relationship).

The significance of these, and that of proof stress, depends on design. Creep strength was originally employed to estimate rupture strengths, but its use is now restricted to designs where the maintenance of clearances, e.g. for rotating components, is essential. '*m* Value' is employed for calculating creep and creep rupture behaviour in complex designs.

Proof stress is used at those (lower) temperature ranges where the proof stress divided by the appropriate design factor is lower than the creep rupture strength divided by the appropriate design factor. Rupture elongation specified as a minimum for the particular application is a safeguard against brittle failure in creep, and a measure of the likely high strain fatigue behaviour of the material.

COMPARISON OF THE CREEP PROPERTIES OF GENERIC STEEL TYPES

The ranges of 1000-h rupture stresses for the general steel types are plotted as a function of temperature in:

Fig 2.1.28 *1000-h rupture stress as a function of temperature for several steels*

The 1000-h rupture stress may be regarded as the ultimate stress which the material will sustain for 1000 h. Like the ultimate stress obtained from short-term tensile tests,

---

[†]This 'precipitation hardening austenitic steel' should not be confused with the precipitation hardening stainless steels described in Section 2.1.10. It is rather a superalloy similar to A286, see Section 2.1.14.9. The tensile strengths of the precipitation hardening stainless steels would at least equal those of any steels included in Fig 2.1.27.

the 1000-h rupture stress exhibits a more marked decrease with increasing temperature for the ferritic steels than for the austenitic steels. Thus, for example, while the 12% chromium steels are superior to the austenitic stainless steels at 550°C, the austenitic stainless steels are superior at temperatures greater than 600°C.

Considering the ferritic steels, it can be seen, from Fig 2.1.28, that, at a stress level of 200 MN/m², the maximum temperatures for a 1000-h life of the plain carbon steels, the Cr–Mo steels, the creep resisting 12% Cr class II and creep resisting 12% Cr class III steels are 465, 575, 640 and 650°C, respectively. At 500°C the maximum stresses for 1000-h life are 140, 420, 540 and 720 MN/m² for the plain carbon steels, Cr–Mo steels, creep resisting 12% Cr class II and creep resisting 12% Cr class III, respectively. Thus, in general, the creep resistance of the ferritic steels, either in terms of strength at a given temperature or temperature at a given strength, increases through the series: plain carbon steels, Cr–Mo steels, creep resisting 12% Cr class II and creep resisting 12% Cr class III steels.

The creep resistance of the austenitic steels is superior to that of the Cr–Mo steels above about 500°C, and becomes superior to the creep resisting 12% Cr steels around 650°C. Increasing the alloy content of the austenitic steels, as shown by the 25 Cr–20 Ni austenitic steel, increases the creep resistance and, in combination with precipitation hardening additions, leads to the most creep-resistant type of steel, known as the iron-based superalloys described in Section 2.1.14.9.

In order to ascertain the importance of creep, it is necessary to compare the tensile and creep properties. This can be done for a number of steels by consulting:

Table 2.1.42   *Comparison of tensile properties, 100 000-h and 10 000-h rupture stresses and 1% creep strengths in 1 000 000-h for several types of steel*

With the ½ Cr–½ Mo–¼ V steel the 0.2% proof stress may be taken as a working stress safe against creep failures at temperatures up to about 500°C; the creep strains will be around 1%. At 550°C, working stresses should be based on the creep properties.

The Super 12 Cr class III steel at 500°C and higher temperatures has a higher 0.2% proof stress than the creep rupture stresses or creep strength and the working stress should be based on the creep properties; the creep and tensile properties of this alloy are superior to those of the Cr–Mo steel at the same temperatures.

The austenitic stainless steels have a proof stress lower than the 100 000-h rupture stress at temperatures up to 600–650°C and the proof stress may be taken as the working stress, safe from creep failure. At temperatures above 600°C the creep rupture properties of the austenitic steels are superior to those of the super 12 Cr class III steels.

The precipitation hardening austenitic steel has 0.2% proof stress greater than the rupture stress at 550°C and higher temperatures and the creep properties should be used for the working stress levels. In the temperature range 550–700°C the creep properties of this steel are markedly superior to the other types of steel.

2.1.14.4   Carbon and carbon manganese steels

Deoxidation practice has an important influence upon the high temperature properties of carbon and carbon manganese steels; the steels made by different deoxidation methods are distinguished as follows:

(a) semi-killed steels;

(b) silicon killed steels with virtually no aluminium;

(c) (1) aluminium treated steels generally with 0.2% or more soluble aluminium and 0.1–0.3% of silicon or

(2) aluminium killed steels with less than 0.01% soluble aluminium and low silicon.

Tensile and creep properties for these three groups of steels and niobium treated steels are discussed below.

Many components fabricated from steels of this type, such as a pressure vessel, are stress relieved before being put into service. Since prolonged periods at 600°C, or only a few hours at 650°C, have a deleterious effect on the high-temperature properties, the stress relieving of carbon steels is usually performed within the temperature range 580–620°C.

TENSILE PROPERTIES

The tensile properties of these steels are shown in:
Fig 2.1.29   *Variation of tensile properties of carbon and carbon manganese steels with temperature*

The most salient features of the behaviour of proof strength of the four groups of steel are as follows:
A. The semi-killed carbon and carbon manganese steels have the lowest strength at all temperatures, up to 450°C, the low-temperature strength is maintained up to 200°C, decreases in the temperature range 200–400°C and then becomes relatively consistent.
B. The Si-killed carbon and carbon manganese steels have higher strengths than the semi-killed steels at low temperatures; the proof stress varies with temperature in a similar manner to that of the semi-killed steels and the two groups of steel have similar strengths above 300°C.
C. The Al-killed carbon and carbon manganese steels have a greater strength than the semi- and Si-killed steels at lower temperatures; the strength decreases with temperature, although there is an inflection in the strength–temperature curve around 200°C, until the strength of the three types becomes similar at 400°C.
D. The semi-killed carbon manganese niobium steels have the greatest proof stresses at low temperatures and the superiority is maintained up to 400°C; the strength decreases steadily from 100–450°C.

The low-temperature strength of carbon manganese steels can be improved by quenching and tempering. Carbon manganese steels have limited hardenability compared to the low alloy steels and the plate thickness range, 6–38mm, is the largest practicable to maintain properties through the section. The tensile properties of quenched and tempered carbon manganese, and carbon-manganese-niobium steels obtained from commercially prepared plates from 25 to 38mm, thickness are shown in:
Fig 2.1.30   *0.2% proof stress and ultimate tensile strength of quenched and tempered low carbon steels compared with the normalised condition*

CREEP PROPERTIES

The stress rupture behaviour of silicon-killed, aluminium-killed and semi-killed carbon and carbon manganese steels at 400, 450 and 500°C is shown in:
Fig 2.1.31   *Stress-rupture life plots for plain carbon steels*
and stresses to produce rupture and creep in the temperature range 380–450°C in:
Fig 2.1.32   *10 000-h and 100 000-h rupture stress and 1% creep strength for 100 000-h for carbon and carbon manganese steels*

The creep properties of the carbon manganese steels are controlled mainly by the manganese content and the deoxidation methods used in making the steel. As can be seen from Fig 2.1.31, increasing the manganese content of the semi-killed, Si-killed and Al-killed groups of steel shifts the stress-rupture life curves at 400, 450 and 500°C to higher stresses, i.e. manganese improves the stress rupture properties of all the carbon manganese steels. The effect of deoxidation practice is illustrated in Fig 2.1.32. The rupture stress for 10 000 and 100 000 h is greater for the semi-killed C–Mn steels than for the C–Mn–Nb steels and the rupture stresses for the C–MN–Nb steels in turn, are greater than those of the Al–killed steels at all temperatures. The Si-killed steels show a somewhat different behaviour in that, at low temperatures and short times, the rupture stress is as great as that of semi-killed C–Mn steels but at high temperatures and long times the rupture stress approaches that of the Al-killed steels. The effect of deoxidation practice on the creep properties can be attributed to active nitrogen, which decays in Si-killed steels due to the formation of silicon nitride.

See also Section 2.1.4.8 Pressure vessel steels.

2.1.14.5   Low alloy steels

This group of steels has tensile and creep properties superior to those of the plain carbon steels (Section 2.1.14.4) and has better tensile properties than the Cr–Mo group of steels (Section 2.1.14.6), up to 600°C. However, the rupture stresses at temperatures above 500°C are not so great as those of the better Cr–Mo steels.

TENSILE PROPERTIES

The variation of tensile properties (UTS and 0.2% proof stress) with temperature of these low alloy steels at different thicknesses is shown in:

Fig 2.1.33   *Variation of the tensile properties of molybdenum boron steels and Mn–Cr–Mo–V steels with temperature. The minimum specification values for different thicknesses are shown.*

The Mo–B steels were air cooled from a temperature within the range 935–980°C and tempered for 1–3 h in the range 600–700°C. The Mn–Cr–Mo–V steels were cooled from a temperature within the range 850–950°C and tempered at a temperature between 600 and 650°C. The Ni–Cr–Mo–V steels were cooled from a temperature within the range 900–950°C and tempered at 650°C. Since these steels rely on the development of a bainitic microstructure for their strength, cooling rate and section thickness influence the transformation and strength. The minimum specification values for 0.2% proof stress for different section sizes are shown in Fig 2.1.33.

The Mo–B steels have the better tensile properties, but are not recommended for use at temperatures above 400°C because of their low rupture ductility. The tensile properties of the Ni–Cr–Mo–V steels are not reported but the specification value for minimum yield strength is the same for the Mn–Cr–Mo–V and Ni–Cr–Mo–V steels over the temperature range 20–500°C and the specification value for the tensile strengths of both steels at room temperature lies between 555 and 700 MN/m^2 depending on section thickness.

CREEP PROPERTIES

The creep rupture properties of these low alloy steels are shown in:

Fig 2.1.34   *Stress–rupture life plots for low alloy steels*

At low temperatures, the Mo–B steel has the greatest creep rupture stress; at high temperatures and longer times, the rupture stress of the Mo–B steel becomes inferior to those of the Ni–Cr–Mo–V and Mn–Cr–Mo–V steels because of the low rupture ductility. The creep rupture stresses for the Mn–Cr–Mo–V steels are greater than those of the Ni–Cr–Mo–V steels, although it appears from the figure that, at times greater than 10 000 h at 550°C, the Mn–Cr–Mo–V would become inferior.

2.1.14.6 Chromium–molybdenum and chromium–molybdenum–vanadium steels

APPLICATIONS

These three steels have excellent time-dependent high temperature properties and relatively low alloy content, which confers relatively low price and easy weldability. They are among the most extensively used steels for high temperature plant. 1 Cr–½ Mo finds applications where the creep resistance of carbon steel is inadequate. The ½ Cr–½ Mo–¼ V composition optimises creep resistance and cost and is extensively used for steam pipes, pressure vessels, turbine rotors and, in its cast form for turbine casings. The additional chromium in 2¼ Cr–1 Mo confers significant environmental resistance. It is used for superheater tubing because of its increased resistance to oxidation, and in refineries because of its resistance to hydrogen embrittlement.

All of these steels can be used for high temperature bolting applications but here the more highly alloyed 1 Cr–1 Mo–¾ V with boron and zirconium additions has properties which make it suitable for the highest conditions which ferritic steel will withstand. The boron and zirconium additions enhance notch ductility.

TENSILE PROPERTIES

The 0.2% proof stress and ultimate tensile strengths are plotted against temperature in:
Fig 2.1.35 *Variation of tensile properties of Cr–Mo steels with temperature*
Separate curves are shown for the following compositions:

1 Cr–½ Mo–2½ Ni;
1 Cr–½ Mo 'normalised', i.e. oil quenched from 860 and tempered at 680°C for 2 h;
1¼ Cr–½ Mo+V;
1 Cr–1 Mo+V;
7 Cr–½ Mo;
5 Cr–½ Mo with and without Ti addition.

A scatter band is shown in the figure for the 0.2% proof stress and ultimate-strength of the following compositions:

½ Mo, ½ Cr–½ Mo, 1 Cr–½ Mo;
2 Cr–½ Mo, 2¼ Cr–1 Mo, 3 Cr–1 Mo;
5 Cr–½ Mo, 5 Cr–½ Mo–1½ Si, 9 Cr–1 Mo.

The 0.2% proof stress of the Cr–Mo steels decreases with increasing temperature; the decrease being greater in the higher strength steels so that, at 700°C, the proof stress is similar for all the steels of this type. The ultimate tensile strength of the Cr–Mo steels behaves in a fashion similar to the 0.2% proof stress, with increasing temperature.

CREEP PROPERTIES

The creep properties of the Cr–Mo steels are shown in:
Fig 2.1.36 *Stress-rupture plots for Cr–Mo steels*
Fig 2.1.37 *10 000-h and 100 000-h rupture stress for Cr–Mo steels*
In general, the rupture strength for a given time and temperature of creep exposure increases as the chromium content or the molybdenum content of the steel increases and the superiority of the more highly alloyed steels becomes more marked at higher temperatures and longer times.

½ Cr–½ Mo–¼ V, 1 Cr–½ Mo AND 2¼ Cr–1 Mo STEELS

Tensile properties

The 0.2% proof strength and ultimate tensile strengths of the steels, represented by scatter bands which include some 95% of the original data, are shown in:

Fig 2.1.38 *Variation of tensile properties of ½ Cr–½ Mo–¼ V, 1 Cr–½ Mo and 2¼ Cr–1 Mo steels with temperature*

The 0.2% proof stresses decrease with increasing temperature for each group steels. The 2¼ Cr–1 Mo steels exhibit the widest range of strengths and tend to be the strongest, whereas the ½ Cr–½ Mo–¼ V steels with the least wide range of strength tend to be the weakest of the three groups. Similar behaviour is exhibited by the ultimate tensile strengths of the three groups of steel.

Creep properties

The creep properties of the ½ Cr–½ Mo–¼ V, 1 Cr–½ Mo and 2¼ Cr–1 Mo steels, are represented as stress-rupture life plots in:

Fig 2.1.39 *Stress-rupture life plots for ½ Cr–½ Mo–¼ V, 1 Cr–½ Mo and 2¼ Cr–1 Mo steels* and as rupture stresses and creep strengths for specific endurances in:

Fig 2.1.40 *10 000–h and 100 000–h rupture stress and 1% creep strength in 100 000–h for ½ Cr–½ Mo–¼ V, 1 Cr–½ Mo and 2¼ Cr–1 Mo steels*

At short times and lower temperatures the rupture stress of the 1 Cr–½ Mo steels is superior to that of the ½ Cr–½ Mo–¼ V steels; at longer times and higher temperatures the 2¼ Cr–1 Mo has a higher creep rupture stress than the 1 Cr–½ Mo alloys but remains inferior to the ½ Cr–½ Mo–¼ V steels. The 1% creep strength for 100 000 h of the ½ Cr–½ Mo–¼ V steels is superior to that of the 2¼ Cr–1 Mo steel, which is superior to that of the 1 Cr–½ Mo.

2.1.14.7   12% Cr steels

The 12% Cr steels may be divided into three classes:

Class I   12% Cr steels containing molybdenum but no other strong carbide formers or cobalt.

Class II   High strength and creep resisting 12% Cr steels containing molybdenum and other strong carbide formers, vanadium, niobium and tungsten but no cobalt.

Class III High strength and creep resisting 12% Cr steels containing molybdenum and other carbide formers, and cobalt.

APPLICATIONS

The 12% Cr steels are used for applications such as steam and aircraft engine turbine blades which utilise their high strengths at elevated temperatures.

TENSILE PROPERTIES

The 0.2% proof stresses and ultimate tensile strengths for 90% of the available data are shown in:

Fig 2.1.41 *Variation of 0.2% proof stress of 12% Cr steels with temperature* and

Fig 2.1.42 *Variation of ultimate tensile strength of 12% Cr steels with temperature*

The Class II steels have been sub-divided into II a. and II b., the former heat-treated to give high strength at low temperatures. There is an increase in strength with increasing amounts of carbide formers so that the Class III steels are superior to the Class I steels. All three classes maintain their strength up to temperatures around 400°C. However, the stronger steels, Class III, show a more rapid decrease in strength with increasing temperature.

CREEP RUPTURE PROPERTIES

The 1000-h creep rupture strengths for the 12% Cr steels are shown in:

Fig 2.1.43 *1000-h rupture stress for 12% Cr steels*

The Class III steels have the better properties, although at temperatures around 650°C the three classes have similar properties. Typical properties of specific class II and III steels are compared with the superalloys in Section 2.1.14.9.

When comparing the creep rupture strengths of the 12% Cr steels with other materials it must be remembered that, while the short time values may be comparable, the curves, particularly of the higher strength class II materials, are more prone to show an inflection than the corresponding curves for austenitic steels. The long-term rupture strengths may therefore be lower than would be inferred from extrapolation.

2.1.14.8   Austenitic steels

AVAILABLE TYPES

There are two classes of austenitic steel for high temperature applications.
(1) The standard austenitic steels, of which the high temperature versions are distinguished by the addition of an H to the normal AISI three figure number, e.g. 316H, or the figure 49 in the BSI number, e.g. 316S49. The standard steels are modified by the addition of small quantities of boron which improves rupture ductility and in some cases also small quantities of niobium. They include:
    304H, 304LH, 310H, 310LH, 316H, 316LH, 321H, 321LH, 347H and 347LH
(2) The non-standard austenitic steels which have been developed specially for high creep rupture strength at high temperatures. These include:

Esshete 1250	BSC
FV548	Firth Brown
12R72	Sandvik
17.14 Cu Mo	Armco

APPLICATIONS

These steels may be used for applications such as superheater tubes, steam pipes and pressure vessels operating up to about 600°C, the choice between standard and non-standard steels being one of economics, the thinner sections of the non-standard steels in some cases outweighing the (usually) higher materials cost. Only the non-standard steels have rupture strengths adequate to operate at 630°C, but their oxidation resistance at this temperature is inadequate for many applications. Superheater tubes with a body of Esshete 1250 or 12R72 and an outer sheath of 310 have given excellent service.

For service at higher temperatures recourse must be made to superalloys or nickel alloys. However, for relatively low stresses at very high temperatures HK40, the cast version of 310, gives excellent service.

TENSILE PROPERTIES

The variation of tensile properties with temperature of five standard austenitic steels is shown in:
Fig 2.1.44   *Variation of tensile properties of austenitic steels with temperature*
Most of the test-pieces used in determining the tensile properties were solution treated commercially at temperatures between 1050 and 1100°C. The 18 Cr–10 Ni–Ti (AISI Type 321) steels were solution-treated in the laboratory at 1020–1050°C or 1100–1130°C; the latter treatment gave lower proof stress values. The tensile properties of a 25 Cr–20 Ni alloy, with and without Nb stabilisation, are also shown in Fig 2.1.44. The most characteristic feature of the tensile properties of the austenitic alloys is the great difference between 0.2% proof stress and the ultimate tensile strength.

The austenitic steels are frequently used at low temperatures in cold worked condition because of the improved tensile properties in this condition. In short-term tensile tests, this beneficial effect is maintained up to 750°C as shown in:

Fig 2.1.45   *Variation of tensile properties of cold-worked 17 Cr–7 Ni and annealed 19 Cr–10 Ni austenitic steels with temperature*

For comparison, the tensile properties of an iron-based superalloy A286 and two 12% chrome steels are shown in:

Fig 2.1.46   *Variation of tensile properties of 12% Cr steels and A286 with temperature*

The Class II super 12 Cr steels were tested after air-cooling from 1150°C and tempering at 650–700°C whereas the Class III super 12 Cr steels were tested after oil quenching from 1170°C and tempering at 600–640°C. The precipitation hardening austenitic steel was oil quenched from 980°C and aged for 16 h at 720°C.

The Class III super 12 Cr steel has a higher 0.2% proof stress than the Class II steel and retains its strength to higher temperatures. The precipitation hardening steel, which shows little variation of proof stress with temperature, becomes similar in strength to the super 12 Cr steels around 400°C and has higher strength at temperatures above 550°C. The standard austenitic steels have generally much lower proof stresses which become comparable with those of the 12% Cr steels only at the highest temperatures.

CREEP PROPERTIES

The creep properties of four standard austenitic steels are shown in:

Fig 2.1.47   *Stress-rupture life plots for austenitic steels* and

Fig 2.1.48   *10 000-h and 100 000-h rupture stress for austenitic steels*

The creep properties of 310 and two compositions of its cast equivalent HK40 are shown in:

Fig 2.1.49   *Stress–rupture life plots for cast and wrought 25 Cr–20 Ni austenitic steels* and

Fig 2.1.50   *10 000-h rupture stress for 25 Cr–20 Ni and 18 Cr–12 Ni austenitic steels*

Of the wrought steels the 18 Cr–12 Ni–Mo alloy has the greatest rupture stress at short times and low temperatures whereas the 18 Cr–10 Ni–Ti alloys solution-treated at high temperatures become superior to the 18 Cr–10 Ni–Mo steels at longer times and higher temperatures. The 18 Cr–10 Ni–Ti alloys solution-treated at the lower temperatures are less creep resistant than those solution treated at higher temperatures. The niobium stabilised 18 Cr–12 Ni steels become relatively less creep resistant than the other steels at longer times and higher temperatures.

The rupture properties of 310 are inferior to those of the other steels but those of the cast steels are superior and exceed those of the 18 Cr–12 Ni–Mo steel.

Much higher time-dependent properties are, however, available, obtainable from the non-standard austenitic steels developed specially for creep resistance. The creep properties of one of these (Esshete 1250) are compared with those of two 12% chrome steels (see Section 2.1.14.7) and the superalloy A286 (see Section 2.1.14.9) in:

Fig 2.1.51   *Stress–rupture life plots for two 12% Cr steels, A286 and Esshete 1250* and

Fig 2.1.52   *10 000-h and 100 000-h rupture stress and 1% creep strength in 100 000-h for 12% Cr steels, A286 and Esshete 1250*

Esshete 1250 was tested after solution treatment within the temperature range 1050–1150°C. Correct working and heat treatment are essential for this alloy.

The super 12 Cr Class III steel has superior creep properties to the Class II steel at short times and low temperatures; at longer times and higher temperatures the Class II steel becomes superior to the Class III steels. The austenitic steels are superior to the super 12 Cr steels at all times and temperatures. The creep properties of the precipitation hardening austenitic steel, A286, are superior to those of Esshete 1250.

However, A286 develops a very coarse, brittle structure at the centre of large sections.

2.1.14.9   Iron-based superalloys

GENERAL

Superalloys are generally defined as alloys developed specifically for service at elevated temperatures where relatively severe mechanical stressing and/or high surface stability in corrosive environments are prime requirements. Iron-based superalloys may be regarded as developments of stainless steels with improved strength and/or corrosion/oxidation resistance. They may be categorised into precipitation strengthened alloys (A286, Discaloy 24) which, depending on stressing, may be useful at temperatures up to about 650°C and alloys which are basically solid solution strengthened (N155, Incoloy DS, Incoloy 800) which may be used up to about 1000°C.

The nickel content is normally 25% or more in order to ensure an austenitic matrix, although lower nickel can be accommodated by cobalt additions, as in N155. Higher nickel contents when combined with higher chromium are normally associated with higher operating temperature capability (e.g. Incoloy 800 and Incoloy DS).

Iron based superalloys are broadly intermediate in strength and temperature capability between their stainless steel and nickel superalloy counterparts.

PROPERTIES

The physical properties are listed in:
Table 2.1.43   *Physical properties of iron-based superalloys*
The mechanical properties are shown in:
Fig 2.1.53   *0.2% proof strength of iron-based superalloys*
Fig 2.1.54   *Ultimate tensile strength of iron-based superalloys*
Fig 2.1.55   *1000-h stress rupture data of iron-based superalloys*
Fig 2.1.56   *10 000-h stress rupture data of iron-based superalloys*

Corrosion/oxidation resistance

A286 and Discaloy 24 are not used above 800°C and for applications in which significant strength is required, are normally limited to temperatures below 650°C. The solid solution alloys have significantly better oxidation resistance, being intermediate between stainless steels and nickel base superalloys.

Incoloy 800 and Incoloy DS can be used at temperatures up to around 1000°C. N155 has a somewhat lower temperature capability, with adequate oxidation resistance up to around 900°C depending on life requirement and whether the exposure is cyclic or isothermal.

Incoloy DS is more resistant than nickel base superalloys to atmospheres which fluctuate between reducing and oxidising, showing less internal oxidation under these conditions. It is this characteristic, together with good resistance to carburisation, that has resulted in its use in applications requiring resistance to combustion furnace atmospheres.

Incoloy 800 has inferior oxidation resistance to nickel alloys such as Inconel 600, but is a significantly cheaper material and may therefore be more cost-effective.

APPLICATIONS OF IRON-BASED SUPERALLOYS

A286 and Discaloy 24 are used for discs in industrial gas turbine engines and in some older aero gas turbine engines. A286 is used for elevated temperature fasteners, springs and bolting in aerospace and chemical applications and for high strength non-magnetic applications such as electrical generator retaining rings. Its freedom from hydrogen cracking relative to low alloy steels has resulted in applications in the petrochemical industry. N155 is used for exhaust valves and manifolds, casings and sheet metal components such as gas turbine jet pipes. Incoloy 800 and Incoloy DS are used for a wide variety of applications involving relatively low stressed exposure to corrosive environments and high temperature such as muffles, retorts and radiant tubes in industrial heating, heat treatment baskets and trays. Because of its higher silicon content, Incoloy DS has a somewhat greater resistance to carburisation than Incoloy 800 but the silicon has an adverse effect on weldability.

Incoloy 800 is used in the industrial heating field for furnace equipment such as radiant tubes and heat treatment baskets and for electrical element sheathing in domestic appliances. It is also used in steam/hydrocarbon reforming plants and ethylene crackers in the form of tubes, piping, etc.

PROCESSING

Forging

Controlled thermomechanical processing is commonly used to control grain size in the forging of discs in A286 and Discaloy 24. Forging is finished slightly below the solvus temperature of the η phase to restrict grain size. Mechanical properties are dependent on forging size and geometry.

Heat treatment

The precipitation strengthened alloys A286 and Discaloy 24 are solution-treated in the range 900–1020°C, with the lower temperatures being used for a better balance of tensile properties and stress rupture ductility. Age hardening is carried out at 720/730°C to precipitate the γ phase. Incoloy 800 and Incoloy DS are relatively weakly solid solution-strengthened alloys and are annealed at temperatures in the range 1000–1150°C. Incoloy 800H (a higher carbon version) is annealed at the higher temperatures to optimise grain size for creep strength in applications such as tubes and piping used in petrochemical processing. Although N155 is primarily a solid-solution-strengthened alloy, carbide precipitation occurs in the 700–850°C temperature range giving some strengthening. Thus, the 1175°C solution treatment is occasionally followed by a precipitation treatment at around 800°C.

Fabrication

A286 is normally cold rolled in the solution-treated condition and sheet formability is good. The readiness with which it can be cold headed makes it a suitable material for manufacture of fasteners but care must be taken to avoid abnormal grain growth during annealing. N115, Incoloy 800 and Incoloy DS can all be readily cold formed by processes such as drawing, spinning, roll forming, etc. N155 work hardens more rapidly than the Incoloy type alloys.

Weldability

The solid solution-strengthened alloys are readily welded in the annealed condition by fusion and resistance welding processes. With the precipitation strengthened alloys, welding parameters must be carefully controlled to avoid cracking in the weld and heat-affected zone, particularly in the welding of large sections or material under restraint.

2.1.14.10   Internal combustion engine poppet valve materials

GENERAL

Poppet valve materials are steels or titanium–nickel or cobalt base metallic alloys which are suitable for controlling the gas flow in internal combustion engines. Their classification is product-orientated but because the vast majority of valves are in one or other class of steel they are included in this chapter.

For exhaust valves, specially designed high-temperature austenitic and martensitic steels are used. Intake valves are low- to highly-alloyed martensitic and austenitic materials or, in some instances, titanium alloys. The conditions of operation are controlled primarily by the gases during the combustion process or their flow past the valve during the gas intake or exhaust portion of the cycle. These conditions include high temperatures and oxidising or corroding atmospheres. Stresses result from firing, spring seating, and acceleration and deceleration forces.

SELECTION CRITERIA

The selection of valve materials is based on a number of mechanical and physical characteristics of the steels and/or alloys. The mechanical property data function as a guideline in material selection. Final approval is usually made on the basis of engine test results. Of the failure modes, corrosion fatigue or fatigue in the absence of corrosion (underhead stem failures) are most common. Failure due to corrosion (inadequate resistance to burning failure) is next in order of occurrence. Thermal fatigue failure by radial cracking of the heads of exhaust valves is a third failure mode.

STANDARDISATION

The most comprehensive list of valve materials is provided in the *SAE Handbook 1986*, which includes materials ranging from constructional carbon manganese steels to superalloys. The SAE also lists a number of limited usage experimental alloys.

Inlet valves and stems for two-piece exhaust valves are listed NV for low rated valves made from constructional steels, and HNV for high rated valves which may be made from chrome silicon, high chromium martensitic or austenitic steels.

The compositions of exhaust valves range from chrome silicon through high chrome martensitic and austenitic steels to superalloys.

The carbon content of austenitic valve steels is often higher than that usually contained in austenitic steels. This does not lead to problems during operating under conditions normally present in internal combustion engines or when the valve is adequately protected during transport and storage. However, for a new design, conceivably using water injection or when the valve is transported or stored in unsatisfactory conditions, the possibility of corrosion, which would be aggravated by the presence of a weld, should not be overlooked.

PHYSICAL PROPERTIES

Valve alloys have physical properties typical of the steel types from which they are selected. The specific gravities of the more highly alloyed austenitics are higher than the simpler ferritic steels and their thermal expansions are higher.

MECHANICAL PROPERTIES

Inlet valves, low duty exhaust valves and in some cases the stems of high duty exhaust valves may be carbon low alloy steels or high alloy martensitic steels. In these duties, properties related to tensile strength and hardness are important.

A hardness of 30–40 Rockwell C presents a good compromise between wear resistance, strength to resist chipping due to pre-ignition, ductility and impact toughness. Tips are hardened to 45–60 Rockwell C to resist wear.

The heads and fillet radii of high duty exhaust valves operate at higher temperatures, and their application is dominated by their high temperature properties, hardness, thermal (and mechanical) fatigue which are related to tensile strength, ductility and creep.

The effects of temperature on yield strength (0.1% proof stress) tensile strength and hardness are summarised in:

Fig 2.1.57   *Variation in hardness of valve steels with temperature* and

Fig 2.1.58   *Tensile strength (bold lines) and 0.1% proof stress (faint lines) for valve steels plotted against temperature*

The martensitic steels show good characteristics at ambient temperature with properties generally superior to the non-hardening austenitics. However, above 300°C the properties tend to fall off sharply, as shown by Fig 2.1.58 which plots tensile strength against temperature.

The non-hardening austenitics show a far more gradual drop in properties and are consistently better at temperatures in excess of 600°C. The precipitation hardening austenitics maintain superior properties to both the non-hardening austenitics and the martensitic steels over the entire temperature range, although any differences are less pronounced at higher temperatures.

Low creep extension of a valve is also of utmost importance for prolonged life and here the martensitic steels prove inferior to the austenitic type over the entire temperature range.

ENVIRONMENT RESISTANCE

The corrosion of valves is accelerated as a result of both their temperature and environment. Corrosion occurs through three mechanisms:

1. Oxidation
2. Attack by various metal oxides and salts
3. Attack by fuel and lubricant additives or contaminants.

Valve alloys are high in chromium and sometimes contain silicon to give them inherent oxidation resistance. Silicon is sometimes intentionally removed from exhaust valve alloys to prevent certain surface reactions with metal oxides. Diesel engine valve alloys frequently have high chromium contents to afford protection from sulphur attack, particularly in alloys which contain substantial nickel contents. If sulphur is present in high proportions it can lower the safe operating temperature of an austenitic steel by 200°C. Although lead increases the corrosion rate of the valve it helps to form a protective film on the seating faces reducing metal-to-metal contact and also wear. A small aluminium addition to a steel will increase resistance to oxidation caused by vanadium in a fuel. Exhaust valves must resist face burning or head oxidation. In extreme conditions these surfaces may be protected by:

1. Face coatings. These are welded overlays applied to valve faces and intended to develop optimum corrosion and wear resistance at the valve seating surface. Cobalt, nickel, and iron base alloys are usually chosen for the purpose.
2. Head coatings. These coatings are applied to the tops of the heads of exhaust valves to inhibit corrosion. Since hot hardness is not required nickel–chromium alloys are most frequently chosen for this purpose.
3. Aluminising. This is a special case of a protective coating. It comprises a thin layer of aluminium applied to the valve face and sometimes to the valve head. Aluminium, when diffused, alloys with the base material and provides a thin, hard, oxidation-resistant coating.

Sodium cooling, inside a hollow stem, lowers the effective head temperature.

WELDING

For high duty valves the head may be made of an austenitic steel (or a nickel alloy or superalloy) and the stem, or just the tip, of a martensitic steel. In such cases inert gas shielded arc, or better, friction welding may be used. Care must be taken to ensure the difficulties of welding relatively high carbon steels are overcome.

A summary of the characteristics of valve steels is provided in:

Table 2.1.44   *Applications and properties of valve steels*

ACKNOWLEDGEMENT

Much of the information and data in this section are derived from 'SAE Information Report' Engine Poppet Valve Materials—SAE J775 JUL80 published in the *1986 SAE Handbook*. Reprinted with permission © 1986 Society of Automotive Engineers, Inc.

TABLE 2.1.42 Comparison of tensile properties, 100 000 h and 10 000 h rupture stresses and 1% creep strengths in 1 000 000 h for several types of steel

Alloy type	Temperature (°C)	0.2% PS (MN/m²)	UTS (MN/m²)	100 000 h (MN/m²)	10 000 h (MN/m²)	1% in 100 000 h (MN/m²)
Carbon steels	400	150	360	165	220	120
	450	130	320	85	130	48
½ Cr–½ Mo–¼ V	460	196	385	290	360	265
	500	180	360	173	260	155
	550	170	320	88	145	85
Creep resisting 12% Cr Class III	500	680	710	360	560	290
	550	600	640	175	300	140
10 Cr–7 Co + Ni, Mo, V, Nb, W	600	380	440	60	100	—
Austenitic steels	550	72	335	140	215	—
	600	71	310	88	148	—
	650	70	270	45	90	—
	700	70	230	20	55	—
Super alloy 15 Cr–25 Ni precipitation hardening austenitic	550	650	895	410	500	—
	600	640	850	250	345	215
	650	620	800	155	220	130
	700	570	740	60	130	60

Table 2.1.42

TABLE 2.1.43 Physical properties of iron-based superalloys

	Incoloy 800	*Incoloy DS*	*A286*	*N155*
Specific gravity	8.02	7.92	7.92	8.25
Mean coefficient of linear thermal expansion ($\times 10^{-6}$/C) 20–100°C 20–1000°C	15.2 19	15 18.7	16.6 —	14.2 18.4
Thermal conductivity (W/mK) 25°C 650°C	11.7 21.9	— —	12.6 24.6	12.3 21.8
Modulus of elasticity (GPa) 25°C 1000°C	196 125	195 119	200 —	204 140

Table 2.1.43

TABLE 2.1.44 Applications and properties of valve steels

Steel type	3Si 8Cr	2Si 20Cr	14Cr 14Ni	21Cr 4Ni 9Mn	21Cr 12Ni
Applications	Inlet valves in petrol engines and exhaust valves in medium duty diesels. Limiting service temperature 700°C for light duty.	Exhaust valves in petrol engines. Limiting temperature ~ 750°C. More resistant to sulphur and high octane fuels than grade 401S45.	High strength suitable for hollow valves. Most suitable for hard coating Grade 331S42 with Mo addition less prone to embrittlement at high temperature for long times. Limiting temperature 800°C with cooling. Hard facing of seats recommended for use above 700°C.	Wide use for petrol engine exhaust valves. Used in fully heat-treated or as-forged and stress relieved conditions, depending on application. Data are given for grade 349S52. Grades 352S52 and S54, with Nb additions show 15% improvement in 100-h rupture stress. Grades 349S54 and 352S54, with S additions are more machinable but less strong. Good scaling resistance to 900°C. Suitable for extreme conditions, with cooling. Hard facing recommended above 700°C.	Moderate hot strength. Wide use in diesel exhaust valves. Not as suitable as the 21Cr 4Ni 9Mn grades for arduous conditions. Hard facing recommended above 700°C.
Thermal expansion coefficients (10^{-6}/K)	20–100°C: 12.6 20–500°C: 13.2 20–700°C: 13.6	20–100°C: 8.2 20–500°C: 10.8 20–700°C: 12.2	20–100°C: 16.9 20–500°C: 18.0 20–700°C: 18.7	20–300°C: 17.1 20–500°C: 18.1 20–700°C: 18.4 20–800°C: 18.6	20–100°C: 14.7 20–500°C: 17.3 20–800°C: 18.6
Thermal conductivity (W/m per K)	20°C: 17 500°C: 54 700°C: 25	20°C: 11	20°C: 16 500°C: 23 700°C: 26	500°C: 21 600°C: 22 700°C: 23 800°C: 25	20°C: 17
Fatigue strength (10^6 cycles at 725°C MN/m²)	±6	±110	±150	±200	±150
Weldability	Readily welded. May need post-weld heat treatment.	Not recommended.	Readily welded. Post-weld heat treatment unnecessary.	All common processes suitable.	Readily welded. Oxyacetylene not suitable.
Resistance to leaded fuels. Lead Oxide Crucible test[a] at 913°C: wt loss g/dm²h.	55	50	53	18	30

[a] Low alloy steel.

Table 2.1.44

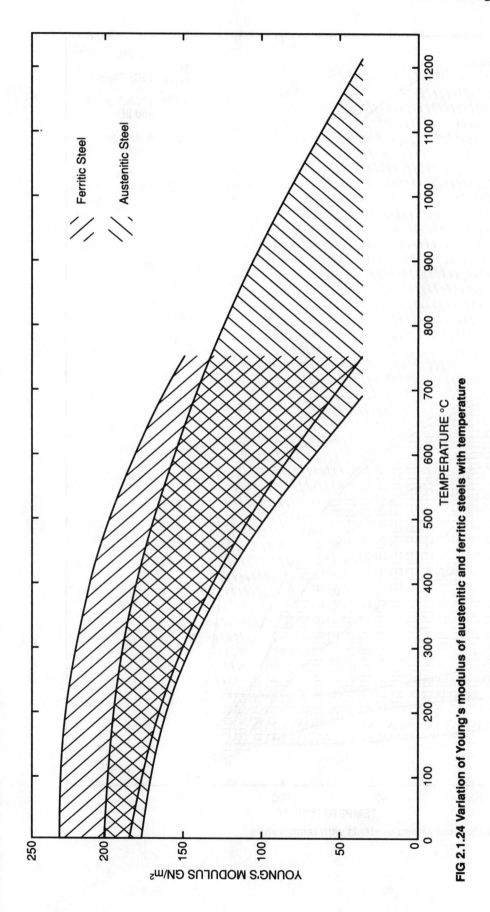

FIG 2.1.24 Variation of Young's modulus of austenitic and ferritic steels with temperature

YOUNG'S MODULUS GN/m²

TEMPERATURE °C

Ferritic Steel

Austenitic Steel

Fig 2.1.24

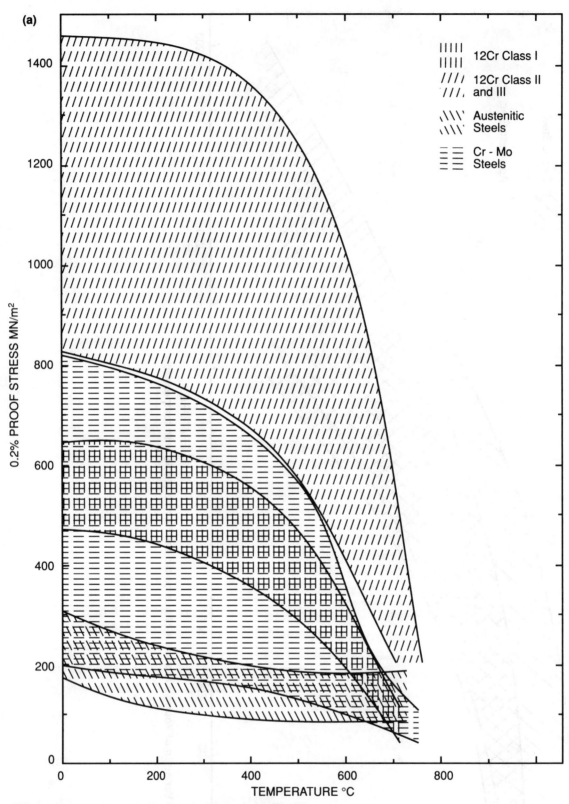

FIG 2.1.25 Variation of 0.2% proof stress of steels with temperature

Fig 2.1.25

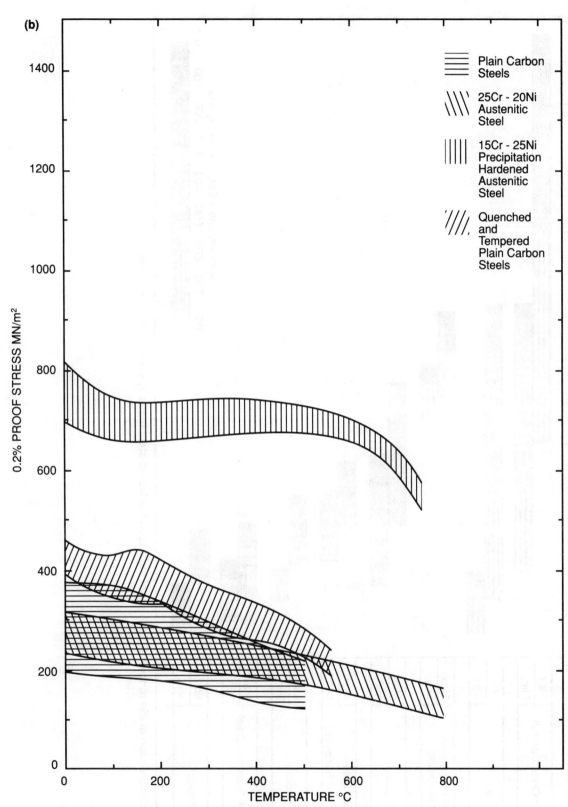

FIG 2.1.25 Variation of 0.2% proof stress of steels with temperature—*continued*

Fig 2.1.25—*continued*

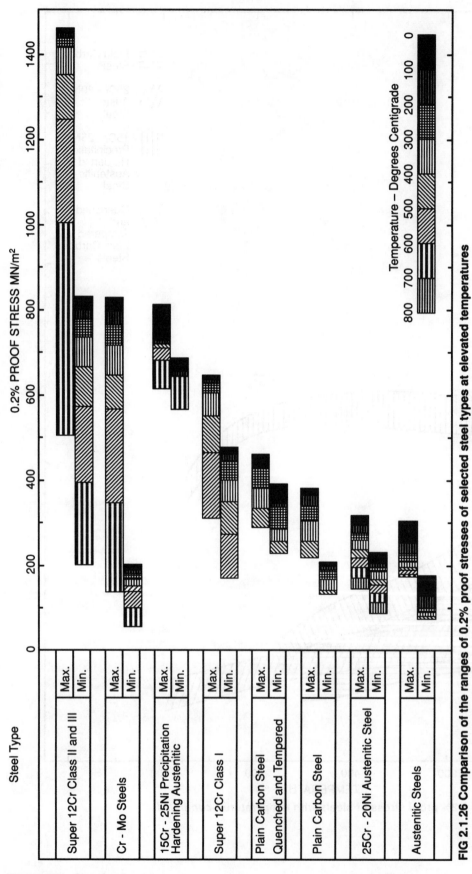

Fig 2.1.26

FIG 2.1.26 Comparison of the ranges of 0.2% proof stresses of selected steel types at elevated temperatures

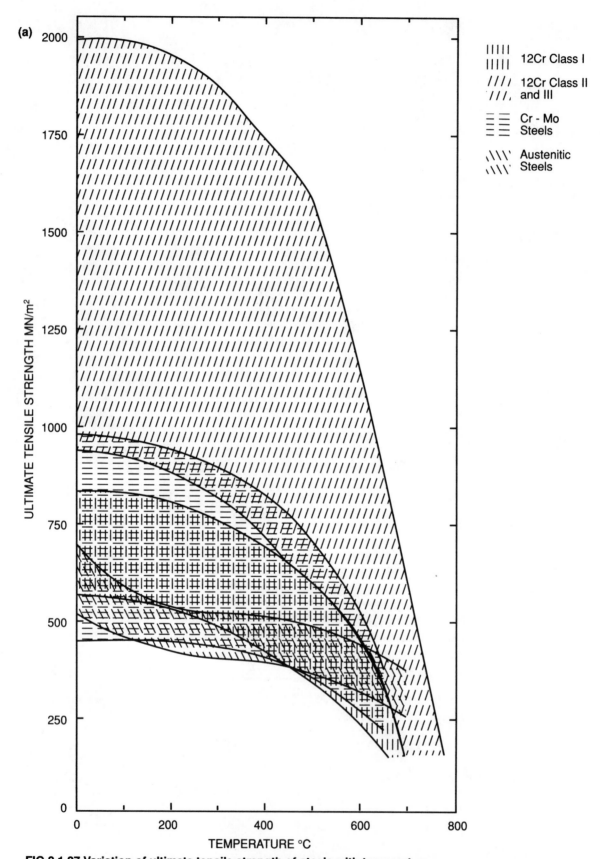

FIG 2.1.27 Variation of ultimate tensile strength of steels with temperature

Fig 2.1.27

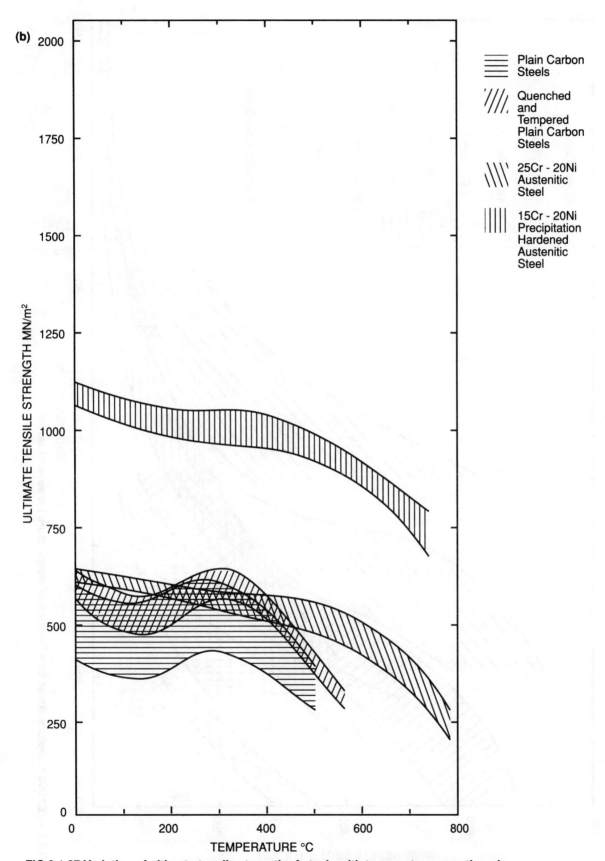

FIG 2.1.27 Variation of ultimate tensile strength of steels with temperature—*continued*

Fig 2.1.27—*continued*

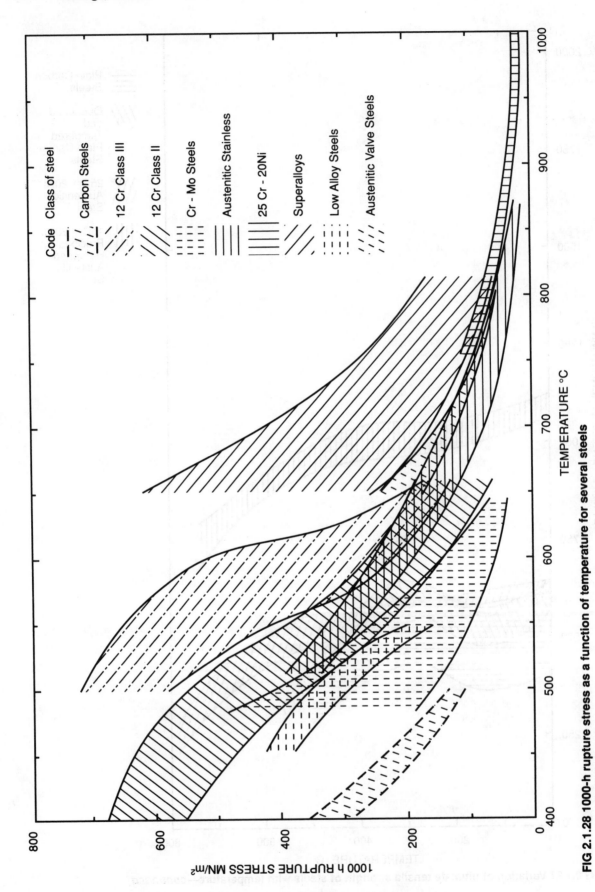

FIG 2.1.28 1000-h rupture stress as a function of temperature for several steels

TEMPERATURE °C

1000 h RUPTURE STRESS MN/m²

Code	Class of steel
	Carbon Steels
	12 Cr Class III
	12 Cr Class II
	Cr - Mo Steels
	Austenitic Stainless
	25 Cr - 20Ni
	Superalloys
	Low Alloy Steels
	Austenitic Valve Steels

Fig 2.1.28

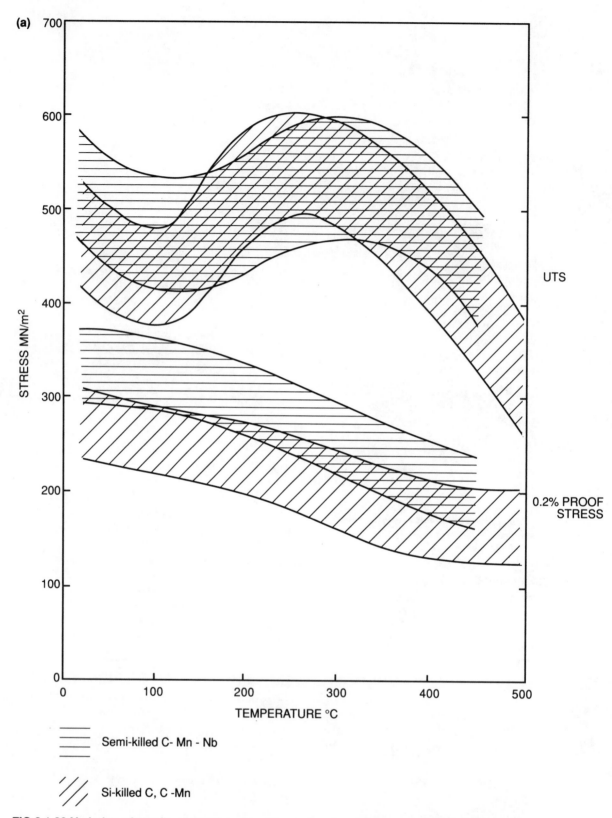

(a)

FIG 2.1.29 Variation of tensile properties of carbon and carbon manganese steels with temperature

Fig 2.1.29

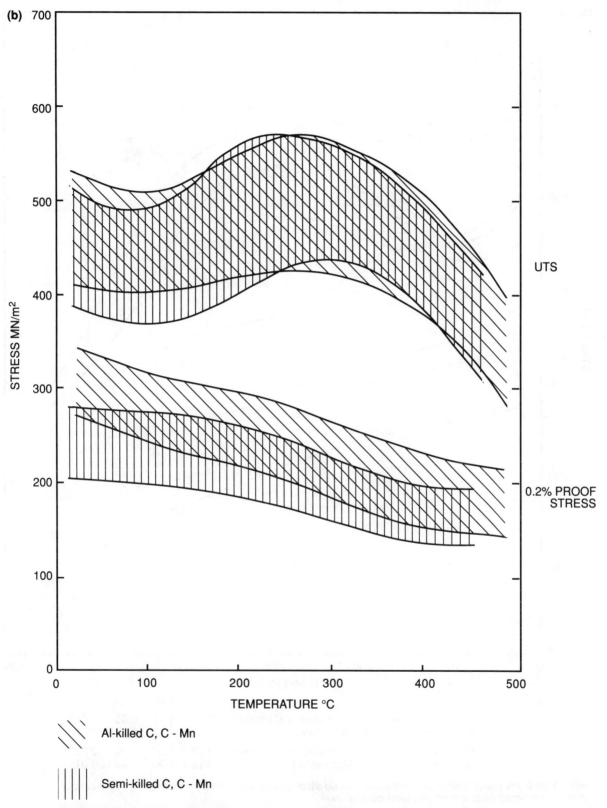

FIG 2.1.29 Variation of tensile properties of carbon and carbon manganese steels with temperature—*continued*

Fig 2.1.29—*continued*

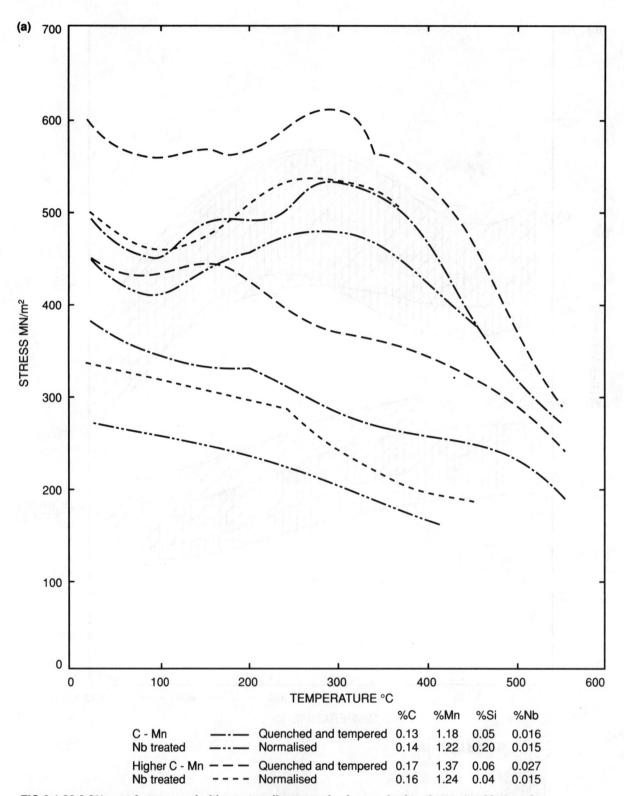

(a)

			%C	%Mn	%Si	%Nb
C - Mn	—·—·—	Quenched and tempered	0.13	1.18	0.05	0.016
Nb treated	—··—··—	Normalised	0.14	1.22	0.20	0.015
Higher C - Mn	— — —	Quenched and tempered	0.17	1.37	0.06	0.027
Nb treated	- - - -	Normalised	0.16	1.24	0.04	0.015

FIG 2.1.30 0.2% proof stress and ultimate tensile strength of quenched and tempered low carbon steels compared with the normalised condition

Fig 2.1.30

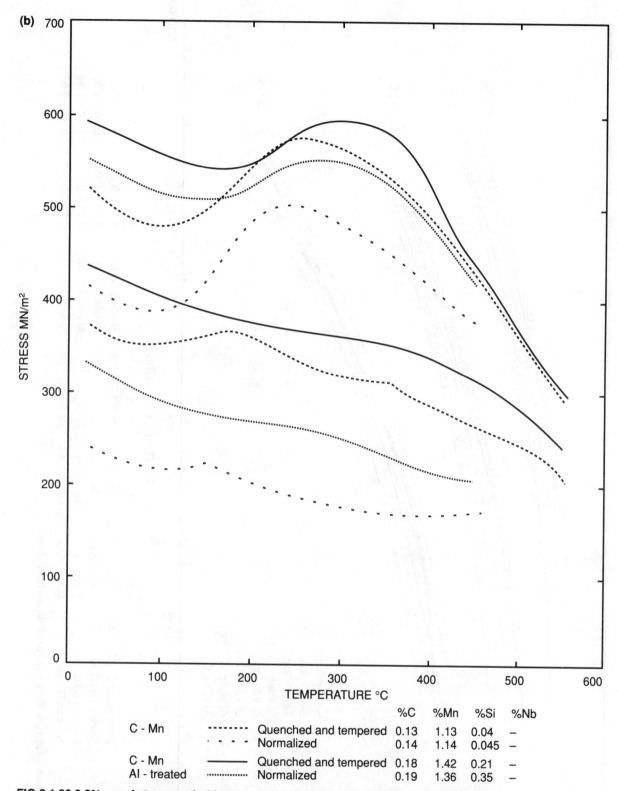

			%C	%Mn	%Si	%Nb
C - Mn	- - - -	Quenched and tempered	0.13	1.13	0.04	–
	- · - ·	Normalized	0.14	1.14	0.045	–
C - Mn	——	Quenched and tempered	0.18	1.42	0.21	–
Al - treated	··········	Normalized	0.19	1.36	0.35	–

FIG 2.1.30 0.2% proof stress and ultimate tensile strength of quenched and tempered low carbon steels compared with the normalised condition—*continued*

Fig 2.1.30—*continued*

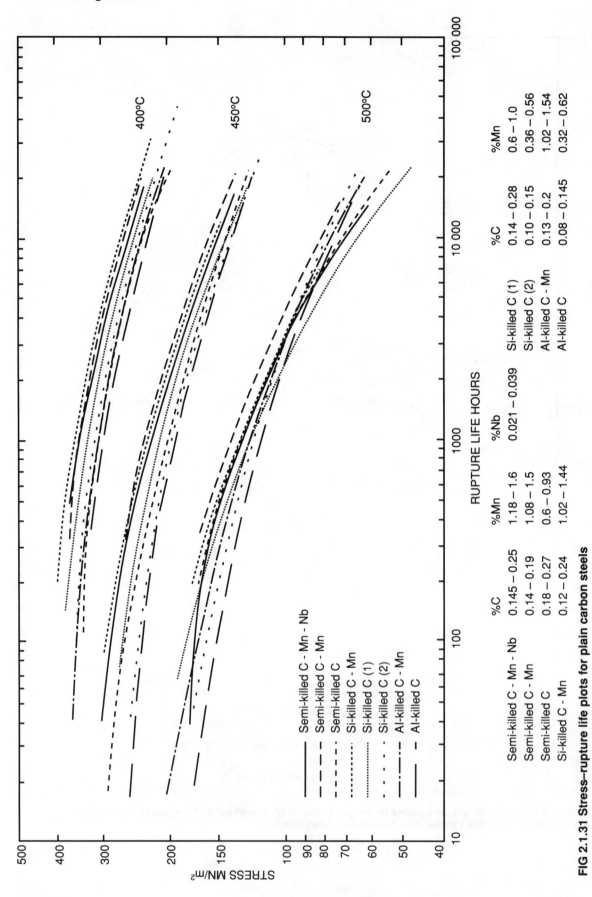

FIG 2.1.31 Stress–rupture life plots for plain carbon steels

Fig 2.1.31

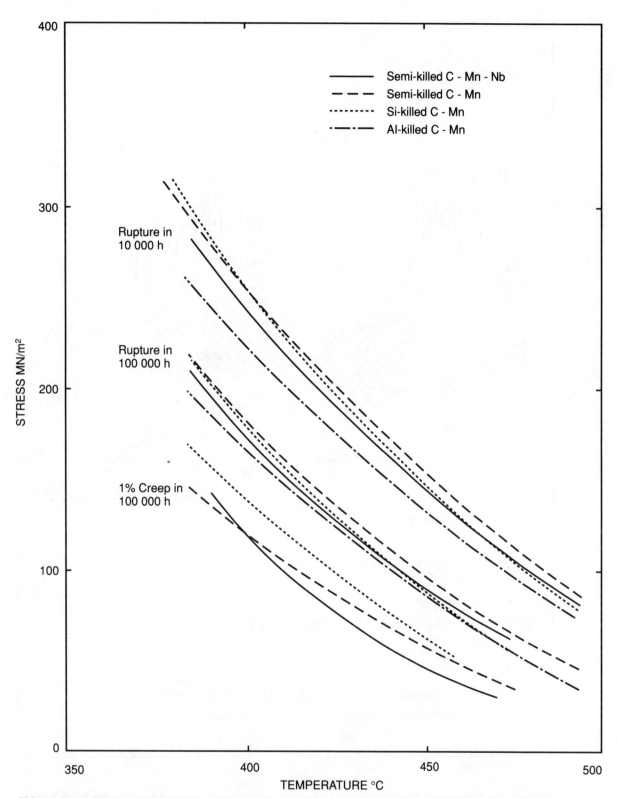

FIG 2.1.32 10 000 h and 100 000 h rupture stress and 1% creep strength for 100 000 h for carbon and carbon manganese steels

Fig 2.1.32

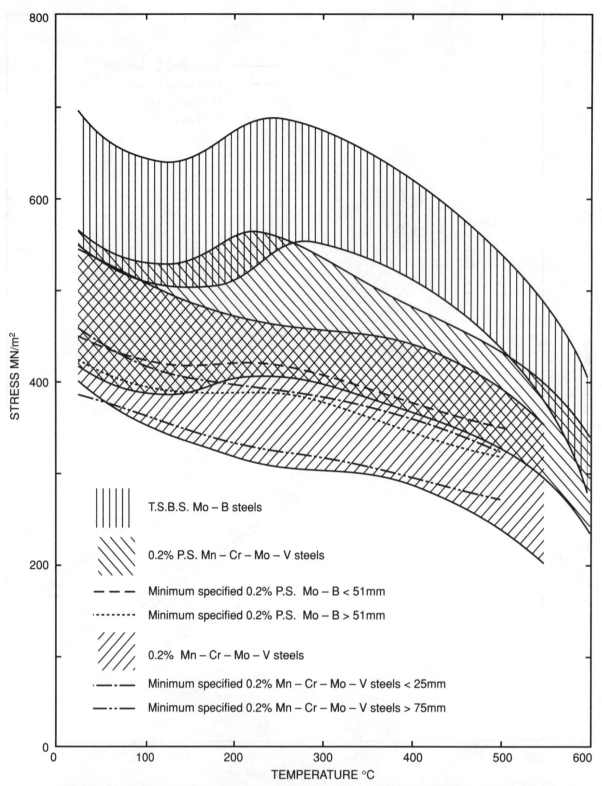

FIG 2.1.33 Variation of the tensile properties of molybdenum boron steels and Mn–Cr–Mo–V steels with temperature. The minimum specification values for different thicknesses are shown

Fig 2.1.33

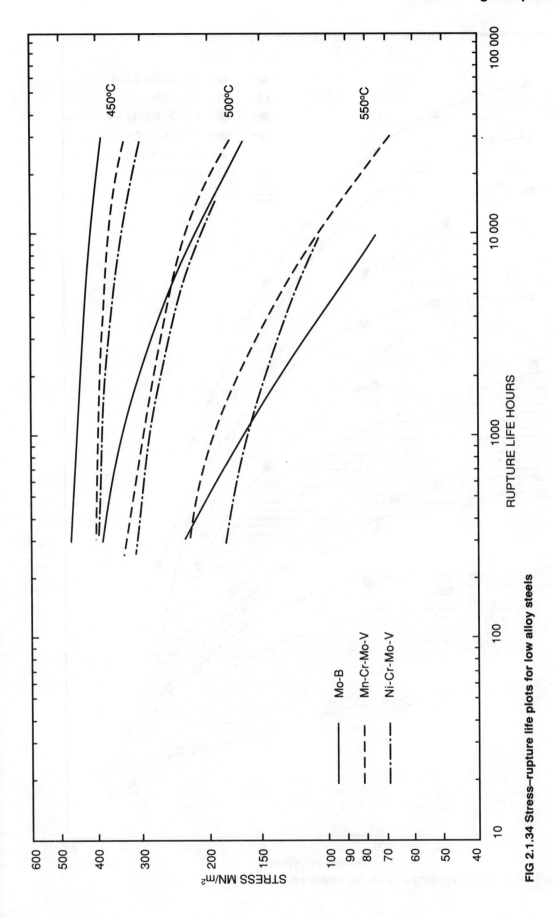

FIG 2.1.34 Stress–rupture life plots for low alloy steels

Mo-B

Mn-Cr-Mo-V

Ni-Cr-Mo-V

Fig 2.1.34

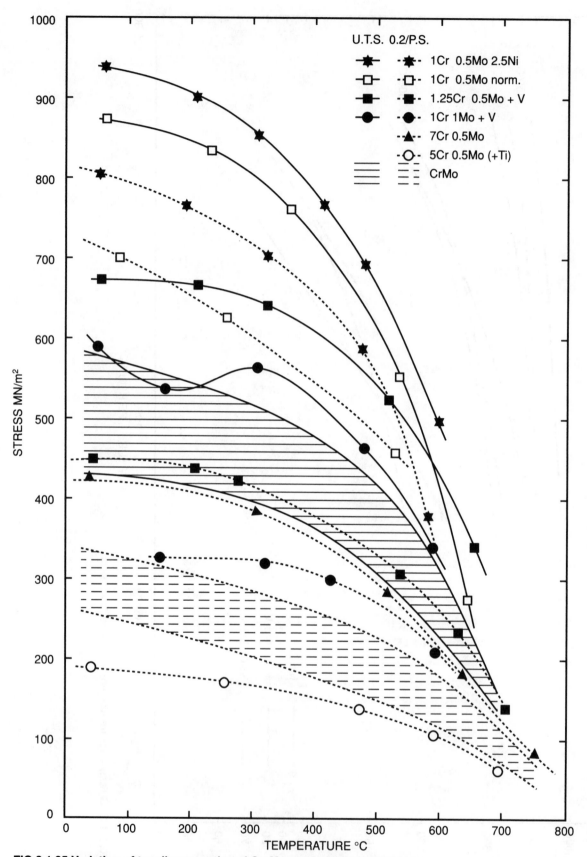

FIG 2.1.35 Variation of tensile properties of Cr–Mo steels with temperature

Fig 2.1.35

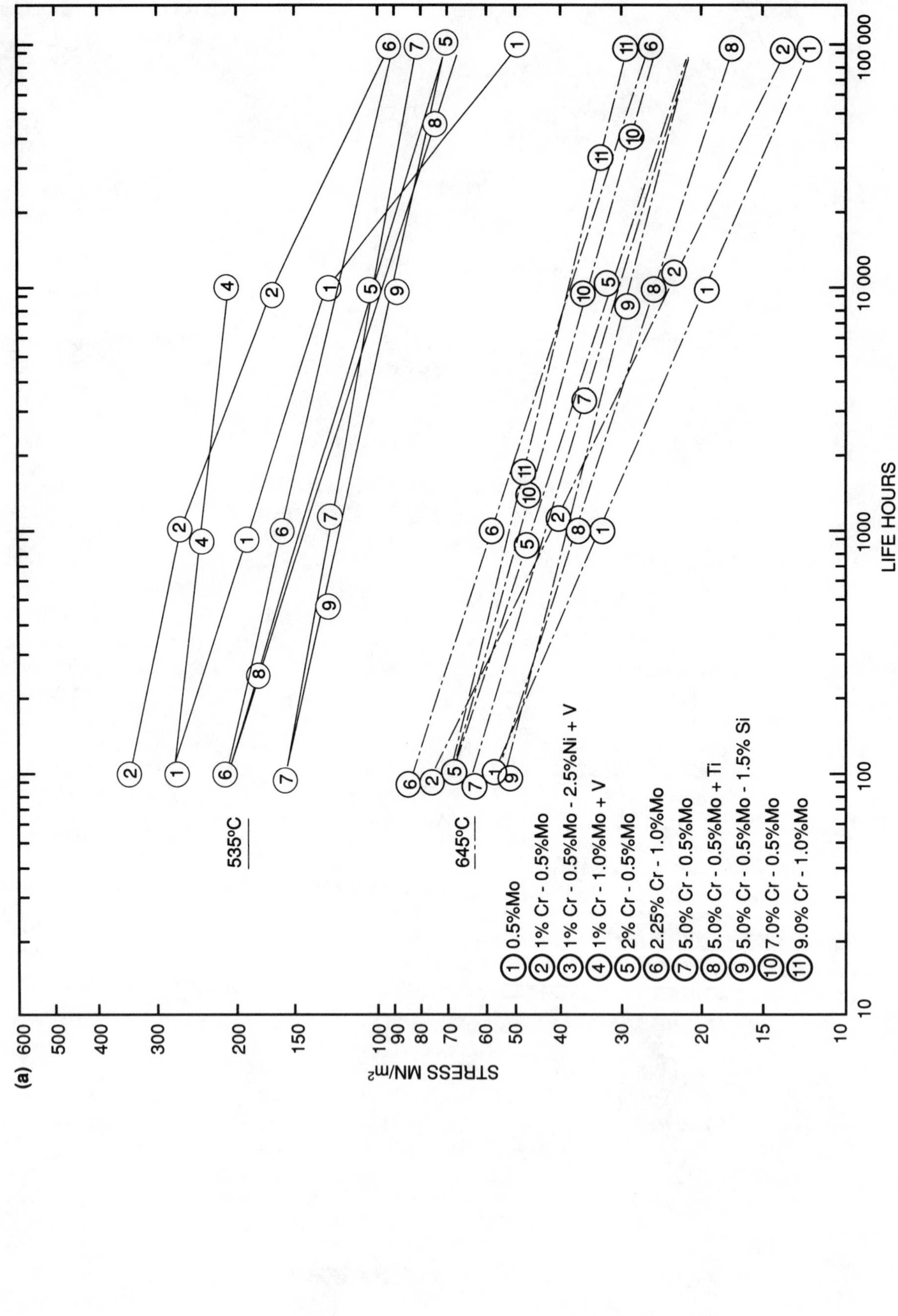

FIG 2.1.36 Stress–rupture plots for Cr–Mo steels

① 0.5%Mo
② 1% Cr - 0.5%Mo
③ 1% Cr - 0.5%Mo - 2.5%Ni + V
④ 1% Cr - 1.0%Mo + V
⑤ 2% Cr - 0.5%Mo
⑥ 2.25% Cr - 1.0%Mo
⑦ 5.0% Cr - 0.5%Mo
⑧ 5.0% Cr - 0.5%Mo + Ti
⑨ 5.0% Cr - 0.5%Mo - 1.5% Si
⑩ 7.0% Cr - 0.5%Mo
⑪ 9.0% Cr - 1.0%Mo

535°C

645°C

(a)

STRESS MN/m²

LIFE HOURS

Fig 2.1.36

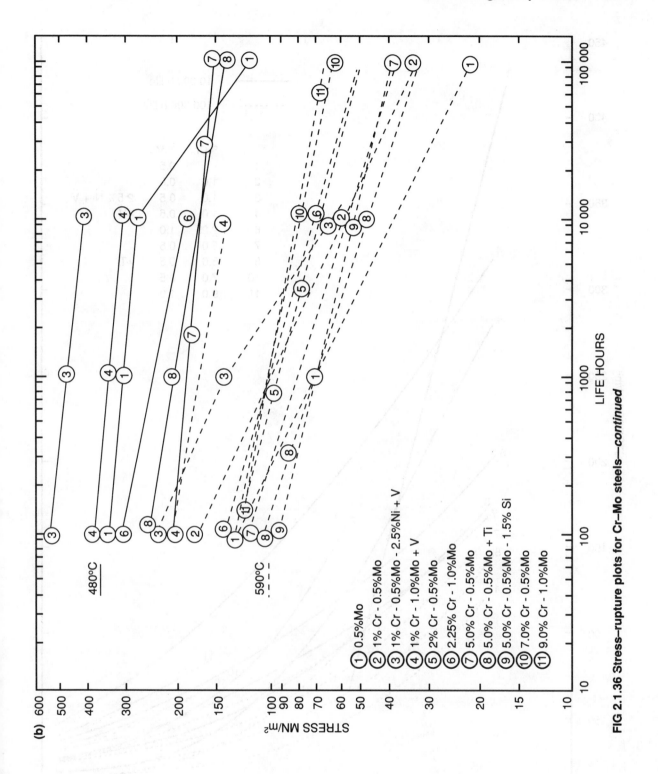

FIG 2.1.36 Stress–rupture plots for Cr–Mo steels—*continued*

Fig 2.1.36—*continued*

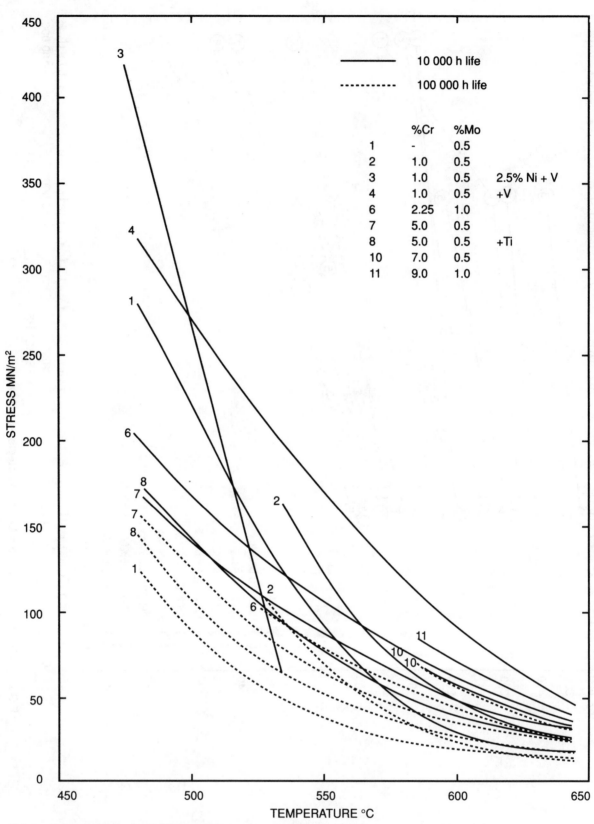

FIG 2.1.37 10 000 h and 100 000 h rupture stress for Cr–Mo steels

Fig 2.1.37

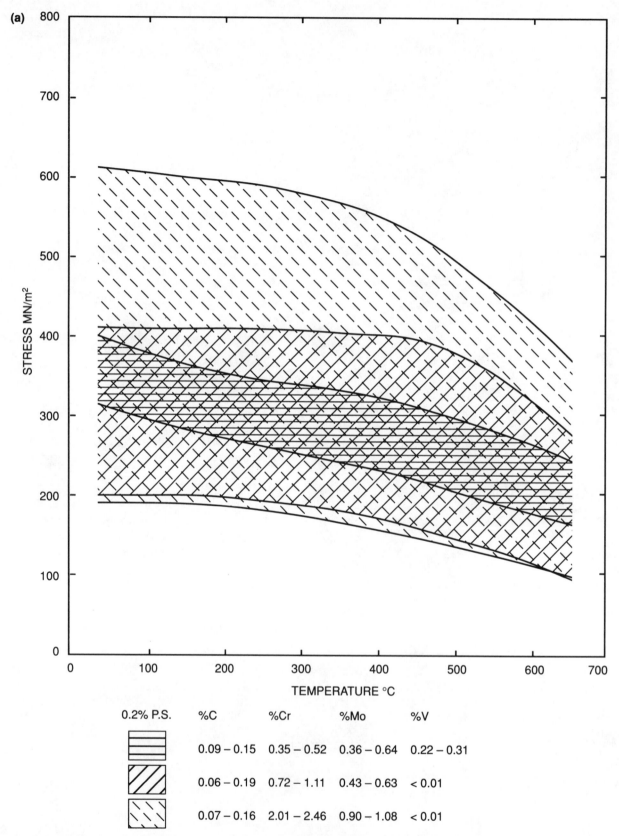

0.2% P.S.	%C	%Cr	%Mo	%V
	0.09 – 0.15	0.35 – 0.52	0.36 – 0.64	0.22 – 0.31
	0.06 – 0.19	0.72 – 1.11	0.43 – 0.63	< 0.01
	0.07 – 0.16	2.01 – 2.46	0.90 – 1.08	< 0.01

FIG 2.1.38 Variation of 0.2% P.S. of ½Cr – ½Mo – ¼V, 1Cr – ½Mo and 2¼Cr – 1Mo steels with temperature

Fig 2.1.38

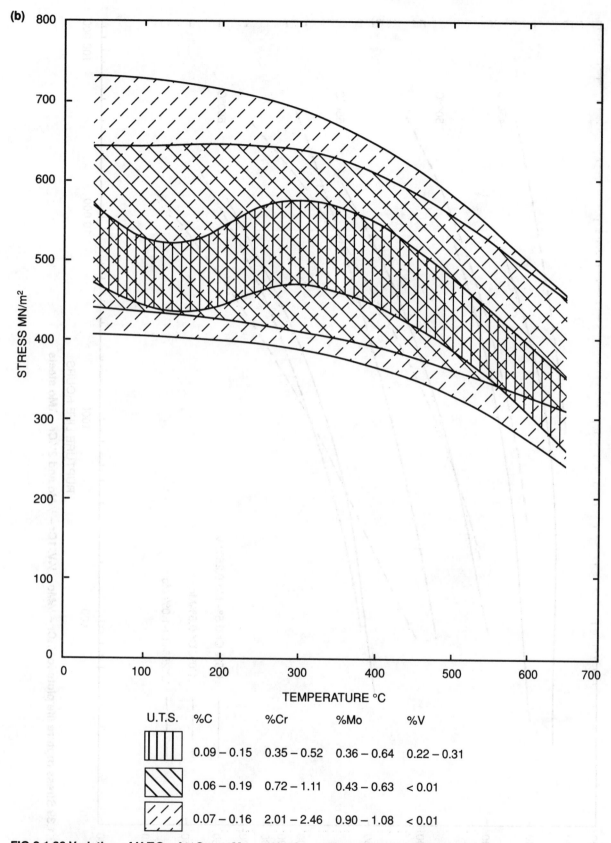

FIG 2.1.38 Variation of U.T.S. of ½Cr – ½Mo – ¼V, 1Cr – ½Mo and 2¼Cr – 1Mo steels with temperature—*continued*

U.T.S.	%C	%Cr	%Mo	%V
	0.09 – 0.15	0.35 – 0.52	0.36 – 0.64	0.22 – 0.31
	0.06 – 0.19	0.72 – 1.11	0.43 – 0.63	< 0.01
	0.07 – 0.16	2.01 – 2.46	0.90 – 1.08	< 0.01

Fig 2.1.38—*continued*

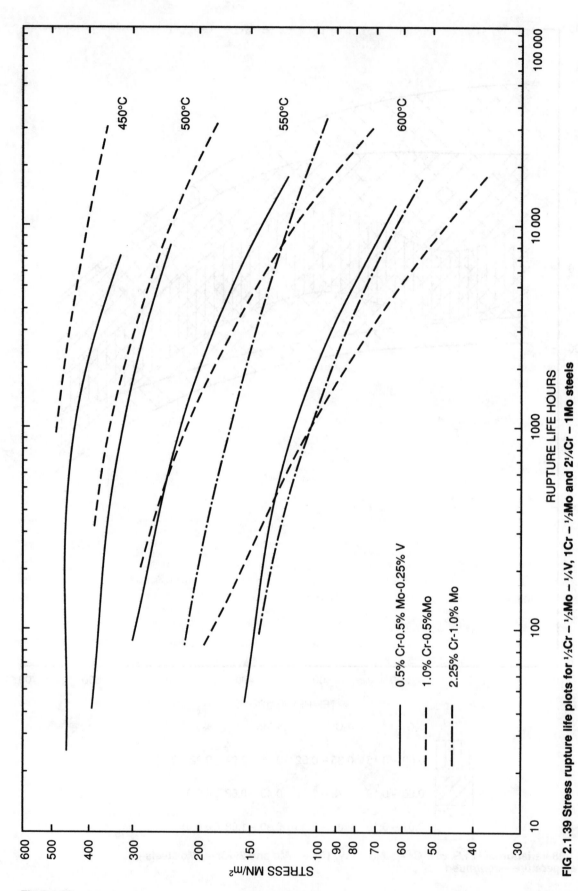

FIG 2.1.39 Stress rupture life plots for ½Cr – ½Mo – ¼V, 1Cr – ½Mo and 2¼Cr – 1Mo steels

Fig 2.1.39

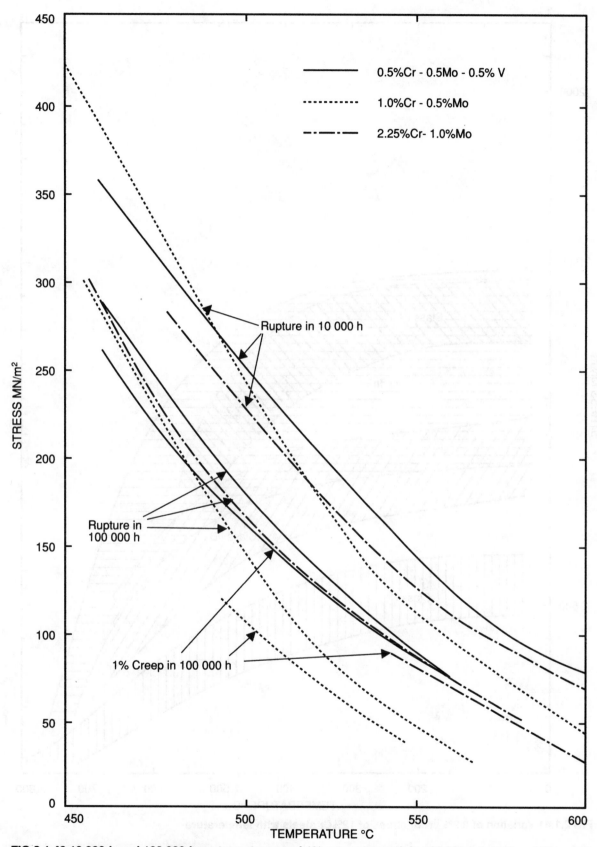

FIG 2.1.40 10 000-h and 100 000-h rupture stress and 1% creep strength in 100 000-h for ½Cr – ½Mo – ¼V, 1Cr – ½Mo and 2¼Cr – 1Mo steels

Fig 2.1.40

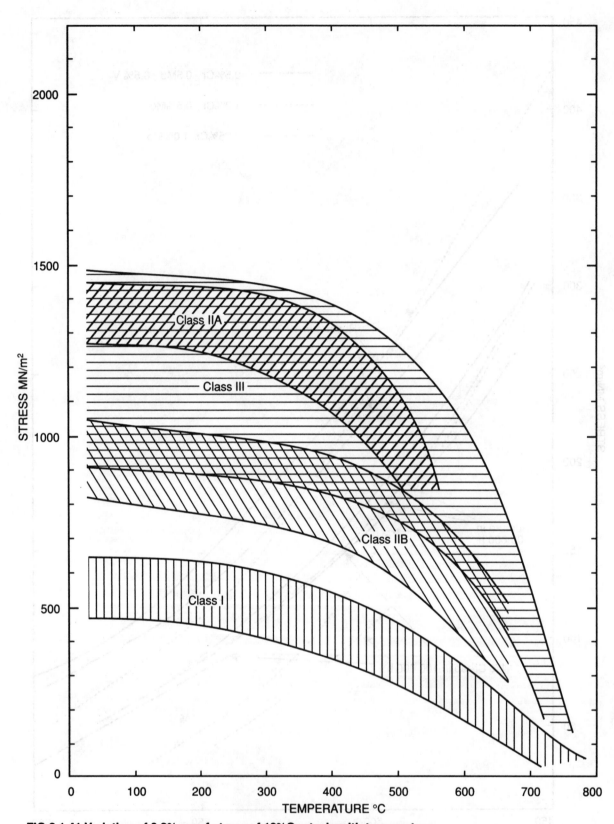

FIG 2.1.41 Variation of 0.2% proof stress of 12%Cr steels with temperature

Fig 2.1.41

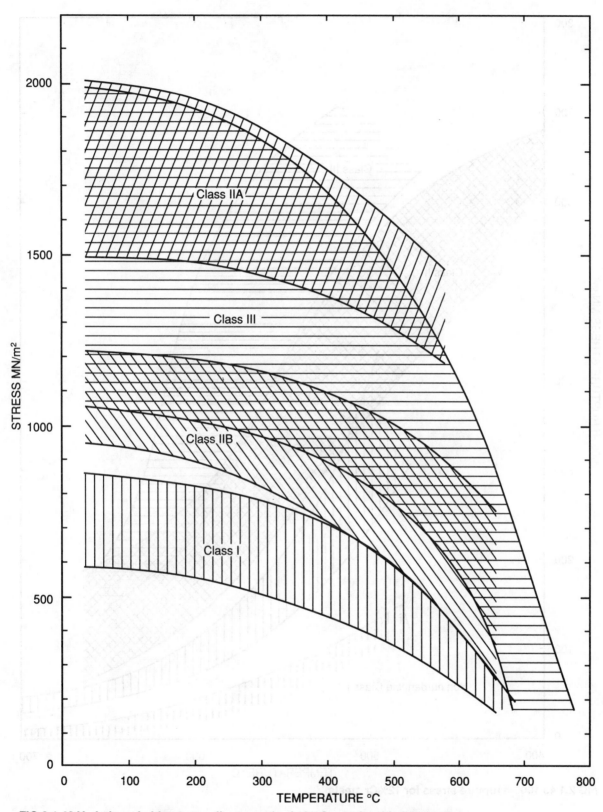

FIG 2.1.42 Variation of ultimate tensile strength of 12%Cr steels with temperature

Fig 2.1.42

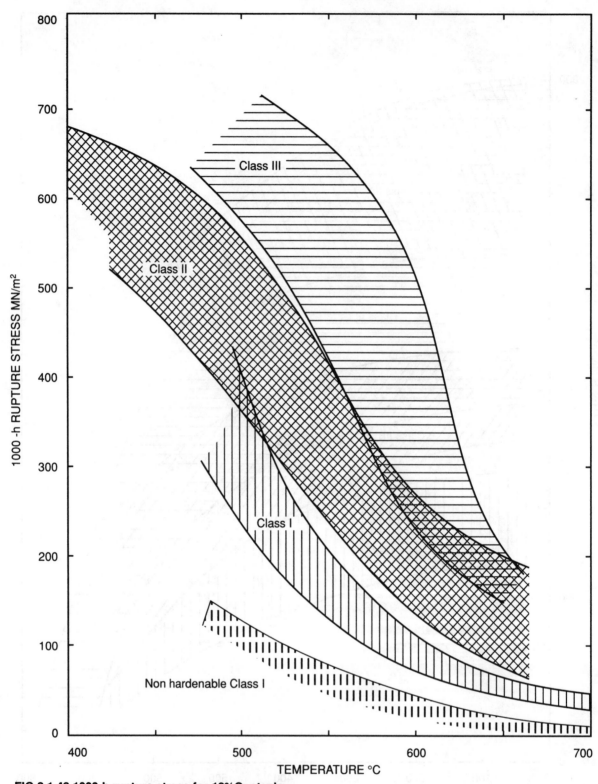

FIG 2.1.43 1000-h rupture stress for 12%Cr steels

Fig 2.1.43

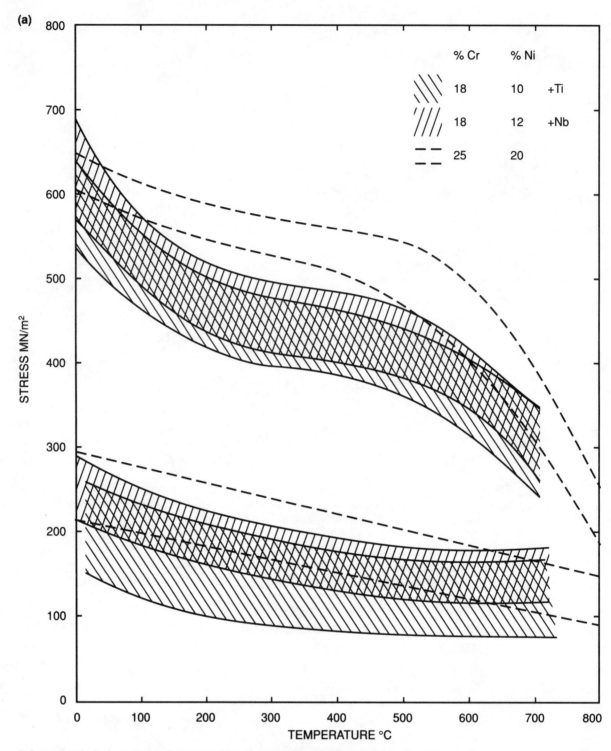

FIG 2.1.44 Variation of tensile properties of austenitic steels with temperature

Fig 2.1.44

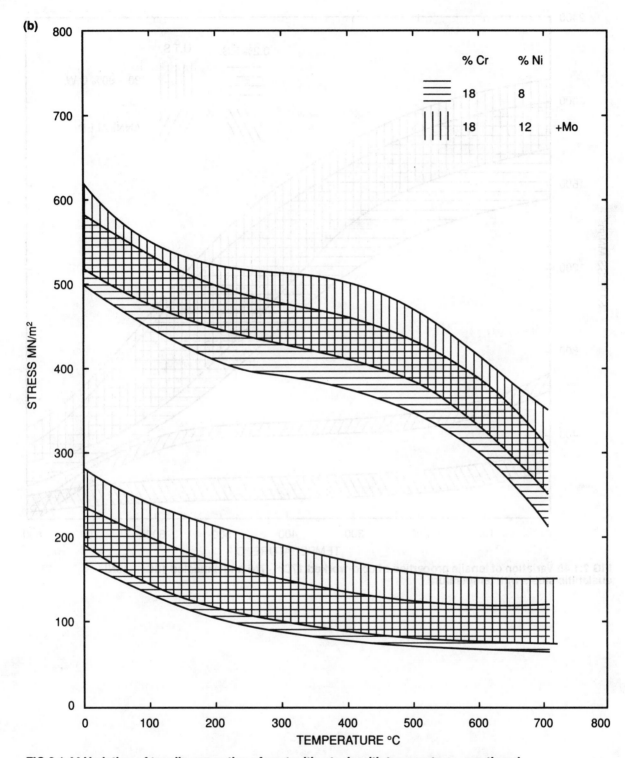

FIG 2.1.44 Variation of tensile properties of austenitic steels with temperature—*continued*

Fig 2.1.44—*continued*

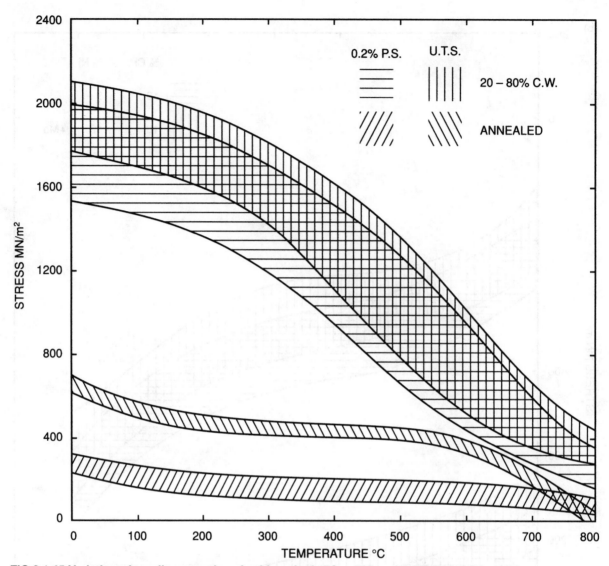

FIG 2.1.45 Variation of tensile properties of cold-worked 17Cr – 7Ni and annealed 19Cr – 10Ni austenitic steels with temperature

Fig 2.1.45

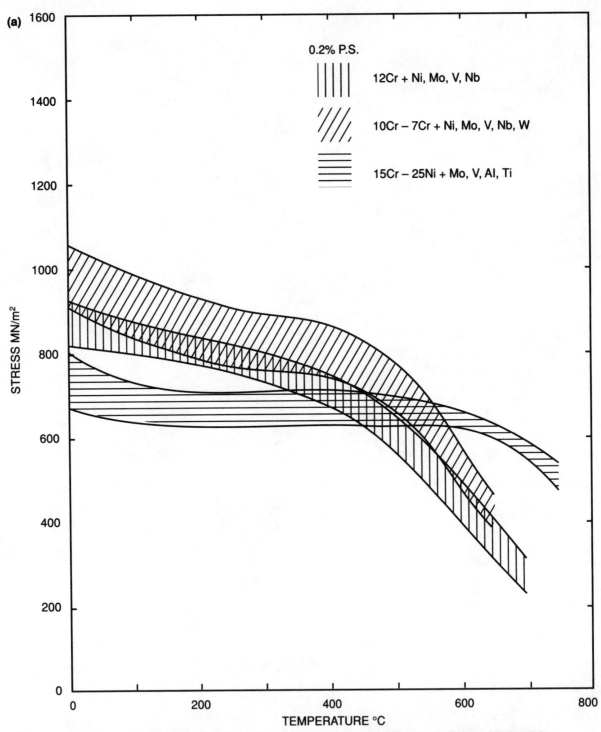

FIG 2.1.46 Variation of tensile properties of 12%Cr steels and A286 with temperature. (a) 0.2% P.S.

Fig 2.1.46

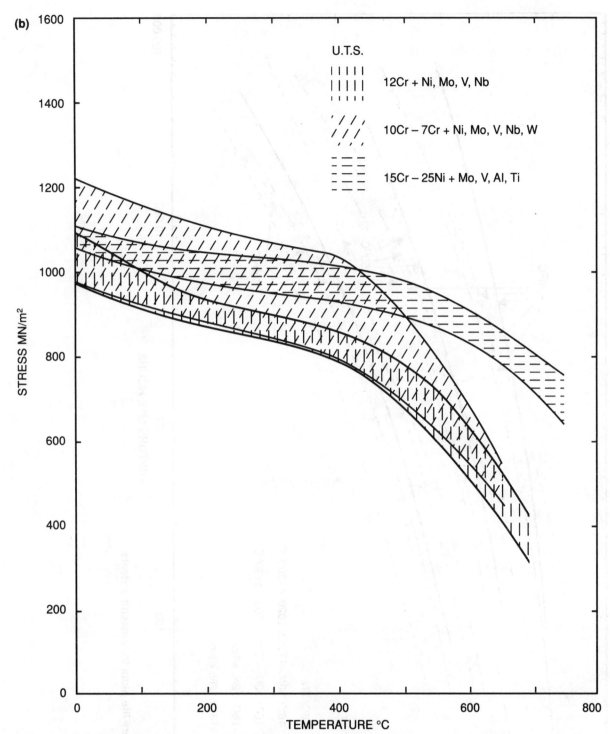

FIG 2.1.46 Variation of tensile properties of 12%Cr steels and A286 with temperature—*continued* (b) U.T.S.

Fig 2.1.46—*continued*

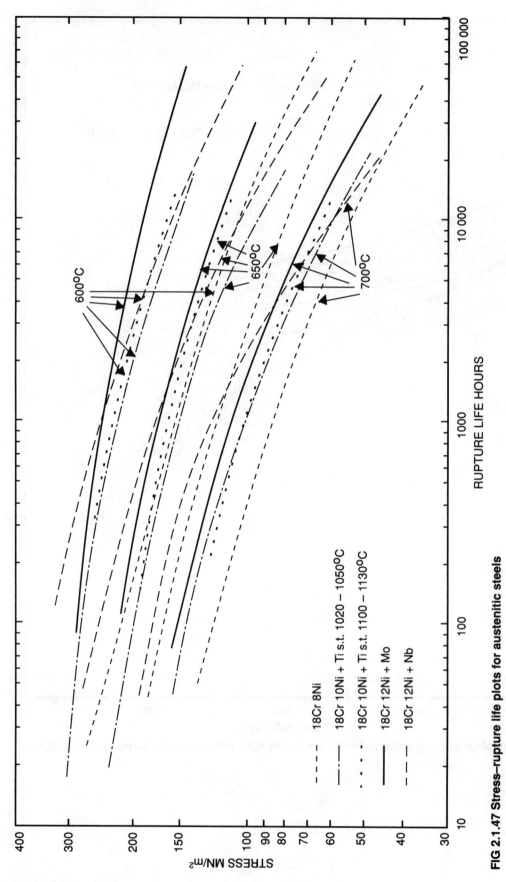

FIG 2.1.47 Stress–rupture life plots for austenitic steels

Fig 2.1.47

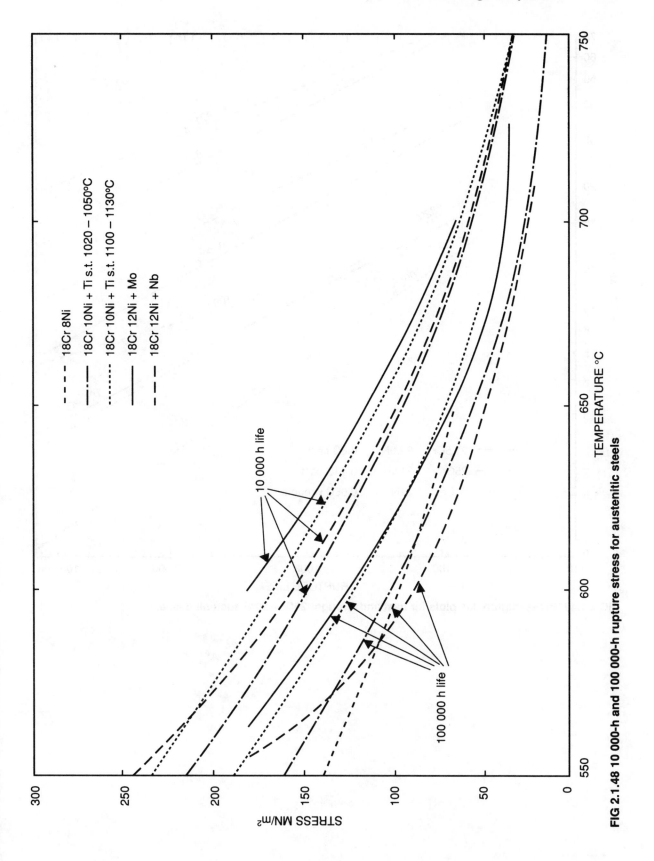

FIG 2.1.48 10 000-h and 100 000-h rupture stress for austenitic steels

Legend:
- - - - 18Cr 8Ni
- · — 18Cr 10Ni + Ti s.t. 1020 – 1050°C
- ······ 18Cr 10Ni + Ti s.t. 1100 – 1130°C
- —— 18Cr 12Ni + Mo
- – – 18Cr 12Ni + Nb

10 000 h life

100 000 h life

TEMPERATURE °C

STRESS MN/m²

Fig 2.1.48

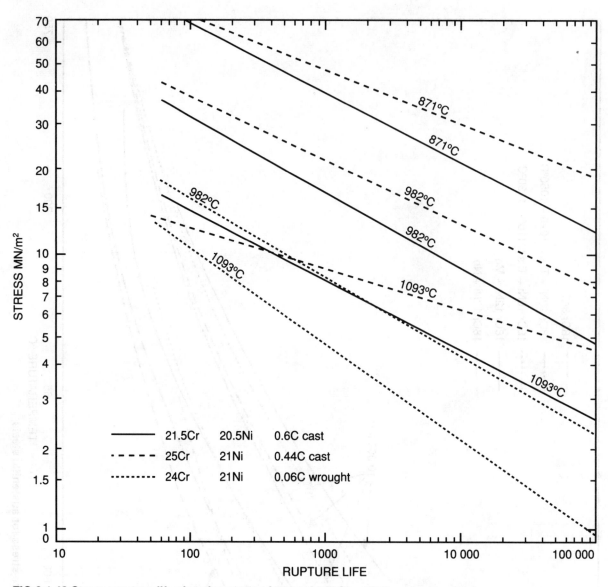

FIG 2.1.49 Stress–rupture life plots for cast and wrought 25Cr – 20Ni austenitic steels

Fig 2.1.49

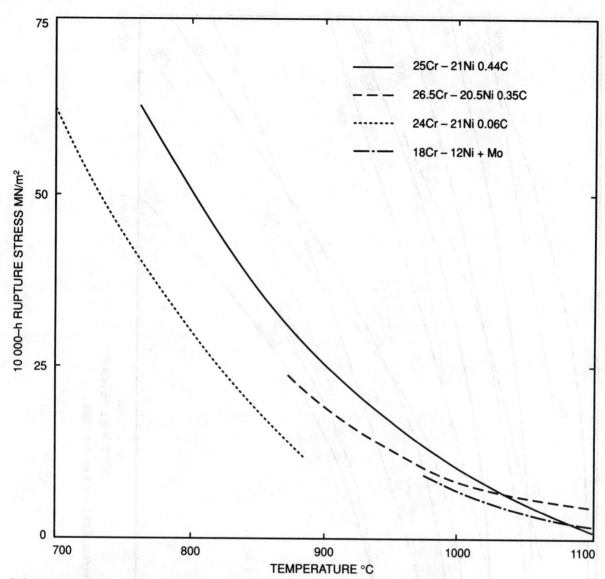

FIG 2.1.50 10 000-h rupture stress for 25Cr–20Ni and 18Cr–12Ni austenitic steels

Fig 2.1.50

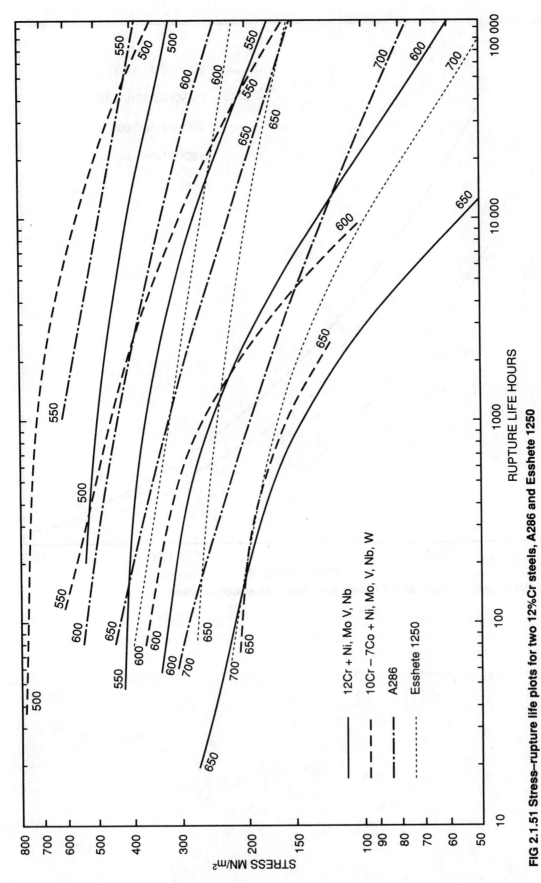

FIG 2.1.51 Stress–rupture life plots for two 12%Cr steels, A286 and Esshete 1250

Fig 2.1.51

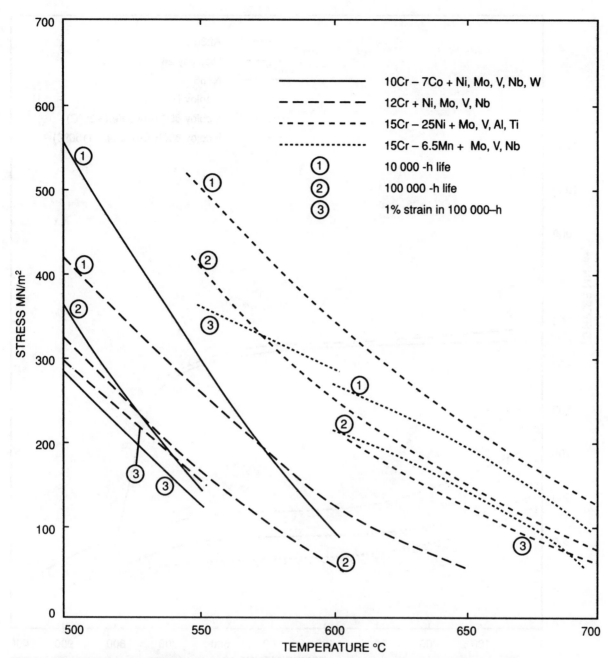

FIG 2.1.52 10 000-h and 100 000-h rupture stress and 1% creep strength in 100 000-h for 12%Cr
steels, A286 and Esshete 1250

Fig 2.1.52

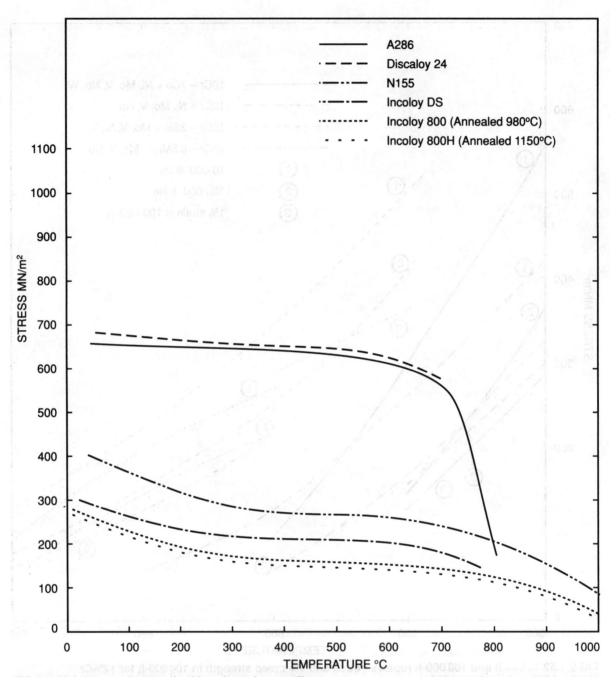

FIG 2.1.53 0.2% proof strength of iron-based superalloys

Fig 2.1.53

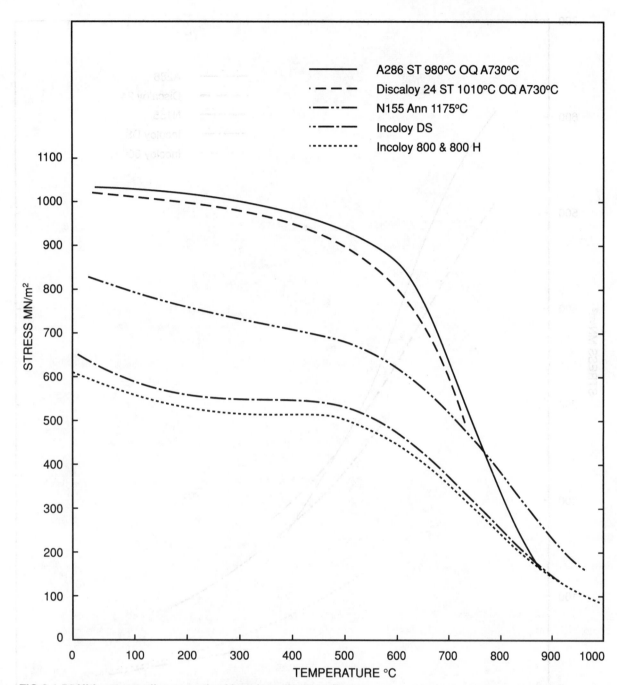

FIG 2.1.54 Ultimate tensile strength of iron-based superalloys

Fig 2.1.54

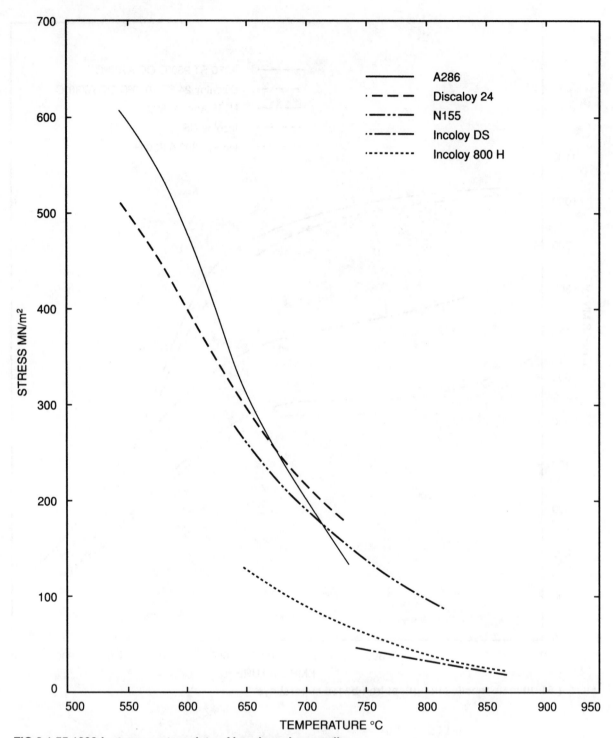

FIG 2.1.55 1000-h stress rupture data of iron-based superalloys

Fig 2.1.55

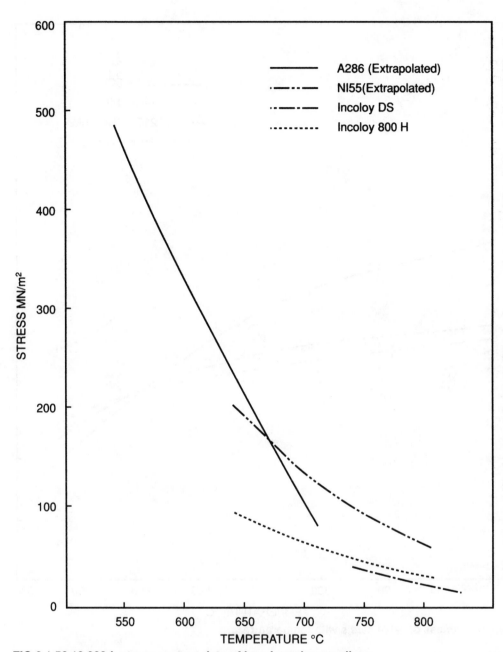

FIG 2.1.56 10 000-h stress rupture data of iron-based superalloys

Fig 2.1.56

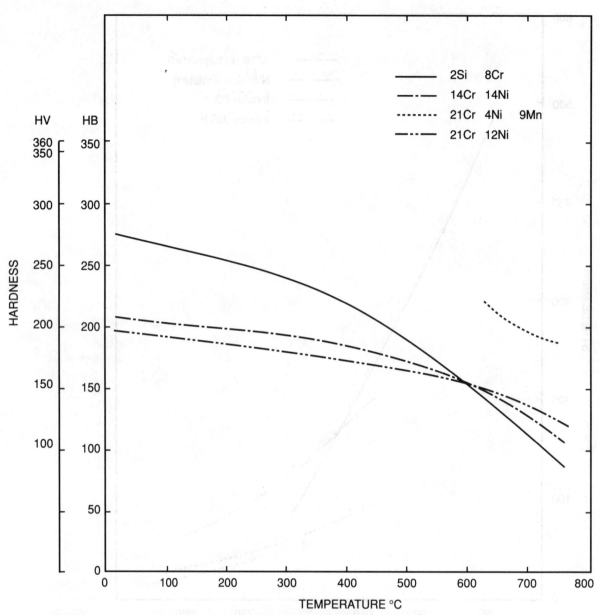

FIG 2.1.57 Variation in hardness of valve steels with temperature

Fig 2.1.57

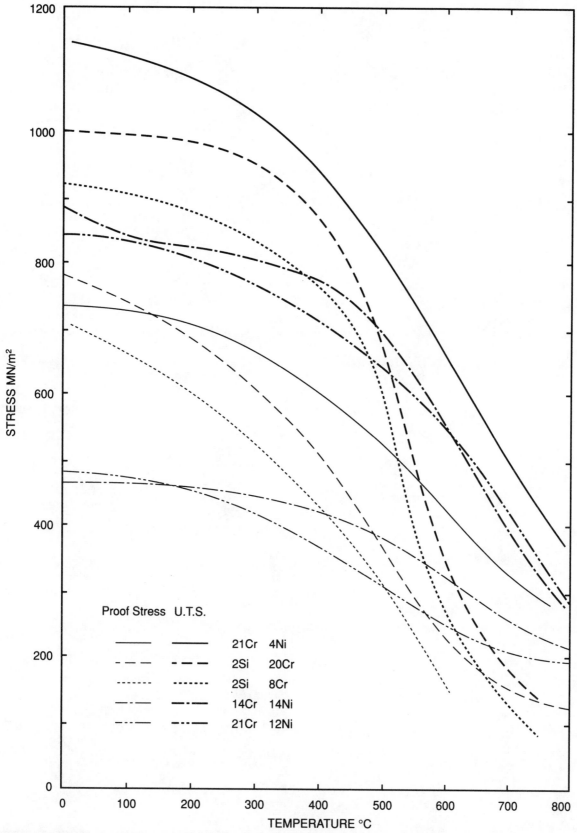

FIG 2.1.58 Tensile strength (bold lines) and 0.1% proof stress (faint lines) for valve steels plotted against temperature

Fig 2.1.58

Cast steels

Contents

List of tables

2.2.1   Introduction

Many of the types of wrought steel described in Chapter 1 of this volume are also available in cast versions made by one or other of the processes described in Vol. 1 Chapter 6. The relative advantages of specifying a casting, as compared with a wrought product of equivalent tensile strength are listed in:

Table 2.2.1   *Relative merits of steel castings as compared with wrought steel products*

In summary, the casting can be made in a wider range of shapes and has better creep resistance but the same component made from a wrought product is usually more consistent, and any defects which may result in failure can be better controlled and oriented so that resistance to fatigue is more consistently high.

Casting standards both in the UK and the USA are divided into:

BS3100 1976, Steel Castings for General Engineering purposes, ASTM 27-83 Steel Castings for General Application, ASTM 148-835 Steel Castings for High Strength Structural Purposes, BS1504 1976 Steel Castings for Pressure Purposes and ASTM 487-83 Steel Castings Suitable for Pressure Purposes.

TABLE 2.2.1 Relative merits of steel castings as compared with wrought steel products

Feature	Advantages	Limitations
Dimensional	Wide range of irregular shapes can be made.	Extended shapes cannot be made.
Strength	Equivalent strength and hardness to wrought product is available in castings.	Properties of a casting are more likely to be influenced by size of section than those of a forging.
Ductility	Ductility of castings is normally isotropic with elongation and fracture toughness the same in all directions.	Working can be so organised that better ductility or fracture toughness is obtainable in a desired direction (but transverse ductility may be lower than for the equivalent casting).
Fatigue properties	Fatigue limit of casting usually about 20% lower than that of equivalent strength forging tested in the longitudinal direction.	Forging tested in transverse direction *may* have lower fatigue resistance than casting.
High temperature time-dependent properties	Creep resistance of casting usually significantly higher than that of equivalent forging. Casting of some alloys can be so organised as to produce higher resistance to creep in a desired direction. This cannot be achieved in a wrought product.	
Structure	Some casting techniques with very high rates of chill can produce very high hardness and strengths (e.g. in amorphous metal strip).	Structure, which may in large castings be coarse and/or segregated, cannot always be refined by heat treatment. If a coarse ingot structure is present it may be possible to refine it by forging. Some working procedures (e.g. warm working or ausforming) can generate very high properties in wrought products.
Defects	Some defects which arise in castings can be directed into innocuous positions by the use of appropriate techniques. Special techniques which can produce castings equal in integrity to forgings exist, but they are usually expensive to carry out.	Defects may occur randomly in size, position and orientation in castings. The defects derived from the ingot in wrought products can be more easily predicted, limited, controlled and detected. The metal working process can orient defects into less harmful directions and may disperse or eliminate them.
Surface finish	Some die castings can have good surface finish.	Surface finish of castings is usually rougher and less regular than cold worked or machined finishes.

Table 2.2.1

2.2.2   Carbon and low alloy steel castings

The most comprehensive range of carbon and low alloy steels is described in BS3100 1976. This is listed with approximate equivalents for BS1504 1976 and the corresponding ASTM equivalents in:

Table 2.2.2   *Chemical compositions and mechanical properties of carbon and low alloy steel castings*

Only limited indications of the application of each grade of steel is provided in the table because it is easily possible to select a cast steel by analogy with the corresponding wrought grade described in Chapter 1 of this volume. It should, however, be possible to select a grade from Table 2.2.2 with properties suitable for almost all requirements.

TABLE 2.2.2 Chemical compositions and mechanical properties of carbon and low alloy steel castings

BS3100 1976	Approx. equivalent grades			Chemical composition % max. unless stated								
Grade	BS1504 1976	ASTM 22/77	487/76	C	Si	Mn	P max.	S max.	Cr	Mo	Ni	Others
A1	161 {430	N1		0.25	0.60	0.90 max.	0.060	0.060	—	—	—	—
A2		N2	1N, 1Q	0.35	0.60	1.0 max.	0.060	0.060	—	—	—	—
A3	{480			0.45	0.60	1.0 max.	0.060	0.060	—	—	—	—
A4	70/40			0.18 to 0.25	0.60	1.20/1.60	0.050	0.050	—	—	—	—
A5			2N, 2Q	0.25 to 0.33	0.60	1.20/1.60	0.050	0.050	—	—	—	—
A6				0.25 to 0.33	0.60	1.20/1.60	0.050	0.050	—	—	—	—
AL1				0.20	0.60	1.10 max.	0.040	0.040	—	—	—	—
AM1				0.15	0.60	0.50 max.	0.050	0.050	—	—	—	—
AM2				0.25	0.60	0.50 max.	0.050	0.050	—	—	—	—
AW1				0.10/0.18	0.60	0.60/1.0	0.050	0.050	—	—	—	—
AW2				0.40/0.50	0.60	1.0 max.	0.050	0.050	—	—	—	—
AW3				0.50/0.60	0.60	1.0 max.	0.050	0.050	—	—	—	—
B1	245			0.20	0.20/0.60	0.50/1.0	0.050	0.050	—	0.45/0.65	—	—
B2	{621		8N, 8Q	0.20	0.60	0.50/0.80	0.050	0.050	1.0/1.50	0.45/0.65	—	—
B3	{622			0.18	0.60	0.40/0.70	0.050	0.050	2.0/2.75	0.90/1.20	—	—
B4	1504 {623			0.25	0.75	0.30/0.70	0.040	0.040	2.50/3.50	0.35/0.60	—	—
B5	{625			0.20	0.75	0.40/0.70	0.040	0.040	4.0/6.0	0.45/0.65	—	—
B6	{629			0.20	1.0	0.30/0.70	0.040	0.040	8.0/10.0	0.90/1.20	—	—
B7			11N, 11Q	0.10/0.15	0.45	0.40/0.70	0.030	0.030	0.30/0.50	0.40/0.60	—	V0.22/0.30
BL1	503 LT60			0.20	0.60	1.0 max.	0.040	0.040	—	0.45/0.65	—	—
BL2				0.12	0.60	0.80 max.	0.030	0.030	—	—	3.0/4.0	—
BT1				—[e]	—[e]	—[e]	0.050	0.050	—	—	—	—
BT2				—[e]	—[e]	—[e]	0.040	0.040	—	—	—	—
BT3				—[e]	—[e]	—[e]	0.030	0.030	—	—	—	—
BW1				0.12/0.18	0.60	0.30/0.60	0.040	0.040	0.60/1.10	0.15/0.25	3.00/3.75	—
BW2				0.45/0.55	0.75	0.50/1.0	0.060	0.060	0.80/1.2	—	—	—
BW3				0.45/0.55	0.75	0.50/1.0	0.060	0.060	0.80/1.2	—	—	—
BW4				0.55/0.65	0.75	0.50/1.0	0.060	0.060	0.80/1.5	0.20/0.40	—	—
BW10				1.00/1.25	1.0	11.0 min.	0.070	0.060	—	—	—	—

[a] Only those ASTM grades which correspond to the BS grades have been included. Other US grades (and US specifications) are available.
[b] HT, Heat treatment at manufacturer's (or purchaser's) direction.
N, Normalised.
T, Tempered.
OQ, Oil quenched.
WQ, Water quenched.
H&T, Hardened and tempered.

[c] At (specified) low temperature.
[d] Angle of bend.
[e] Not specified.

Table 2.2.2

TABLE 2.2.2 Chemical compositions and mechanical properties of carbon and low alloy steel castings—*continued*

Treatment	Tensile strength (MPa)	0.2% proof stress (MPa min.)	E% on 5.65√area (% min.)	Charpy V Notch Impact (J min.)	Comments
HT	430 min.	230	22	25	Carbon steel castings for general purposes.
HT	490 min.	260	18	20	
HT	540 min.	295	14	18	
N, N & T or HT	540/690	320	16	30	Carbon manganese steel castings for general purposes.
N, N & T or HT	620/770	370	13	25	
OQ or WQ & T	690/850	495	13	25	
	430 min.	230	22	20[c]	Carbon steel castings for low temperature.
	340/430	185	22	120[d]	Carbon steel castings with high magnetic permeability.
	400/490	215	22	120[d]	
	460 min.	—	12	25	Carbon steel castings for case hardening.
N or N & T	620 min.	325	12	—	Carbon steel castings resistant to wear and suitable for case hardening.
N or N & T	690 min.	370	8	—	
N & T	460 min.	260	18	20	C–Mo steel castings for elevated temperatures.
N & T or H & T	480 min.	280	17	30	Cr–Mo steel castings for elevated temperatures.
"	540 min.	325	17	25	
"	620 min.	370	13	25	
"	620 min.	420	13	25	
"	620 min.	420	13	—	
N & T or H & T	510 min.	295	17	120[d]	Cr–Mo–V steel castings for elevated temperatures.
HT	460 min.	260	18	20[c]	Steel castings for low temperatures.
HT	460 min.	280	20	20[c]	
H & T	690/850	495	11	35	Castings with higher tensile strengths.
H & T	850/1000	585	8	25	
H & T	1000/1160	695	6	20	
WQ & T or H & T	1000 min.	—	7	20	Alloy steel castings for case hardening.
"	—	—	—	—	1% Cr steel castings to resist abrasion.
"	—	—	—	—	
"	—	—	—	—	
WQ	—	—	—	—	Austenitic manganese steel castings.

Table 2.2.2—*continued*

2.2.3   Cast stainless steels

2.2.3.1   Introduction

Stainless steel casting alloys are available to match the more commonly used wrought stainless steels in composition, corrosion resistance and, allowing for the difference between castings and forgings, mechanical properties and structure.

2.2.3.2   Specifications

So far as British and German Specifications are concerned, cast stainless steels are allocated a specification number which includes the characteristic AISI three figure identification number of the corresponding wrought steel.

There are two British Standards, 3100 1976 for general engineering purposes and 1504 1976 for pressure purposes, with marginal differences. The first three figures are followed by a 'C' for casting and then two further figures which indicate slight compositional variations. The use of the easily recognised three figures simplifies the choice when design or economic considerations require a cast rather than an already identified wrought component.

In the USA however, the US Alloy Castings Institute uses a system of numbers and letters which bear no relationship to the AISI wrought steel numbers. There is first a three figure number related to the application, typically 351, 743 or 744. This is followed by a letter C which recognises that there is a significant chromium content. A second letter ranges from A to N as the proportion of major alloying constituents chromium and nickel increases. A figure which is 100 times the percentage carbon content follows and then a further letter or letters indicating other alloying additions. All the stainless steels specified in BS 3100 for general engineering purposes, BS 1504 for pressure purposes and the ASTM (Steel Founder's Society of America) designations are listed in:

Table 2.2.3   *Specification numbers (BS3100, BS1504 and Steel Founder's Society of America) and compositions of cast stainless steels*
and
Table 2.2.4   *Mechanical properties, structure and applications of stainless steels casting alloys*

2.2.3.3   Characteristics

In general, the properties of a stainless steel casting resemble those of the corresponding wrought steel. Corrosion resistance is very closely related but it should be remembered that castings are more liable to segregation and that this particularly may adversely affect local resistance to environment particularly in the non-heat-treated condition.

Mechanical properties (see Table 2.2.4) are also very similar to the worked materials, and the tensile properties of the non-hardening steels are to all intents and purposes identical. Similar properties to wrought steels can also be obtained in the martensitic and precipitation hardening stainless steels, but care should be exercised when heat treating for higher properties.

In general, a cast material should have a similar structure to the equivalent worked material, but there is always a tendency in alloying an austenitic steel for casting to choose a composition higher in ferrite because this lessens the risk of fissuring during solidification. If a fully austenitic structure is required alloys higher in nickel must be used, and (usually) the casting must be solution-treated.

Most of the higher carbon stainless steels are sensitised by heating to temperatures between 425 and 870°C. 316, 347, 310 and CK20 are less susceptible.

The strengths quoted for the US steels are higher that those in BS 3100 and 1504. This is probably because the British figures are guaranteed minimum and the American are typical values.

The compositions of the British and US steels are mostly identical. A noted exception is the 25/20 type (310 and CK20) where the British steels have significantly higher alloy content. This stems from successful use as a reformer tube material.

In addition to the standard steel equivalents the US specifications include a number of steels with added copper which are claimed to have exceptional corrosion resistance to aqueous chemical environments. These include CF16F, CN7M, CN7MS, CO4MCU, Ilium P and Ilium PD.

TABLE 2.2.3 Specification numbers (BS3100, BS1504 and Steel Founder's Society of America) and compositions of cast stainless steels

BS 3100 1976	BS 1504 1976	Steel Founder's Society of America	C	Si	Mn	P max.	S max.	Cr	Mo	Ni	Others
Austenitic steels											
302C25			0.12 max	1.5 max.	2.0 max.	0.040	0.040	17.0/21.0	—	8.0 min.	—
302C35		CF20	0.2 max	2.0 max.	2.0 max.	0.060	0.060	17.0/22.0	1.5 max.	6.0 to 10.0	—
304C12	304C12	CF3	0.03 max.	1.5 max.	2.0 max.	0.040	0.040	17.0/21.0	—	8.0 min.	—
304C15	304C15	CF8	0.08 max.	1.5 max.	2.0 max.	0.040	0.040	17.0/21.0	—	8.0 min.	—
309C30			0.5 max.	2.5 max.	2.0 max.	0.060	0.060	22.0/27.0	1.5 max.	10.0/14.0	—
		CH10	0.10 max.	2.0 max.	1.5 max.	0.040	0.040	22.0/26.0	—	12.0/15.0	—
		CH20	0.20 max.	2.0 max.	1.5 max.	0.040	0.040	22.0/26.0	—	12.0/15.0	—
309C32		CE30	0.20/0.45	1.5 max.	2.5 max.	0.040	0.040	24.0/28.0	1.5 max.	11.0/14.0	—
309C35		CK20	0.20/0.50	1.5 max.	2.0 max.	0.040	0.040	24.0/28.0	1.5 max.	11.0/14.0	—
309C40			0.2 max.	2.0 max.	2.0 max.	0.060	0.060	25.0/30.0	1.5 max.	8.0/12.0	—
310C40	310C40		0.30/0.50	1.5 max.	2.0 max.	0.040	0.040	23.0/27.0	—	19.0/22.0	—
310C45			0.5 max.	3.0 max.	2.0 max.	0.060	0.060	24.0/27.0	—	19.0/22.0	—
311C11			0.08 max.	3.0 max.	2.0 max.	0.060	0.060	22.0/27.0	—	17.0/22.0	—
315C16	315C16	CF3M	0.03 max.	1.5 max.	2.0 max.	0.040	0.040	17.0/23.0	1.0/1.75	23.0/28.0	—
316C12	316C16	CF8M	0.08 max.	1.5 max.	2.0 max.	0.040	0.040	17.0/21.0	2.0/3.0	8.0 min.	—
316C16	316C71		0.08 max.	1.5 max.	2.0 max.	0.040	0.040	17.0/21.0	2.0/3.0	10.0 min.	—
316C71	317C12		0.03 max.	1.5 max.	2.0 max.	0.040	0.040	17.0/21.0	2.0/3.0	10.0 min.	—
317C16		CG8M	0.08 max.	1.5 max.	2.0 max.	0.040	0.040	17.0/21.0	3.0/4.0	10.0 min.	—
318C17	318C17		0.08 max.	1.5 max.	2.0 max.	0.040	0.040	17.0/21.0	3.0/4.0	10.0 min.	Nb8 × C to 1.0
330C11	330C11		0.35/0.55	1.5 max.	2.0 max.	0.040	0.060	13.0/17.0	2.0/3.0	33.0/37.0	—
330C12			0.75 max.	1.5 max.	2.0 max.	0.060	0.060	13.0/17.0	1.5 max.	30.0/40.0	—
331C40			0.35/0.55	3.0 max.	2.0 max.	0.060	0.040	17.0/21.0	1.5 max.	37.0/41.0	—
331C60			0.75	3.0 max.	2.0 max.	0.040	0.060	15.0/25.0	1.5 max.	36.0/46.0	—
334C11			0.75 max.	1.5 max.	2.0 max.	0.060	0.040	10.0/20.0	1.5 max.	55.0/65.0	—
347C17	347C17	CF8C	0.08 max.	2.0 max.	2.0 max.	0.040	0.040	17.0/21.0	—	8.5 min.	Nb8 × C to 1.0
		CF16	0.16 max.	2.0 max.	1.5 max.	0.170	0.040	18.0/21.0	1.5 max.	9.0/12.0	Se 0.2/0.35
		CN7M	0.07 max.	1.5 max.	1.5 max.	0.040	0.040	22.0/25.0	2.0/3.0	27.5/30.5	Cu 3.0/4.0
		CN7MS	0.07 max.	2.5 to 3.5	1.0 max.	0.030	0.040	18.6/20.0	2.5/3.0	22.0/25.0	Cu 1.5/2.0
Martensitic steels											
410C21		CA15	0.15 max.	1.0 max.	1.0 max.	0.040	0.040	11.5/13.5	—	1.0 max.	—
		CA15M	0.15 max.	0.6 max.	1.0 max.	0.040	0.040	11.5/14.0	0.15/1.0	1.0 max.	—
420C24		CA40	0.25 max.	2.0 max.	1.0 max.	0.060	0.060	12.0/16.0	—	—	—
420C29	420C29		0.20 max.	1.0 max.	1.0 max.	0.040	0.040	11.5/13.5	—	1.0 to 4.2	—
425C11	425C29	CA6NM	0.10 max.	1.0 max.	1.0 max.	0.040	0.040	11.5/13.5	0.60 max.	3.4 to 4.2	—
		CB7Cu	0.07 max.	1.0 max.	0.7 max.	0.035	0.030	14.0/15.5	—	4.5 to 5.5	Cu 2.5/3.5 Nb 0.2/0.35
Ferritic steels											
452C11		CC50	1.0 max.	2.0 max.	1.0 max.	0.060	0.060	25.0/30.0	1.5 max.	4.0 max.	—
452C12			1.0/2.0	2.0 max.	1.0 max.	0.060	0.060	25.0/30.0	1.5 max.	4.0 max.	—
		CB30	0.03 max.	1.5 max.	1.0 max.	0.040	0.040	18.0/21.0	—	2.0 max.	—
Duplex steels											
		CD4MC	0.04 max.	1.0 max.	1.0 max.	0.040	0.040	25.0/26.5	1.75/2.25	4.75/6.00	Cu 3
		IliumP[a]	0.2					28	2	8	Cu 3
		IliumPD[a]	0.06					27	2.5	5	Co 7
		Feralium 255–3SC[a]	0.05					25	3	6	Cu22 N0·18
		Fermanel[a]	0.06					27	3.1	8.5	Cu 1.0

[a]Manufacturer's designation.

Table 2.2.3

TABLE 2.2.4 Mechanical properties, structure and applications of stainless steels casting alloys

| Specification | | | BS minimum values | | | Steel Founder's Society typical values | | | Structure | | Applications |
BS3100 1976	BS1504 1976	Steel Founder's Society of America	Tensile strength R_m (MPa)	1.0% proof stress $R_{p1.0}$ (MPa)	Elongation A% $L_o=5.65\sqrt{S_o}$	Tensile strength R_m (MPa)	0.2% proof stress $R_{p0.2}$ (MPa)	Elongation A% min L = 50mm	As cast	Solution treated	
Austenitic steels											
302C25			480	240	26	—	—	—			
302C35		CF20	—	—	—	530	250	50	Austenite + chromium carbides	Austenite (sensitised by heating)	General
304C12	304C12	CF3	430	215	26	552	262	55	Austenite + ferrite	Austenite (not sensitised)	Where cannot be heat-treated for welding. General.
304C15	304C15	CF8	480	240	26	530	255	55	Austenite + carbides + ferrite	Austenite + ferrite (sensitised)	General corrosion resistant.
309C32		CH10	560	—	3	607	348	38	Austenite + carbides +ferrite	Austenitic + ferrite (not badly sensitised)	
309C35		CH20	510	—	7						
309C40		CE30	—	—		655	310	18	Austenite + ferrite + carbides		Sulphuric and sulphurous acids.
310C40	310C40	CK20	450	—	7	524	262	37	Austenite + chromium carbides	Austenite	High temperature corrosion resistance.
310C45			—	—							(Mainly reformer tubes).
311C11			—	—							
315C16	315C16		480	240	26						
316C12		CF3M	430	215	26	552	262	55	Austenite + ferrite		
316C16	316C16	CF3M	480	240	26	552	262	50	Austenite + ferrite		
316C71	316C71		510	260	26						Parts where post-welding heat treatment is impossible. General engineering use.
317C16	317C12		430	215	22						
318C17	318C17	CC8M	480	240	22	565	305	45	Austenite + ferrite	Austenite + ferrite	
330C12	330C11		480	240	18						
330C40			450	—	3						
331C40			—	—							
331C60			450	—							
334C11			—	—							
347C17	347C17	CF8C	480	240	22	530	262	39		Austenite + ferrite (not sensitised)	Aircraft and general.
		CF16F				530	275	52			Special chemical environments.
		CN7M				475	217	48			
		CN7MS				475	217	48			

Table 2.2.4

TABLE 2.2.4 Mechanical properties, structure and applications of stainless steels casting alloys—*continued*

Martensitic steels

410C21	CA15	540	370	15				Tempered martensite[b]	Jet engine and general engineering components.
420C24	CA40							Tempered martensite[b]	Aircraft and chemical components.
420C29		690	465	11				Tempered martensite[b]	Aircraft and chemical components.
425C11	CA6NM	770	620	12				Tempered martensite[b]	Seawater turbine impellers and components.
425C29	CB7Cu								Corrosion resistant applications.

Ferritic steels

452C11	CC50	690	465		655	414	15c	Ferrite	Mining, paper, fibre
452C12	CB30	770	620		655	415	15	Ferrite	Heat treating, oil refining.

Duplex steels

	CD4MCu / Ilium P[a]				745	562	35	2 Phase	General. Resists corrosion/erosion in strong acid.
					670	365	10	2 Phase	
	Feralium 255-3SC[a] / Fermanel[a]	725-825		18-20	480			2 Phase	
		700-850		25-35	420-500			2 Phase	

a Manufacturer's designation.
b As hardened and tempered.
c >2.0% Ni   0.15% N

Table 2.2.4—*continued*

2.2.4   Investment castings

The complex shapes, high dimensional tolerances, excellent surface finish and good mechanical properties obtainable by investment casting has led to the application of this process to many steel, nickel and cobalt alloys. It is of special advantage to the production of tool steel components, because the cast carbide structure confers a resistance to abrasion superior to that obtainable with wrought or powder compacted materials of equivalent composition. Commercially available investment castings (including, for completeness, nickel and cobalt alloys in addition to steels) are listed in:

Table 2.2.5   *Investment casting alloys commercially available*

Table 2.2.6   *British standard for investment castings showing some comparable materials and common names.*

TABLE 2.2.5 Investment casting alloys commercially available

Pic spec	Type	C	Ni	Cr	Mo	Others	En	AISI	BS'S series	DTD	BS 3100	CS 3146	Nr	Code	AFNOR
(1)	*General engineering and structural steels*														
2	Mild steel	0.15	—	—	—	—	2 32	1012 1015 1016	91 14	—	4239 16178	CLA9	0401	C15	XC12
3A	Mild steel	0.20	—	—	—	—	3A	1020 1021 1022	1 21	5199	592A	CLA1A	0402	C22	XC18
12	1% Ni–steel	0.50	1.0	—	—	—	12	—	—	—	—	—	—	—	—
19	1% Cr–Mo steel	0.40	—	1.1	0.22	—	19	4137 4140	3T50 81	5112 5219	—	CLA3	7225	42CrMo4	42CD4
19A	1% Cr–Mo– steel	0.30	—	1.1	0.35	—	19A	4135 4137	81	5219	—	CLA3	7220	34CrMo4	30CD4
24	1½ Ni–Cr–Mo steel	0.35	1.5	1.25	0.30	—	24	4340	119 95	—	1458	CLA3	6582	34CrNi-Mo6	35NCD6
24A	1½ Ni-Cr-Mo steel	0.40	1.5	1.25	0.30	—	24A	4340	—	—	—	CLA4	—	—	35NCD6
25	2½ Ni-Cr-Mo steel	0.30	2.5	0.6	0.6	—	25	—	96 97	5072	—	CLA5	5736	28(36)NiCr10	—
30B	4¼ Ni-Cr-MO steel	0.35	4.2	1.2	0.3	—	30B	—	28	—	1458	—	5864	35NiCr-18	30NCD16
31	1% Cr-1%C steel	1.0	—	1.3	—	—	31	51100 52100	135 136	—	—	—	3505	100 Cr6	100C6
34R	2% Ni-Mo steel	0.35	1.7	—	0.3	—	—	4640	—	—	—	—	—	—	—
44B	C-V Spring steel	0.85	—	—	—	V-0.25	44B	1090	—	—	—	—	1243	Mk 75	—
44D	C-V Spring steel	1.0	—	—	—	V-0.25	44C	1095	—	—	—	—	1274	Mk 101	—
47	1% Cr-V Spring steel	0.55	—	1.0	—	V-0.15	47	6150	204	—	1956A	CLA12	2241 8159 7561	50CrV4 50CrV4 42CrV6	—
320A	2% Ni-Cr-Mo steel	0.28	2.0	2.0	0.20	—	320	—	—	—	—	—	5920 6590	118CrNi8 30CrNiMo8	
325	2% Ni-Cr-Mo steel	0.20	2.0	0.5	0.3	—	325	4320	—	—	—	—			
1140	3% Cr-Mo steel	0.25	0.2	3.0	0.6	—	29A	—	106	5072 5229	1458	CLA5A/B	1.7765	GS-35 CrMo V10.4	—
1141	3% Cr-Mo steel	0.25	0.2	3.0	0.6	—	29B	—	—	5172	1461	CLA4 CLA7			—
(2)	*Surface hardening steels* c/h = Case hardening														
2	C-c/h	0.15	—	—	—	—	32C	1015 1016	14	—	1617	CLA9	0401 1141	C15 CK 15	XC15 XC12F
33	3% Ni-c/h	0.15	3.2	—	—	—	33	4815	15	5239	4240	CLA10	—	—	—
34	2% Ni-c/h	0.17	1.7	—	0.3	—	34	4617	—	—	—	—	—	—	15ND8
34c	2% Ni-c/h	0.12	1.7	—	0.3	—	—	4615	—	—	—	—	—	—	15ND8
34e	2% Ni-c/h	0.10	1.7	—	0.3	—	—	—	—	—	—	—	—	—	—
36	3% Ni-Cr-c/h	0.10	3.3	1.0	—	—	36A	9310	—	—	—	—	5752	14NiCr-14	14NC12
36t	3% Ni-Cr-Mo c/h	0.18	3.3	1.0	0.3	—	36C	9310	107	—	4241	—	—	—	—
37	5% Ni-c/h	0.13	5.0	—	—	—	37	2515	67	—	—	—	5680	12Ni19	—
39b	4¼% Ni-Cr-Mo-c/h	0.15	4.3	1.4	0.3	—	39B	—	82	—	—	—	5860	14NiCr 18	12NC15
40b	3% Cr-Mo (Nitriding)	0.25	0.2	3.0	0.6	—	40B	—	106	5249	1461	CLA11	8519	31CrMoV9	30CD12
8620	Low alloy c/h	0.20	0.5	0.75	0.15	—	362	8620	—	—	—	—	—	—	20NCD2
9	'60' Carbon Flame Harden	0.60	—	—	—	—	8/9	1060	70 79	—	1760b	CLA1-C CLA8	0601	C60	XC55
(3)	*Corrosion-resistant steels*														
	Rust-resistant steels														
56a	13% Cr	0.10	—	13.5	0.2	)a	56A	410	61	—	1630A	ANC1-A	4006	X10Cr13	Z12C13
56b	13% Cr	0.14	—	12.5	—	)	56B	410	61	—	1630B	ANC1-B	4021	X20Cr13	—
56c	13% Cr	0.20	—	12.5	—	)	56C	420	62	—	—	—	4021	X20Cr13	Z20C13
56d	13% Cr	0.28	—	12.8	—		56D	420	—	271	—	ANC1-C	4034	X40Cr13	Z30C13
57	18% Cr-Ni	0.12	2.0	17.5	—	—	57	431	80	—	—	ANC2	4057	X22CrNi-17	Z15CN16-2
	Stainless steels														
58	18/11/3	0.12	11.0	18.6	3.0	)a	58J	316	110 520	5279	1632B	ANC4-B	4436	X5CrNiMo18.12	Z8ND18-12
58a	18/11/3	0.08	11.0	18.6	3.0	)	58J	316	520	5279	1632B	ANC4-B	4436	X5CrNiMo 18.12	Z8CND 18-12
58b	18/9/Nb	0.08	9.0	18.6	—	) No. 11	58B	347	130	189 5269	1631B	ANC3-B Nb	4550	X10CrNiNb 18.19	Z10CNNb 18.10
3001	High chrome ferritic	0.15	—	21.5	—	—	61	51442	—	—	—	—	—	—	—
1180	1% C-18.0Cr	1.0	—	18.0	—	—	—	440c	—	—	—	—	1.4124	—	Z100CD17
1181	Cutlery	0.65	—	16.5	—	—	—	440a	—	—	—	—	—	—	—
17/4	Precipitation hardening stainless	0.05	4.0	17	—	Cu2.5	—	AMS 5355a	—	5299	—	—	—	—	ZCNU17-04

Table 2.2.5

TABLE 2.2.5 Investment casting alloys commercially available—*continued*

Pic spec	Type	Composition %					Comparable materials								
		C	Ni	Cr	Mo	Others	BS970 En	AISI	BS'S series	DTD	BS 3100	CS 3146	DIN Nr	Code	AFNOR
(4)	*Heat-resistant steels*														
2512	25/12/3	0.25	13.0	25.0	—	W3.25	—	309	—	—	1648E	ANC6-B	—	—	Z25CNWS 22.12
											4238EC				
2513	25/12	0.30	14.0	25.0	—		—	ASTM 297-HH	—	—	1648E 4238EC	ANC6-A	4837	GX35-CrNiSi 25.12	—
2515	25/20	0.10	21.0	24.0	—	—	—	310	—	—	1648F	ANC5-A	4848	GX40-CrNiSi 25.20	Z15CNS 25.20
													4849	GX15-CrNiSi 25.20	—
PE10	PE10	0.04	BAL	20.0	6.0	W2.5 Nb-6.5	—	—	—	—	—	—	—	—	—
(5)	*Special applications*														
For magnetic uses															
1	Low carbon iron	0.07	—	—	—	—	—	—	—	—	1617A	—	—	—	—
2	Mild steel	0.10	—	—	—	—	2	1008	—	—	1617B	—	0301	C10	XC10f XC69
56SL	Low carbon 13% chrome	0.05	—	13.0	—	—	—	—	—	—	—	—	—	—	—
5050	50% Ni-Fe	0.05	49	—	—	—	—	—	—	—	—	—	—	—	—

a Free machining versions also available.

Pic spec	Type	Composition %					AISI tool steel	BS4659 tool steel	Nearest comparable materials DIN (nearest composition) Nr	Symbol	AFNOR
		C	Cr	Mo	W	V					
(6)	*Tool steels*										
31	1% C-chrome	1.0	1.35	—	—	—	L1	BL3	2067	100Cr6	Y100C6
44b	Carbon-vanadium	0.85	—	—	—	0.25	W2 (1090)	(EN44b)	1248	Mk75	
44d	Carbon-vanadium	1.0	—	—	—	0.25	W2 (1095)	BW2 (EN44c)	2833	100V1	Y105V
									1274	Mk101	
47	1% Chrome-V	0.55	1.0	—	—	0.15	L2 (6150)	(EN50)	2241	50CrV4	Y50CV4
									8150	50CrV4	
									7561	42CrV6	
1035	10% W-Cr-V	0.40	2.8	—	10.5	0.5	H21-H22	BH21	2581	X30WCrV 9.3	Z30WCV9
1180	18%Cr–1%C	1.0	18.0	0.4	—	—	(440c)	—	—	—	—
1181	Cutlery	0.65	16.5	0.4	—	—	(440a)	—	—	—	—
1230	Carbon-2%W	1.0	—	—	2.0	0.3	F1-F2	BF1	2442	115W8	100WC10
1231	1% C-Cr-W	1.0	1.0	—	2.0	—	07	—	2550	60WCrV7	—
1232	0.7% C-Cr-W	0.7	2.2	—	1.6	—	—	—	—	—	—
1233	3% W-Cr-V	0.50	1.5	—	3.3	0.25	S1	BS1	2542	45WCrV7	—
1234	4% W-Cr-V	1.2	1.0	—	4.5	0.3	F3	—	2562	142WV13	—
1235	1% C-Cr-W	1.0	1.4	—	0.6	—	—	—	2067	100Cr6	Y100C6
1236	13% Cr-V	1.5	13.0	0.7	—	1.0	D2	BD2A	2379	X155CrVMo12.1	Z160CDV12
1512	12% Cr-Mo-V	1.5	12.0	0.75	—	0.25	D2	BD2	2601	X165CrMoV12	Z160CDV12
1841	18.4-1H.S.S.	0.75	4.0	0.75	18.0	1.25	T1	BT1	3255	B18	18.0-1
1842	18.4-2H.S.S.	0.80	4.0	0.75	18.0	2.0	T2	BT2	3357	C18	18.0-2
2120	12% CrWear/Die	2.2	12.0	0.4	—	—	D3	BD3	2080	X210Cr12	Z200C12
4056	5% Cr-Mo-V	0.44	5.0	1.3	—	1.2	H13	BH13	2344	X40CrMoV51	—
9550	Mn-C-Cr-W	0.9	0.5	—	0.5	Mn1.3	01	BO1	2419	105WCr6	90MCW5
9551	Mn-C-Cr	1.0	0.7	—	—	Mn1.3	02	BO2	2127	105MnCr4	—
9552	Mn-C-Cr-W	0.9	0.9	—	0.3	Mn1.3	L5	—	2127	105MnCr4	—

Data courtesy of P. I. Castings (Altrincham) Ltd. UK.

Table 2.2.5—*continued*

TABLE 2.2.6 British standard for investment castings showing some comparable materials and common names
(a) BS3146 Part 1 Carbon and low alloy steels

| Alloy | Pic spec. | Composition % | | | | Mechanical properties | | | | Comparable materials | | | | Common name |
		C	Ni	Cr	Mo	Tensile strength, R_m (MPa)	Elong%	Impact J	Hardness Hb 30	BS3100 (BS1504) BS970	DIN (Nearest composition)	ASTM	AFNOR	
CLA1-A	3A	0.25 max.	—	—	—	430 min.	15	—	121–174	592A(101A) EN3A	1155-GS-Ck25 0402-C22	A216-WC-A A27-60-30	CC20	Mild steel
CLA1-B	—	0.35 max.	—	—	—	500 min.	13	—	143–183	592B(101B)	0501-C35	A216-WC-B	CC35	Mild steel
CLA1-C	8	0.45 max.	—	—	—	550 min.	11	—	163–207	592C(101C) EN8	0503-C45	A148-80-40	CC45	'40' Carbon steel
CLA2	—	0.18–0.25	Mn 1.20–1.70 Ni	—	—	550–700	13	40	152–201	1456-A EN14-A	0555-GS62 0559-GS62.3	A148-80-50	20M5	1½% Mn steel
CLA3	19 or 24	—	—	—	—	700–850	11	35	201–255	1458-A	7218-GS-25CrMo4	A148-105-85	35 NC D6	Low alloy steel
CLA4	24 or 40 b	—	—	—	—	850–1000	11	20	248–302	1458-B	7361-32CrMo12	A148-120-95	30 CD 12	Low alloy steel
CLA5-A	1141	—	—	—	—	1000 min.	9	40	269–321	1458-C	7755-GS-35CrMo-V10.4	A148-150-125	30 CD 12	High strength steel
CLA5-B	1140	—	—	—	—	1160 min.	5	15	341–388	—	7755-GS-35CrMo-V10.4	A148-175-145	30 CD 12	High strength steel
CLA7	1141	0.25 max.	—	2.5–3.5	0.35–0.60	620–770	14	25	179–223	1461-(623) EN29-R	8519-31CrMoV9	A217-WC9	30 CD 12	Chrome–moly steel
CLA8	8 or 9	0.35–0.45	—	—	—	540 min.	15	—	Water Quench 500HV min.	1760A EN8 or 9	1191-GS-Ck45	—	CC45	Flame hardening steel
CLA9	2	0.10–0.18	—	—	—	495 min.	15	25	148–217	4239 EN32	0301-C10	—	XC15	Case-hardening steel
CLA10	33	0.10–0.18	2.75–3.50	—	—	700 min.	14	40	—	4240(503) EN33	—	A352-LC3	10 N 12	Ni-case-hardening steel
CLA11	40b	0.20–0.30	—	2.9–3.5	0.40–0.70	850–1000	8	20	248–302	1461 EN40-B	—	—	—	Nitriding steel
CLA12-A	47	0.45–0.55	—	0.80–1.20	—	700 min.	8	—	207 min.	1956-A 1956-B	GS-25CrMo4 GS-34CrMo4 (similar comp'n)	—	—	1% Cr abrasion resistant steel
CLA12-C	—	0.55–0.65	—	0.80–1.50	0.20–0.40	—	—	—	341 min.	1956-C	GS-42CrMo4	—	—	1% Cr–Mo abrasion resistant steel
CLA13	—	0.12–0.20	1.5–2.0	0.30 max.	0.20–0.30	700 min.	14 min.	40 min.	—	—	—	—	—	Ni–Mo case-hardening steel

Data courtesy of P.I. Castings (Altrincham) Ltd. Revised to accord with BS3146 (1974).

Table 2.2.6

TABLE 2.2.6 British standard for investment castings showing some comparable materials and common names—continued
(b) BS3146 Part 2 High alloy steels, nickel and cobalt alloys

Alloy	Plc spec.	C	Si	Mn	P	S	Ni	Cr	Mo	Nb	W	Co	Al	Fe	Tensile strength, R_m (MPa, Min.)	Elong % (Min.)	U.K. specs.	DIN (Nearest composition)	AISI (ASTM)	Common name
		Composition % (max. unless stated)													Mech. props		Related specifications			
ANC1-A	56A	0.15	0.2–1.2	0.2–1	0.035	0.035[b]	1	11.5–13.5							540	15[b]	1630–A EN56–A	4006 X 10 Cr 13	410 (A296–CA 15)	13% chrome rust-resisting
ANC1-B	56B	0.12–0.20	0.2–1.2	0.2–1	0.035	0.035[b]	1	11.5–13.5							620	13[b]	1630–B	4021 X 20 Cr 13	410 (A296–CA 15)	13% chrome rust-resisting
ANC1-C	56D	0.20–0.30	0.2–1.2	0.2–1	0.035	0.035[b]	1	11.5–13.5							695	11[b]	1630–C	4027 GX 25 Cr 14	420 (A296–CA 40)	13% chrome rust-resisting
ANC2	57 57A	0.12–0.25	0.2–1	0.2–1.0	0.035	0.035	1.5 3.00	15.5–20							850–1000	8	EN57 S80	4059 GX 22 CrNi 17	431	'18/2' Cr–Ni rust-resisting
ANC3-A	58X	0.12–	0.2–2	0.2–2	0.035	0.035[b]	8–12	17 min.							460	20[b]	1631–A EN58–A	3955 GX 12 CrNi 18.11 4308 GX 6 CrNi 18.9	302 (A296.CF8)	'18/9' stainless steel
ANC3-Bnb	58B	0.12–	0.2–2	0.2–2	0.035	0.035[b]	8.5–12	17–20		8×C 1.1					460	20[b]	1631–BNb	4552 GX 10 CrNiNb 18.9	347 (A296.CF8.C)	'18/9' stainless steel
ANC4	—	0.08	0.2–1.5	0.2–2	0.035	0.035[c]	11–14	18–20	3–4						500	12[c]	1632–A EN58J	4436 X 5CrNiMo 18.12	316 (A296.CG8.M)	'18/12/3' stainless steel
ANC4-B	58 58D	0.08	0.2–1.5	0.2–2	0.035	0.035[c]	10 min.	17–20	2–3						500	12[c]	1632.B EN58H	4408.G X 5CrNiMo 18.10	316 (A296.CF8.M)	Austenitic stainless steel
ANC4-CNb	—	0.12–1.5	0.2–2	0.2	0.035	0.035[c]	10 min.	17–20	3	8×C 1.1					500	12[c]	1632–CNb EN58H	4580.G X 10CrNiMoNb 18.10	316 (A296.CF10.M)	Austenitic stainless steel
ANC5-A	2515 2518	0.5	0.2–3	0.2–2	0.035	0.035	17–22.	22–27									1648.F	4848 GX 40CrNiSi 25.20 4849.GX 15 CrNiSi 25.20	(A351.CK20)	'25/20' heat-resistant steel
ANC5-B	—	0.5	0.2–3	0.2–2	0.035	0.035	36–46.	15–25									1648.H	4865.GX40NiCrSi 36.16	(A297.HU)	'37/20' heat-resistant steel
ANC5-C	—	0.75–	0.2–3	0.2–2	0.035	0.035	55–65.	10–20									1648.K		(A297.HW) (A297.HK)	'55/18' heat-resistant steel
ANC6-A[a]	2513	0.15–0.3	0.75–2	0.2–1	0.035	0.035	10–15	20–25							460	17	1648.E	4837.GX35CrNiSi 25.12	(A297.CH20) (A447.1/2)	'25/12' heat resistant steel H.R. Crown One (Firth Vickers Patent)
ANC6-B[a]	2512	0.15–0.35	0.75–2	0.2–1	0.035	0.035	10–15	20–25			2.5–3.5				460	17	1648.E EN55			'25/20' heat-resistant steel H.R. Crown Max. (Firth Vickers Patent)
ANC6-C[a]	—	0.05–0.5	0.75–2	0.2–1	0.035	0.035	10–18	20–25			2.5–3.5				460	17				'25/12/3' heat-resistant steel
ANC8		0.08–0.15	0.2–1	0.2–1			Bal.	18–22	Ti 0.2–0.6				0.3	5			DTD 703B	NiCr20Ti 24630		Nimocast 75 Nimonic 75

[a] If the castings are to be welded the structure should contain between 3 and 12% ferrite. To achieve this: 8.5 < [(Cr + W + 1.5 Si) − (30 × Ni + 0.5 Mn)] < 12.5.
[b] Where a free machining grade is specified (SO.3) no minimum elongation is specified.
[c] Where a free machining grade is specified (SO.3) the minimum elongation shall be 10.

Table 2.2.6—continued

TABLE 2.2.6 British standard for investment castings showing some comparable materials and common names—*continued*
(b) BS3146 Part 2—*continued*

Alloy	C	Si	Mn	S	P	Ni	Cr	Mo	Nb	Ti	W	Co	Al	Fe	Mg	R_m MPa min.	Elong % min.	U.K. specs.	DIN (Nearest composition)	AISI (ASTM)	Common name
ANC9	0.04–0.1	0.2–1	0.2–1			Bal.	18–22			2.2–3		2	0.8–1.6	2				DTD 736B	NiCr20TiAl 2.4631		Nimocast 80 / Nimonic 80
ANC10	0.05–0.13	0.2–1	0.2–1			Bal.	18–21		2–2.7			15–18	1–1.6	2				DTD 747B	NiCr20Co18Ti 2.4632		Nimocast 90 / Nimonic 90
ANC11	0.27–0.4	0.2–0.45	0.2–0.5			Bal.	18–23	9.5–11		0.3		9–11	0.2	1							C242 (Rolls Royce Patent)
ANC13	0.4–0.55	0.5–1	0.5–1			9.5–11.5	24.5–26.5				7–8	Bal.		2							X40–Delloro–Stellite Patent
ANC14	0.2–0.3	0.2–1	0.2–1			1.75–3.75	25–29	5–6				Bal.		3		650	6				Stellite 8 (Patent)
ANC15	0.02–0.12	0.5–1.2	0.5–1.2	0.03		Bal.	25–29	26–30						4–7							Hastelloy B
ANC16	0.05–0.15	0.5–1.2	0.5–1.2	0.03		Bal.	15.5–17.5	16–18		3.75–5.25				4–7							Hastelloy C
ANC17	0.05–0.12	8.5–10	0.5–1.2	0.03		Bal.							Cu 2–4	2							Hastelloy D
ANC18-A	0.1–0.3	0.5–1.5	0.5–1.5	0.05		Bal.							28–34	3	0.07–0.13			BS3071 NA.1	2.4360 (17743)	Federal QQ.N.288 A	Monel (H. Wiggin Patent)
ANC18-B	0.05–0.15	2.5–3	0.5–1.5	0.05		Bal.							28–34	3	0.07–0.13			BS3071 NA.2		QQ.N.288 B	Monel H (H. Wiggin Patent)
ANC18-C	0.05–0.15	3.5–4	0.5–1.5	0.05		Bal.							28–34	3	0.07–0.13			BS.3071 NA.3		QQ.N.288 D	Monel S (H. Wiggin Patent)
ANC19	0.06	0.1–0.4	0.1–0.5	0.015		Bal.	19–21	5.5–6.5	6.2–7[d]		2–3	2	0.2	2–4							
ANC20-A	0.07	0.2–2	0.2–1	0.025	0.025	3–6	12.5–15.5	0.5–2.5	0.5				1–3.5	Bal.		950–1200	12				
ANC20-B	0.07	0.2–2	0.2–1	0.025	0.025	3–6	12.5–15.5	0.5–2.5	0.5				1–3.5	Bal.		1250–1500	18				
ANC21	0.05	0.75	0.75	0.05	0.05	4.75–6	25–27	1.75–2.25				N 0.1	2.75–3.25	Bal.		700	18				

[d] Nb + Ta

Data courtesy of P.I. Castings (Altrincham) Ltd.

Table 2.2.6—*continued*

TABLE 2.2.6 British standard for investment castings showing some comparable materials and common names—continued
(c) BS3146 Part 3 Vacuum melted alloys

Composition % (max. unless stated) / Mechanical properties

Alloy	C	Si	Mn	P	S	Ni	Cr	Mo	W	Co	Ti	Al	B	Fe	Nb	Zr	Ta	Cu	N	Tensile UTS (MPa min.)	Elong% (min.)	Stress (MPa)	Time (h) min.	Temp. (°C)
VMA 1–A	0.03	0.1	0.1	0.01	0.01	16–17.5	0.25	4.4–4.9	—	9.5–11	0.15–0.6	0.07–0.15	—	Bal.	—	—	—	—	—	1540–1850	5	—	—	—
VMA 1–B	0.03	0.1	0.1	0.01	0.01	18–19	0.25	4.9–5.2	—	8.5–9.5	0.5–0.7	0.07–0.15	—	Bal.	—	—	—	—	—	1760–1930	5	—	—	—
VMA 2	0.06–0.14	0.5	0.5	—	0.015	Bal.	5–6.5	1.5–2.5	10–11.5	—	0.5	5.6–6.2	0.015–0.025	0.5	1–2[a]	0.08–0.16	—	0.5	—	750	4	750	50	850
VMA 3	0.1	0.6	0.6	—	—	Bal.	20–23	9–10.5	—	—	2.4–2.8	0.7–1	—	0.5	—	—	—	—	—	—	—	120	30	980
VMA 4	0.55–0.65	0.4	0.1	—	0.015	9–11	22.5–24.25	—	6.5–7.5	Bal.	0.15–0.35	—	0.01	1.5	—	0.1–0.6	3–4	0.2	—	725	2	—	—	—
VMA 5	0.04–0.08	0.4	0.6	—	—	Bal.	19–21	5.6–6.1	—	19–21	1.9–2.4[b]	0.3–0.6[b]	—	0.7	—	—	—	—	—	620	12	150	23	980
VMA 6A	0.08–0.2	0.5	0.25	—	0.015	Bal.	12–14	3.8–5.2	—	1	0.5–1.2	5.5–6.5	0.005–0.015	2.5	1.8–2.8[a]	0.05–0.15	—	0.5	—	755	2.5	150	30	980
VMA 6B	0.08–0.2	0.25	0.25	—	0.015	Bal.	12–14	3.8–5.2	—	1	0.5–1.2	5.5–6.5	0.005–0.015	1	1.8–2.8[a]	0.05–0.15	—	0.2	—	755	4	150	45	980
VMA 6C	0.03–0.07	0.25	0.25	—	0.015	Bal.	11–13	3.8–5.2	—	1	0.4–1	5.5–6.5	0.005–0.015	0.5	1.5–2.8[a] / 0.15	0.05	—	0.2	—	775	6	190	50	815
VMA 7A	0.1–0.2	0.3	0.25	—	—	Bal.	14–17	4.5–6	—	—	1.5–2.5[c]	2.5–3.5[c]	0.02–0.07	8–12	—	—	—	—	—	—	—	240	30	815
VMA 7B	0.1–0.2	0.3	0.25	—	—	Bal.	14–17	4.5–6	—	—	2–3[c]	3.25–4[c]	0.02–0.07	3.5–5	—	—	—	—	—	—	—	130	30	980
VMA 8	0.12–0.18	0.2	0.2	—	—	Bal.	14.5–16.5	7.6–9	—	9–10.5	3.4–2.8	3.9–4.4	0.004–0.008	0.5	—	—	—	—	—	—	—	170	35	930
VMA 9	0.05–0.1	0.3	0.2	—	0.015	Bal.	16–20	3–5	—	16–20	2.5–3.25	2.5–3.25	0.003–0.01	2	—	—	—	0.1	—	—	—	150	23	980
VMA 10	0.05–0.1	0.2	0.15	—	0.015	Bal.	14–15.25	3.9–4.5	—	14.25–15.25	3–3.7	4–4.6	0.012–0.020	0.5	—	0.05	—	0.1	0.05	—	—	200	20	930
VMA 11	0.05–0.11	0.2	0.2	—	0.015	Bal.	13.5–16.5	4–5	—	20–24	2–2.7	4–4.8	0.01–0.02	1	—	—	—	0.1	—	—	—	230	30	950
VMA 12	0.15–0.2	0.2	0.2	—	0.015	Bal.	8–11	2–4	—	13–17	4.5–5	5–6	0.01–0.02	1	—	0.03–0.09	—	0.2	V 0.7–1.2	—	—	450	20	700
VMA 13	0.02–0.10	0.35	0.35	—	0.015	50–55	17–21	2.8–3.3	—	1	0.65–1.15	0.4–0.8	0.006	Bal.	4.75–5.5	0.1	—	0.3	—	860	5	220	30	980
VMA 14	0.12–0.17	0.2	0.2	—	0.015	Bal.	8–10	2.25–2.75	9–11	9–11	1.25–1.75	5.25–5.75	0.01–0.02	1	—	0.03–0.08	1.25–1.75	0.1	—	—	—	110	50	1040
VMA 15	0.13–0.17	0.2	0.2	—	0.015	Bal.	8–10	0.5	9–11	9–11	1.25–1.75	5.25–5.75	0.01–0.02	0.5	—	0.03–0.08	2.25–2.75	0.1	Ha 1.2–1.6	—	—	150	27	980
VMA 16A	0.15–0.2	0.3	0.2	—	0.015	Bal.	15.7–16.3	1.5–2	2.4–2.8	8–9	3.2–3.7	3.2–3.7	0.005–0.015	0.5	0.6–1.1	0.05–0.15	1.4–2	—	—	—	—	150	23	980
VMA 16B	0.09–0.13	0.3	0.2	—	0.015	Bal.	15.7–16.3	1.5–2	2.4–2.8	8–9	3.2–3.7	3.2–3.7	0.007–0.012	0.5	0.6–1.1	0.03–0.08	1.4–2	—	—	—	—	—	—	—

a Nb + Ta.
b 2.4 < Al + Ti < 2.8.
c 5.6 < Al + Ti < 6.5.

Table 2.2.6—*continued*

Cast irons

Contents

List of tables

List of figures

2.3.1 Introduction to cast iron

The term 'cast iron' covers materials with a hardness range of 100–750 HB, ductilities (measured in terms of elongation at fracture) from 0 to 17%, and strengths from 155 to 730 MN/m². This wide spectrum of properties is controlled by three main factors:

(i)   the chemical composition of the iron;
(ii) the rate of cooling of the casting in the mould (which depends in part on the section thicknesses in the casting);
(iii) the type of graphite formed (if any).

A description of the main types of cast iron, their advantages and limitations are given in:

Table 2.3.1   *Characteristics of cast irons*

TABLE 2.3.1 Characteristics of cast irons

Material	Description	Advantages	Limitations
a.1 Grey Cast Iron. (Flake Graphite Cast Iron)	Most common type. Structure consists of graphite flakes in a matrix of ferrite (soft iron), pearlite (alternate lamellae of ferrite and hard cementite–iron carbide) or mixtures of both depending on composition and section size. Strength depends on amount and size of graphite—fewer smaller flakes give higher strength but greater casting difficulties. Normally specified according to tensile strength which would be obtained in a separately cast 30mm diameter bar from the same melt as the casting and not chemical composition. Inoculation (as is used in the production of Meehanite cast iron) improves homogeneity of graphite and allows iron which would otherwise solidify white to be graphitic, but increases cost.	Cheapest material for metal castings especially for small quantity production. Very easy to cast—much narrower solidification temperature range than steel. Low shrinkage in mould due to formation of graphite flakes. Good machinability, faster material removal rates but poorer surface finish with ferritic matrix and vice versa for pearlitic matrix. Graphite acts as a chip breaker and tool lubricant. Very high damping capacity. No difference in notched and un-notched fatigue strength. Good dry bearing qualities due to graphite (only true of pearlitic matrix). After formation of protective scale resists corrosion in many common engineering environments.	Brittle (low impact strength) due to sharp ends of graphite flakes severely limits use for critical applications. Graphite acts as a void and reduces strength. Maximum recommended design stress is ¼ of ultimate tensile strength. Maximum fatigue loading limit = ⅓ fatigue strength. (Fatigue strength = ½UTS for low strength iron; ⅓UTS for high strength iron.) Changes in section size will cause variation in machining characteristics (due to variation in microstructure). Higher strength irons more expensive to produce. Tensile strength range 155–400 MPa (compressive strength 3 times tensile). Tensile modulus 70–140 GPa (lower for higher graphite contents). Hardness increases from 100 to 350 HV as ferritic becomes pearlitic.
a.2 Low Alloy Grey Cast Iron	Grey cast irons with additions of Ni, Cu, Cr, Mo, V or Sn with the intention of ensuring a fine pearlitic matrix, and/or carbides and/or finer graphite form resulting in improved and more reliable engineering properties compared with unalloyed grey cast iron, especially in thick sectioned castings (for detailed effect of each alloy addition see Section 2.3.8).	Enables castings formerly produced in unalloyed grey cast iron to be used in higher duty applications without redesign or need for costly materials. Reduction in section sensitivity. Improvement in strength, corrosion, heat, and wear resistance or combination of these properties.	Higher cost. Alloy additions can cause foundry problems with re-use of scrap (runners, risers, etc.) and interrupt normal production. Increase in strength does not bring corresponding increase in fatigue strength. Cr, Mo, V are carbide stabilisers which improve strength and heat resistance but impair machinability.
a.3 Acicular Cast Iron (Flake Graphite)	Grey cast iron alloyed with Ni and Mo to give a microstructure of fine ferrite needles and fine pearlite with normal cooling. Precise structure depends on alloy content and section size.	Tougher than martensitic irons. Harder and stronger than pearlitic cast irons. Hardness of 320 HB with better machinability than unalloyed irons of same hardness due to absence of free cementite. High strength and uniformity of structure in heavy sections plus wear resistance responsible for use as die block material.	Softer than martensitic cast irons. Phosphorus content must be kept below 0.1%. Lower fatigue endurance limit than pearlitic grey cast iron. Lower carbon content means decreased fluidity and increased shrinkage risk which mean extra feeding requirements and lower yield.

Table 2.3.1

TABLE 2.3.1 Characteristics of cast irons—*continued*

Material	Description	Advantages	Limitations
b.1 Nodular Cast Iron (Spheroidal Graphite Cast Iron)	Ductile iron produced by inoculation of the melt with nickel magnesium and cerium compounds which gives a structure of graphite nodules (near spheroidal) in a matrix of pearlite, ferrite or mixture of both depending on composition and section size. Commonly a homogenising heat treatment is employed followed by: (i) fast air cooling to produce a fully pearlitic (strong, abrasive wear resistant) structure or (ii) slow cooling to produce a fully ferritic (soft machinable) structure Normalising improves properties but not fatigue strength.	Strength and ductility more typical of steel but with castability, damping capacity and machinability more typical of grey cast iron. Much better thermal shock resistance than grey cast iron. Can replace steel fabrications, castings and even forgings in some applications at lower total cost. Much shorter heat treatment times than malleable cast iron.	Higher cost than grey cast iron due to need for greater control over charge materials, cost of inoculation and generally low yield. Recommended design stress = ½ (0.1% proof stress). Fatigue design stress = ⅓ fatigue strength. Tensile strength range 380–700 MPa. Tensile modulus range 150–180 GPa.
b.2 Low Alloy Nodular Cast Iron	Nodular cast irons with additions of Ni, Cu, Cr, Mo, V or Sn with the intention of ensuring a fine pearlitic matrix, and/or carbides resulting in improved engineering properties compared with unalloyed nodular cast iron, especially in thick sectioned castings.	Similar improvement in properties to those listed for low alloy grey cast iron. Alternative to quenching and tempering unalloyed nodular iron.	As for low alloy grey cast iron (a.2).
b.3 Acicular Cast Iron (Nodular Graphite)	Nodular cast iron alloyed with Ni and Mo to give a microstructure of fine ferrite needles and fine pearlite with normal cooling. Precise structure depends on alloy content and section size.	Similar improvement in properties to those listed for acicular flake graphite cast iron. (a.3).	As for acicular flake graphite cast iron. (a.3).
c Austenitic Flake and Nodular Graphite Cast Iron, e.g. Ni-Resist Nicrosilal	Alloy additions—principally nickel but also copper and manganese—in sufficient quantities suppress transformation to ferrite (or martensite). Austenitic structure is stable at room temperature and below (if alloying content is high enough). Cr added to improve hardness, stiffness and erosion resistance. Cu replaces some of nickel content for corrosion resistance at lower cost. High silicon grades (Nicrosilal) developed for high temperature oxidation resistance. Most compositions produced in both flake and nodular iron except low Ni high Cu (flake only) and high Mn low temperature grade (nodular only). Nodular irons have higher strength and ductility. Flake irons have lower production costs and better machinability.	Non-magnetic. Wide range of grades for specific applications: (i) Resistance to oxidation and growth up to 800–950°C. (ii) Thermal shock resistance under cyclic heating conditions. (iii) Valve and pump parts in mildly corrosive and erosive process industries where low alloy steel is unsuitable and high cost stainless steel, phosphor bronze or gunmetal cannot be justified. (iv) Cryogenics.	High cost due to high nickel content and poor castability compared with grey and nodular cast iron. Copper-containing grades cannot be used in food processing, soap or rayon production industries. Not suitable for strong acid media.
Austempered ductile iron (Nodular graphite)	Special heat treatment of as-cast nodular iron.	Improvements in tensile strength, fatigue strength and elongation depending on austempering temperature (see Section 2.3.3.2).	Higher cost as a result of complex heat treatment. Not all heat treatment facilities capable of performing required heat treatment cycle.

Table 2.3.1—*continued*

TABLE 2.3.1 Characteristics of cast irons—*continued*

Material	Description	Advantages	Limitations	
d.1 White Heart Malleable Cast Iron	High temperature (850–900°C) heat treatment in oxidising medium for 5–6 days of white iron castings causes decomposition of carbides and formation of graphite aggregates (spider-like formation) in a ferritic (or in centre of heavy section—pearlitic) matrix. Castings have a decarburised skin.	Higher carbon content than other types of malleable iron gives better castability, especially thin sections. Decarburised layer improves weldability and provides soft ductile surface to absorb local impact blows. Marked increase in shock resistance above 100°C. Used in furnaces up to 450°C. Can be galvanised and does not suffer galvanising embrittlement. Can be bronzed or bronze welded. Very good machinability—high metal removal rates.	Long heat treatment time. Decarburised layer unsuitable surface for wear resistance. Welded casting should be annealed.	The need for the as cast structure to be graphite-free restricts the maximum section size of castings to about 38mm. Most malleable iron castings weigh less than 5 kg. and have a maximum section size of 25mm or less although parts in excess of 100 kg. are made.
d.2 Blackheart Malleable	White iron with carbon content of 2–3% is heated in a neutral atmosphere for 40–60 h at 850–875°C, cooled to 690°C at 4–5°C/h and then air-cooled to produce graphite aggregates in a ferritic matrix.	Best combination of machinability and strength for any ferrous material. Lower cost than nodular cast iron.	Not suitable for wear resistant applications unless surface treated. Long heat treatment cycle times compared with ferritic nodular cast iron.	
d.3 Pearlitic Blackheart Malleable Cast Iron	Variation of above but with pearlitic matrix produced by rapid cooling after annealing or by alloy additions (e.g. 0.5% Mn) by appropriate heat treatment. Structures can be made to vary from lamellar pearlite to spheroidised carbides to tempered martensite.	Good wear resistance. Highest strength of malleable irons. Can be hardened. Wide range and combination of properties possible by control of matrix microstructure.	Difficult to weld. Longer heat treatment cycle times compared with nodular cast iron.	
e.1 Unalloyed White Cast Iron	Silicon and carbon contents are lowered so that no graphite is formed—all carbon is combined as free iron-carbide (cementite) or in iron carbide lamellae in pearlite matrix. This structure exhibits a 'white' fracture surface when broken. White cast iron can also be produced from grey iron composition by increasing the cooling rate through the incorporation of metal chills (hence chilled iron) into the mould or through casting thin sections	Very hard (400–600 DPN), abrasion resistant material. Low cost compared with competitive materials. (e.g. manganese steel).	Very brittle—must be subjected to compressive loads only unless supported. All white irons require careful running and feeding if shrinkage in castings is to be avoided and hence yields are lower than for grey irons. White irons are unmachinable and are finished by grinding when necessary.	
e.2 Low Alloy White Cast Iron	Addition of alloying elements (usually Cr and Ni) in sufficient quantity to improve properties but not to produce martensitic white cast iron.	Improved toughness and wear resistance.	Extra cost must be justified by high performance or longer service.	
e.3 Martensitic White Cast Iron (e.g. Ni-Hard)	Additions of Cr and Ni in sufficient quantity to produce a structure of cementite and martensite. (Cr and Mo can give the same structure.)	Higher hardness and toughness than other types of white iron. Stable at high temperatures (e.g. 480–540°C) due to presence of Cr. Lower carbon compositions have higher toughness but lower hardness.	Cost of higher alloy content must be justified by improved performance or longer service. Stress relieving heat treatment necessary for optimum properties.	

Table 2.3.1—*continued*

TABLE 2.3.1 Characteristics of cast irons—*continued*

Material	*Description*	*Advantages*	*Limitations*
e.4 High Chromium White Cast Iron	2.8–28% Cr alloy gives partially austenitic structure (with consequent work hardening qualities) with tough, hard carbide. Martensitic and partially austenitic white cast iron can also be obtained with 2.75–3.25% Mo and 15–18% Cr and high carbon.	Abrasion resistance similar to martensitic (Ni-Hard) white iron but with higher toughness strength and corrosion resistance which are responsible for wet grinding applications.	High cost.
f.1 5–7% Silicon Flake Cast Iron	Addition of 5–7% silicon produces fine undercooled graphite in a ferritic matrix.	Good corrosion resistance to mineral acids. Resistant to high temperature oxidation and growth due to: (i) Surface passivation conferred by high silicon content. (ii) Fully ferritic matrix avoids growth associated with pearlite decomposition. (iii) Raising of transformation temperature from ferrite to austenite and thus (in some situations) avoiding phase transformation induced stresses. Stronger than 14–15% silicon irons (see below). Nodular grade has better elevated temperature properties.	Very brittle at room and elevated temperatures. Poor thermal shock resistance. Care needed during casting and subsequent processing to avoid cracking.
f.2 High Silicon Flake Cast Iron	Addition of 14–15% silicon produces silico-ferritic solid solution with dispersed graphite. Carbon content is reduced to avoid kish graphite (large clumps).	Outstanding corrosion resistance, especially to mineral acids and oxidation resistance. Hard (450–500 HB). Used in acid industries as pipes, stills and vats in locations where strength is not needed.	Extremely brittle (comparable with ceramic materials) with low strength and poor thermal shock characteristics. Casting care needed to avoid cracking during cooling.

Table 2.3.1—*continued*

2.3.2   Grey cast iron

2.3.2.1   Properties

The room temperature mechanical properties of the cast irons are given in:

Table 2.3.2   *Mechanical properties of grey cast irons*

It should be noted that the actual properties of castings are dependent on section thickness, and strengths will be lower in thicker sections. This effect is illustrated in:

Fig 2.3.1   *Variation of tensile strength with section thickness.*

Typical physical properties of these irons are given in:

Table 2.3.3   *Typical physical properties of grey cast irons*

Grey cast irons may be used at elevated temperatures up to about 500°C.

TABLE 2.3.2 Mechanical properties of grey cast irons

DIN 1691 (y)	BS1452 Grade	ASTM A48 Class[d]	0.1% proof strength (MPa)	Tensile strength (MPa)	Tensile elongation (%)	Compressive strength (MPa)	Impact strength (J)[a]	Brinell hardness (HB)[b]	Elastic modulus (GPa)	Modulus of rigidity (GPa)	Shear strength (MPa)	Fatigue limit[c] Unnotched (MPa)	Notched (MPa)
GG 14–18	150	20 B	98	150	0.6–0.75	600	8–14	145–180	100	40	173	68	68
GG 18+	180	25 B	117	180	0.5–0.70	672	8–14	155–220	109	44	207	81	79
GG 22	220	30 B	143	220	0.4–0.63	768	19–26	165–245	120	48	253	99	94
GG 26	260	40 B	169	260	0.6	864	19–26	165–295	128	51	299	117	108
GG 30	300	45 B/50 B	195	300	0.5	960	16–47	180–300	135	54	345	135	122
GG 35	350	55 B	228	350	0.5	1080	16–47	200–310	140	56	403	149	129
GG 35–40	400	60 B	260	400	0.5	1200	16–47	200–330	145	58	460	152	127

a Unnotched 20 mm diameter bar.
b Typical range. For grey cast iron hardness is not simply related to tensile strength but varies with composition and section size.
c Wohler. Unnotched, 8.4 mm dia. bar. Notched, 8.4 mm dia. at root of notch; 45° V-notch 3.4 mm deep, root radius 0.25 mm.
d New equivalents.

Table 2.3.2

TABLE 2.3.3 Typical physical properties of grey cast irons

BS1452 grade	Density (Mg/m³)	Specific heat (J/kg per K)			Coefficient of thermal expansion × 10⁻⁶ K⁻¹		Thermal conductivity (W/m per K)				Electrical resistivity (μΩm)
		20–200°C	20–400°C	20–600°C	20–200°C	20–400°C	100°C	200°C	400°C		
150	7.05	265	400	445	11.0	12.5	52.5	51.5	49.5		0.80
180	7.10	330	440	480	11.0	12.5	51.5	50.5	48.5		0.78
220	7.15	420	465	495	11.0	12.5	50.1	49.1	47.1		0.76
260	7.20	460	505	535	11.0	12.5	48.8	47.8	45.8		0.73
300	7.25	460	505	535	11.0	12.5	47.4	46.4	44.4		0.70
350	7.30	460	505	535	11.0	12.5	45.7	44.7	42.7		0.67
400	7.30	460	505	535	11.0[a]	12.5[a]	44.0	43.0	41.0		0.64

Since the specified tensile strengths may be obtained from a range of chemical compositions, the physical property values given are typical.
[a] Grade 400 may be supplied with either pearlite or acicular matrix. The values given are for pearlite iron. For acicular iron the values are 15.0 and 16.5 × 10⁻¹ K⁻¹, respectively, for 20–200 and 20–400°C.

Table 2.3.3

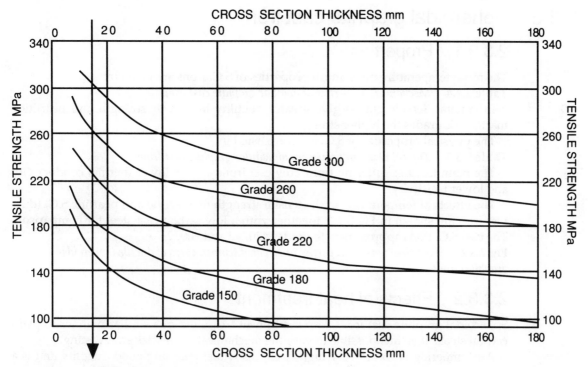

Equivalent to 30mm dia. bar

Example An iron which will give a tensile strength of 220 N/mm² in a 30mm diameter test bar should give a tensile strength of approximately 220 N/mm² in the centre of a casting whose ruling section is 15mm in thickness and approximately 147 N/mm² where the cross section is 100 mm in thickness

FIG 2.3.1 Variation of tensile strength with section thickness (Source BSI-1452-1977)

Fig 2.3.1

2.3.3   Spheroidal graphite cast iron

2.3.3.1   Properties

The room temperature mechanical properties of S.G. irons are given in:

Table 2.3.4   *Mechanical properties of nodular graphite cast irons*

In general, ferritic grades give greater ductility but lower strength and pearlitic/martensitic grades the higher strengths.

The physical properties of S.G. irons are listed in:

Table 2.3.5   *Typical physical properties of nodular graphite cast irons*

The main features differing from grey cast irons are the lower thermal conductivity and lower resistivity.

The effects of temperature on the tensile strength of a ferritic and pearlitic S.G. (ductile) iron are shown in Fig 2.3.2, together with a low carbon cast steel for comparison. The two S.G. irons approximate to grades 420/12 and 900/2:

Fig 2.3.2   *Short time tensile properties at elevated temperatures of ductile (spheroidal graphite) irons*

2.3.3.2   Effects of heat treatment

Spheroidal graphite cast irons respond to heat treatment in a similar manner to plain carbon steels. It is thus possible to vary the mechanical properties after casting.

Austempering achieves very high strengths whilst retaining good ductility and is a controlled transformation heat treatment. 'Austempered Ductile Irons' as they are called, can replace steels in many engineering applications. Typical properties are illustrated in:

Fig 2.3.3   *Tensile strengths and elongations in austempered ductile iron compared with other ductile irons*

Fig 2.3.4   *Properties of austempered ductile irons*

Fig 2.3.5   *Comparing contact-fatigue strengths,* and

Fig 2.3.6   *Bending fatigue strength*

TABLE 2.3.4 Mechanical properties of nodular graphite cast irons

DIN 1693[d]	BS2789 Grade	ASTM A536 Grade[d]	0.2% proof strength (MPa)	Tensile strength (MPa)	Tensile elongation (%)	Compressive strength (MPa)	Impact strength (J)	Brinell hardness (HB)	Elastic modulus (GPa)	Modulus of rigidity (GPa)	Fatigue limit (MPa)[c]	Normal matrix Structure
	350/22L40	—	220	350	22	600	12[a]	–160	160	62	180	Ferrite
	350/22	—	220	350	22	—	17[a]	–160	—	—	—	Ferrite
	400/18L20	60–40–18	250	400	18	700	12[a]	–179	160	62	200	Ferrite
	400/18	60–40–18	250	400	18	800	14[a]	–179	165	65	220	Ferrite
	420/12	65–45–12	270	420	12	800	80[b]	–212	165	65	220	Ferrite
	450/10	65–45–12	320	450	10	—	—	160–221	—	—	—	Ferrite/Pearlite
	500/7	80–55–06	320	500	7	900	60[b]	170–241	170	66	240	Ferrite/Pearlite
GGG 38	600/3	80–55–06	370	600	3	1000	40[b]	192–269	175	68	260	Pearlite/Ferrite
GGG 42	700/2	100–70–03	420	700	2	1100	30[b]	229–302	180	70	280	Pearlite
GGG 50	800/2	120–90–02	480	800	2	1200	20[b]	248–352	185	72	300	Pearlite or tempered
GGG 60	900/2	—	600	900	2	1300	—	302–359	185	72	—	Tempered Martensite

[a] Charpy V-notch test pieces tested at 20°C except for 350/22L40 (–40°C) and 400/18L20 (–20°C).
[b] Unnotched 10 mm square test bars.
[c] Wohler, unnotched.
[d] New equivalents.

Table 2.3.4

TABLE 2.3.5 Typical physical properties of nodular graphite cast irons

Type	BS2789 Grades	Density (Mg/m³)	Specific heat [a] (J/kg per K)			Coefficient of thermal expansion × 10⁻⁶ K⁻¹		Thermal [b] conductivity (W/m per K)	Electrical [c] resistivity (μΩm)
			20–200°C	20–400°C	20–600°C	20–200°C	20–500°C		
Ferritic	350/22–420/12	7.1	460	505	535	10.0–11.8	10.0–13.2	32–40	0.50–0.55
Ferritic/pearlitic	450/10–600/3	7.1–7.2	460	505	535	11.0–12.0	11.5–12.0	31–39	0.55–0.65
Pearlitic	700/2–800/2	7.2	460	505	535	11.8–12.5	12.5–13.4	31–35	0.55–0.70

Nodular cast irons (like grey cast irons) are specified on tensile properties, which may be obtained from a range of compositions. Physical properties are more dependent on composition and microstructure and the values given in the table are therefore typical.

[a] Specific heat is less dependent on composition than other physical properties. These values may therefore be used for all grades.

[b] Thermal conductivity is heavily dependent on total silicon plus nickel content and for a ferritic iron would fall from 38 to 32 W/m per K as Si + Ni increased from 2.5 to 5.5%. The values given are typical throughout the range 0–500°C.

[c] Electrical resistivity is most affected by silicon content. For a ferritic iron, resistivity can be as low as 0.40 μΩm for a nickel-free iron with 1.5% silicon or as high as 0.70 μΩm at 3.5% Si.

Table 2.3.5

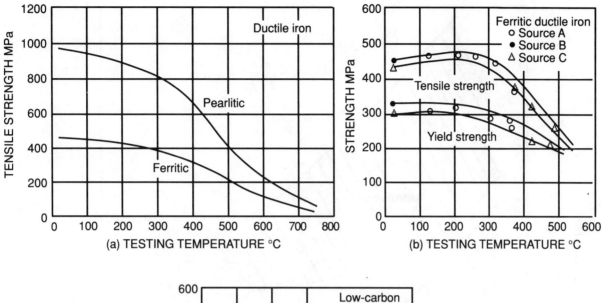

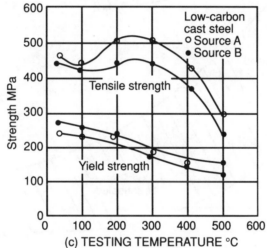

(a) and (b) Properties of ductile iron. (c) Tensile and yield strengths of cast steel, for comparison.

FIG 2.3.2 Short time tensile properties at elevated temperatures of ductile(spheroidal graphite) irons

(Source: *Metals handbook* Vol. I, 1978—ASM, Int., Materials Park, Ohio 44073, USA)

Fig 2.3.2

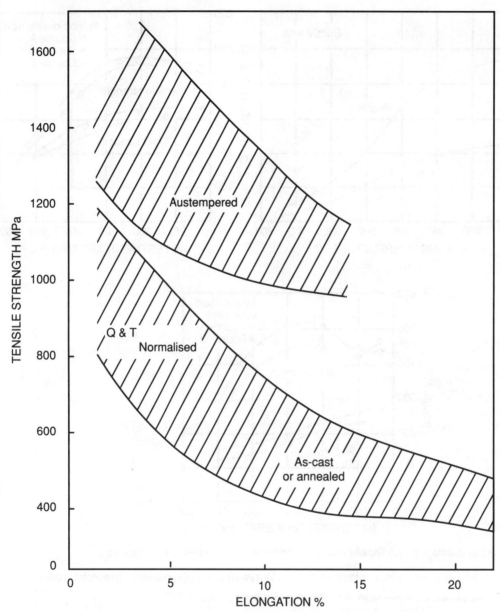

FIG 2.3.3 Tensile strengths and elongations in austempered ductile iron compared with other ductile irons

(Source: BCIRA, The Cast Metals Technology Centre, Alvechurch, Birmingham, UK)

Fig 2.3.3

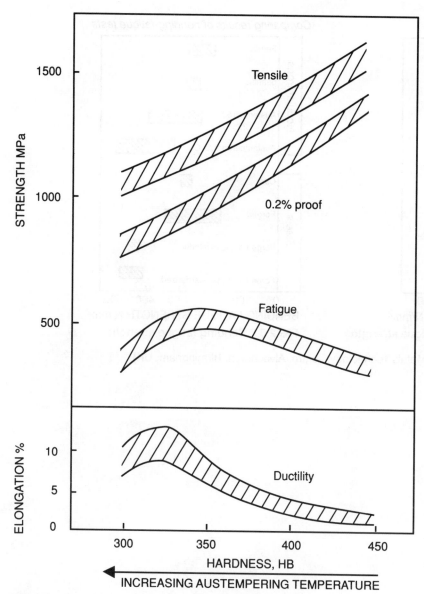

FIG 2.3.4 Properties of austempered ductile irons

(Source: BCIRA, The Cast Metals Technology Centre, Alvechurch, Birmingham, UK)

Fig 2.3.4

Comparing results of bending-fatigue tests

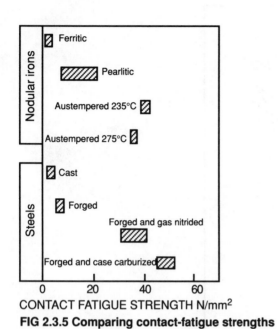

FIG 2.3.5 Comparing contact-fatigue strengths

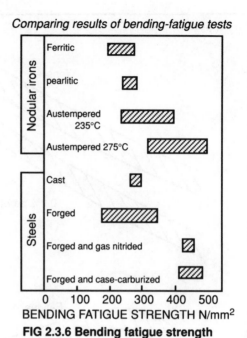

FIG 2.3.6 Bending fatigue strength

(Source: BCIRA, The Cast Metals Technology Centre, Alvechurch, Birmingham, UK)

Fig 2.3.5 and Fig 2.3.6

2.3.4   Malleable cast iron

2.3.4.1   Description

Malleable cast irons are available in three basic types, i.e. 'whiteheart', 'blackheart' and 'pearlitic'. These are obtained by different heat treatments and variations in alloy chemistry. Whiteheart and blackheart have a ferritic matrix and differ mainly in that whiteheart has a heavily decarburised surface. Whiteheart malleable iron is not used in the USA where blackheart malleable iron is known as 'ferritic malleable'.

2.3.4.2   Properties

(i) The room temperature mechanical properties of malleable cast irons are listed in Table 2.3.6. Note that ferritic grades have a specified MAXIMUM hardness.

Table 2.3.6   *Mechanical properties of malleable cast irons*

Typical physical properties of malleable cast irons are given in:

Table 2.3.7   *Typical physical properties of malleable cast irons*

(ii) Elevated temperature properties

Malleable irons are normally used for ambient temperature applications. Strength falls off over 300°C.

TABLE 2.3.6 Mechanical properties of malleable cast irons

BS 6681 Grade	Type	Test bar diameter (mm)	0.2% proof strength (mPa)	Tensile strength (mPa)	Elongation %	Brinell hardness (HB)	Elastic modulus (GPa)	Modulus of rigidity (GPa)	Fatigue ratio (% UTS)	Impact[a] strength (J)
W 35–04	Whiteheart	9	—	340	5	230 max.				Notched: 3–4
		12	—	350	4					
		15	—	360	3					
W 38–12	Whiteheart	9	170	320	15	200 max.	165 to 186	69 to 83	40 to 60	Unnotched decarburised layer: 41–54
		12	200	380	12					
		15	210	400	8					
W 40–05	Whiteheart	9	200	360	8	220 max.				Core: 11–14
		12	220	400	5					
		15	230	420	4					
W 45–07	Whiteheart	9	230	400	10	220 max.				
		12	260	450	7					
		15	280	480	4					
B 30–06	Blackheart	12–15	—	300	6	150 max.	158 to 173	62 to 69	40 to 60	9.5–19.0
B 32–10	Blackheart	12–15	190	320	10					
B 35–12	Blackheart	12–15	200	350	12					
P 45–06	Pearlitic	12–15	270	450	6	150–200	158 to 179	62 to 83	40 to 60	—
P 50–05	Pearlitic	12–15	300	500	5	160–220				
P 55–04	Pearlitic	12–15	340	550	4	180–230				
P 60–03	Pearlitic	12–15	390	600	3	200–250				
P 65–02	Pearlitic	12–15	430	650	2	210–260				
P 70–02	Pearlitic	12–15	530	700	2	240–290				

[a] Izod test on 10 mm square bars. Notched bars except where stated. Values for Whiteheart malleable iron vary widely due to the extensive decarburised surface.

Table 2.3.6

TABLE 2.3.7 Typical physical properties of malleable cast irons

Type	Density (Mg/m^3)	Specific heat [a] (J/kg per K)		Coefficient of thermal expansion $\times 10^{-6}$ K^{-1}		Thermal [b] conductivity (W/m per K)	Electrical [c] resistivity ($\mu\Omega$ m)
		0–100°C	0–600°C	0–100°C	0–600°C		
Whiteheart	7.30–7.70	460	—	10.0	12.5	42–46	0.24–0.26
Blackheart (ferritic)	7.20–7.45	500	690	10.0	12.0	46–49	0.27–0.37
Pearlitic	7.20–7.45	500	690	10.0	12.5	39–45	0.34–0.41

See Table 2.3.5 for footnotes.

Table 2.3.7

2.3.5   White cast irons

2.3.5.1   Description

White cast iron is the generic term for those cast irons which solidify with the carbon as carbide. It includes plain unalloyed types which have a pearlitic or martensitic matrix plus low alloy (usually nickel plus chromium) and higher alloy (high chromium plus molybdenum, nickel and copper) types which are martensitic. The alloy grades are usually heat-treated to develop the optimum combination of abrasive wear resistance and toughness which is sought in these materials.

2.3.5.2   Properties

These materials are extremely hard and brittle and are normally used under compressive loading conditions (i.e. abrasive/sliding wear). Apart from hardness, mechanical properties are therefore normally measured by transverse rupture testing. Typical mechanical properties are given in:

Table 2.3.8   *Mechanical properties of abrasion resistant white cast irons*

and physical properties in:

Table 2.3.9   *Typical physical properties of abrasion resistant white cast irons*

TABLE 2.3.8 Mechanical properties of abrasion resistant white cast irons

Type	Hardness[a]			Transverse rupture strength (MPa)	Elastic modulus (GPa)
	Normal condition (HV)	As cast (HV)	Annealed (HV)		
Unalloyed	550–650	—	—	310–460	165–190
Unalloyed and low alloy grades	370–430 370–430 205–255	— — —	— — —	Typically 390–540	Typically 200
Nickel chromium grades	485–540 540–600 485–540 540–600 600–655	— — — — —	— — — — —	Ranges from 480 to 850 dependent on C, Ni and Cr contents	Typically 165 to 180
High chromium grades	655 710 655 655 655 655 710	485 540 485 430 485 485 485	430 485 485 430 485 430 485	Typically 720–920 dependent on C and Cr content and heat tratment.	Depends on microstructure after heat treatment. May vary from 165 to 220

[a] Hardness values are for material in the heat-treated condition unless otherwise indicated and minima. Castings supplied will usually have higher hardness.

TABLE 2.3.9 Typical physical properties of abrasion resistant white cast irons

Type	Density (Mg/m³)	Specific heat (J/kg per K)		Coefficient of thermal expansion × 10⁻⁶ K⁻¹		Thermal conductivity (W/m per K)	Electrical resistivity (μΩm)
		100°C	400°C	20–200°C	20–400°C		
Unalloyed and low alloy grades	7.6–7.8	560–600	710–740	10–11	15–16	15–32	0.53
Nickel chromium grades	7.6–7.8	a	a	8–10[c]	11.5–12.6[c]	20–30	0.80
High chromium grades	7.6–8.0	a	a	d	d	d	0.6–1.0[b]

The values given are typical and may vary with composition and heat treatment.
[a] Similar to unalloyed and low alloy grades.
[b] Estimated values, dependent on alloy composition and heat treatment.
[c] Depends on carbon content. Higher values result from lower carbon contents.
[d] Similar to nickel, chromium and low alloy grades.

Table 2.3.8 and Table 2.3.9

2.3.6   Austenitic cast irons

2.3.6.1   Description

Austenitic cast irons are a group of alloys which achieve an austenitic matrix from high additions of nickel sometimes in conjunction with manganese and copper. The main reason for alloying, is to obtain enhanced corrosion and/or oxidation resistance. This may be further enhanced by additions of chromium. The austenitic matrix also enables non-magnetic grades to be obtained.

Both flake graphite and nodular (ductile) forms may be produced, the latter having correspondingly higher mechanical properties.

2.3.6.2   Mechanical properties

The room temperature mechanical properties of austenitic cast irons are given in:
Table 2.3.10   *Mechanical properties of austenitic cast irons*

The long-term tensile properties of grade S2C are shown in Fig 2.3.7 with results for unalloyed ferritic and pearlitic ductile iron for comparison. The better creep resistance of the austenitic iron can be seen:
Fig 2.3.7   *Composite tensile, rupture and creep strength data at various test temperatures for (a) ferritic, (b) pearlitic and (c) austenitic nodular iron*

2.3.6.3   Physical properties

Typical physical properties of austenitic cast irons are given in:
Table 2.3.11   *Physical properties of austenitic cast irons*

TABLE 2.3.10 Mechanical properties of austenitic cast irons

Type	0.2% proof strength (MPa)	Tensile strength (MPa)	Tensile elongation (%)	Elastic modulus (GPa)	Brinell hardness (HB)	Impact[a] strength (J)
Flake Graphite	—	170–240	1–2	85–113	140–220	—
	—	170–240	1–3	85–113	140–220	—
	—	190–240	1–3	98–113	120–215	—
Spheroidal Graphite	210–260	370–490	7–20	112–133	140–230	4–20
	210–260	370–490	7–20	112–133	140–200	10–20
	210–260	370–490	7–20	112–133	140–230	4–10
	170–250	370–440	20–40	85–112	130–170	20–30
	210–240	420–470	25–45	120–140	150–180	15–25
	210–260	370–470	7–18	92–130	130–200	4–20
	210–260	370–470	7–20	92–120	130–180	—
	200–260	390–460	15–25	140–150	130–170	15–25

[a] Charpy V-notch samples tested at room temperature. Property varies with chromium content.

TABLE 2.3.11 Physical properties of austenitic cast irons

Type	Density (Mg/m^3)	Specific heat (J/kg per K)	Coefficient of thermal expansion $\times 10^{-6}$ (K^{-1})	Thermal conductivity (W/m per K)	Electrical resistivity ($\mu\Omega$m)	Magnetic permeability (H/m)
Flake graphite	7.3	461	18.7	38–42	1.4–1.7	1.3
	7.3	486	18.7	38–42		1.3
	7.3	465	12.4	38–42		Magnetic
Spheroidal graphite	7.4	—	18.7	12.6	Typically 1.0	1.3
	7.4	—	18.7	12.6		1.3
	7.4	—	18.7	12.6		1.3
	7.4	—	18.4	12.6		1.03
	7.4	—	14.7	12.6		1.03
	7.4	—	12.6	12.6		Magnetic
	7.6	—	12.1	12.6		Magnetic
	7.3	—	18.2	12.6		1.02

Table 2.3.10 and Table 2.3.11

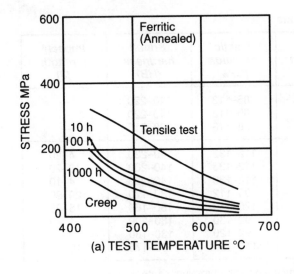

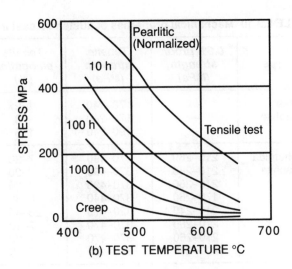

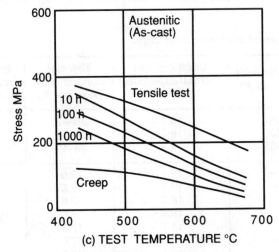

The curve labeled 'Creep' shows the stress-temperature combination that will result in a creep rate of 0.0001%/h.

FIG 2.3.7 Composite tensile, rupture and creep strength data at various test temperatures for (a) ferritic, (b) pearlitic and (c) austenitic nodular iron.

Compositions: ferritic and pearlitic irons: 3.6C–0.5Mn–2.7Si–0.7Ni–0.08Mg
Austenitic iron: 2.9C–2.1Mn–2.6Si–23Ni–0.08Mg

Fig 2.3.7

2.3.7   Silicon cast irons

2.3.7.1   Description

Unalloyed cast irons may contain up to 3% silicon. At higher levels enhanced oxidation or corrosion resistance may be obtained at the expense of ductility.

At moderate levels of silicon (4–7%) good oxidation resistance is obtained whilst retaining good mechanical properties. This is usually the result of careful balance of composition and may be improved further by making spheroidal graphite (ductile) grades.

For corrosion resistance, much higher silicon contents are used (around 14–16%) and these alloys are extremely brittle.

2.3.7.2   Properties

The room temperature mechanical properties and typical physical properties of silicon cast irons are given in:

Table 2.3.12   *Typical mechanical and physical properties of silicon cast irons*

2.3.7.3   Application

Medium silicon cast irons are used mainly for high temperature applications such as exhaust manifolds.

TABLE 2.3.12 Typical mechanical and physical properties of silicon cast irons

Type	Tensile strength (MPa)	Compressive strength (MPa)	Impact strength (J)[b]	Brinell hardness (HB)	Density (Mg/m³)	Coefficient of thermal expansion[c] × 10⁻⁶ (K⁻¹)	Thermal conductivity (W/m per K)	Electrical resistivity (μΩm)
High silicon iron[a]	90–180	690	0.1–3.0	480–520	7.0	12.4–13.1	—	0.50
Medium silicon grey iron	170–310	620–1040	20–31	170–250	6.8–7.1	10.8	37	—
Medium silicon ductile iron	415–690	—	7–155	140–300	7.1	10.8–13.5	—	0.58–0.87
Silicon molybdenum ductile iron	600–660	—	—	220–240	6.85	10.0–11.8	20–25	—

[a] High silicon irons are extremely brittle. Strength values are given for guidance only.
[b] Unnotched.
[c] 20–100°C.

Table 2.3.12

2.3.8 Optimisation of grey cast iron

Grey cast iron is capable of being tailored to meet the needs of particular applications by varying the composition and processing conditions and thereby the reconstructive end properties. Table 2.3.13 gives in matrix form the relationship between these variables—more details are given in the notes following:

Table 2.3.13 *Influence of composition and processing variables on the properties of grey cast irons*

TABLE 2.3.13 Influence of composition and processing variables on the properties of grey cast irons

Grey cast irons — Relationship of Specifiable items. To desired Characteristics →

			Strength			4. Deflection	5. Modulus of E	Heat resistance			9. Density	10. Hardness	11. Damping capacity	Thermal		14. Electrical resistance	15. Magnetic properties	16. Machinability	17. Wear resistance	18. Hardenability	19. Cost
			1. Tensile	2. Fatigue	3. Impact			6. Scaling	7. Growth	8. Creep				12. Expansion	13. Conductivity						
MICROSTRUCTURE	Section size	A	X	X	—	—	X	O	O	O	X	X	X	O	O	X	X	X	X	X	X
	Graphite	B	X	X	X	—	X	X	X	X	X	X	X	O	X	X	X	X	X	—	X
	Pearlite	C	X	X	—	—	—	O	X	—	X	X	X	O	X	X	X	X	X	X	X
	Ferrite	D	X	X	—	—	—	O	X	—	X	X	X	O	X	X	X	X	X	X	X
	Phosphide	E	X	X	X	X	X	X	O	—	X	X	—	O	X	X	X	X	X	X	—
	Carbides	F	—	O	—	—	—	O	O	X	—	X	—	O	X	—	X	X	X	X	X
COMPOSITION	C	G	X	X	X	—	X	X	X	X	X	X	X	O	X	X	X	X	X	—	X
	Si	H	X	O	—	—	—	X	X	—	X	X	X	O	X	X	O	X	X	—	—
	Mn	I	X	O	—	—	—	O	X	—	X	X	X	O	X	O	X	—	X	X	—
	P	J	X	X	X	X	X	X	O	—	X	X	—	O	X	X	X	X	X	X	X
	S	K	—	O	—	—	—	O	O	O	X	X	—	O	—	O	O	—	—	X	—
	Mo	L	X	—	X	—	—	O	X	X	X	X	X	O	X	O	—	X	X	X	X
	Cr	M	X	—	O	—	—	X	X	X	X	X	X	O	X	O	X	X	X	X	X
	V	N	X	—	O	—	—	O	X	X	X	X	X	O	X	O	—	X	X	X	X
	Ni	O	X	—	O	—	—	O	O	O	X	X	X	O	X	X	X	X	X	X	X
	Cu	P	X	—	O	—	—	O	O	O	X	X	X	O	X	O	X	X	X	X	X
	Sn	Q	X	—	X	—	—	O	O	O	X	—	X	O	—	O	—	X	X	X	X
	Al	R	—	—	O	—	—	X	O	O	X	—	—	O	X	X	—	—	—	—	X
	Ti	S	—	—	O	—	—	O	O	O	X	—	—	O	—	O	—	—	—	—	X
PROCESSING	Inoculation	T	X	O	O	O	—	O	O	O	O	—	X	O	O	X	X	X	X	—	X
	Stress relieving	U	O	—	—	X	X	O	O	O	O	X	—	O	O	O	O	X	—	—	X
	Annealing	V	X	X	X	X	X	O	X	—	X	X	—	O	X	X	X	X	X	X	X
	Normalising	W	—	—	—	—	—	O	—	—	—	X	—	O	—	—	—	X	—	—	X
	Pearlitising	X	—	—	—	—	—	O	X	—	—	X	X	O	X	X	X	X	X	X	X
	Surface treatments including hardening	Y	X	X	X	X	X	—	O	—	O	X	—	O	—	X	—	X	X	—	X

Key: X = See Notes; O = no effect; — = no data.

Table 2.3.13

Notes to Table 2.3.13

1.   TENSILE STRENGTH

1-A   **Section size**

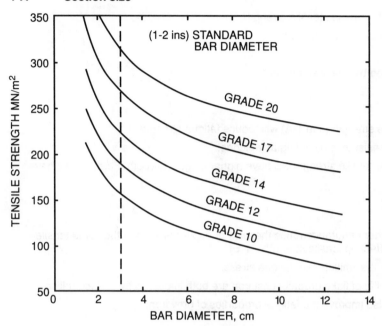

BS GRADE	ASTM CLASS	DIN
10	20–25	GG14–18
12	25–30	GG18+
14	30–35	GG22
17	40	GG26
20	45	GG30

The graph shows the reduction in strength with increasing test bar size for several grades of unalloyed grey cast iron.

1-B   **Graphite**
The greater the amount of graphite, the lower the tensile strength.

1-C   **Pearlite**
The % of pearlite is directly proportional to the tensile strength.

1-D   **Ferrite**
Tensile strength is inversely proportional to the % ferrite in the microstructure.

1-E   **Phosphide**
The presence of phosphide is detrimental to tensile strength.

1-G   As for graphite, see 1-B.

1-H   Silicon being a graphitizer will affect tensile strength as indicated in 1-B.

1-I, L, M, N, O and P   See Table 1.

Table 1 Effect of alloy additions of tensile strength of grey cast iron

	Molybdenum	Chromium		Vanadium	Nickel	Copper	Manganese
Up to approx. % of:	1.0	0.5	1.0	0.35	3.0	1.5	0.5–1.0
Approximate % increase in tensile strength for each 1% alloy	40–45	30	20	30–80	5–10	8–10	0–10

1-J   See 1-E.

Table 2.3.13—*continued*

Notes to Table 2.3.13—*continued*

1-Q	**Tin** 0.1% additions to hypoeutectic irons raises the tensile strength by 7% and by 25% in hypereutectic irons.
1-T	Inoculation refines structure and increases tensile strength.
1-V	The effect of annealing is to reduce tensile strength.
1-X	See 1-C.
1-Y	Tensile strength can be increased by heat treatments.

2	**FATIGUE**
2-A	**Section size** by reducing tensile strength (see 1-A) will reduce fatigue strength.
2-B	**Graphite** flakes reduce fatigue resistance by acting as notches.
2-C,D	**Pearlitic** irons by virtue of higher tensile strength will have higher fatigue strength. Ferritic irons have higher endurance ratio.
2-E	See 2-J.
2-G	See 2-B.
2-J	High phosphorus irons have higher endurance ratios than low phosphorus irons of the same tensile strength (but note detrimental effect on tensile strength—1-E).
2-V	Annealing approximately halves the limiting safe fatigue stress.
2-Y	The fatigue limit and the fatigue life of flake graphite cast iron are both improved by surface rolling. Surface hardening treatments also improve the fatigue properties of grey iron.

3	**IMPACT**
3-B	Impact strength is inversely proportional to graphite content.
3-E	The presence of phosphide reduces impact resistance.
3-G	See 2-B.
3-J	See 3-E.
3-L	Impact strength is proportional to molybdenum content.
3-Q	0.1% tin addition may lower the impact strength of hypoeutectic irons by 10%.
3-V	Annealing has the effect of increasing the impact strength.
3-Y	Surface treatments such as surface rolling will improve impact resistance.

4	**DEFLECTION**
4-E	Phosphide has the effect of reducing deflection in grey cast irons.
4-J	See 4-E. Irons of 1% phosphorus or more will show only 50–80% the deflection of those with less than 0.2% phosphorus.
4-U	Stress relieving cast iron increases the extent to which it will deflect.
4-V	Annealing also improves deflection of grey cast irons.
4-Y	Surface treatments greatly reduce the deflection of grey cast irons.

5	**MODULUS OF ELASTICITY**
5-A	If section size is large enough to cause increased graphite formation then elasticity modulus will be decreased (see 5-B).
5-B	The modulus of elasticity is affected greatly by the graphite content, and is lower in high graphite irons than in low graphite irons.
5-E	Increasing phosphide content tends to increase the value for Young's Modulus.

Table 2.3.13—*continued*

Notes to Table 2.3.13—*continued*

5-G	Low total carbon values will increase the modulus of elasticity.
5-J	See 5-E.
5-U	Stress relieving reduces the modulus of elasticity.
5-V	Annealing grey irons has the effect of reducing the modulus of elasticity.
5-Y	Surface treatments such as rolling or hardening will increase the modulus of elasticity.

6	**SCALING**
6-B	Grey irons with low graphite content tend to have higher resistance to scaling.
6-E	See 6-J.
6-G	See 6-B.
6-H	Additions of silicon greatly improve scaling resistance particularly silicon contents between 3.8 and 6.5%.
6-J	There is a tendency for high phosphorus irons to be more resistant to scaling than low phosphorus irons.
6-M	Chromium additions of 1% or more improve scaling resistance, 15% chromium being particularly suitable for temperatures up to 950°C.
6-R	Aluminium additions improve an iron's resistance to scaling, values between 1% and 7% being particularly advantageous.

7	**GROWTH**
7-B	Growth may be caused by carbon deposition on graphite flakes in reducing atmospheres.
7-C	The breakdown of pearlite and ferrite and graphite is the main cause of growth. One method of prevention is to stabilise the pearlite structure by alloy additions.
7-D	The presence of a fully ferritic matrix reduces growth to very low values.
7-G	See 7-B.
7-H	Additions of silicon below 1.6% stabilises any pearlite present considerably. Silicon from 2.0% to 3.5% does not entirely prevent pearlite formation, but makes it less stable, without conferring any oxidation resistance.
7-I	Manganese contents of up to 1.5% confer marked stability in growth up to 450°C.
7-L	Molybdenum is a powerful pearlite stabiliser, thus providing high resistance to growth in the temperature range 450–650°C.
7-M	0.5–2.0% chromium reduces growth to very low values in the temperature range 500–750°C.
7-N	Vanadium increases pearlite stability, and thus resistance to growth in the temperature range 450–650°C.
7-V	Growth will occur after annealing on account of pearlite breakdown.
7-X	Pearlitising a grey iron will reduce its resistance to growth unless it is stabilised by alloy additions (see 7-C).

8	**CREEP**
8-B	Creep resistance is inversely proportional to graphite content in grey iron.
8-F	The presence of carbides in a cast iron matrix greatly improves its creep resistance.
8-G	See 8-B.
8-L	Molybdenum additions give great resistance to creep between 450 and 650°C.
8-M	The presence of chromium greatly improves creep resistance in irons.
8-N	Vanadium increases resistance to creep within the temperature range 450–650°C.

Table 2.3.13—*continued*

Notes to Table 2.3.13—*continued*

9	DENSITY
9-A	The larger the section, the slower the rate of cooling and hence the greater the amount of graphite. The larger the amount of graphite the lower the density of the cast iron.
9-B to 9-E	The specific gravities of the main grey iron constituents are given as follows:

Ferrite	7.86
Pearlite	7.78
Graphite	2.25
Phosphide Eutectic	7.32

The specific gravity of the grey iron varies according to the relative amounts of these structural constituents.

9-G	See 9-A.
9-H to 9-S	The density of a grey iron will be changed according to the amount and density of any alloy additions.

	Si	Mn	Mo	Cr	C	Ni	Cu	Sn	Al	Ti
Specific Gravity	2.33	7.43	10.22	7.19	6.1	8.90	8.96	7.70	2.69	4.51

9-V	Full annealing of a pearlite grey iron reduces the specific gravity by 0.1–0.2 due to the further deposition of graphite caused by the breakdown of pearlite to ferrite.
10	HARDNESS
10-A	Large sections of grey iron have larger graphite flakes, and therefore are softer than smaller sections.
10-B	The presence of large graphite flakes makes grey irons softer than irons with small graphite flakes.
10-C	Pearlite has a hardness of between 175 and 330 Brinell, depending on the interlamellar spacing.
10-D	The hardness of ferrite varies depending on silicon content, e.g. silico-ferrite (0.82% Si)–88 BHN, silico-ferrite (3.4% Si)–150 BHN.
10-E	The presence of phosphide eutectic tends to increase the hardness of grey iron as it has a hardness of between 400 and 600 BHN.
10-F	Carbides increase the hardness of grey irons because of their extremely high hardness.
10-G	See 10-B.
10-H	The presence of silicon reduces hardness.
10-I	Manganese increases hardness when the amount present is above that required to neutralise the sulphur content.
10-J	See 10-E. Phosphorus also acts as a centre for the deposition of carbides of iron, vanadium, chromium and molybdenum.
10-K	The hardness of cast irons tends to increase with increasing sulphur content.
10-L	Molybdenum additions increase hardness.
10-M	Chromium increases hardness.
10-N	Vanadium increases hardness.
10-O	Nickel decreases hardness.
10-P	Cu being a pearlite stabiliser increases hardness.
10-U	Stress relieving grey irons slightly reduces hardness.
10-V	Grey irons are considerably softened by being fully annealed.
10-W	Normalising gives more homogeneous hardness.
10-X	Pearlitising increases hardness.
10-Y	Grey irons may be surface hardened by induction or flame hardening.
	Surface rolling also hardens to a lesser extent by work hardening.

Table 2.3.13—*continued*

Notes to Table 2.3.13—*continued*

---

11 DAMPING CAPACITY

11-A Affects damping capacity via effect on graphite (see 11-B).

11-B Increase in graphite content increases damping capacity.

11-C, D Pearlitic irons generally have lower damping capacity than ferrite because of decreased graphite content.

11-G See 11-B.

11-H Silicon being a graphitiser will affect damping capacity as indicated in 11-B.

11-I, L, M, N, O, P As Mn, Mo, Cr, V, Ni and Cu are pearlite stabilisers effect will be as indicated in 11-B.

11-Q Tin refines graphite structure and will therefore reduce damping capacity.

11-T Refines graphite and hence reduces damping capacity.

11-X See 11-C.

12 THERMAL EXPANSION
No significant effects.

13 THERMAL CONDUCTIVITY

		B	C	D	E	F	G	H	I	J	K	L	M	N	O	P	R
Increase In →		Graphite	Pearlite	Ferrite	Phosphide	Carbides	C	Si	Mn	P	S	Mo	Cr	V	Ni	Cu	Al
Effect on Thermal Conductivity	13	↑	↓	↑	↓	↓	↑↓*	↓	↓	↓	—	↑	↑↓*	—	↓	↓	↓

*13-G Increase in carbon content decreases conductivity if carbon is present as increased cementite. Reverse effect if more graphite is formed.

*13-M Chromium increases thermal conductivity of cast iron unless it causes iron carbide to form in which case conductivity decreases.

13-V, X See B, C and D.

14 ELECTRICAL RESISTANCE

14-A Section size can affect microstructure and this in turn affects resistivity (see 14-C and D).

14-B Coarse flake graphite microstructures give highest resistivity.

14-C, D Ferritic irons have lower resistivity than pearlitic irons. Coarse pearlite gives a lower resistivity than fine pearlite.

14-E See 14-J.

14-G Increase in carbon content increases resistivity by increasing graphite content (see 14-B).

14-H Increase in silicon increases resistivity by increasing graphite content (see 14-B) and because silicon-ferrite has higher resistivity than pure ferrite.

14-J (Phosphorus) Up to 0.2% no effect. Above 0.2% influences graphite structure to produce increase in resistivity (effect is small compared with Si and C).

14-O, R Ni and Al raise resistivity with increasing additions.

14-T Inoculation reduces resistivity (see 14-B).

14-V Annealing affects microstructure and hence affects resistivity (see 14-C and D).

Table 2.3.13—*continued*

Notes to Table 2.3.13—*continued*

14-X	See 14-C and D.
14-Y	Martensitic structures have higher resistance.

15	MAGNETIC PROPERTIES
15-A	Section size affects microstructure and this in turn affects magnetic properties (see 15-B, C and D).
15-B	Coarse graphite lowers remanent magnetism. As graphite becomes finer there is tendency to increased hysteresis loss.
15-C, D	Ferrite gives low hysteresis loss (lower with larger grain size) and high permeability whereas pearlite gives a high hysteresis loss and low permeability.
15-E	In a ferritic matrix phosphide eutectic increases hysteresis loss and reduces maximum permeability. Effect in pearlitic matrix is masked.
15-F	Free cementite causes low values of magnetic induction, permeability and remanent magnetism, together with increased coercive force and hysteresis loss.
15-G	See 15-B.
15-I	Manganese is, in general, prejudicial to high magnetic induction and permeability.
15-J	See 15-E.
15-M	Chromium reduces magnetic induction, permeability and remanent magnetism, and increases coercive force and hysteresis loss.
15-O	Nickel decreases the permeability and increases hysteresis loss.
15-P	Copper reduces permeability and increases hysteresis loss.
15-T	Refines graphite (see 15-B).
15-V	Annealed ferritic irons have highest magnetic-induction and permeability. Remanent magnetism and coercive force tend to show lower values in the annealed state.
15-X	See 15-C.

16	MACHINABILITY
16-A	Section size has marked effect on machinability via modifications of the microstructure (see 17-B, C, D). For the same chemical composition thin sections can be hard and unmachinable and thick sections can be soft and very machinable.
16-B	Graphite flakes act as discontinuities which facilitate chip breakage and as a lubricant reducing friction and tool wear. Coarseness of graphite determines minimum c.l.a. that can be achieved (see below). Increase in graphite content weakens cast iron and therefore reduces power requirements.

Graphite		Minimum possible cla
Size	*Type*	μm
1	A	0.2–0.25
5–7	B	0.05–0.08
8	D	0.04–0.05

16-C	Pearlitic cast irons are much less machinable than ferritic but surface finish is improved. Power requirements and tool wear will increase as the interlamellar spacing of the pearlite decreases (hardness and strength increasing). Relatively small differences in pearlite microstructure can cause very marked differences in machinability.
16-D	Ferritic irons are very machinable and high metal removal rates can be achieved. A good surface finish may be difficult to achieve due to smearing of ferrite. Regions of ferrite in a predominantly pearlitic matrix may cause increased tool wear by setting up vibration.

Table 2.3.13—*continued*

Notes to Table 2.3.13—*continued*

16-E	Above 0.7% phosphorus a continuous hard phosphide network is formed and this will impair machinability causing premature tool wear. Ternary form of phosphide eutectic with a hardness of 760–870 DPN cf. pseudo binary form 420–560 DPN is to be avoided.
16-F	The presence of free carbides (800 DPN hardness) are detrimental even in small amounts.
16-G	See 16-B.
16-H	Si being a graphitiser improves machinability.
16-I	See 16-L.
16-J	See 16-E.
16-L, M, N, O, P	The elements being pearlite stabilisers will reduce machinability and increase the power requirements of the machine tools. In addition Mo, Cr and V being carbide formers will further impair machinability.
16-Q	Additions of Sn (0.05–0.1%) produce more homogenous structure and thereby improve machinability.
16-T	Inoculation by producing a more homogeneous structure improves machinability.
16-U	Machining can itself induce residual stresses and these should be relieved by heat treatment for critical engineering components.
16-V	Annealing to produce a ferritic structure greatly improves machinability (see 16-D).
16-X	See 16-C.
16-Y	Grinding or honing are usually the only metal removal operations which follow surface treatments.

17	WEAR
17-A	Section size affects microstructure and this in turn affects wear rate (see 17-B, C, D).
17-B	Graphite flakes of type A size 4–5 randomly dispersed are believed to give the best resistance to sliding wear. Graphite acts as a lubricant in dry sliding situations.
17-C, D	Fully pearlitic structures are preferred for maximum wear resistance. The presence of free ferrite may cause galling and pick-up.
17-E	Phosphide eutectic when present as a continuous hard network gives improved resistance in marginally lubricated situations and multiplies the effect of free ferrite.
17-F	Free carbides, provided they are supported by a pearlitic matrix, will give increased load bearing capacity and reduce wear.
17-G	See 17-B.
17-I	See 17-L below.
17-J	Above 0.8% a continuous hard (400–600 HB) network is formed. Phosphorus improves the load-bearing capacity of grey iron by increasing the mild–severe wear transition load. Phosphorus reduces running-in wear – maximum reduction for 1.0–1.25% phosphorus. Nullifies the deleterious effects of free ferrite.
17-L, M, N, O, P, Q	These elements being pearlite stabilisers or carbide will affect wear resistance via microstructural variation.
17-T	Innoculation affects wear resistance by refining distribution of graphite (see 17-B)
17-V, X	See 17-C and D.
17-Y	There are a wide range of surface treatments designed to improve wear resistance and reduce friction suitable for cast iron (see Section 1-C–3).

18	HARDENABILITY
18-A	Sections of greater than 10–12.7mm will require low alloy additions if satisfactorily hardened castings are to be obtained. Hardenability is inversely proportional to section size.
18-C, D	Pearlitic irons show better hardening response than ferritic irons—require much shorter heating times.
18-E	See 19-J.

Table 2.3.13—*continued*

Notes to Table 2.3.13—*continued*

18-F	The presence of any free carbide in the initial structure before hardening greatly increases risk of cracking.
18-H	Irons high in silicon show poorer hardening response than low silicon irons.
18-I, K	Mn reduces initial quenching speed and should be at least 0.5% in excess of that required to balance sulphur content (i.e. $1.7 \times S\% + 0.3$).
18-J	Phosphorus content should be kept below 0.08% in irons to be hardened to reduce the risk of cracking by avoiding the formation phosphide–carbide complex.
18-L, M, N	Additions of Mo, Cr and V increase hardenability and are commonly used in simple castings such as piston rings. However care is required for more complex castings as these elements are potent carbide formers (see 18–F).
18-O, P	Additions of 1–2% Ni or Cu increase depth of hardness for a given section and decrease the critical quenching speed. Used for thick sections.
18-V, X	Annealed ferritic irons are more difficult to harden than pearlitic irons.

19	COST
19-A	As a general rule casting cost is directly proportional to weight. However cores or tight tolerances will increase costs disproportionately to weight.
19-B	High graphite content irons are generally cheaper to produce than low carbon irons due to the improved fluidity which carbon confers thus reducing the need for additional feeding of the casting.
19-C	Pearlitic irons which require the use of alloying additions or heat treatment to ensure ferrite free microstructure will be more expensive than mixed microstructure irons.
19-D	Fully ferritic microstructures throughout a casting are not generally achieved in the as cast structure and annealing heat treatments are needed which will increase cost.
19-F	Irons with free carbides if produced by alloy additions will be more expensive than unalloyed irons.
19-J	Phosphorus increases the fluidity of cast irons and hence reduces need for feeding of castings and lowers costs. Reduction of phosphorus content to very low levels for impact strength and reliability will also increase costs.
19-L, M, O, P, Q, R, S	Alloy additions increase costs directly due to high cost per unit weight and also indirectly as alloy content in runners and risers may prohibit re-use in subsequent melts.
19-T	Inoculated irons are slightly more expensive than un-inoculated.
19-U, V, W, X, Y	All these post casting processes will evidently increase the cost of a finished component but the additional expenditure is usually small in comparison with the total cost.

Table 2.3.13—*continued*

2–4

Copper and its alloys

Contents

List of tables

List of figures

2.4.1   General characteristics

2.4.1.1   Introduction

The advantages and limitations of copper alloys are listed in:
Table 2.4.1   *Characteristics of copper and its alloys*

2.4.1.2   Classification of copper alloys

Copper-based alloys are usually classified in terms of one of the main alloying elements, but each class can contain other elements which have considerable influence on specific properties. The main classes of alloys are referred to by traditional names, but in some cases misnomers have arisen over the years, and this is especially true of the bronzes.

The two main categories of copper alloys are brass and bronze. Brasses are essentially copper–zinc alloys to which other elements may be added (and which may give their name as a prefix, e.g. aluminium brass). There are basically two types of brass, alpha brasses which are cold workable alloys and alphabeta or beta brasses which are essentially hot workable.

True bronzes are copper–tin alloys. Other elements added in small quantities are often pre-fixed as in phosphor–bronze. However, manganese bronze is really a brass, and aluminium bronze contains neither tin nor zinc.

The following alphabetical glossary lists most common names of the copper-based alloys.

Admiralty Brass	Copper–zinc–tin alloy.
Alloy D	An aluminium bronze containing Fe.
Alloy E	An aluminium bronze containing Fe and Ni.
Aluminium Brass	Copper–zinc alloy with Al (and As).
Aluminium Bronze	Copper plus aluminium with smaller amounts of Fe, Ni and/or Mn. (A misnomer now accepted.)
American free-machining Brass	Cu–30%Zn–3%Pb, leaded brass.
Architectural Bronze	Cu–40%Zn–3%Pb or Cu–43%Zn–1%Pb leaded brass.
Basis Brass	Cu–35%Zn (the lowest copper content for cold working.)
Beryllium Bronze	Also called beryllium–copper. Cu–Be alloys containing 2% Co with 0.2–0.4%Be.
Best English Brass	Cu–30%Zn brass.
Brass	Generic name for the Cu–Zn alloy series, i.e. copper plus zinc plus other minor elements including Pb, Mn, Al, Sn.
Cadmium Bronze	Cu–Cd alloy, better known as Cadmium Copper.
Cap Copper	Cu up to 10% Zn brass (originally drawn to make percussion caps).
Cartridge Brass	Deep-drawable quality of Cu–30%Zn.
Cast Brass	Usually based on Cu–40%Zn alloy.
Clock Brass	Cu–34/39%Zn–1/3%Pb leaded brasses.
Commercial Bronze	Cu–10%Zn (i.e. a brass).
Common Brass	Cu–37%Zn (sometimes known also as a Basis Brass).
Conductivity Bronze	Cu–2%Sn with P addition.
Constantan	Cu–45%Ni alloy with very low temperature coefficient of electrical resistivity.
Copper (Commercial)	An alloy of copper and oxygen (sometimes also containing phosphorus).

Copperoid	Copper steel (0.2%Cu min) or zinc - coated copper or an American bearing alloy.
Cupro-Nickel	Generic name for Cu plus up to 30% Ni with minor additions.
Deep-drawing Brass	Cartridge Brass, i.e. Cu–30%Zn.
Engraving Brass	Cu–36%Zn–2%Pb, Cu–38%Zn–1%Pb leaded brasses.
Extra-high-leaded Brass	Cu–36%Zn–3%Pb.
Forging Brass	Cu–39%Zn–2%Pb.
Free Cutting Brass	Cu–36%Zn–3%Pb.
Free Cutting Muntz Metal	Cu–39%Zn–1%Pb.
Free-Machining Brass	Cu–40%Zn–3%Pb.
German Silver	Copper–zinc–nickel alloys (see Nickel Silver).
Gilding Metal	Cu–5/20%Zn brasses including '80/20 Gilding Metal'.
Gunmetal	Generic name for copper–tin–zinc alloys.
HCC	High conductivity copper.
High-leaded Brass	Cu–36%Zn–2%Pb.
Inhibited Admiralty Brass	Admiralty brass with As added as inhibitor.
Iron Aluminium Bronze	Cu–Al–Fe alloy.
Jewellery Bronze	A brass with about 14%Zn.
Leaded Commercial Brass	Cu–10%Zn–2%Pb.
Leaded Muntz Metal	Cu–39%Zn–1%Pb.
Leaded Nickel Silver	Cu–Zn–Ni alloy plus Pb.
Low Brass	Cu–20%Zn
Medium Leaded Brass	Cu–34%Zn–1%Pb.
Muntz Metal	Cu–40%Zn brass.
Naval Brass	Cu–38%Zn–1%Sn.
Nickel Brass	Cu–42%Zn–10%Ni–2%Pb.
Nickel Bronze	A nickel-bearing tin bronze.
Nickel–Iron–Aluminium Bronze	Cu–Al alloy with Fe and Ni.
Nickel Silvers	A series of cold workable Cu–Zn–Ni alloys (containing no silver).
OFHC	Oxygen-free high conductivity copper.
Phosphor Bronze	A phosphorus-deoxidised tin bronze usually containing 3 or 5% Sn.
Phosphor Copper	An alloy (used as a hardener) of copper and phosphorus.
Red Brass	Cu–15%Zn.
Rivet Brass	Cu–36%Zn–2%Pb.
Silicon Bronze	Cu–Si alloy which may contain other minor elements Mn or Fe.
Silver Copper	Silver bearing tough-pitch or oxygen-free copper.
Sulphur Copper	Cu–S free-machining copper.
Tellurium Copper	Cu–Te free-machining copper.
Yellow Brass	Cu–33%Zn (cast Yellow Brass Cu–38%Zn– 1%Sn–1%Pb).
Yellow Metal	Cu–40%Zn brass.
2/1 Brass, 2 in 1 Brass	Cu–33%Zn.

The room temperature properties of copper and copper alloys are given in Table 2.4.2 which also gives typical British Standard designations—for international equivalents see Table 2.4.3.

Table 2.4.2   *Room temperature properties of copper and copper alloys*
Table 2.4.3   *International specifications for copper and copper alloys*

2.4.1.3   Influence of alloying elements on the properties of copper and its alloys

The variation of properties of copper and its alloys as elements are added is shown in:
Fig 2.4.1   *Variation of properties of copper and its alloys as elements are added*

ELECTRICAL CONDUCTIVITY

Most pure copper is used for electrical purposes, and in general the electrical conductivity of copper increases with increasing purity. Only silver increases conductivity, but the deleterious effect of other elements varies from the slight effect of cadmium to the very serious effect of phosphorus. Thus silver and cadmium are added to alternator windings to reduce creep in service without substantially affecting electrical performance.

THERMAL CONDUCTIVITY

The same principle applies in that the thermal conductivity of copper increases with increasing purity.

STRENGTH

Copper alloys can be strengthened by the same mechanisms used to strengthen other metals, for example,

(a)   Solid solution strengthening where the alloying elements are soluble in the face-centred cubic alpha phase of copper and form a homogeneous structure, e.g. single phase copper-zinc (alpha brass).

(b)   Presence of a second phase where the amount of alloying element exceeds the solid solubility in copper producing a duplex structure of a monogeneous solid solution plus the (usually) harder second phase, e.g. duplex copper–zinc (alpha-beta brass).

(c)   Precipitation hardening where the alloying element is retained in supersaturated solid solution by rapid cooling from a high temperature. The solid solution is then decomposed by allowing the excess solute to form a precipitate of a second phase on aging at a lower temperature at which its solubility in copper is reduced. The size and form of such precipitates can be controlled during heat treatment to produce the optimum effect, e.g. Copper-Beryllium.

Mechanisms (a) and (b) are most frequently used to harden copper alloys. The effects of certain elements are much greater than others, for example tin is a more effective strengthener than zinc. However, tin has a more deleterious effect on the electrical and thermal conductivities than zinc and it is also more expensive.

More details of the effects of alloying elements on copper and each type of copper alloy are given in the following sections.

TABLE 2.4.1 Characteristics of copper and its alloys

Advantages	Limitations
High conductivity of electrical grades superior to all metals except silver on a volume basis and aluminium on a weight basis.	High cost relative to other common metals.
High thermal conductivity.	Conductivity reduced by small quantities of other elements present.
Excellent ductility permits easy working.	High casting temperatures of the metal and its alloys.
Wide range of copper base alloys, most types having good ductility and malleability in the annealed condition and are particularly suitable for tube forming, hot forming, spinning, deep drawing, etc.	High temperature properties of the metal impose limitations on its use.
Mechanical properties of copper—strength, creep resistance and fatigue performance are improved by alloying (but conductivity is impaired).	The 'gassing' reaction of copper with oxygen requires precautions when temperatures exceed 700°C.
Good corrosion resistance to potable water and to atmospheric and marine environments and can be further improved by alloying.	Toxic—therefore must not be used in contact with foodstuffs, e.g. food processing plant.
Antifouling properties in marine environments.	Some alloys prone to stress corrosion, and other forms of attack e.g. dezincification of brasses.
Useful biocidal properties of the metal and salts.	
Wide range of alloys with special properties, e.g. very high damping capacity.	
Mechanical and electrical properties retained at cryogenic temperatures.	
Weldability of alloys good by appropriate processes.	
Non-magnetic, except some Cu–Ni alloys.	

Table 2.4.1

TABLE 2.4.2 Room temperature properties of copper and copper alloys

Alloy class	Type	Chemical composition (wt %)	Typical BS specification	Hot formability	Cold formability	Cold reduction between anneals	Thermal treatment	Machinability (free cutting brass = 100)	Electrical conductivity % IACS	Thermal conductivity (W/m per K)	Thermal expansion coefficient (10^{-6} K^{-1})	UTS (MN/m²) Soft	UTS (MN/m²) Hard
A. Commercially pure copper	Oxygen-free high conductivity copper	Cu + Ag 99.95	C103	Good	Exc.	95% max.	A		100–101.8	395			
	Electrolytic tough-pitch copper	Cu + Ag 99.90	C101	Good	Exc.	90% max.	A		100–100.8				
	Fire-refined tough-pitch high conductivity copper	Cu + Ag 99.90	C102	Good	Exc.	90% max.	A	20	100–100.5	395	17	215 to 254	261 to 441
	Fire-refined tough-pitch	Cu + Ag 99.85	C104	Good	Exc.	85% max.	A		85–95	330–380			
	Phosphorus-deoxidised copper (low resid. P)	Cu + Ag 99.90 P 0.005–0.012	—	Good	Exc.	95% max.	A		85–98	330–390			
	Phosphorus-deoxidised copper (high resid. P)	Cu + Ag 99.90 P 0.013–0.050	C106	Good	Exc.	95% max.	A		70–90	290–365			
B. Low alloyed coppers	Phosphorus-deoxidised arsenical copper (Cu–DPA)	Cu + Ag 99.70	C107	Good	Exc.	85% max.	A	25	35–45	150–190	17	215–235	264–372
	Silver-bearing oxygen-free copper (Cu–OFS)	Cu + Ag 99.95	—	Good	Exc.	95% max.	A	20	99–101	395	18	215–254	225–372
	Copper–sulphur Copper–selenium Copper–tellurium	Cu bal S 0.2–0.5 or Te 0.3–0.8	C109 C111	Good	Good	70% max.	A	75–85	92–98	369–373	17		
C. High copper alloys	Copper–beryllium	Cu bal Be 1.7–1.9 W and Ni 0.2	CB101	See Table 4	See Table 4		PH	See Table 4	18(a) 25(b) 38(c)	84–105	17	460 540	1200–1500
	Copper–silicon–manganese (Silicon bronze)	CuSi3Mn1	CS101				A	30	7	36	18	350	770

Notes to Table 2.4.2 (pages 00 to 00)

1. To convert specific gravity to density in kg/m³ multiply by 10^3.

2. Electrical conductivities quoted for 20°C. In case of beryllium copper (a) = solution-treated, (b) = aged to maximum hardness, (c) = aged to maximum conductivity. 100% IACS ≡ 5.857×10^7 Ω^{-1} m^{-1}.

3. Impact strengths quoted for the range of material tempers, i.e. minimum value corresponds to hardest temper condition, maximum value corresponds to softest temper condition.

4. Fatigue strengths quoted for annealed materials only. Corresponding UTS values given in brackets. All values are quoted for 25 mm diameter rod.

Table 2.4.2

TABLE 2.4.2 Room temperature properties of copper and copper alloys—*continued*

0.1% Proof stress (MN/m²)		Young's Modulus (GN/m²)		Elongation (%)		Impact strength (J)	Fatigue strength (MN/m²) Annealed 25 mm dia. rod	Specific gravity	Characteristics	Further reference
Soft	Hard	Soft	Hard	Soft	Hard					
49 to 78	138 to 324	117	117 to 132	35 to 50	6 to 30	34 to 61	67	8.9	Best conductivity. Max. ductility. Soft. Expensive.	Tables 2.4.4 & 2.4.5
									High conductivity. High ductility. Suffers gassing in hydrogen atmospheres.	
									Cheaper than above types, slightly lower conductivity. Tough, but ductile and malleable. Suffers gassing.	
									Cheapest, toughest copper. Low conductivity.	
									Not susceptible to hydrogen embrittlement. Slightly lower conductivity than the above.	
									Not susceptible to hydrogen embrittlement. Low conductivity relative to above.	
49–77	138–324	117	117–132	45–48	6–30	—	—	8.9	Enhanced corrosion resistance and softening temperature.	Tables 2.4.6 & 2.4.7
49 77	138–324	117	117–132	45–48	6–30	—	—	8.9	Greater resistance to softening and creep than unalloyed copper. Not embrittled in reducing atm.	
Similar to copper						—	—	8.9	High conductivity applications requiring free-machining. Sulphur preferred to tellurium or selenium.	
125–185	770–1100	—	124–131	40–50	2–4	—	—	8.25	High strength and corrosion resistance.	Tables 2.4.8 & 2.4.9
92	620	113	—	70	5	—	130	8.53	Good corrosion resistance, moderate strength, high toughness. Can be hot or cold worked.	

Notes to Table 2.4.2 (continued)

5. Soft means fully annealed except when value is underlined, when the value is for hot formed material.

6. A　= annealing
   QH = quench hardening
   PH = precipitation hardening.

7. An arbitrary index has been used where a standard of 100 has been assigned to free-cutting brass (BS2874 CZ121).

General — absence of data is indicated in the columns by a dash (—). Where an asterisk (*) appears, data is available but special qualification is required, and the user is directed to the more detailed information on the material in this section.

Table 2.4.2—*continued*

TABLE 2.4.2 Room temperature properties of copper and copper alloys—*continued*

Alloy class	Type	Chemical composition (wt %)	Typical BS specification	Hot formability	Cold formability	Cold reduction between anneals	Thermal treatment	Machinability (free cutting brass = 100)	Electrical conductivity % IACS	Thermal conductivity (W/m per K)	Thermal expansion coefficient (10^{-6} K^{-1})	UTS (MN/m²) Soft	UTS (MN/m²) Hard
D. Brasses (copper–zinc alloys)	CuZn10 Commercial Bronze) (α)	Cu bal Zn10	CZ101	Good	Exc.	90% max.	A	25	44	188	18.2	250–	370–
	CuZn30 (Cartridge Brass) (α)	Cu bal Zn30	CZ106	Fair	Exc.	90% max.	A	30	28	117	20	310–340	450–600
	CuZn40 (α–β)	Cu bal Zn40	CZ109	Exc.	Limited	40% max.	A	45	27	117	20.9	350–	380–
	CuZn40Pb3 (Free-Machining Brass)	Cu bal Zn40Pb3	CZ121	Good	Limited	20% max.	A	100	28	121	21	411	450–637
	Aluminium Brass	Cu bal Zn20Al2	CZ110	Fair	Good	80% max.	A	30	23	100	18	350–420	—
	Admiralty Brass	Cu bal Zn28Sn1	CZ111	Fair	Good	80% max.	A	30	25	109	19	330–390	—
	Naval Brass (non-leaded)	Cu bal Zn38Sn1	CZ112	Exc.	Fair	40% max.	A	40	26	117	20	380–390	450–510
	(leaded)	Pb2 added			Good	25% max.		70					
E. Copper–tin alloys (Tin, Bronze, Phosphor–Bronze)	3% Phosphor–Bronze	Cu bal Sn 3–4.5 P 0.02–0.40	Pb101	Limited	Exc.	80% max.	A	20	15–25	60–120	17	320–340	390–650
	5% Phosphor–Bronze	Cu bal Sn 4.5–5.5 P 0.02–0.40	Pb102	Limited	Exc.	75% max.	A	20	13–18	63–96	17	340–360	420–700
	9% Phosphor–Bronze	Cu bal Sn 7.5–9.0 P 0.02–0.40	Pb104	None	Good	60% max.	A	20	10–14	46–63	17	410–420	470–830
F. Copper–tin–zinc alloys (gunmetals)	Admiralty Gunmetal	Cu88Sn10 Zn2 Ni, Pb	G1				A					270–340 as cast	—
	Leaded Gunmetals	Cu85Sn5 Zn5Pb5 Ni	LG2				A					200–270 as cast	—

Notes to Table 2.4.2 (see pages 1080 and 1081)

Table 2.4.2—*continued*

TABLE 2.4.2 Room temperature properties of copper and copper alloys—*continued*

0.1% Proof stress (MN/m²)		Young's Modulus (GN/m²)		Elongation (%)		Impact strength (J)	Fatigue strength (MN/m²) Annealed 25 mm dia. rod	Specific gravity	Characteristics	Further reference
Soft	Hard	Soft	Hard	Soft	Hard					
60–	185–	124	117–	50–	2–10	154 (annealed)	—	8.8	Excellent cold working. Not susceptible to dezincification or stress-corrosion. Golden.	
92–160	216–463	114	97–114	60–70	10–20	21–120	115	8.5	Optimum strength and ductility for copper–zinc. Suitable for extreme cold deformation.	
123	150–	101	94–101	40–45	20–25	24–52	145	8.4	Alpha-beta brass. Excellent hot working properties.	
170	200–280	—	95	25–32	8–15	—	132–156	8.5	Used for applications requiring extensive machining.	Tables 2.4.10–2.4.16
123–138	—	109	—	55–60	—	—	—	8.35	May contain As (0.02–0.06) to protect against dezinc. Very good corrosion and erosion resistance in clean water.	
108	—	109	—	57–65	—	50–151	—	8.55	Contains As as above. Good corrosion resistance in clean sea-water or moderately polluted river water if slow moving (2m/s).	
138	230–340	102	—	40	20–35	21–51	147	8.4	Good hot working properties. Better resistance than CuZn40 to marine corrosion and mildly aggressive environments.	
									Lead added to improve machinability. CuZn38Sn1Pb1 used for casting (e.g. plumbing fitments).	
90–110	216–510	119	109	50	2–35	61 (cold wkd)	—	8.85	Good cold workability, strength and hardness. Good resistance to corrosion and stress corrosion.	
110–150	230–550	121	105	50	2–38	70–140	137	8.85	Higher strength and hardness than 3% phosphor-bronze.	Tables 2.4.14–2.4.16
154	324–720	111	88–105	55–65	2–35	—	147	8.80	Excellent wear resistance, fatigue strength, bearing properties. High strength, hardness and corrosion resistance.	
130 as cast	—	—	—	13–25 as cast	—	—	—	—	Excellent castability and corrosion resistance in marine conditions. Zinc retards gas absorption.	Tables 2.4.17 & 2.4.18
100 as cast	—	—	—	11–15 as cast	—	—	—	—	Most widely used grade. Better machinability, cheaper to cast, pressure tightness.	

Table 2.4.2—*continued*

TABLE 2.4.2 Room temperature properties of copper and copper alloys—*continued*

Alloy class	Type	Chemical composition (wt %)	Typical BS specification	Hot formability	Cold formability	Cold reduction between anneals	Thermal treatment	Machinability (free cutting brass = 100)	Electrical conductivity % IACS	Thermal conductivity (W/m per K)	Thermal expansion coefficient (10^{-6} K^{-1})	UTS (MN/m²) Soft	Hard
G. Copper–aluminium alloys	5% Aluminium Bronze	Cu bal Al 4.0–6.5	CA101	Fair	Good	60% max.	A	20	15–18	75–84	17	370–410	440–640
	7% Aluminium Bronze (Alloy D)	Cu bal Fe 2.0–3.5 Al 6.0–8.0	CA106	Good	Fair	25% max.	A	30	12–14	58–71	15	470–640	540–640
	Iron–Aluminium Bronze	Cu bal Al 8.5–10 Fe 2.0–4.0 Mn, Ni	CA103	Good	Limited	15% max.	QH	30	12–14	58–67	15	610–640	690–640
	Nickel–Iron Aluminium Bronze	Cu bal Al 9–11 Fe 4–6 Ni 4–6 + Mn	CA104	Good	Limited	10% max.	QH	20	7–9	37–46	15	735	784
H. Copper–nickel alloys (Cupro-nickels)	90/10 Copper–Nickel–Iron	Cu bal Ni 9–11 Fe 1–2 Mn 0.3–1.0	CN102	Good	Exc.	80% max.	A	20	9	50	17	310–320	410
	70–30 Copper–Nickel–Iron	Cu bal Ni 29–32 Mn 0.5–1.5 Fe 0.4–1.0	CN105	Good	Good	50% max.	A	20	5	29	16	350–410	490–850
I. Copper–nickel–zinc alloys (Nickel silvers)	10% Nickel Silver	Cu bal Ni 9–11 Mn 0.05 Zn 25	NS103	None	Good	70% max.	A	25	8.5	37–46	16	380–420	450–590
	18% Nickel Silver	Cu bal Ni 17–19 Mn 0–0.7 Zn 20	NS106	None	Good	70% max.	A	25	6	25	15	401–440	470–680
	Leaded 10% Nickel Brass	Cu bal Ni 9–11 Pb 1–2.5 Mn 0–0.5 Zn 42	NS101	Exc.	Fair	20% max.	QH	80	7	33	19	460–480	510–610
	Leaded 15% Nickel Silver	Cu bal Ni 17–19 Zn 19 Pb 0.5–1.5 Mn 0.07	NS113	None	Fair	30% max.	A	70	6	25	16	450	490–590

Notes to Table 2.4.2 (see pages 1080 and 1081)

Table 2.4.2—*continued*

TABLE 2.4.2 Room temperature properties of copper and copper alloys—*continued*

0.1% Proof stress (MN/m²)		Young's Modulus (GN/m²)		Elongation (%)		Impact strength (J)	Fatigue strength (MN/m²) Annealed 25 mm dia. rod	Specific gravity	*Characteristics*	*Further reference*
Soft	Hard	Soft	Hard	Soft	Hard					
138	400–540	125	115	50	15–30	68	—	8.2	As, Ni, Mn added to improve corrosion resistance which is generally good. Good cold workability.	Tables 2.4.19, 2.4.20 & 2.4.21
200	—	120	—	40	20–30	86–150	—	7.8	Improved strength. Good corrosion resistance and hot working properties.	
260–340	—	117	—	15–18	10	21–47	—	7.6	High corrosion and oxidation resistance, retaining strength at moderately high temperature. Good hot working properties.	
350–450	—	130–140	—	15	12	14–27	250	7.6	Good corrosion and oxidation resistance. High strength, fatigue, wear resistance. Good hot workability and high temperature strength retention.	
120–140	185–460	135	127	38	12–14	—	—	8.9	Iron and manganese added to cupronickels to improve resistance to high-velocity waters. Relatively insensitive to stress corrosion. Good weldability.	Tables 2.4.22, 2.4.23 & 2.4.24
155–170	370–570	151	143	40–	16–18	155 (annealed)	151	8.95	Almost insensitive to stress corrosion. Retains strength at moderately elevated temperatures. Readily weldable.	
123–185	308–555	117	—	48–52	6–25	—	—	8.6	Good corrosion resistance to rural and marine atmospheres. Good cold-working properties.	Tables 2.4.25 & 2.4.28
150–230	370–590	130	134	40–45	3–25	127 (annealed)	115	8.75	Good corrosion resistance and spring properties. Used in telecommunication equipment.	
260–270	310–370	—	127	20–22	15–20	—	—	8.5	Excellent hot working properties, very good machinability and tarnish resistance.	
—	370–480	127	131	—	7–15	29	—	8.8	Good corrosion resistance and machinability.	

Table 2.4.2—*continued*

TABLE 2.4.3 International specifications for copper and copper alloys

Class	Alloy	ISO	UK (BS)	US (ASTM)	Sweden (SIS)	France (NF)	Japan[a] (JIS)	FRG (DIN)
A. Commercially pure copper	Oxygen-free h.c. copper	Cu-OF	C103	OF	Cu-OF	Cu/c1	—	SE-Cu (2.0070)
	Electrolytic tough pitch copper	Cu-ETP	C101	ETP	Cu-ETP	Cu/a1	—	E-Cu (2.0060)
	Fire refined t.p.h.c. copper	Cu-FRHC	C102	FRHC	—	Cu/a2	—	E-Cu (2.0060)
	Fire refined t.p. copper	Cu-FRTP	C104	FRTP	Cu-FRTP	Cu/a3	—	F-Cu (2.0080)
	Phos. deox. copper (low P)	Cu-DLP	—	DLP	—	—	—	—
	Phos. deox. copper (high P)	Cu-DHP	C106	DHP	Cu-DHP	Cu/b	—	SF-Cu (2.0090)
B. Low-alloyed copper	Phos. deox. arsenical copper	Cu-DPA	C107	DPA, 142	—	—	—	SB-Cu (2.0150)
	Silver bearing t.p. copper	Cu-LSTP	—	STP, 113–116	145030	—	—	E-CuAg (2.1202)
	Silver bearing o.f. copper	Cu-OFS	—	OFS, 104, 105, 107	—	—	—	SE-CuAg
	Copper–sulphur	CuS	C111	—	—	—	—	—
	Copper–tellurium	CuTe	C109	DPTE, OFTE OFPTE, 145	—	—	—	E-CuTE (2.1545) SE-CuTe (2.1546)
C. High copper alloys	Copper–cadmium	CuCd	C108	—	—	—	—	—
	Copper–cadmium–tin	CuCdSn	—	—	—	—	—	—
	Copper–chromium	CuCr	—	—	—	—	—	—
	Copper–beryllium (1.7% Be)	CuBe1.7 CoNi	CB101	—	—	—	—	—
	Copper–beryllium (2% Be)	CuBe2CoNi	—	—	—	—	—	—
	Copper–cobalt–beryllium	CuCo2Be	—	—	—	—	—	—
	Copper–nickel–silicon	CuNi2Si	—	—	—	—	—	—
	Copper–silicon–manganese	CuSi3Mn1	CS101	—	—	—	—	—
D1. Brasses	95/5 Brass	CuZn5	CZ125	210	—	—	RBs P1 R1 W1	CuZn5 (2.0020)
	90/10 Brass	CuZn10	CZ101	220	—	U-Z10	RBs P2, R2, W2, T2	CuZn10 (2.0230)
	85/15 Brass	CuZn15	CZ102	230	145112	U-Z15	RBs P3, W3, R3, T3	CuZn15 (2.0240)
	80/20 Brass	CuZn20	CZ103	240	145114	—	RBs P4, W4, R4	CuZn20 (2.0250)
	72/28 Brass	CuZn28	—	260	—	—	—	CuZn28 (2.0265)
	70/30 Brass	CuZn30	CZ105 CZ106	—	—	U-Z30	Bs P1, W1, R1, T1	—
	67/33 Brass	CuZn33	CZ107	268, 270	145124	U-Z33	Bs P2A, W2, R2A, T2	CuZn33 (2.0280)
	63/37 Brass	CuZn37	CZ108	272, 274	145150	U-Z36	Bs P2B, R2B	CuZn36/37 (2.0320)
	60/40 Brass	CuZn40	CZ109	280	—	U-Z40	Bs P3, BF, W3, R3, T3	CuZn40 (2.0360)

Table 2.4.3

TABLE 2.4.3 International specifications for copper and copper alloys—*continued*

Category	Material	Designation	Code	No.	U- code	No.	PbBs/PB code	German no.
D2. Leaded brass	90/10 Brass (leaded)	CuZn9Pb2	—	314	—	—	—	—
	67/33 Brass (leaded)	CuZn9Pb2	CZ118	340	—	—	—	—
	63/37 Brass (leaded)	CuZn36Pb2	CZ119	353	U–Z36Pb2	5140	PbBs P11, W11, R11	CuZn36Pb1 (2.0330)
	63/37 Brass (leaded)	CuZn36Pb3	CZ124	360	U–Z36Pb3	—	—	CuZn36Pb3 (2.0375)
	62/38 Brass (leaded)	CuZn38Pb1	—	370	—	5165	PbBs P12, R12, W12	CuZn38Pb1 (2.0370)
	60/40 Brass (leaded)	CuZn40Pb1	CZ123	365–368	—	5163	PbBs P1, R1, W1	CuZn38Pb1 (2.0370) CuZn40 (2.0360)
	60/40 Brass (leaded)	CuZn39Pb2	CZ120 CZ122	377	U–Z39Pb1	5168	PbBs P13, R13, W1	CuZn39Pb2 (2.0380)
	60/40 Brass (leaded) (Free-Machining Brass)	CuZn40Pb3	CZ121	—	U–Z39Pb2	5170		CuZn40Pb3 (2.0405)
	57/43 Brass (leaded)	CuZn43Pb1	—	—	—	5272	—	CuZn44Pb2 (2.0410)
D3. Special brasses	Aluminium Brass	CuZn20Al2	CZ110	687	CuZn22Al2	5217	BsTF 2, 3, 4	CuZn20Al (2.0460)
	Admiralty Brass	CuZn38Sn1	CZ111	442–445	CuZn29Sn1	5220	BsTF 1	CuZn28Sn1 (2.0470)
	Naval Brass	CuZn38Sn1	CZ112	462, 464, 465, 466, 467, 482 485	—	—	NBs P1, B1, P2, B2, BsTP	CuZn39Sn (2.0530)
E. Copper–tin alloys	Copper–2% Tin	CuSn2	—	505	—	5420	—	CuSn2 (2.1010)
	3% Phosphor Bronze	CuSn4	PB101	511	-	—	PB P1, B1, R1, W1	—
	5% Phosphor Bronze	CuSn5	PB102	510	U–E5P	—	As above	—
	7% Phosphor Bronze	CuSn6	PB103	519	U–E7P	5428	PB P2, B2, R2, W2	CuSn6 (2.1020)
	9% Phosphor Bronze	CuSn8	PB104	521	—	5431	PB P3, SP, SR, R3, SPS, SRS, B3, W3	CuSn8 (2.1030)
	10% Phosphor Bronze	CuSn10	CT1	524	U–E9P	5431	—	—
F. Copper–tin–zinc alloys (gunmetals)	88/10/2 Admiralty Gunmetal	—	G1	—	—	—	—	—
	Nickel Gunmetal (7Sn2Zn5Ni)	—	G3, G3–WP	—	—	—	—	—
	83/3/9/5 Leaded Gunmetal (3Sn9Zn5Pb2Ni)	—	LG1	—	—	—	—	—
	85/5/5/5 Leaded Gunmetal	—	LG2	—	—	—	—	—
	87/7/3/3 Leaded Gunmetal	—	LG4	—	—	—	—	—

Table 2.4.3—*continued*

TABLE 2.4.3 International specifications for copper and copper alloys—*continued*

Class	Alloy	ISO	UK (BS)	US (ASTM)	Sweden (SIS)	France (NF)	Japan[a] (JIS)	FRG (DIN)
G. Copper–aluminium alloys (aluminium bronzes)	5% Aluminium Bronze	CuAl5	CA101	606, 608	—	U–A6	—	CuAl5, CuAl5As
	8% Aluminium Bronze	CuAl8	—	—	—	U–A8	—	CuAl8
	7% Aluminium–Iron Bronze	CuAl8Fe3	CA106	614	—		—	CuAl8Fe
	9% Aluminium–Mn Bronze	CuAl9Mn2	—	—	—		—	CuAl9Mn
	Iron–Aluminium Bronze	CuAl10Fe3	CA103	623	—		ABP1, ABB1	CuAl10Fe
	Ni–Fe–Al–Bronze	CuAl10Fe5Ni5	CA104	632			ABP5	CuAl10Ni CuAl11Ni
	Ni–Fe–Al–Bronze	CuAl9Ni6Fe3	CA105	628, 630	—		ABP4	—
H. Copper–nickel alloys (cupro-nickels)	95/5 Copper–Nickel–Iron	CuNi5Fe1Mn	CN101	704	—		<u>CN</u> P1, TF1, TF1S	CuNi5Fe
	90/10 Copper–Nickel–Iron	CuNi10Fe1Mn	CN102	706	—	AS ISO		CuNi10Fe
	80/20 Copper–Nickel	CuNi20	CN104	—	—			—
	80/20 Copper–Nickel–Iron	CuNi20Mn1Fe	—	710	—		<u>CNTF</u> 2, 25	CuNi20Fe
	75/25 Copper–Nickel	CuNi25	CN105	—	—			CuNi25
	70/30 Copper–Nickel–Iron	CuNi30Mn1Fe	CN107	71	—		<u>CN</u> P3, TF3, TF35	CuNi30Fe
	55/45 Copper–Nickel	CuNi44Mn1	—	—	—	CuNi44 FeMn	—	CuNi44
I. Copper–nickel–zinc alloys	10% Nickel–Silver	CuNi10Zn27	NS103	745	—	U–Z28N9	<u>NS</u> P4, R4, W4	—
	12% Nickel–Silver	CuNi12Zn24	NS104	757	5243 CuNi12Zn24	—	<u>NS</u> P3, B3,	CuNi12Zn24
(Nickel Silvers)	15% Nickel–Silver	CuNi15Zn21	NS105	—	—	U–Z22N15	As above	—
	12% Nickel–Silver	CuNi18Zn20	NS106	752	5246 CuNi18Zn20	U–Z22N18	<u>NS</u> P2, B2, R2, W2	CuNi18Zn20
	18% Nickel–Silver	CuNi18Zn27	NS107	770	—	U–Z27N18 ×85	<u>NS</u> SP, BS, SR, WS, SPS, SRS	—
	Leaded 10% Nickel Brass	CuNi10Zn42Pb2	NS101	—	—	U–Z45N19		CuNi10Zn42Pb
	Leaded 18% Nickel Silver	CuNi18Zn19Pb1	NS113	794	—		PbNSB	CuNi18Zn19Pb

[a] Japan—underlined part of specification is a prefix for all subsequent sets of digits, e.g. <u>CNTF</u> 2, 25 ≡ CNTF 2, CNTF 25.

Table 2.4.3—*continued*

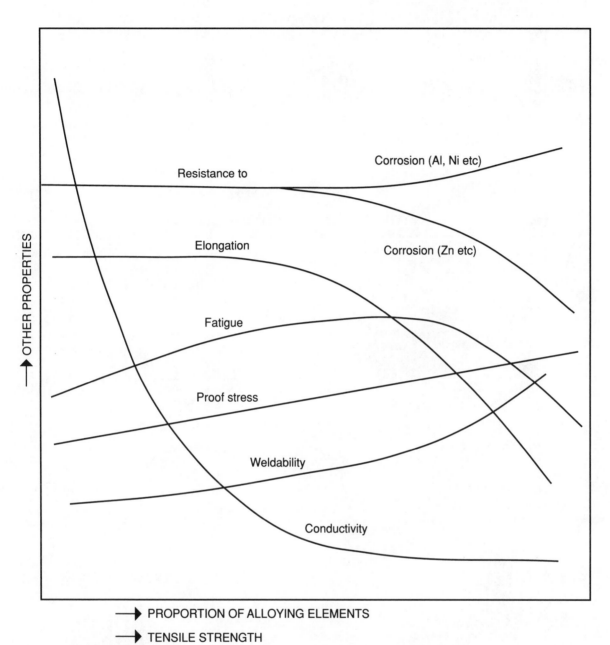

FIG 2.4.1 Variation of properties of copper and its alloys as elements are added

Fig 2.4.1

2.4.2   Commercial coppers and high copper alloys

The comparative characteristics of commercially pure coppers are summarised in:
Table 2.4.4   *Comparative characteristics of different types of commercially pure copper*
The effects of low alloy (less than 1.0%) additions on the properties of copper are shown in:
Table 2.4.5   *Effect of low alloy additions on the properties of copper*
Figure 2.4.2 illustrates the effect of low alloy (<1.0%) elements on the electrical conductivity of copper:
Fig 2.4.2   *Effect of some elements on electrical conductivity of copper at 20°C*
Figure 2.4.3 shows their effect on thermal conductivity:
Fig 2.4.3   *Effect of some elements on thermal conductivity of copper*
The common names of the commercial coppers, their BS and ISO specification numbers and their more important properties are listed in:
Table 2.4.6   *Basic properties of copper and low alloyed copper of high conductivity*
and their characteristics and applications in:
Table 2.4.7   *Comparative characteristics of copper and low alloyed coppers of high conductivity*
The common names of the high copper alloys used mainly for their conducting properties, their BS and ISO specification numbers and their more important properties are listed in:
Table 2.4.8   *Basic properties of high copper alloys*
and their characteristics and applications in:
Table 2.4.9   *Comparative characteristics of high copper alloys*

TABLE 2.4.4 Comparative characteristics of different types of commercially pure copper

Type	Purity (Cu+Ag min)	IACS	Advantages	Limitations	Typical uses
1. Oxygen-free high conductivity	99.95	100.0–101.8	Highest electrical and thermal conductivities. Freedom from gassing when heated in atmospheres containing hydrogen. Maximum ductility and malleability.	More expensive than other types. Soft.	Electrical applications demanding highest conductivity. Foils.
2. Electrolytic tough-pitch	99.90	100.0–100.8	Free from impurities and inclusions. Very high electrical and thermal conductivities. Excellent ductility and malleability.	Generally more expensive than fire refined types. Suffer gassing in hydrogen atmospheres.	Chemical plant. Extended surface heat exchangers. Fine tubes and wire Electrical applications.
3. Fire refined tough-pitch high conductivity	99.90	100.0–100.5	Usually cheaper than electrolytic types but nearly as good electrical and thermal conductivities. Tough but good ductility and malleability in annealed condition.	Some impurities and inclusions may be present. Suffers gassing.	Wires, cables, tubes bus-bars, chemical plant, plain heat exchangers.
4. Fire refined tough-pitch	99.85	85–95	Cheapest and toughest type, but still has good ductility and malleability in annealed state plus good stiffness when hard drawn or cold worked.	Lower electrical and thermal conductivity. Not always homogeneous. Suffers gassing.	Chemical plant when very high conductivity not essential. Drawn and spun products. Architectural uses.

Table 2.4.4

TABLE 2.4.5 Effect of low alloy additions on the properties of copper

Copper with up to 1% alloying addition		Strength				Hardness	Elasticity		Ductility	Heat resistance			Thermal			Electrical conductivity	Magnetic properties	Machinability	Surface finish	Corrosion resistance
		Tensile	Proof	Fatigue	Impact	Hardness	Modulus of elasticity	Limit of proportionality	Ductility	Oxide scaling	Creep	Softening temperature	Specific heat	Expansion	Conductivity	Electrical conductivity	Magnetic properties	Machinability	Surface finish	Corrosion resistance
Alloy addition ↓		1	2	3	4	5	6	7	8	9	10	11	12	13	14	15	16	17	18	19
As 0.15–0.5%	A	0	0	0		0	0	0	0		↑	↑	0	0	↓	↓	0	0	0	↑
Ag 0.03–0.12%	B	0	0	0		0	0	0	0		↑	↑	0	0	0	0	0	0	0	
Cd 0.2–1.0%	C	↑	↑	↑		↑	0	↑	0		↑	↑	0	0	0	↓	0			
Cr 0.5–1.0%	D	↑	↑			↑	0		0	↑		↑	0	0	↓	↓	0	↑	↑	↑
Se	E	0	0	0		0	0	0	0			↑			↓	↓	0	↑	↑	
Te 0.3–0.8%	F	0	0	0		0	0	0	0		↑	↑	0	0	↓	↓	0	↑	↑	
S 0.2–0.5%	G	0	0			0	0	0	0				0	0	↓	↓	0	↑	↑	
Zr 0.1–0.2%	H	↑	↑						0		↑	↑				↓				

↑ = Increase.   ↓ = Decrease.
0 = No effect.   Blank = No data.

A.   Arsenic additions improve hot strength, but the effect on electrical conductivity is severe (35–45% IACS). Not suitable for electrical applications.

B.   Silver improves hot strength without degrading electrical conductivity.

C.   Cadmium improves cold and hot strength with a relatively small effect on electrical conductivity (80–95% IACS).

D.   Chromium improves cold and hot strength, with a relatively small effect on electrical conductivity (80–85% IACS). Can be used up to 350°C without undue impairment of properties. Hardenable by precipitation heat treatment.

E, F, G.   Selenium, tellurium and sulphur additions improve the machinability of copper (machinability ratings Cu 20, Cu-S 75, Cu-Te 85). The improved machinabilities do not impair mechanical properties and the reduction in electrical conductivity is slight (90–98% IACS). Sulphur is preferred to tellurium or selenium.

H.   Zirconium additions improve cold and hot strength with a slight reduction in electrical conductivity (85–90% IACS). Hardenable by precipitation heat treatment.

Table 2.4.5

TABLE 2.4.6 Basic properties of copper and low alloyed coppers of high conductivity

Group	Usual name	BS ref. (ISO)	Basic chemical composition	Specific gravity	Electrical conductivity (% IACS)[b]	Thermal conductivity (W/m per K)	Thermal expansion coefficient (10^{-6} K^{-1})
Commercial coppers	Electrolytic tough pitch h.c. copper	C101[a] (Cu ETP)	99.90 Cu (min.)	8.9	101.5–100	410	17
Wrought	Fire-refined tough pitch h.c. copper	C102 (Cu FRHC)	99.90 Cu (min.)	8.9	101.5–100	410	17
	Oxygen-free h.c. copper	C103 (Cu OF)	99.95 Cu (min.)	8.9	101.5–100	410	17
	Oxygen-free h.c. copper (special)	C110	99.95 Cu (min.)	8.9	101.5–100	410	17
	Oxygen-free silver bearing copper	(Cu OFS)		8.9	101–100	410	18
	Fire-refined tough pitch non-arsenical copper	C104 (Cu FRTP)	99.85 Cu (min.)	8.9	95–89	330–380	17
	Tough pitch arsenical copper	C105	99.20 Cu (min.) 0.40 As	8.9			17
	Phosphorus deoxidised non-arsenical copper	C106 (Cu DHP)	99.85 Cu (min.)	8.9	90–70	290–365	17
	Phosphorus deoxidised arsenical copper	C107	99.20 Cu (min.) 0.40 As	8.9	50–35	150–190	17
	Copper-cadmium (cadmium-copper)	C108	99.5–98.8 Cu 0.5–1.2 Cd	8.9	83–87	315–355	16.6
	Copper tellurium	C109	99.7–99.3 Cu 0.3–0.7 Te	8.9	92–98	369–375	17
	Copper-sulphur	C111	99.7–99.4 Cu 0.3–0.65 S	8.9	93–95	370–375	17
Castings	High conductivity copper	HCC1	99.9 Cu	8.9	90	372	17

[a] Within C101 (tough pitch) for electrical conductors there are grades with specified minimum Ag contents as follows:–
grade Cu-Ag-1: 0.01%, grade Cu-Ag+3: 0.06%, grade Cu-Ag+4: 0.019%, grade Cu-Ag-5: 0.14%.

[b] Electrical conductivities quoted for 20°C.

[c] Soft means fully annealed.

Table 2.4.6

TABLE 2.4.6 Basic properties of copper and low alloyed coppers of high conductivity—*continued*

Modulus of elasticity (GN/m^2)		Mechanical properties								Limitation creep approx. service temperature
		Tensile strength (MN/m^2)		0.1% Proof stress (MN/m^2)		Elongation (%)		Fatigue^d (MN/m^2)	Impact^e (J)	
Soft^c	Hard	Soft^c	Hard	Soft^c	Hard	Soft^c	Hard			Range (°C)
117	117	210–215	250–400	50–70	135–300	35–50	30–6	66	34–60	−200–80
117	117	210–215	250–400	50–70	135–300	35–50	30–6	66	34–60	−200–80
117	117	210	240–400	50–70	135–300	40–60	30–6	66	34–60	−200–80
117	117	210	240	50	135	60	30			−200–80
117	117	215–235	240	50–70	135–350	40–55	30–6	66	34–60	−200–80
117	120	240–270	240–400	50–75	140–325	40–55	30–6	66	34–60	−200–80
117	125	240–270	260–420	50–75	140–325	35–50	30–6	66	34–60	−200–120
117	120	240–280	260–440	50–75	140–350	35–50	30–6	66	34–60	−200–80
117	120	215–235	260–380	50–75	135–350	40–50	30–6			−200–120
117	117	270–300	500–650	60–80	220–400	40–50	10–2		40	−200–80
117	117	215–235	240–400	50–75	135–300	35–50	30–6			−200–80
117	117	215–235	240–400	50–75	135–300	35–50	30–6			−200–80
Conductivity test only specified										−200–80

^{d} Fatigue strengths quoted for annealed materials.
  Most values are for 25 mm rods and are an approximate guide only.

^{e} Impact strengths cover a range of tempers.
  Minimum value corresponds to hardest temper condition. Maximum value corresponds to soft temper.

Table 2.4.6—*continued*

TABLE 2.4.7 Comparative characteristics of copper and low alloyed coppers of high conductivity

BS ref.	Type	Available forms[a]	Thermal treatment[b]	Hardness	Formability[c] Hot	Cold	Maximum reduction between anneals	Machinability[d]
C101	Electrolytic tough pitch	S.B. W	A	45–115	G	E	90%	20
C102	Fire-refined high conductivity	S.B. W.	A	45–115	G	E	90%	20
C103	Oxygen-free high conductivity	S.B. W.	A	45–115	G	E	95%	20
C104	Fire-refined tough pitch	S.B. P.W.	A	45–115–	G	E	85%	20
(Cu-OFS)	Oxygen-free, silver-bearing	S.B.	A	45–115	G	E	90%	20
C105	Tough pitch arsenical	P.	A	45–115	G	E	85%	20
C106	Phosphorus deoxidised	PST.	A	45–115	G	E	95%	20
C107	Phosphorus deoxidised arsenical	PT.	A	45–115	G	E	85%	25
C108	Cadmium copper	P.W.	A	60–140	G	E	90%	25
C109	Copper tellurium	B.	A	50–100	G	G	70%	75–85
C111	Copper sulphur	B.	A	50–100	G	G	70%	75–85
C110	Oxygen-free h.c.		A	45–115	G	E	90%	20
HCC1	Cast h.c. copper	C.	—	45	—	—	—	20

[a] S, sheet, strip. B, bars, rods, sections. W, wire. P, plate. T, tube. C, castings. Sd, Sand castings. Ch, chill castings. F, forgings.
[b] A, anneal. QH, quench hardened (W condition). PH Precipitation hardened (WP condition).
[c] G = good, F = fair, P = poor, E = excellent, L = limited.
[d] An arbitrary index of 100 is assigned to free-cutting brass.
[e] Special care due to risk of 'gassing'.

Table 2.4.7

TABLE 2.4.7 Comparative characteristics of copper and low alloyed coppers of high conductivity—*continued*

Joining[c]			Limitations	Typical uses
Soldering	Brazing	Welding		
E	G[e]	P	Gassing risk. Soft. Expensive. Creeps readily. Welding difficult.	High conductivity favours electrical wiring, cables, equipment.
E	G[e]	P	Gassing risk. Soft. Expensive. Creeps readily. Welding difficult.	High conductivity favours electrical wiring cables, equipment.
E	G	P	Soft. Expensive. Creeps readily. Welding difficult.	Highest ductility and conductivity.
E	G[e]	P	Lower conductivity than above. Gassing risk. Creeps readily.	Toughest, less costly.
E	G	F	Creeps less readily.	Good resistance to softening and slightly better to creep.
E	G[e]	P	Gassing risk. Creeps readily. Fusion welds often porous.	Tough. Slightly better at raised temperatures steam locomotive fire boxes and stay bolts.
E	G	G	Creeps readily.	No gassing risk permits fusion welded and brazed assemblies.
E	G	G	Creeps readily.	Welded and brazed assemblies.
G	G	G	Cd. fume risk when casting or welding.	Good strength and conductivity favour overhead lines and traction catenaries.
G	G	P	Low strength for fittings. Welding difficult.	High conductivity applications requiring free machining.
G	F	P	Low strength for fittings. Welding difficult.	Preferred to C109 for high conductivity by machining operations.
—	—	—	Expensive.	Special uses for electronics purposes especially in vacuum.
—	F	P	Difficult to cast gas free.	Cast fittings of high conductivity.

Table 2.4.7—*continued*

TABLE 2.4.8 Basic properties of high copper alloys

Group	Usual name	BS ref. (ISO)	Basic chemical composition	Specific gravity	Electrical conductivity (% IACS)[b] (a)		Thermal conductivity (W/m per K)	Thermal expansion coefficient (10^{-6} K^{-1})
Copper Chromium Wrought	Copper chromium Chromium–copper	CC101 (Cu Cr1)	99.7–98.8 Cu 0.3–1.4 Cr	8.9	W PH	32–37 75–85	166 315–340	17
	Copper–Chromium Zirconium	CC102 (Cu Cr 1Zr)	99.5–98.4 Cu 0.3–1.4 Cr. 0.02–0.2 Zr	8.9	W PH	45 80	166 330–340	17
	Copper–Zirconium	—	99.8 Cu 0.1 Zr	8.9	W PH	45 90	~160 ~300	17
Castings	Cast Copper-Chromium	CC1–TF	99 Cu 1 Cr	8.9	TF	~80	312	17
Wrought	Beryllium–Copper Beryllium Bronze	CB101 Cu 1.7 Be Co Ni	97.8 Cu 1.7 Be 0.05–0.4 (Co + Ni)	8.25	O PH	22–32 17–24	~150 108–130	17
	Beryllium–Copper Beryllium Bronze	— (Cu 2 Be Co Ni)	97.5 Cu 2 Be 0.2–0.6 (Co + Ni)	8.25	W PH	~17 20–30	~85 ~105	17
	Copper–Cobalt, Beryllium	C112 Cu Co 2 Be	97 Cu 2–2.8 Co 0.4–0.7 Be 0.5 (Ni + Fe)	8.75	W PH	~22 45–52	~130 230	17
	Copper–Silicon Silicon Bronze	CS101 (Cu Si 3 Mn)	95.5 Cu 3.5 Si 1 Mn	8.6		7	42	17
	Cast Silicon Bronze	—	94 Cu 3.5 Si 1Mn 1.5 Fe Zn	8.6				
	Copper–Nickel– Phosphorus	C113	98.5 Cu 1 Ni 0.2 P	8.8				
	Copper–Nickel Silicon	— (Cu Ni 2 Si)	97 Cu 2 Ni 0.8 Si	8.9	W TF	~17 26–40	~80 ~95	16
	Conductivity Bronze	—	98 Cu 1 Sn (1Cd)	8.9		55–75	230	17

[a] W = Quench hardened. PH = Precipitation hardened. TF = Solution-treated and precipitation-treated.

[b] See footnotes for Table 2.4.6.

Table 2.4.8

TABLE 2.4.8 Basic properties of high copper alloys—continued

Modulus of elasticity (GN/m²)		Mechanical properties								Approx. service temperature (Limitation creep)
		Tensile strength (MN/m²)		0.2% Proof stress (MN/m²)		Elongation (%)		Fatigue[d] (MN/m²)	Impact[e] (J)	
Soft[c]	Hard	Soft[c]	Hard	Soft[c]	Hard	Soft[c]	Hard			Range (°C)
110	140–160	230–250	380–410			25–30	15–18			−200–350
~110	~150	230–250	380–410			25–30	15–18			−200–350
~110	~150	200–225	400–415			25–30	15–18			−200–350
		(200)	(400)			(15)	(5)			−200–350
125	135	510–560	570–1400	230	400–1100	50	5–20		130 30	−200–250
125	135	490–510	550–1400	180–230	400–1100	40–50	2–30		140 40–50	−250–200
120	140	310–320	360–900	120–150	280–800	25–30	2–12		14–16	−250–200
105	105	390–400	440–850	140–160	350–720	45–55	3–20	110–230	80–160	−250–200
130	145–158	300	440–780	120	380–650	30–35	7–14			−250–200
135	1270	260–280	450–750	90–100	400–700	35–45	3–7	140–200		−200–200

Table 2.4.8—continued

TABLE 2.4.9 Comparative characteristics of high copper alloys

BS ref.	Type	Available forms[a]	Thermal treatment[b]	Hardness	Formability[c]		Maximum reduction between anneals	Machinability[d]
					Hot	Cold		
CC101	Copper Chromium	SPBF	QH	65–170	G	W.G PH.F	W75% PH. 35%	30
CC102	Copper–Chromium–Zirconium	SBFW	QH	90–150	G	W.G. PH.F		30
—	Copper–Zirconium	SPBW	QH	60–150	G			30
CC1TF	Cast Copper–Chromium	Sd.	QH	125	—	—	—	30
CB101	Beryllium–Copper	SPBFW	QH	95–380	G	W. G PH L	W.40% PH.10%	W. 30 PH. 20
—	Beryllium–Copper	SPBFW	QH	105–400	G	W. G QH. L	W. 4% PH.10%	W. 30 PH. 20
C112	Copper–Cobalt–Beryllium	SPBFW	QH	60–230	G	W. G QH. F	W. 50% PH. 30%	W. 30 PH. 30
CS101	Copper–Silicon	SPBFTW	A	75–210	G	G	75%	30
—	Cast Silicon Bronze	Sd.	—	70–115	—	—	—	30
—	Copper–Nickel–Silicon	SBFTW	QH	70–210	E	W. G QH. F	W. 75% PH. 20%	W. 20 PH. 30
—	Conductivity Bronze	W	A	55–145	F	G	80%	20

See Footnotes for Table 2.4.7.

Table 2.4.9

TABLE 2.4.9 Comparative characteristics of high copper alloys—*continued*

Joining[c]			Limitations	Typical uses
Soldering	Brazing	Welding		
G	F	F	Heat treatment necessary making costs higher. Precise working and heat treatment control essential.	Electrical connectors, switch gear parts, commutator bars and end rings. Spot and seam welding electrodes, disc brakes, oxygen jet tips.
G	F	F		Electronic items. Electrical components, commutator bars and end rings at raised temperatures. Electrical items, commutator and end rings. Switch gear parts. Raised temperature terms.
G	F	F		
G	F	F		Cast components for use with Cu–Cr and Cu–Zr assemblies. Electrode holders.
G	G	F	Health risks to foundry workers. Heat treatment necessary adding to costs. Heat treatments and working are critical, requiring careful control and liaison with makers.	Springs, contacts, connectors, HF plugs. Pressure elements, Bourdon tubes, non-sparking tools, resistance welding electrodes, cryogenic items, cams, bushings, chains, etc., for non-magnetic components, wear-resistant items and special bearings guides, etc.
G	G	F		
G	G	F		
G	E	E	Silicon content wears in machinery operations. More costly than brasses.	Connectors, spring contacts, circuit breakers, clips. Bearings and bushings. Nozzles for gas burners. Die-casting machine plunger tips.
G	G	E		
G	G	E		Welded assemblies for chemical plant, picking equipment, sewage plant, marine pipes and fittings, fasteners, hot water tanks, hydraulic tubes. Cast components joined to sheet and tube assemblies of CS101. Good corrosion resistance to sea-water. Contacts, switchgear parts, reinforcing rings for rotor ends. Fasteners, catenary supports, etc., for electric railways. Ball & roller bearing cages, springs.
G	F	F		Catenary wires for railways. Contact shoes. Electrode holders for arc furnaces and resistance welding machines.

Table 2.4.9—*continued*

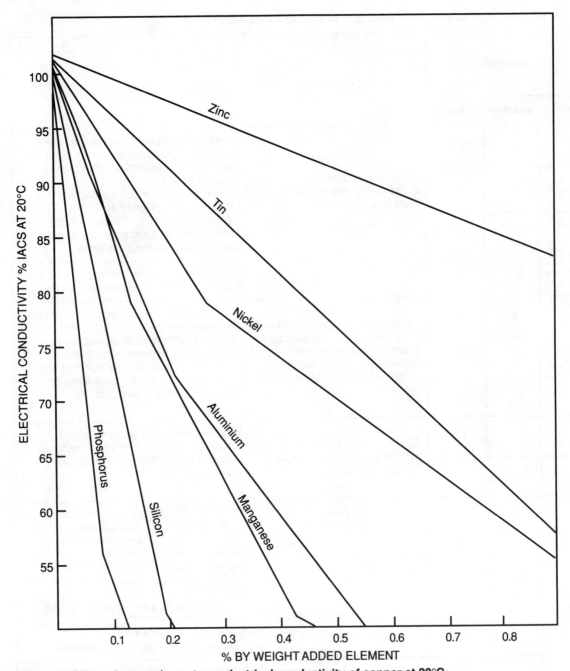

FIG 2.4.2 Effect of some elements on electrical conductivity of copper at 20°C

Fig 2.4.2

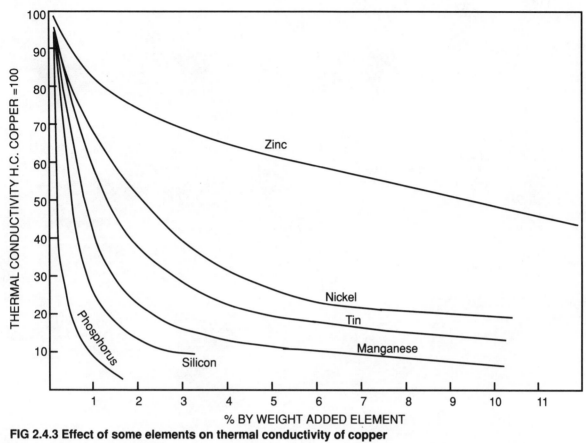

FIG 2.4.3 Effect of some elements on thermal conductivity of copper

Fig 2.4.3

2.4.3 Brasses

Brasses are the most numerous and most widely used of the copper alloys because of their low cost, easy (and inexpensive) fabrication and machining and their resistance (compared with steel) to aggressive environments. They are, however, generally inferior in strength to bronzes and must not be used in environments which cause dezincification.

Increase in zinc increases strength and hardness but reduces ductility and conductivity. Up to about 37% Zn gives single phase alloys but a higher zinc content gives duplex (alpha+beta) alloys which are readily worked hot but are less readily cold worked than single phase brasses.

Lead additions improve machinability, the 'standard' being 61% Cu, 36% Zn, 3% Pb with an arbitary value of 100 for the machinability index of other alloys. Al increases the strength and corrosion resistance especially of the duplex alloys. As additions increase the resistance of brass to dezincification corrosion. Sn increases strength and corrosion resistance slightly. Fe increases slightly the strength of duplex brasses but is used in the high tensile brasses chiefly for its grain-refining effect. Mn increases strength and is used largely in the high tensile brasses. Si improves resistance to corrosion particularly for marine purposes.

The main types of brasses are listed in:

Table 2.4.10 *Effect of alloy addition on the properties of copper–zinc alloys*

More detailed information on the influence of alloy additions on the properties of 60% Cu–40% Zn—a most commonly employed brass—is given in:

Table 2.4.11 *Effect on properties of alloy additions to 60–40 brass*

The common names of the brasses, their BS and ISO specification numbers and their more important properties are listed in:

Table 2.4.12 *Basic properties of copper–zinc alloys (brasses)*

and their general characteristics and applications:

Table 2.4.13 *Comparative characteristics of copper–zinc alloys (brasses)*

TABLE 2.4.10 Effect of alloy addition on the properties of copper–zinc alloys (see footnotes)

Copper–zinc alloys (Alpha brasses)		Tensile	Proof	Fatigue	Impact	Hardness	Modulus of elasticity	Limit of proportionality	Ductility	Oxide scaling	Creep	Softening temperature	Specific heat	Expansion	Conductivity	Electrical conductivity	Magnetic properties	Machinability	Surface finish	General	Stress corrosion	Dezincification	Zinc equivalent[a]	Cost
Composition		1	2	3	4	5	6	7	8	9	10	11	12	13	14	15	16	17	18	19	20	21	22	23
Cap Copper – Gilding Metal Zn (0–10%)	A	↑	↑			↑	↓	↑	0			↓	0	0	↓	↓	0	0·		0	0	0	1	↓
Gilding Metal – Cartridge Brass Zn (10–30%)	B	↑	↑			↑	↓	↑	↑			↑	0	0	↓	↓	0	↑		↓	↓	↓	1	↓
Cartridge Brass – Basis Brass Zn (30–37%)	C	↑	↑	↓	↓	0	↓	↑	↓			0	0	0	0	↓	0	↑		↓	↓	↓	1	↓
Aluminium Brass Al (1.8–2.5%)	D	0	0			0	↓	0	0			0	0	0	↓	↓	0	0		↑		↑	6	
Admiralty Brass Sn (0.9–1.3%)	E	0	0			0	↓	0	0			0	0	0	↓	↓	0	0		↑			2	↑
As (0.02–0.06%)	F																					↑		
Pb (0.5–1.5%)	G	0	0		↓	0	↓		↓			0	0	0	0	0	0	↑	↑	↓			1	

Key: ↑ = Increase.   ↓ = Decrease.   0 = No effect.   Blank = No data.

A, B & C The effect of increasing zinc content is indicated, and three ranges of zinc concentration have been selected as in each range different effects are concerned, e.g. ductility (A8, B8 and C8) increases with increased zinc content to a maximum at 30% Zn above which ductility is reduced as small particles of beta phase occur owing to segregation. In each case the effect of increasing zinc content on properties is indicated by comparison with the lowest zinc content within that range.

D, E, F, G The effects of the alloy additions are indicated by comparison with the properties of 70% Cu 30% Zn brass. Arsenic may be added to all the alpha brass alloys to inhibit corrosion by dezincification. Arsenic additions to 30% Zn brass are employed in the manufacture of heat-exchanger tubes.

[a] The zinc equivalent indicates the potency of an element in promoting the formation of the β phase in a copper zinc alloy. In the case of aluminium brass, with 76% copper, 22% zinc and 2% aluminium, aluminium has a zinc equivalent of 6, thus the effective zinc content is $(2 \times 6) + 22 = 34$ parts. The alloy contains 76 parts of copper $(100 - [2 + 22])$ thus the effective zinc concentration is:

$$\frac{34}{34 + 76} \times 100 = 31\% \text{ Zn}$$

which is insufficient to produce a duplex α + β phase structure. The zinc equivalents are only approximate.

Table 2.4.10

TABLE 2.4.11 Effect on properties of alloy additions to 60–40 brass

Copper–zinc alloys (2) (Alpha + beta brasses)		Strength				Hardness	Elasticity		Ductility	Heat resistance			Thermal			Electrical conductivity	Magnetic properties	Low temperature properties	Machinability	Surface finish	Cold formability	Hot formability	Corrosion resistance			Zinc equivalent	Cost
		Tensile	Proof	Fatigue	Impact		Modulus of elasticity	Limit of proportionality		Oxide scaling	Creep	Softening temperature	Specific heat	Expansion	Conductivity								General	Stress corrosion	Dezincification		
		1	2	3	4	5	6	7	8	9	10	11	12	13	14	15	16	17	18	19	20	21	22	23	24	25	26
Zn (30–40%)	A	↑	↑	↑	↓	↑	↑	↑	↓				0	0	0	0	0	0	↑	↑	↓	↑	↓	↓	↓	1	↓
Sn (1%)	B	↑	↑	0		↑	↑	↑	0				0	0	↓	↓	0		0	0	0	0	↑		0	2	↑
Mn (1.25%)	C	↑	0			0	0		↓					0	↓	↓	0		↑	↑	↓	↑	↓			0.5	
Ni (5.5–9.5%)	D	↑							0			↑									0	0	↑			1.2	
Pb (0.3–3.0%)	E	0	0		↓	↓	↓		↓				0	0	0	0			↑	↑	0	↓				1	
As (0.02–0.1%)	F																								↑		
Al (0.5–3.0%)	G	↑	↑			↑	↓	↑	↓	↑				↓	↓	↓			↑	↑	0	0	↑		↑	6	
Fe (0.1–2.0%)	H	↑	↑																							0.9	
Si (0.6%)	I	↑	↑	↑		↑																	↑	↑	↑		

Key: ↑ = Increase. ↓ = Decrease. 0 = No effect. Blank = No data.

A. The properties of a 60% Cu, 40% Zn brass are indicated by comparison with properties of 70% Cu, 30% brass used as the reference material in Table 2.4.10 for alpha brasses. The effects of alloying element additions are indicated by comparison with 60% Cu, 40% Zn brass.

B. Brass containing 61% Cu, 38% Zn, 1% Sn is commonly known as naval brass.

C. Brass containing 57.75% Cu, 41% Zn, 1.25% Mn is commonly known as manganese brass, not to be confused with manganese bronze which is another name for some high tensile brasses which contain other elements in addition to manganese.

D. Brasses containing up to 10% nickel with sufficient zinc to retain an alpha plus beta phase structure (nickel has a negative zinc equivalent) are known as nickel brasses. Copper–zinc–nickel alloys with single phase alpha structure are known as nickel silvers.

E. The addition of lead to all the brasses has the effect of improving the machinability, and an alloy containing 61% Cu, 36% Zn, 3% Pb, known as free-cutting brass is used as the standard for a machinability index, with an arbitrary value of 100. Higher machinability ratings have been achieved with certain leaded proprietary brass alloys.

F. Arsenic additions have the effect of increasing the resistance of the brasses to dezincification corrosion.

G. Aluminium additions have the effect of increasing the strength of the duplex brasses as well as increasing their corrosion resistance. Aluminium is the most strengthening addition in the high tensile brasses.

H. Iron has the effect of slightly increasing the strength of the duplex brasses but is generally only used in the high tensile brasses mainly for its grain refining effect. A typical high tensile brass contains 54% Cu, 40% Zn, 1% Sn, 1% Pb, 1% Fe, 1.5% Mn, 1% Al.

I. The addition of silicon improves resistance to corrosion. An alloy containing 80% Cu, 4% Si, 16% Zn is used for certain marine applications.

Table 2.4.11

TABLE 2.4.12 Basic properties of copper–zinc alloys (brasses)

Group	Usual name	BS Ref. (ISO)	Basic chemical composition	Specific gravity	Electrical Conductivity[a, b] (% IACS)	Thermal conductivity (W/m per K)	Thermal expansion coefficient ($10^{-6} K^{-1}$)
Brasses– Copper–zinc	Gilding Metal	CZ125 (—)	95 Cu 5 Zn	8.85	56	230	17.5
Alloys: Wrought forms	Commercial Bronze	CZ101 (Cu Zn 10)	90 Cu 10 Zn	8.85	44	188	18.2
	Red Brass	CZ102 (Cu Zn 15)	85 Cu 15 Zn	8.75	37	160	18.3
	Low Brass	CZ103(W) (Cu Zn 20)	80 Cu 20 Zn	8.65	32	138	18.2
	Leaded 80–20 Brass	CZ104 (Cu Zn 20)	79 Cu 20 Zn 1 Pb	8.8	31	138	18.5
	Cartridge or Deep Drawing Brass	CZ106(W) (Cu Zn 30)	70 Cu 30 Zn	8.55	27	117	20
	70/30 Arsenical Brass	CZ105	70 Cu 30 Zn 0.04 As	8.5	27	117	20
	2/1 Brass	CZ107 (Cu Zn 33)	67 Cu 33 Zn	8.45	27	117	19
	Common Brass or Basic Brass	CZ108 (Cu Zn 37)	63 Cu 37 Zn	8.4	26	117	19
	Muntz Metal	CZ109 (Cu Zn 40)	60 Cu 40 Zn	8.4	28	117	20.8
	Aluminium Brass	CZ110 (Cu Zn 20 Al 2)	Cu 77 21 Zn Al 2 As 0.04	8.6	27	100	18
	Admiralty Brass	CZ111	Cu 72 27 Zn Sn 1 As 0.04	8.55	25	109	19
	Naval Brass	CZ112	Cu 63 36 Zn Sn 1	8.4	26	117	20
	Special Naval Brass	CZ113(W)	Cu 59 40 Zn 0.60–1.25 Sn	8.4	23	106	20
	High Tensile Brass	CZ114	58 Cu 37 Zn 0.6 Sn 1 Pb 0.8 Fe 1.1 Mn. 1.5 max. Al	8.5	26	117	19

See footnotes for Table 2.4.6.

Table 2.4.12

TABLE 2.4.12 Basic properties of copper–zinc alloys (brasses)—*continued*

| Modulus of elasticity (GN/m²) | | Mechanical properties | | | | | | Fatigue^d (MN/m²) | Impact^e (J) | Approx. service temperature |
| | | Tensile strength (MN/m²) | | 0.2% proof stress (MN/m²) | | Elongation (%) | | | | |
Soft^c	Hard	Soft^c	Hard	Soft^c	Hard	Soft^c	Hard			Range (°C)
120	110	250	350	100	240–360	50–75	20–3			−200–200
124	117	250–280	370–480	90–100	200–350	50	2–12	—	154	−250–200
124	120	250	360–400	110–130	240–400	35	5–10	110–140	132	−250–200
121	110	310	410–550	120–140	280–440	35–50	5–10	105–160	75	−250–200
120	105	310	400–480	120	240–400	35–45	5–10			−250–200
120	105	350	390–550	140	230–480	50–60	6–35	75–150	30–80	−250–20
120	105	330–370	380–570	120–150	230–480	50–60	8–35	75–150	30–80	−250–200
120	105	340–370	390–600	140–160	250–540	50–60	3–35	85–140	40–60	−250–200
113	110	360–370	390–600	140	280–550	50–55	3–35	95–145	40–70	−200–200
104	100	380	430–520	160	320–450	40	3–25	105–150	47–66	−200–180
112	100	340–380	480–520	120	360–460	50	20–10	100		−200–180
112	100	330–390	460–560	110	390–460	50–60	20–10	120–185	80–150	−200–180
109	100	340–390	400–510	138	230–400	40–45	10–35	147	21–50	−200–150
110	—	350–400	—	150–170	—	20–25	—			−200–150
98		460–490		195–240		20–15				−180–100

Table 2.4.12—*continued*

TABLE 2.4.12 Basic properties of copper–zinc alloys (brasses)—*continued*

Group	Usual name	BS Ref. (ISO)	Basic chemical composition	Specific gravity	Electrical Conductivity[a, b] (% IACS)	Thermal conductivity (W/m per K)	Thermal expansion coefficient (10^{-6} K^{-1})
Alloys: Wrought form—*continued*	High Tensile Brass (for soldering)	CZ115	57.5 Cu 40 Zn 0.8 Sn 1 Pb 0.8 Fe	8.5	26	117	19
	High Tensile Brass	CZ116	66 Cu 37.5 Zn 0.8 Fe 4.5 Al 1.1 Mn	8.5	26	117	19
	Leaded Brass (64 Cu 1 Lead)	CZ118	64 Cu 35 Zn 1 Pb	8.5	26	117	19
	Leaded Brass Clock Brass	CZ119	62 Cu 36 Zn 2 Pb	8.5	26	117	19
	Leaded Brass (Clock Brass)	CZ120 (Cu Zu 39 Pb 2)	59 Cu 39 Zn 2 Pb	8.4	27	117	20
	Leaded Brass Free-cutting brass 'Architectural bronze'	CZ121	57 Cu 39 or 40 Zn, 3 Pb or 4 Pb	8.5	28	117	21
	Leaded Brass Forging Brass	CZ122	58 Cu 40 Zn 2 Pb	8.4	27	117	20
	60/40 Brass (leaded Muntz Metal)	CZ123 (Cu Zn 40 Pb)	60 Cu 39.5 Zn 0.5 Pb	8.4	27	119	20
	Leaded Brass Free-cutting Brass	CZ124	62 Cu 35 Zn 3 Pb	8.5	26	117	19
	Leaded Brass	CZ128	60 Cu 38 Zn 2 Pb	8.5	26	117	20
	Leaded Brass	CZ129	60 Cu 39 Zn 1 Pb	8.4	27	117	20
	Leaded Brass (for sections)	CZ130	56.5 Cu 40 Zn 3 Pb 0.5 Al	8.5	28	114	20
	Leaded Brass (for sections)	CZ131	62 Cu 36 Zn 2 Pb	8.5	27	117	20
	Commercial 'bronze'	—	89 Cu 9 Zn 2 Pb (for sections)	8.8	42	180	18
	Special 70/30 Arsenical Brass	CZ126	70 Cu 30 Zn 0.05 As	8.5	26		20

Table 2.4.12—*continued*

TABLE 2.4.12 Basic properties of copper–zinc alloys (brasses)—*continued*

Modulus of elasticity (GN/m^2)		Tensile strength (MN/m^2)		0.2% proof stress (MN/m^2)		Elongation (%)		Fatigue^d (MN/m^2)	Impact^e (J)	Approx. service temperature
Soft^c	*Hard*	*Soft^c*	*Hard*	*Soft^c*	*Hard*	*Soft^c*	*Hard*			Range (°C)
97		460–490		195–240		20–15				−180–100
100		540–600		290–320		15–12				−180–180
105		330	420–530	120	280–480	45	25–10		36	
102	102	340	420–500	150	300–460	45	28–10	105		−200–180
98	98	400	450–560	220	300–450	30–35	20–15	155(c)	21	−200–100
97	97	420–440	470–650	200	380–500	25–32	15–8	135–160		−200–180
98	98	400	420–450	220	300–450	30–35	15–25		21	−200–100
100	100	360–380	420–440	160	300	35	20		21	−200–120
101	101	360–400	440–520	150	300–450	40	25–12	100–140	20	−200–120
100		350–380				22–25				−200–120
100		350–380				25–28				−200–120
97.5		350				20				−200–120
102		330–350				22–28				
120		270	330–400	70	240–350	45–50	12–25	140–160		
120	110	330–370	380–550	120–150	230–280	50–60	8–35		30–80	

Table 2.4.12—*continued*

TABLE 2.4.12 Basic properties of copper–zinc alloys (brasses)—*continued*

Group	Usual name	BS Ref. (ISO)	Basic chemical composition	Specific gravity	Electrical Conductivity[a, b] (% IACS)	Thermal conductivity (W/m per K)	Thermal expansion coefficient (10^{-6} K^{-1})
Alloys: Wrought form—*continued*	Aluminium–Nickel–Silicon Brass	CZ127	83 Cu 13 Zn 1.2 Ni 1 Si 1 Al	8.7	~ 36		20
	Dezincification-resistant brass	CZ132	62 Cu 36 Zn 22 Pb 0.12 As	8.4	~ 18		20
	Naval Brass (replacing CZ113)	CZ133	60.5 Cu 39 Zn 75 Sn	8.5	26	117	19
	Naval Brass (high lead)	CZ134	60.5 Cu 37 Zn 0.75 Sn1.75 Pb	8.5	26	117	19
	High tensile Silicon Brass	CZ135 (Cu Zn 37 Mn 3 Al Si)	58.5 Cu 37 Zn 1.5 Al 2.5 Mn 0.8 Si	8.5	26	117	19
	Manganese Brass	CZ136	57.5 Cu 38.5 Zn 3 Pb 1 Mn	8.5	26	117	19
Brass Castings:	Brass for sand castings	SCB1	70 Cu 25 Zn 3 Pb Sn 2 Ni Fe	8.5	18	81	19
	Brass for sand castings	SCB2 (Cu Zn 33 Pb 2)	65 Cu 33 Zn 2 Pb	8.4	20	90	20
	Naval Brass for sand castings	SCB4	62 Cu 36 Zn 1–1.5 Sn	8.4	18	81	21
	Brass for brazable castings	SCB6	85 Cu 15 Zn 0.15 As	8.6	25	111	19
	Brass for die castings	DCB1	60 Cu 40 Zn 0.5 Al	8.3	18	81	21
	Brass for die castings	DCB3 (Cu 40 Zn Pb)	60 Cu 40 Zn 2 Pb 0.5 Al Ni Fe Mn	8.3	18	81	21
	Brass for pressure die castings	PCB1	58 Cu 40 Zn 2 Pb Al Fe Sn	8.3	18	81	21
	High tensile Brass	HTB1 (Cu 35 Zn Al Fe Mn)	60 Cu 35 Zn 1 Fe 2 Al 2 Mn	8.3	22	87	21

Table 2.4.12—*continued*

TABLE 2.4.12 Basic properties of copper–zinc alloys (brasses)—continued

Modulus of elasticity (GN/m²)		Tensile strength (MN/m²)		0.2% proof stress (MN/m²)		Elongation (%)		Fatigue[d] (MN/m²)	Impact[e] (J)	Approx. service temperature Range (°C)
Soft[c]	Hard	Soft[c]	Hard	Soft[c]	Hard	Soft[c]	Hard			
120										
112		280				30				
~ 100		350–400	—	150–170	—	20–25	—			−180–100
~ 100		350–400	—	150–170	—	15–20				−180–100
~ 100		550	—	270	—	12				−180–180
~ 100		350–380	—		—	20–25	—			−180–180
Sand	Chill	Sand	Chill	Sand	Chill	Sand	Chill			
~	120	170–200	—	80–110	—	18–38	—			−100–120
~	120	190–220	—	70–110	—	11–28	—			−100–120
~	120	250–310	—	70–110	—	18–35	—			−100–120
~	120	170–190	—	80–110	—	18–35	—			−100–120
~	120	—	280–370	—	90–120	—	20–45			−100–120
~	120	—	300–400	—	90–120	—	13–40	110		−100–120
~	120	—	280–370	—	90–120	—	25–45			−100–120
~	120	470–500	500–600	170–270	210–280	18–20	20–23	140	26	−100–150

Table 2.4.12—continued

TABLE 2.4.13 Comparative characteristics of copper–zinc alloys (brasses)

BS ref.	Type	Available forms[a]	Thermal treatment[b]	Hardness	Formability[c]		Maximum reduction between anneals	Machinability[d]
					Hot	Cold		
CZ125	Gilding Metal	S	A	65–120	G	E	90%	25
CZ101	Commercial 'bronze'	SW	A	60–125	G	E	90%	25
CZ102	Red Brass	SW	A	80–135	F	E	85%	25
CZ103	Low Brass	SBW	A	75–130	F	E	85%	30
CZ104	Leaded 80/20 Brass	B	A	75–100	F	E	85%	30
CZ106	Cartridge or Deep Drawing Brass	SBPW	A	65–170	F	E	90%	30
CZ105	70/30 Arsenical Brass	SP	A	65–170	F	E	90%	30
CZ107	2/1 Brass	SW	A	65–175	F	E	85%	30
CZ108	Common or Basic Brass	STW	A	65–170	G	G	65%	30
CZ109	Muntz Metal	BFP	A	85–150	E	L	40%	45
CZ110	Aluminium Brass	STP	A	70–175	F	G	80%	30
CZ111	Admiralty Brass	T	A	65–175	F	G	80%	30
CZ112	Naval Brass	BFP	A	90–150	E	F	40%	40
CZ113	Special Naval Brass	B	A	90–155	E	L	20%	30
CZ114	High tensile brass	BF	A	120–140	E	L	20%	80
CZ115	High tensile brass (for soldering)	BF	A	120–140	E	L	20%	70
CZ116	High tensile brass	BF	A	165–175	E	L	20%	65
CZ118	Leaded Brass	S	A	65–135	L	G	65%	70

See footnotes for Table 2.4.7.

Table 2.4.13

TABLE 2.4.13 Comparative characteristics of copper–zinc alloys (brasses)—*continued*

Joining[c]			Limitations	Typical uses
Soldering	*Brazing*	*Welding*		
E	E	G		Ammunition Caps
E	E	G		Golden colour favours architectural and decorative uses, jewellery and enamelled items. Brazed assemblies such as heat exchangers. Pressed parts for domestic and automobiles.
E	E	G		
E	E	G		
E	E	G		
E	E	G	Stress corrosion and dezincification risk. Expensive.	Cartridge cases and many deep drawn items, freshwater heat exchangers and header tanks. Hardware. Cold headed fasteners, etc.
E	E	G	(As reduces dezincification corrosion)	
E	E	G		Good general working alloys for pressings (not deep drawn), spinnings, cold headed items fasteners and hardware, electrical components, wire brushes.
E	E	G		
E	G	G		Hot formed parts, extrusions and forgings. Condenser tube plates.
E	G	G		Heat exchangers, oil tanker heating coils marine uses requiring corrosion–erosion conditions.
E	G	G		Heat exchangers, especially oil refineries and power stations.
E	G	G		Marine heat exchangers, underwater items, forgings.
E	G	G	BS withdrawn—replaced by CZ133 (1986)	Underwater items, especially marine.
G	G	P		Architectural sections and fixings. Forgings. Pump components. Pressure items.
E	G	F		
F	G	P		
E	G	P	Cold working may cause cracking.	Headed or blanked items, keys, printers matrices, instrument and clock parts.

Table 2.4.13—*continued*

TABLE 2.4.13 Comparative characteristics of copper–zinc alloys (brasses)—*continued*

BS ref.	Type	Available forms[a]	Thermal treatment[b]	Hardness	Formability[c]		Maximum reduction between anneals	Machinability[d]
					Hot	Cold		
CZ119	Clock Brass	STB	A	70–130	F	F	50%	75
CZ120	Clock Brass	S	A	90–145	E	L	20%	85
CZ121	Free-cutting Brass 'Architectural Bronze'	BF	A	120–160	G	L	20%	100
CZ122	Forging Brass	BF	A	90–145	E	L	20%	85
CZ123	60/40 Brass	SP	A	85–130	G	L	35%	60
CZ124	Free-cutting leaded Brass	B	A	75–165	F	L	35%	100
CZ128	Leaded Brass	BF	A	80–140	F	F	50%	65
CZ129	Leaded Brass	BF	A	85–160	G	L	30%	65
CZ130	Leaded Brass sections	BP	A	125–160	E	L	20%	90
CZ131	Leaded Brass sections	BP	A	75–160	E	L	20%	90
—	Commercial 'bronze' sections	B	A		L	G	75%	80
CZ126	Special 70/30 Arsenical Brass	T	A	65–170	L	E	90%	30
CZ127	Al–Ni–Si Brass	T	A	70–130	L	E	85%	30
CZ132	Dezincification resistant brass	SFBT	A	70–130	G	F	50%	75
CZ133	Naval Brass (replacing CZ113)	B	A	90–115	E	L	20%	30
CZ134	Naval Brass (high lead)	B	A	90–115	E	L	20%	
CZ135	High tensile silicon brass	BF	A					
CZ136	Manganese Brass	BF	A					
CZ137	Leaded 60/40 Brass	B	A	85–130	G	L	30%	60

Table 2.4.13—*continued*

TABLE 2.4.13 Comparative characteristics of copper–zinc alloys (brasses)—*continued*

Soldering	Brazing	Welding	Limitations	Typical uses
	Joining[c]			
E	G	P	Cold working may cause cracking.	Machined or blanked parts. Spark plug parts. Engraved plates. Door and window fixings.
E	G	P	Cold working may cause cracking.	Clock and instrument fittings. Decorative items. Dials.
E	G	P		Machined parts especially repetition work, bushes, fittings and trim, terminals and fasteners.
E	G	P		Door and window furniture. Forgings. House fittings.
E	G	F		Condenser tube plates, marine fittings and hot forged parts. Decorative items.
E	G	P	Cold working may cause cracking.	Machined items on automatic lathes, etc.
E	G	P	Cold working may cause cracking.	Forged and machined parts. Locks and keys.
E	G	P	Cold working may cause cracking.	
E	G	P	Cold working may cause cracking.	
E	G	P	Cold working may cause cracking.	
E	G	P	Cold working may cause cracking.	
E	E	G		Standard for condensers and other corrosive situations (As inhibits dezincification).
G	G	G		Special purposes.
E	E	G	Accurate control of composition essential.	Wrought items to resist dezincification corrosion. Machinable.
E	G	G	Not inhibited.	Marine items, especially for underwater use.
E	G	F		
E	G	F		Marine items—similar to CZ123.

Table 2.4.13—*continued*

TABLE 2.4.13 Comparative characteristics of copper–zinc alloys (brasses)—_continued_

BS ref.	Type	Available forms[a]	Thermal treatment[b]	Hardness	Formability[c]		Maximum reduction between anneals	Machinability[d]
					Hot	Cold		
Brass Castings SCB1	Brass for sand castings	Sd	—	45–60	—	—	—	80
SCB3	Brass for sand castings	Sd	—	45–65	—	—	—	80
SCB4	Naval Brass for sand castings	Sd	—	50–75	—	—	—	50
SCB6	Brass for sand (brazable) castings	Sd	—	45–60	—	—	—	30
DCB1	Brass for die castings	Ch	—	60–70	—	—	—	50
DCB3	Brass for die castings	Ch	—	60–70	—	—	—	80
PCB1	Brass for pressure die castings	Ch	—	60–70	—	—	—	50
HTB1	High tensile brass	SdCh	—	100–150	—	—	—	30

Table 2.4.13—_continued_

TABLE 2.4.13 Comparative characteristics of copper–zinc alloys (brasses)—*continued*

Joining[c]			Limitations	Typical uses
Soldering	Brazing	Welding		
G	P	P		Pressure taps, valves and fittings. Hardware and ornamental items. Centrifugal castings.
G	P	P		General purpose water and gas fittings, electrical items. Centrifugal castings.
G	P	F		Marine fittings, heat exchanges, fasteners, mechanical items.
E	G	F		Marine fittings and other items to be brazed (e.g. end plates). Jewellery, etc.
G	P	P		Accurate castings for general purposes, gas and plumbing fittings. Automobile parts.
G	P	P		Accurate castings for general purposes, food and pickling plant, especially when machined.
G	P	P		Accurate castings for general purposes, especially when machined.
G	F	P	Normal temperature purposes only. Expensive.	Stressed components—marine propellers, hydraulic items, pumps. Axle boxes.

Table 2.4.13—*continued*

2.4.4   Copper–tin alloys (bronzes) and copper–tin–zinc alloys (gunmetals)

The influence of alloying elements on the properties of bronzes (copper–tin alloys) is shown in:

Table 2.4.14   *Effect of alloy additions on properties of copper–tin alloys*

(The effect for gunmetals, copper–tin–zinc alloys which are also bronzes was given in Table 2.4.10.)

The common names of bronzes, and their more important properties are listed in:

Table 2.4.15   *Basic properties of copper–tin alloys (bronzes)*

and their general characteristics and applications in:

Table 2.4.16   *Comparative characteristics of copper–tin alloys (bronzes)*

The corresponding data for gunmetals are listed in:

Table 2.4.17   *Basic properties of copper–tin–zinc alloys (gunmetals)*, and

Table 2.4.18   *Comparative characteristics of copper–tin–zinc alloys (gunmetals)*

TABLE 2.4.14 Effect of alloy additions on properties of copper–tin alloys (see footnotes)

Copper–tin alloys (Tin Bronze)		Tensile	Proof	Fatigue	Impact	Modulus of elasticity	Limit of proportionality	Ductility	Oxide scaling	Creep	Softening temperature	Specific heat	Expansion	Conductivity	Electrical conductivity	Magnetic properties	Low temperature properties	Machinability	Surface finish	Wear resistance	Cold formability	Hot formability	General	Stress corrosion	Cost	Cast (C) or wrought (W)
		1	2	3	4	5	6	7	8	9	10	11	12	13	14	15	16	17	18	19	20	21	22	23	24	25	26
Sn (0–5%)	A	↑↑	↑↑	↑↑		↑↑	↑	↑	↓		↑↑↑	↑	0	0	↓↓↓	↓↓↓			0			0	↓				W
Sn (5–10%)	B	↑↑	↑↑	0		↑↑	↓	↑	↑			0	0	0	↓↓	↓			0			↓↓	↓↓↓				W + C
Sn (10–30%)	C			↓↓↓			↓↓↓		↓↓↓						↓	↓					↑↑	↓↓↓	↓↓↓				C
P (0.02–0.4%)	D	↑	↑				↑								↓↓↓	↓↓↓											
Pb (0–5%)	E	↓	↓		↓	↓			↓										↑↑		↑↑	↓	↓				
Ni (0–10%)	F	↑	↑			↑			↑		↑	↑									↑			↑			

Key: ↑ = Increase.   0 = No effect.   ↓ = Decrease.   Blank = No data.

A. Plain tin bronzes of up to 5% tin are often referred to as 'low tin' (up to 3½%) or ordinary quality bronzes and are used where extensive fabrication is required.

B. Tin bronzes of between 5 and 10% tin become progressively more difficult to fabricate by deformation due to the potency of tin as a hardening agent. Alloys with more than 8% tin cannot be worked cold due to their lack of ductility, and also due to the presence of a brittle constituent of the alloy structure. However, prolonged heating at 700°C causes this constituent to be dissolved enabling the alloys to be formed into wire or sheet.

C. Alloys containing more than 10% tin are invariably used in a cast form, and are used in specific applications.
  i. 12–18% tin—hard wearing for use under large compressive loads.
  ii. 20–30% tin (Bell metal)—low damping capacity has good sonority for use in bells.
  iii. 30–40% tin (speculum)—high reflectance (better than chromium) for use in mirrors, optical instruments and decorative purposes where the high hardness imparts considerable scratch resistance.

D. Phosphorus is a usual constituent of the general engineering bronzes A and B. It is added primarily to act as a deoxidiser to prevent the formation of particles of tin oxide during casting, but the improvement of elastic modulus makes phosphor bronze a suitable material for springs, and phosphorus also improves stability.

E. Lead additions to the engineering bronzes improve machinability, for which purpose additions of 0.5 to 1.0 are usually sufficient. Further additions improve the anti-friction characteristics, but the strength, elongation and shock resistance are reduced. Bronzes containing up to 10% lead are called 'lead bronzes' and are used for bearing purposes. Bronzes containing more than 20% lead are termed 'plastic bronzes' and are suitable for uses as non-lubricated bearings or where a degree of self alignment is required. The addition of up to 1% nickel to bronzes containing lead assists the uniform distribution of lead globules.

F. Nickel additions of up to 5% to bronze have the effect of improving the strength and improving the castability. Bronzes containing 5 to 10% nickel are hardenable by heat treatment but such alloys usually contain zinc and so are more accurately described as nickel-gunmetals than nickel-bronzes. These nickel-gunmetals are hardenable by precipitation heat treatment, and are characterised by their excellent pressure tightness and high tensile strength.

These alloys have an improved thermal stability due to the presence of nickel. When more than 10% nickel is present the corrosion resistance is improved and increased strength and hardness values can be achieved by the addition of silicon from 1–5%. These high nickel alloys are used for valve faces and similar fittings for use in corrosive waters or steam, and are also used in bearing applications as rollers, sleeves and bushes.

Table 2.4.14

TABLE 2.4.15 Basic properties of copper–tin alloys (bronzes)

Group	Usual name	BS Ref. (ISO)	Basic chemical composition	Specific gravity	Electrical conductivity[a, b] (% IACS)	Thermal conductivity (W/m per K)	Thermal expansion coefficient (10^{-6} K^{-1})
Wrought	Phosphor Bronze (4%)	PB101 (Cu Sn 4)	96 Cu 4 Sn + P	8.85	15–25	62	17
	Phosphor Bronze (5%)	PB102 (Cu Sn 5)	95 Cu 5 Sn + P	8.85	12–18	64	17
	Phosphor Bronze (7%)	PB103 (Cu Sn 6)	93.5 Cu 6.5 Sn + P	8.8	10–15	67	17
	Phosphor Bronze (9%)	PB104 (Cu Sn 8)	91 Cu 9 Sn + P	8.8	10–14	46	17
	Coinage Bronze	—	96 Cu 2.5 Sn 1.5 Zn	8.8	15–25		17
Cast		BS1400					
	Tin Bronze	CT1	90 Cu 10 Sn P	8.8	11	50	18
	Nickel Tin Bronze	CT2	86 Cu 12 Sn 2 Ni P	8.8	9	45	19
	Phosphor Bronze	PB1	89 Cu 10 Sn 1 P	8.8	9	47	18
	Phosphor Bronze	PB2	88 Cu 12 Sn P	8.8	9	45	19
	Phosphor Bronze	PB4	89 Cu 10 Sn 1 P Pb	8.8	10	47	18
	Leaded Bronze	LB1	76 Cu 9 Sn 15 Pb	9.1	11	47	19
	Leaded Bronze	LB2	80 Cu 10 Sn 10 Pb	9.0	10	47	19
	Leaded Bronze	LB4	86 Cu 5 Sn 9 Pb	9.0	17	71	19
	Leaded Bronze	LB5	75 Cu 5 Sn 20 Pb	9.2	14	71	19
	Leaded Phosphor Bronze	LPB1	88 Cu 7 Sn 4 Pb P Zn Ni	8.8	11	47	18

For footnotes see Table 2.4.6.

Table 2.4.15

TABLE 2.4.15 Basic properties of copper–tin alloys (bronzes)—*continued*

Modulus of elasticity (GN/m²)		Mechanical properties								Approx. service temperature
		Tensile strength (MN/m²)		0.2% proof stress (MN/m²)		Elongation (%)		Fatigue[d] (MN/m²)	Impact[e] (J)	
Soft[c]	Hard	Soft[c]	Hard	Soft[c]	Hard	Soft[c]	Hard			Range (°C)
122	112	295–330	340–620	130	250–580	40–50	4–35	90–240	60	−200–150
124	108	340–370	450–750	130	250–600	45–55	4–40	85–265	140	−200–150
120	98	340–390	440–695	150	200–600	50–60	3–35	195–260		−200–160
114	97	400–430	530–900	150–170	320–700	55–65	4–30	120–230		−200–160
			450–550		350–450		8–10			
Sand	Chill	Sand	Chill	Sand	Chill	Sand	Chill			
122	—	240	—	130	—		—			
120		280	300	160	180	12	8–10			
120		220	310–330	130	170	3	2–6	110		
120		220	270–310	130	170	5	3–5			
120		190	270–330	100	140–160	3	2–7			
120		170	200–230	80	130	4	3–9			
120		190	220–280	80	80–160	5	3–6	80–150	11	−200–150
100		160	200–230	60	80–130	7	5–9	90	11	
120		160	170–190	60	80–100	5	5–8			
120		190	220–270	80	130	3	2–5			

Table 2.4.15—*continued*

TABLE 2.4.16 Comparative characteristics of copper–tin alloys (bronzes)

BS ref.	Type	Available forms[a]	Thermal treatment[b]	Hardness	Formability[c] Hot	Formability[c] Cold	Maximum reduction between anneals	Machinability[d]
PB101	4% Phosphor Bronze	PSWB	A	75–200	L	E	80%	20
PB102	5% Phosphor Bronze	PSWB	A	80–210	L	E	75%	20
PB103	7% Phosphor Bronze	PSWB	A	85–245	L	G	70%	20
PB104	9% Phosphor Bronze	SW	A	90–250	L	G	60%	20
—	Coinage Bronze	S	A	65–180	P	E	75%	20
Castings CT1	Tin Bronze	Sd. Ch.	A	70–130	—	—	—	45
CT2	Nickel Tin Bronze	Sd. Ch.	A	75–150	—	—	—	25
PB1	Phosphor Bronze	Sd. Ch.	A	70–145	—	—	—	45–50
PB2	Phosphor Bronze	Sd. Ch.	A	75–140	—	—	—	45–50
PB4	Phosphor Bronze	Sd. Ch.	A	70–150	—	—	—	75
LB1	Leaded Bronze	Sd. Ch.	A	50–90	—	—	—	65–75
LB2	Leaded Bronze	Sd. Ch.	A	65–90	—	—	—	65–75
LB4	Leaded Bronze	Sd. Ch.	A	55–80	—	—	—	65–75
LB5	Leaded Bronze	Sd. Ch.	A	45–70	—	—	—	80
LPB1	Leaded Phosphor Bronze	Sd. Ch.	A	60–110	—	—	—	65

For footnotes see Table 2.4.7.

Table 2.4.16

TABLE 2.4.16 Comparative characteristics of copper–tin alloys (bronzes)—*continued*

Joining[c]			Limitations	Typical uses
Soldering	Brazing	Welding		
E	E	F		Tube plates, pressure plant, bellows and diaphragms. Masonry fixings and bolts. Clips and springs. Electrical contacts.
E	E	F		Tubes and tube plates, bellows, etc. Masonry fixings and fasteners. Springs including electrical contacts. Textile and paper-making plant.
E	E	F		Tubes for acid water. Chemical plant and textile machine items. Bushings, gears, springs and clips. Diaphragms and Bourdon tubing.
G	G	F		Wire in hard condition used in mesh for Foudrinier paper machines.
G	E	G		Coins, medals, tokens—die stamped/coined.
E	F	F		General purpose sand castings. Acid resistant castings. Boiler feed water items. Cast components to resist impingement corrosion.
E	E	F		
E	E	F	Less tolerant to impurities in castings than PB4.	High grade gears, bearings and bushes when lubrication is adequate. Pumps and gears in corrosive conditions.
E	E	F		Gears and worms subject to shock loads. Heavy duty bearings.
E	E	P		Widely used for less onerous duties than PB1.
G	F	P	Not to be used for fittings in contact with potable waters or foodstuffs due to P6 pickup.	Bearings for light–moderate duties with hard to soft steel shafts in less than perfect alignment or lubrication conditions. Useful when abrasive pesticides present. Low impact loads or pounding require adequate backing. Can be continuous cast.
E	F	P		
E	F	P		
G	F	P		Steel backed bearings for car engines, etc. Some self lubrication.
E	E	F		Lighter duty bearings and bushes than PB1, PB2 and PB4. Poor lubrication conditions assisted by P6 content.

Table 2.4.16—*continued*

TABLE 2.4.17 Basic properties of copper–tin–zinc alloys (gunmetals)

Usual name	BS Ref. (ISO)	Basic chemical composition	Specific gravity	Electrical conductivity[a, b] (% IACS)	Thermal conductivity (W/m per K)	Thermal expansion coefficient (10^{-6} K^{-1})
Bell Metal		75–85 Cu 25–15 Sn	8.7 to 8.8			
Statuary Bronze	—	76–80 Cu 10–12 Sn 5–10 Pb 1–5 Zn				
Admiralty (88/10/2) Gunmetal	G1	88 Cu 10 Sn 2 Zn	8.8	10–11	47	18
Nickel Gunmetal	G3	86 Cu 7 Sn 2 Zn 5 Ni	8.8	10–11	47	18
Nickel Gunmetal (Fully heat treated)	G3–TF	86 Cu 7 Sn 2 Zn 5 Ni	8.8	10–11	47	18
Leaded Gunmetal	LG1	84 Cu 3 Sn 8 Zn 5 Pb	8.8	12–16	81	18
Leaded (85–5–5–5) Gunmetal	LG2	85 Cu 5 Sn 5 Zn 5 Pb	8.85	10–15	71	18
Leaded Gunmetal	LG4	87 Cu 7 Sn 3 Zn 3 Pb Ni	8.8	10–13	61	18

For footnotes see Table 2.4.6.

Table 2.4.17

TABLE 2.4.17 Basic properties of copper–tin–zinc alloys (gunmetals)—*continued*

Modulus of elasticity (GN/m^2)		Mechanical properties						Fatigue^d (MN/m^2)	Impact^e (J)	Approx. service temperature
		Tensile strength (MN/m^2)		0.1% proof stress (MN/m^2)		Elongation (%)				
Sand	Chill	Sand	Chill	Sand	Chill	Sand	Chill			Range (°C)
		—	—	—	—	—	—			—
		—	—	—	—	—	—			—
105		270–340	230–310	130–170	130–170	13–20	3–8	90–150		−50–200
105		280–340	—	140–160	—	16–25	—			−50–200
110		430–490	—	280–320	—	3–5	—	110		−50–180
105		180–200	180–200	80–90	80–90	11–13	2–8			−50–180
110		200–220	260–270	100–110	110–120	13–14	6–15	70–110	25–26	−50–180
105		250–320	250–340	130–140	130–160	16–20	5–15	80	25–26	−50–180

Table 2.4.17—*continued*

TABLE 2.4.18 Comparative characteristics of copper–tin–zinc alloys (gunmetals)

BS ref.	Type Castings	Available forms[a]	Thermal treatment[b]	Hardness	Formability[c]		Maximum reduction between anneals	Machinability[d]
					Hot	Cold		
G1	Admiralty Gunmetal	Sd. Ch.	—	70–130	—	—	—	45
G3	Nickel Gunmetal	Sd. Ch.	—	70–130	—	—	—	45–50
G3–TF	Nickel Gunmetal (heat treated)	Sd. Ch.	QH	160–180	—	—	—	50
LG1	Leaded Gunmetal	Sd. Ch.	—	55–80	—	—	—	45
LG2	'85 three fives' Gunmetal	Sd. Ch.	—	65–95	—	—	—	65–75
LG4	Leaded Gunmetal	Sd. Ch.	—	70–95	—	—	—	65–75

For footnotes see Table 2.4.7.

Table 2.4.18

TABLE 2.4.18 Comparative characteristics of copper–tin–zinc alloys (gunmetals)—*continued*

Joining[c]			Limitations	Typical uses
Soldering	Brazing	Welding		
E	E	G		Good resistance to impingement attack by natural and sea-waters. Useful when fatigue conditions to be met. Feed water pumps, etc. Components for high velocity waters. G3 resists attack by S-contaminated atmospheres.
E	E	G		
E	E	G		
E	E	G	Unsuitable for valves, etc., in contact with potable waters, etc., due to risk of Pb pick-up.	
E	E	G		Lightly loaded, non-critical bearings. Very light duty gears with low loads. Items to be pressure tight in thin sections.
E	E	G		

Table 2.4.18—*continued*

2.4.5   Other copper alloys (aluminium bronzes, copper–nickel alloys and nickel silvers)

2.4.5.1   Aluminium bronzes

The effect of alloy additions on the properties of copper aluminium alloy is given in:
Table 2.4.19   *Effect of alloy additions on properties of copper–aluminium alloys*
    The common names of the aluminium bronzes, their BS and ISO specification numbers and their more important properties are listed in:
Table 2.4.20   *Basic properties of copper–aluminium alloys (aluminium bronzes)*
and their general characteristics and applications in:
Table 2.4 21   *Comparative characteristics of copper–aluminium alloys (aluminium bronzes)*

2.4.5.2   Copper–nickel alloys

The effect of alloy additions on copper–nickel alloys is given in:
Table 2.4.22   *Effect of alloy additions on properties of copper–nickel alloys*
    The common names of the cupronickel alloys, their BS (ISO) specification numbers and their more important properties are listed in:
Table 2.4.23   *Basic properties of copper–nickel alloys (cupro-nickel)*
and their general characteristics and applications in:
Table 2.4.24   *Comparative characteristics of copper–nickel alloys (cupro-nickel)*

2.4.5.3   Nickel–silver alloys

The common names of the nickel silvers, their BS (and ISO) specification numbers and their more important properties are listed in:
Table 2.4.25   *Basic properties of copper–zinc–nickel alloys (nickel silvers)*
and their general characteristics and applications in:
Table 2.4.26   *General characteristics of copper–zinc–nickel alloys (nickel silver)*

TABLE 2.4.19 Effect of alloy additions on properties of copper–aluminium alloys (see footnotes)

Copper–aluminium alloys (Aluminium Bronze)		Strength				Hardness	Elasticity		Ductility	Heat resistance			Thermal		Conductivity	Electrical conductivity	Magnetic properties	Low temperature properties	Machinability	Surface finish	Wear resistance	Cold formability	Hot formability	Corrosion resistance		Cost	Cast (C) or wrought (W)
		Tensile	Proof	Fatigue	Impact		Modulus of elasticity	Limit of proportionality		Oxide scaling	Creep	Softening temperature	Specific heat	Expansion										General	Stress corrosion		
		1	2	3	4	5	6	7	8	9	10	11	12	13	14	15	16	17	18	19	20	21	22	23	24	25	26
Al (0–7%)	A	↑↑↑	↑↑↑	↑↑↑		↑↑	0		↑			↑↑	↑	0	↓↓↓	↓↓↓			0		↑	↓	↓				W
Al (7–10%)	B	↑↑	↑↑	↑↑		↑↑	↓					0	0	0	↓	↓			0		↑↑						W + C
Fe (0–5%)	C	↑	↑	↑	0	0	↓		↓			↑	0	0	0	0			↑			0	0	↑			W + C
As (0–0.4%)	D																							↑			
Ni (0–6%)	E	↑	↑	↑							↑													↑			
Mn (0–3%)	F	↑	↑	↑					↑														↑				
Pb (0–1.5%)	G																		↑				↓				
Si (0–2.2%)	H	↑	↑		↓				↓										↑				↑	↑			

Key: ↑ = Increase.   0 = No effect.   ↓ = Decrease.   Blank = No data.   1 arrow indicates slight effect. 2 arrows indicate moderate effect. 3 arrows indicate large effect.

A & B. Copper–aluminium alloys are used in two forms; single phase alloys containing up to about 7% aluminium and duplex alloys (two phases) containing between about 7 and 11% aluminium. In many respects the single phase alloys are similar to the single phase brasses and tin bronzes, although the amounts of zinc, tin and aluminium necessary to produce particular properties are different in each case. Aluminium increases corrosion resistance especially in marine atmospheres. The duplex copper–aluminium alloys are not similar to duplex copper–zinc or copper–tin since their properties can be modified by heat treatment in a manner similar to that used for hardening and tempering of steel. An 11% aluminium alloy, which is the most common composition used for heat treatment, shows an increase of about 25% in tensile strength accompanied by an increase of about 70% in elongation as a result of a heat treatment involving quenching from 900°C followed by heating at 550°C for 2 h. This type of treatment gives the improvement in properties by the formation of a particular distribution of the second phase in the matrix, but heat treatment is not commonly used for copper–aluminium alloys except in particular situations where the application demands this enhancement of properties. The duplex alloys are most commonly used in the cast or hot worked conditions where the strength and ductility are adequate, but where the other properties such as corrosion resistance or wear resistance are of greater importance.

C.   The addition of iron is generally confined to the duplex alloys containing more than 7% aluminium, where its main function is to inhibit grain growth. The need for this is that the duplex alloys are generally hot worked and during multiple working operations involving reheating there is a tendency for the grains to enlarge, resulting in reduced strength and ductility. It does not prevent eutectoid decomposition of beta phase (see E).

D.   Arsenic is added to the single phase alloys, containing less than 7% aluminium where it increases the corrosion resistance with a slight increase in strength. The use of too much arsenic (more than about 0.4%) renders the alloy brittle.

E.   Nickel is normally added in conjunction with iron to the duplex alloys containing more than 7% aluminium, where the two elements improve the corrosion and oxidation resistance as well as increasing the strength, wear and fatigue resistance. It also avoids the deleterious decomposition of beta phase to the $\alpha + \gamma_2$ (δ) eutectoid. This eutectoid is brittle and reduces corrosion resistance especially in marine atmospheres.

F.   Manganese is added to duplex alloys to improve the hot workability and weldability by extending the lower temperature limit of the single phase structure which is present at hot working temperatures. This results in a better structure at room temperatures after cooling from the working temperatures, giving improved strength and ductility properties. Manganese reduces the beta phase eutectoid temperature and in so doing delays the decomposition to eutectoid (see E).

G.   Lead is sometimes added to improve machinability.

H.   Additions of silicon to single phase alloys produces an increase in strength with improvement of machinability, hot workability and corrosion resistance.

Table 2.4.19

TABLE 2.4.20 Basic properties of copper–aluminium alloys (aluminium bronzes)

Group	Usual name	BS Ref. (ISO)	Basic chemical composition	Specific gravity	Electrical conductivity[a, b] (% IACS)	Thermal conductivity (W/m per K)	Thermal expansion coefficient (10^{-6} K^{-1})
Copper Aluminium Wrought	Aluminium Bronze	CA102	93 Cu 7 Al	8.2	14–15	~80	17
	Aluminium Bronze	CA103	88 Cu 9 Al (4 Fe + Ni)	7.8	12–14	~60	16
	Aluminium Bronze	CA104 (Cu Al 10 Fe 5 Ni 5)	80 Cu 10 Al 5 Fe 5 Ni	7.6	7–9	~50	16
	(Alloy E)	CA105	82 Cu 9 Al 6 Ni 3 Fe 1 Mn	7.6	7–9	~50	16
	(Alloy D)	CA106	90 Cu 7Al 2.5 Fe	7.9	14–15	~70	16
	DGS. 1044 DGS. 8453 Bars only	—	92 Cu 6 Al 2 Si	8.0	14		16
Castings							
	Cast Aluminium Bronze	AB1 (Cu Al 10 Fe 3)	88 Cu 9.5 Al 2.5 Fe	7.6	13	61	17
	Cast Aluminium Bronze	AB2 (Cu Al 10 Fe 5 Ni 5)	80 Cu 10 Al 5 Fe 5 Ni	7.6	8	42	16
	Cast Aluminium Silicon Bronze	AB3	92 Cu 6 Al 2 Si	7.7	8	45	18
	CMA1	CMA1	73 Cu 8 Al 13 Mn 3 Fe 3 Ni	7.5	3–4	14	19
	CMA2	CMA2 (obsolete BS)	72 Cu 9 Al 13 Mn 3 Fe 3 Ni	7.5	3–4	14	19

For footnotes see Table 2.4.6.

Table 2.4.20

Other copper alloys (aluminium bronzes, copper–nickel alloys and nickel silvers)   1133

TABLE 2.4.20 Basic properties of copper–aluminium alloys (aluminium bronzes)—*continued*

| Modulus of elasticity (GN/m^2) | | Mechanical properties | | | | | | Fatigue[d] (MN/m^2) | Impact[e] (J) | Approx. service temperature |
| | | Tensile strength (MN/m^2) | | 0.2% proof stress (MN/m^2) | | Elongation (%) | | | | |
Soft[c]	Hard	Soft[c]	Hard	Soft[c]	Hard	Soft[c]	Hard			Range (°C)
125	114									−250–300
120	105	580–640	650–690	260–340		15–18	10		21–27	−250–300
125–135	115	700–750	760–796	360–450		12–18	10–12	250	14–28	−250–350
120	110	710–	750	280–300	330	20				−250–350
110	100	470–540	560–640	200–230	250	35–40	20–30		80–150	−250–300
		Sand	Chill	Sand	Chill	Sand	Chill			
		500–590	540–620	170–200	200–270	18–40	18–40	200	41	−150–250
		640–700	650–740	250–300	250–310	13–20	13–200	220	24	−150–400
		450–500		180–190		20–30			38	−200–300
		650–730	670–740	280–340	310–370	18–25	27–40	230	41	−150–400
		740–820	—	380–470	—	9–20	—			−150–400

TABLE 2.4.21 Comparative characteristics of copper–aluminium alloys (aluminium bronzes)

BS ref.	Type	Available forms[a]	Thermal treatment[b]	Hardness	Formability[c]		Maximum reduction between anneals	Machinability[d]
					Hot	Cold		
Wrought CA102	7% Aluminium Bronze	PTW	A	130–150	G	G	20%	25
CA103	9% Aluminium–Iron Bronze	BF	QH	150–190	G	L	15%	30
CA104	10% Aluminium Bronze + Fe + Ni	BF	QH	190–210	G	L	10%	20
CA105	'Alloy E'	SPBF	QH	190–220	G	L	10%	20
CA106	'Alloy D'	PB	A	150–165	G	F	25%	30
Cast AB1	10% Aluminium Bronze + Fe	S.D.C.	ST	90–160	—	—	—	30
AB2	10% Aluminium Bronze + Fe + Ni	S.D.C.	ST	140–190	—	—	—	30
AB3	Aluminium–Silicon Bronze	S		160–210	—	—	—	35
CMA1	Copper–Manganese–Aluminium Bronze	S		200–260	—	—	—	20

For footnotes see Table 2.4.7.

Table 2.4.21

TABLE 2.4.21 Comparative characteristics of copper–aluminium alloys (aluminium bronzes)—*continued*

Soldering	Brazing	Welding	Limitations	Typical uses
Joining[c]				
P	F	G		Condensers, heat exchangers for acids and salt solutions. Marine non-magnetic items. Chains, racks, coins and jewellery.
P	F	G	Cold working limited.	Chemical plant, pumps and valves. Cryogenic uses. Non-sparking tools. Heavy duty bearings, gears and slides. Dies. Fasteners.
P	F	G	Cold working limited.	As for CA103. Marine fittings and davits. Exhaust fans.
P	F	G	Cold working limited.	As for CA103 and CA104. Components for superheated steam.
P	F	G	Cold working limited.	As above. Welded assemblies.
P	P	G	Special techniques necessary for casting.	Sand and die-castings (Group B). High strength and corrosion resistant items for marine and chemical equipment. Valves.
P	P	F	Special techniques necessary for casting.	Sand and die-castings (Group B) as for AB1. Heavy gears, glass moulding dies, paper-making plant. Raised temperature items.
P	P	F	Special foundry technique necessary.	As for AB1. Bearings especially heavy plant.
P	P	G	Special foundry techniques.	Marine propellers, heavy and intricate items for corrosive conditions. Wear resisting components. Raised temperature items.

Table 2.4.21—*continued*

TABLE 2.4.22 Effect of alloy additions on properties of copper–nickel alloys (see footnotes)

Copper–nickel alloys (Cupro–Nickel)		Strength					Elasticity			Heat resistance			Thermal										Corrosion resistance			
		Tensile	Proof	Fatigue	Impact	Hardness	Modulus of elasticity	Limit of proportionality	Ductility	Oxide scaling	Creep	Softening temperature	Specific heat	Expansion	Conductivity	Electrical conductivity	Magnetic properties	Low temperature properties	Machinability	Surface finish	Wear resistance	Cold formability	Hot formability	General	Stress corrosion	Cost	Cast (C) or wrought (W)
		1	2	3	4	5	6	7	8	9	10	11	12	13	14	15	16	17	18	19	20	21	22	23	24	25	26
Ni (0–15)	A	↑	↑			↑	↑	↑	↓	↑		↑			↓	↓			0					↑			W
Ni (15–30)	B	↑	↑			↑	↑		0	↑					↓	↓			0					↑			W
Ni (40–45)	C	0	0				↑		0						↓												
Fe (2%)	D																							↑			
Mn (20%)	E	↑	↑			↑																					

Key: ↑ = Increase.   0 = No effect.   ↓ = Decrease.

A. Alloys containing up to 15% Ni have a reddish colour, the 15% Ni alloy being the lowest nickel content required to produce a white colour. The low nickel alloys can be regarded essentially as toughened copper, with superior corrosion resistance.

B. Alloys with up to 30% Ni have improved corrosion resistance and superior mechanical properties.

C. Alloys containing about 40–45% Ni are used as resistance materials due to their low conductivity and very low temperature coefficient of resistance.

D. Iron additions are made to alloys with about 30% Ni to improve corrosion resistance especially to impingement attack and sea water attack.

E. Copper–nickel–manganese alloys, characteristically 60% Cu, 20% Mn, are heat-treatable with properties similar to those of copper beryllium. Used for spring applications.

Alloys with greater than 50% nickel comprise the monel metals whose main features are their corrosion and oxidation resistance coupled with good mechanical properties. Since these alloys are nickel based, they are considered in the section on nickel alloys.

Table 2.4.22

TABLE 2.4.23 Basic properties of copper–nickel alloys (cupro–nickel)

Group	Usual name	BS Ref. (ISO)	Basic chemical composition	Specific Gravity	Electrical Conductivity[a, b] (% IACS)	Thermal conductivity (W/m per K)	Thermal expansion coefficient (10^{-6} K^{-1})
Copper–Nickel	95/5 Cupro–Nickel	CN101	94 Cu 5 Ni 1 Fe	8.9	14	62	17
Wrought	90/10 Cupro–Nickel	CN102	88 Cu 10 Ni 2 Fe 1 Mn	8.9	9	50	17
	80/20 Cupro–Nickel	CN104	80 Cu 20 Ni 0.5 Mn	8.95	6	38	16
	75/25 Cupro–Nickel	CN105	75 Cu 25 Ni Mn	8.95	5	29	16
	70/30 Cupro–Nickel	CN107	69 Cu 30 Ni 1 Fe 1 Mn	8.95	5	29	16
	66/30/2/2 Cupro–Nickel	CN108	66 Cu 30 Ni 2 Fe 2 Mn	8.9	5	24	16
	Hiduron alloy 130		Cu bal 14.5 Ni	8.5	10.0	46.1	16.4
	Hiduron alloy 191		Cu bal 14.5 Ni 4.5 Mn	8.5		25	16
	Marinel		Cu bal 14.5 Ni 4.5 Mn	8.5		25	16
	Hidural 5		Cu bal 2.0 Ni 0.4–0.8 Si 0.05–0.3 Mn	8.87	40	190–232	16.4
Castings		**BS1400**					
	Cast copper–nickel–chromium	CN1	66 Cu 30 Ni 1 Fe 1 Mn 2 Cr Si Zr	8.8	5	23	18
	Cast copper–nickel–niobium	CN2	67 Cu 30 Ni 1 Fe 1 Mn 1 Nb Si	8.8	5	23	18

For footnotes see Table 2.4.6.

Table 2.4.23

TABLE 2.4.23 Basic properties of copper–nickel alloys (cupro–nickel)—*continued*

Modulus of elasticity (GN/m²)		Mechanical properties						Fatigue^d (MN/m²)	Impact^e (J)	Approx. service temperature
		Tensile strength (MN/m²)		0.1% proof stress (MN/m²)		Elongation (%)				
Soft^c	Hard	Soft^c	Hard	Soft^c	Hard	Soft^c	Hard			Range (°C)
135	120	280–320	380	90–130	350	35–40	10	115–170	170	−200–150
138	130	280–330	420	120–140	350–380	25–40	12–14	95–170	58–150	−200–150
149	136	320–340	400–560	150	340–420	40	12–20	120–200	60–100	−200–150
149	141	340–370	450–580	140–160	290–380	35–42	3–15	200–270	140	−200–200
155	146	310–420	500–590	150–170	420–460	30–45	7–18	140–245	148	−200–200
		770–850				10		270	14–20	−196–300
148		725				18			40	−196–300
145		930				13				−196–300
		620		340		21		220	90	−196–300
		Sand	Chill	Sand	Chill	Sand	Chill			
		480–540	—	300–320	—	18–25	—		45	
		480–540	—	300–320	—	18–25	—		45	

Table 2.4.23—*continued*

TABLE 2.4.24 Comparative characteristics of copper–nickel alloys (cupro–nickel)

BS ref.	Type	Available forms[a]	Thermal treatment[b]	Hardness	Formability[c]		Maximum reduction between anneals	Machinability[d]
					Hot	Cold		
CN101	5% Cupro–Nickel	PST	A	63–100	G	E	80%	20
CN102	10% Cupro–Nickel	PSTB	A	80–165	G	E	80%	20
CN104	20% Cupro–Nickel	PS	A	85–165	G	E	70%	20
CN105	25% Cupro–Nickel	PS	A	90–170	G	G	50%	20
CN107	30% Cupro–Nickel	PSTB	A	95–185	G	E	40%	20
CN108	60–30–2–2 Cupro-Nickel	T	A	105–140	G	E	40%	20
	Hiduron alloy 130 14.5 Ni				X	X		
	Hiduron alloy 191				√			
	Marinel	Bar Forgings			√			
	Hidurel 5	Bar Forgings Strip			X	√		
Castings CN1	30% Cupro–Nickel (+ CR) (+ Cr)	Sand Cast	—	100	—	—	—	20
CN2	30% Cupro–Nickel (+ N6)	Sand Cast	—	110	—	—	—	20

For footnotes see Table 2.4.7.

Table 2.4.24

TABLE 2.4.24 Comparative characteristics of copper–nickel alloys (cupro–nickel)—_continued_

Joining[c]			Limitations	Typical uses
Soldering	Brazing	Welding		
P	G	G	Expensive (but cost-effective).	Marine fire mains, cooling circuits and sanitary services. Ammunition driving bands.
P	G	G	Expensive (but cost-effective).	Marine corrosion resistant assemblies. Desalination plant. Medals, etc.
P	G	G	Expensive (but cost-effective).	Marine assemblies for slow-moving sea water. Pressings and deep drawn items. Electrical resistance wires.
P	G	G		Coins and medals. Resistance wires. Fasteners.
P	G	G	High cost but essential uses.	Tubes, tube plates and fittings for marine condensers and desalination plants. Feed water fittings. High velocity waters.
P	G	G	High cost but essential uses.	Items to resist corrosion–erosion in sea-water with silt.
√	√			Airframe components. Naval and marine engineering components.
√	√	√		General corrosion resistant applications in marine environments.
				Fasteners for severe duties in offshore environments.
√	√	X		Engine components. Non magnetic naval applications.
P	F	G	Sand casting only: skill required. Expensive.	Items used in conjunction with wrought cupro–nickel alloy in corrosive environments, especially sea-water. CN1 outstanding in impingement conditions.
P	F	G		

Table 2.4.24—_continued_

TABLE 2.4.25 Basic properties of copper–zinc–nickel alloys (nickel silvers)

Group	Usual name	BS Ref. (ISO)	Basic chemical composition	Specific gravity	Electrical conductivity[a, b] (% IACS)	Thermal conductivity (W/m per K)	Thermal expansion coefficient (10^{-6} K^{-1})
Copper–Zinc–Nickel	Leaded Nickel Brass (10%)	NS101	45 Cu 43 Zn 10 Ni 2 Pb Mn	8.5	7	33	19
Wrought	Leaded Nickel Brass (14%)	NS102	40 Cu 42 Zn 14 Ni 2 Pb 2 Mn	8.5	6	32	18
	Nickel Silver (10%)	NS103	63 Cu 27 Zn 10 Ni Mn	8.6	8.5	37	16
	Nickel Silver (12% Special)	NS104	64 Cu 24 Zn 12 Ni Mn	8.6	8	30	16
	Nickel Silver (15%)	NS105	64 Cu 21 Zn 15 Ni Mn	8.65	7	26	16
	Nickel Silver (18%)	NS106 (Cu Ni 18 Zn 20)	62 Cu 20 Zn 18 Ni Mn	8.75	5.5	22	16
	Nickel Silver (18% Special)	NS107 (Cu Ni 18 Zn 27)	55 Cu 27 Zn 18 Ni Mn	8.7	5.5	21	17
	Nickel Silver (20%)	NS108	62 Cu 18 Zn 20 Ni Mn	8.75	6	21	17
	Nickel Silver (25%)	NS109	57 Cu 18 Zn 25 Ni Mn	8.6	5	21	17
	Leaded Nickel Silver (10%) Sections	NS111	63 Cu 25 Zn 10 Ni 2 Pb Mn	8.6	8	36	16
	Leaded Nickel Silver (15%) Sections	NS112	64 Cu 2 Zn 15 Ni 1 Pb Mn	8.65	7	27	16
	Leaded Nickel Silver (18%) Sections	NS113	62 Cu 19 Zn 18 Ni 1 Pb Mn	8.75	5	26	16
Castings	Cast Nickel Silver (Sand Cast)	—	55 Cu 22 Zn 12 Ni 9 Pb 2 Sn	8.6			17
	Cast Nickel Silver (Sand Cast)	—	65 Cu 4 Zn 25 Ni 1 Pb 5 Sn	8.7			17
	Cast Nickel Silver (Sand Cast)	—	62 Cu 2 Zn 1 Mn 20 Ni 4 Pb 4 Sn	8.7			17

[a] For footnotes see Table 2.4.6.

Table 2.4.25

TABLE 2.4.25 Basic properties of copper–zinc–nickel alloys (nickel silvers)—*continued*

Modulus of elasticity (GN/m²)		Tensile strength (MN/m²)		0.2% proof stress (MN/m²)		Elongation (%)		Fatigue[d] (MN/m²)	Impact[e] (J)	Approx. service temperature
Soft[c]	Hard	Soft[c]	Hard	Soft[c]	Hard	Soft[c]	Hard			Range (°C)
130	127	450–480	510–610	260–280	310–370	20–22	12–18			−200–200
120	117	370–430	460–620	150–200	320–520	50	6–25	110–180		−200–200
120	117	380–420	450–590	125–185	300–550	40–50	6–18	140–190		−200–200
123	130	380–440	450–700	140–150	450–700	40–50	5–15	175–220		−200–200
127	132	400–450	470–630	160–220	360–570	35–45	6–15			−200–200
133	137	400–440	470–680	150–230	370–580	40–45	3–20	160–220	125	−200–200
134	140	400–440	530–800	160–210	580–690	40–45	2–15	110–225		−200–200
~130		380–400	500–700	130–145	400–600	50–55	4–8			−200–200
~130		390–400	600–710	130–145	450–630	50–55	4–8			−200–200
~130		380–400	—	150–200	—	40–50	—			−200–150
~130		400–420	—	160–200	—	35–45	—			−200–150
~130	134	440–460	—	150–180	—	25–35	—		29	−200–150
		200–270	—		—	25–10	—	—		−200–100
		340–350	—		—	25–15	—	—		−200–150
		280–400	—		—	25–15	—	—		−200–150

Table 2.4.25—*continued*

TABLE 2.4.26 Comparative characteristics of copper–zinc–nickel alloys (nickel silvers)

BS ref.	Type	Available forms[a]	Thermal treatment[b]	Hardness	Formability[c]		Maximum reduction between anneals	Machinability[d]
					Hot	Cold		
NS101	Leaded Nickel Brass (10%	BF	QH	100–180	E	F	20%	80
NS102	Leaded Nickel Brass (14%)	B	A	100–180	E	F	15%	80
NS103	10% Nickel Silver	S	A	70–210	L	G	70%	25
NS104	12% Nickel Silver	S	A	70–210	L	E	85%	25
NS105	15% Nickel Silver	S	A	90–205	L	E	80%	25
NS106	18% Nickel Silver	S	A	80–210	L	E	70%	25
NS107	18% Special Nickel Silver	SW	A	90–240	L	G	65%	30
NS108	20% Nickel Silver	PSW	A	90–205	L	E	60%	25
NS109	25% Nickel Silver	PSW	A	90–200	L	E	60%	25
NS111	Sections only corresponding to NS103, 105 and 106 but BSs withdrawn 1986	B	A	70–180	F	G	30%	75
NS112		B	A	70–180	F	G	30%	75
NS113		B	A	70–180	L	F	30%	75
—	Cast Nickel Silvers	Sd. Ch.	—	~70	—	—	—	~25

For footnotes see Table 2.4.7.

Table 2.4.26

TABLE 2.4.26 Comparative characteristics of copper–zinc–nickel alloys (nickel silvers)—*continued*

Joining[c]			*Limitations*	*Typical uses*
Soldering	*Brazing*	*Welding*		
G	E	F		Architectural metalwork-grilles, handrails, balustrading, screens, window and door fittings. Shop fronts, trim, lighting fittings, clock and watch parts. Machined items for instruments. Latch keys.
G	E	F	Relatively expensive but little maintenance required.	
G	E	F		Decorative items, tableware, pressed and spun articles, jewellery. Electrical contacts and resistance wire. Terminals and clips. Instrument and camera parts. Dials and nameplates. Small cryogenic tubes, etc.
G	E	F		
G	E	F	Relatively expensive but little maintenance required.	Electrical resistance wires for moderate temperatures. Telephone selector items in exchange equipment. Bellows and pressure sensitive devices. Musical instrument keys and springs, etc. Medical equipment parts. Rivets and screws.
G	E	F		
G	E	F		Springs and clips for electrical equipment—relays, contactors (telephones), connectors.
G	E	F		Higher grade decorative items, jewellery, tableware, etc. Doors, kickplates OK in major buildings.
G	E	F		Highest quality decorative and tableware—may not require final electroplating.
G	E	P	Relatively expensive but little maintenance required.	Long-lasting decorative and architectural sections.
G	E	P		
G	E	P		
F	G	P		Statuary and other decorative items. Small parts for machines in pharmaceutical industry.

Table 2.4.26—*continued*

2.4.6   Non-standard alloys

There are several groups of copper base alloys which are widely used for specific purposes:

(i)   Copper head alloys for bearings see Table 2.4.27.
(ii) Porous bronze bearings: pressed and sintered powder mixtures 90% Cu 10% Sn, often with additions of graphite.
(iii) High manganese alloys (see Table 2.4.28).
(iv) Memory alloys.

Memory alloys are basically of copper and zinc with aluminium within the range 55–80% copper and 2–8% aluminium in the wrought form. Such alloys are able to exist in two distinct configurations above and below a transformation temperature. Actual compositions can be selected for transition temperatures between –70 and +130°C: details must be obtained from the manufacturers.

Compositions and characteristics of typical non-standard alloys are listed in:

Table 2.4.27   *Copper–lead alloys for bearings (lead bronzes)*
Table 2.4.28   *Copper–manganese alloys*

TABLE 2.4.27 Copper–lead alloys for bearings (lead bronzes)

Type	Composition (Wt %)					Tensile Strength (MN/m^2)	Characteristics
	Cu	Pb	Sn	Ag	Others		
80/20 Cu–Pb	78	20	—	—	1.2–5 Ni	140	Bonded to steel shells for bearings. Alloys up to 45% lead are in use.
74/24 Cu–Pb	74	24	2	—	—	140	Good conductivity.
70/30 Cu–Pb	69	30	—	0.6	—	140	Withstands 'pounding'.
60/40 Cu–Pb	59	40	—	1	—	—	Lead distribution improved by powder metallurgical processing.
55/45 Cu–Pb	55	45	1			120	

Table 2.4.27

TABLE 2.4.28 Copper–manganese alloys

Wrought copper in manganese alloy

Type	Composition (Wt%)				Tensile Strength (MN/m^2)		Elongation (%)		Characteristics
	Cu	Mn	Ni	Al	Soft	Hard	Soft	Hard	
Cu–Mn–Ni (Manganin)	Bal.	12	2–4	—	420	—	30		High specific resistance and low temperature coefficient of electrical resistance.
Cu–Mn–Al	85	13	—	2	430	680	30	10	Resistance alloy family having similar properties to manganin (above).

Copper–manganese–aluminium damping alloys—comparative characteristics of different types

Type	Composition (Wt%)				Advantages	Limitations	Typical uses	Trade name and source
	Cu	Mn	Al	Ni				
General purpose high damping casting alloy	58	40	2	—	Good vibration damping. Good castability capable of both hot and cold forming. Good corrosion resistance and pleasant colour.	Expensive and tough to machine.	Components of pneumatic compressors, road drills and machine beds where high damping is required.	'Incramute' (International Copper Research Association Inc.)
Marine high damping casting alloy	30	63.5	4	2.5	Good vibration damping. Excellent sea-water and general corrosion resistance. Can be cast accurately with skill.	Very expensive and tough to machine.	Marine propellers and similar marine components. Chemical process plant stirrers etc.	'Sonoston' (Stone Manganese Marine Inc.)

Table 2.4.28

2.4.7   Corrosion resistance of copper and its alloys

Corrosion resistance of all grades of copper is good, hence their wide usage for roofing and for contact with most waters. The metal develops adherent protective coatings, initially of oxide but subsequently thickening to give a familiar green patina on roofs and the dark brownish colour of bronze statuary. The composition of patina varies depending on specific atmospheric conditions, but it is primarily basic copper sulphate with copper carbonate and, in marine atmospheres, some chloride. Testing over 20 years has given the following annual corrosion rates in µm.

Rural	0.37
Marine	0.75
Industrial	1.0

When pitting occurs, the corrosion rate is from 2 to 5 times greater.

Because copper is largely unaffected by potable water it is widely used for calorifiers and tubes for carrying domestic and industrial water, but some natural waters are sufficiently acid to cause cupro-solvency when the pH values range downwards from the neutral point of pH7. Waters with a high content of carbon dioxide, with sulphates and chlorides, may dissolve sufficient copper to cause green discolouration in the presence of soap. Pitting of tube walls may be encountered when the interior is contaminated by a carbon film along the tube, residual from annealing during manufacture. Damage by corrosion/erosion to tube walls by entrained gas bubbles may occur when high water speeds are encountered; hence avoidance of high velocity and turbulence in the presence of corrodants is necessary.

Sea-water causes only negligible corrosion of copper at water speeds up to 1 m/s, but at higher speeds the protective film is removed. Corrosion is increased in the presence of ammonia or moist hydrogen sulphide which may be present as a result of bacterial activity, particularly in stagnant and polluted waters. Copper withstands attack by many substances including alcohols (beer and spirits), asphalt and creosote, most oils and petrol, sugars and molasses, acetone, trichloroethylene, formaldehyde, freons, turpentine and varnishes.

Copper alloys have useful corrosion resistance roughly equal to that of the metal itself but some compositions give additionally protective films, for example, when they contain aluminium oxide.

Comparative data on the resistance of cast copper alloys to sea-water attack is given in Tables 2.4.29 and 2.4.30.

Table 2.4.29   *Resistance of cast coppers to impingement attack and general corrosion in sea-water*

Table 2.4.30   *Resistance of copper alloys and stainless steels to general corrosion, crevice corrosion and impingement attack in sea-water*

For more details of the corrosion resistance of copper and its alloys, see Volume 1, Chapter 3. See also Table 2.4.32 in Section 2.4.9.

1150   Copper and its alloys

TABLE 2.4.29 Resistance of cast coppers to impingement attack and general corrosion in sea-water

Alloy	Composition % (Bal. Cu)					Depth of impingement attack (mm)		General corrosion weight loss (mg/cm² per day)	
						28-day Jet impingement (20°C)	14-day Brownsdon & Bannister (20°C)	Water in slow motion	Water speed (10 m/s)
	Al	**Fe**	**Ni**	**Mn**	**Zn**				
Aluminium Bronze	8.2	1.7	—	—	—	0.04	0.19	0.15	0.17
Nickel Aluminium Bronze	8.2	2.9	4.3	2.4	—	0.00	0.32	0.04	0.10
Nickel Aluminium Bronze	8.8	3.8	4.5	1.3	—	0.00	0.28	0.04	0.16
Manganese Aluminum Bronze	7.6	2.8	3.1	10.0	—	0.01	0.24	0.04	0.11
High Tensile Brass	0.8	0.8	0.2	0.5	37.0	0.03	0.08	0.09	0.73
	Sn	**Zn**	**Pb**						
Gunmetal	9.7	1.4	0.6			0.02	0.32	0.14	0.74
Gunmetal	5.1	5.0	4.3			0.23	0.39	0.22	1.66

Table 2.4.29

TABLE 2.4.30 Resistance of copper alloys and stainless steels to general corrosion, crevice corrosion and impingement attack in sea-water

Alloy	General corrosion rate (mm/year)	Crevice corrosion (mm/year)	Corrosion/erosion resistance limit (m/s)
Wrought alloys:			
Phosphorus deoxidised copper C106 or C107	0.04	<0.025	1.8
Admiralty brass CZ111	0.05	<0.05	3.0
Aluminium brass CZ110	0.05	0.05	4.0
Naval brass CZ112	0.05	0.15	3.0
HT brass CZ115	0.18	0.75	3.0
90/10 copper–nickel	0.04	<0.04	3.0
70/30 copper–nickel	0.025	<0.025	4.6
5% aluminium bronze CA101	0.06	<0.06	4.3
8% aluminium bronze CA102	0.05	<0.05	4.3
9% aluminium bronze CA103	0.06	0.075	4.6
Nickel aluminium bronze CA104	0.075		
Aluminium silicon bronze DGS 1044	0.06	<0.075	
17% Cr stainless steel 430	<0.025	5.0	>9
Austenitic stainless steel 304	<0.025	0.25	>9
Austenitic stainless steel 316	0.025	0.13	>9
Monel	0.025	0.5	>9
Cast alloys:			
Gunmetal LG2	0.04	<0.04	3.7
Gunmetal G1	0.025	<0.025	6.1
High tensile brass HTB1	0.18	0.25	2.4
Aluminium bronze AB1	0.06	<0.06	4.6
Nickel aluminium bronze AB2	0.06		
Manganese aluminium bronzes CMA1/CMA2	0.04	3.8	4.3
Austenitic cast iron (AUS 202)	0.075	0	>6
Austenitic stainless steel 304	<0.025	0.25	>9
Austenitic stainless steel 316	<0.025	0.125	>9
3% or 4% Si Monel	0.025	0.5	>9

Table 2.4.30

2.4.8   Effect of temperature on the properties of copper and its alloys

The effect of temperature on the electrical and thermal conductivities of tough pitch high conductivity and deoxidised copper is given in:

Fig 2.4.4   *Effect of temperature on electrical and thermal conductivities of tough pitch high conductivity and deoxidised (0.4% phosphorus) copper*

Figure 2.4.5 shows the creep of copper and highly conducting alloys over the temperature range 0–1000°C. The variation in hardness with temperature of some low alloy coppers is given in Fig 2.4.6.

Fig 2.4.5   *Curves showing creep of copper and highly conducting alloys*

Fig 2.4.6   *Hardness values as influenced by softening temperatures of typical tough pitch high conductivity copper, silver copper, chromium copper and zirconium copper*

Creep curves for a range of copper alloys are given in:

Fig 2.4.7   *Stress for 10–5% creep per hour of copper alloys*

Fig 2.4.8   *Creep data for two hot rolled aluminium bronzes*

Maximum permissible stress levels over the temperature range 0–300°C are given in Fig 2.4.9 for a range of copper–tin–zinc alloys (gunmetals):

Fig 2.4.9   *Permissible stress for five standard gunmetals (G1, LG1, LG2, LG3, and G4 of BS1400) and two gunmetals with nickel additions*

Short term tensile test data for wrought aluminium bronze alloys over the temperature range 0–600°C is given in:

Fig 2.4.10   *Short-term tensile test data up to 550°C for wrought aluminium bronze CA104 (9.6% Al, 5% Fe, 4.9% Ni)*

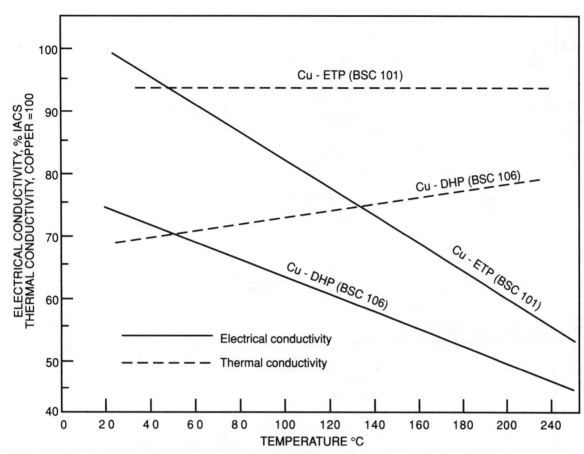

FIG 2.4.4 Effect of temperature on electrical and thermal conductivities of tough pitch high conductivity and deoxidised (0.04% phosphorus) copper

Fig 2.4.4

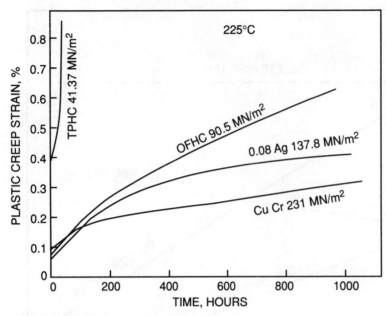

TPHC =Tough Pitch High Conductivity Copper
OFHC =Oxygen Free High Conductivity Copper

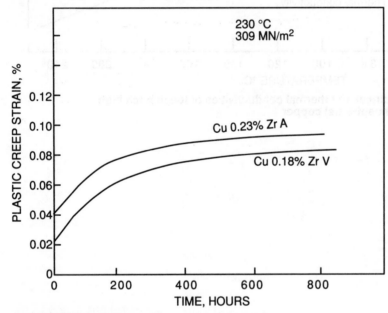

A =Air melted and cast
V =Vacuum melted and cast

FIG 2.4.5 Curves showing creep of copper and highly conducting alloys

Fig 2.4.5

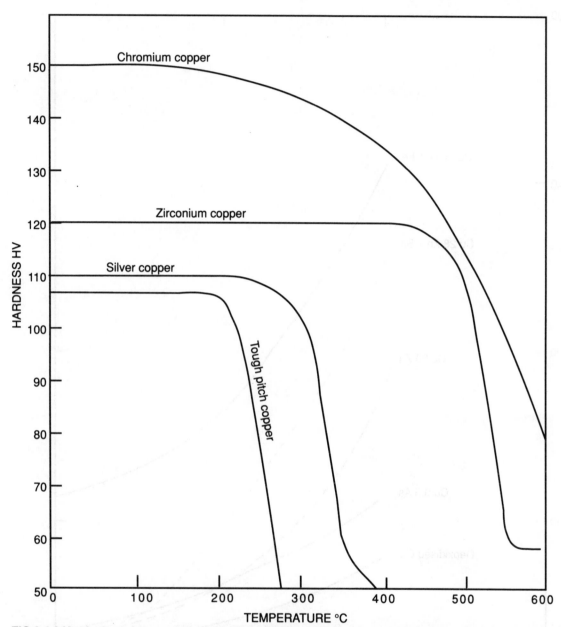

FIG 2.4.6 Hardness values as influenced by softening temperatures of typical tough pitch high conductivity copper, silver copper, chromium copper and zirconium copper

Fig 2.4.6

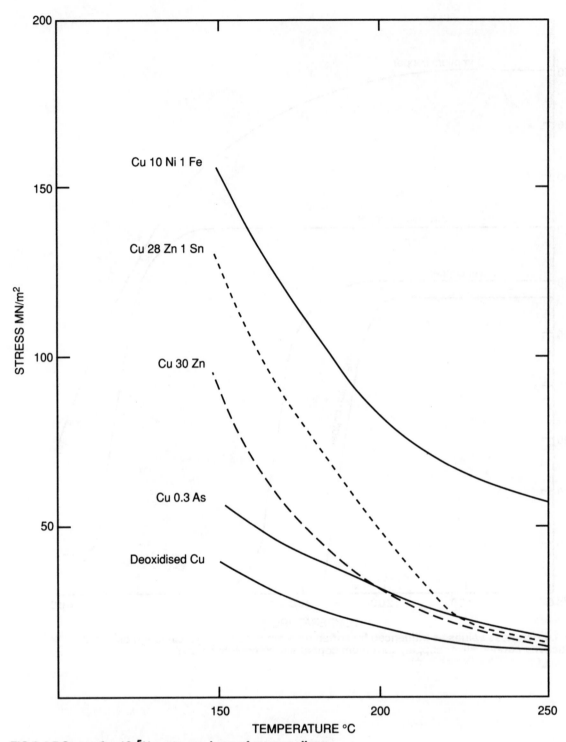

FIG 2.4.7 Stress for 10^{-5}% creep per hour of copper alloys

Fig 2.4.7

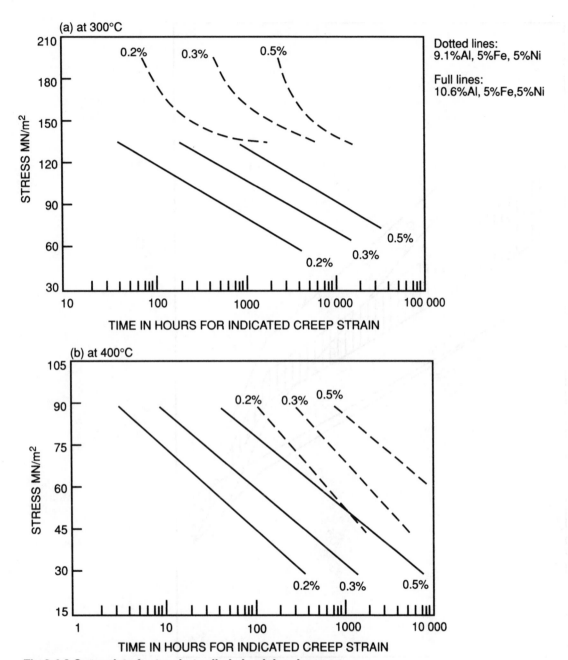

Fig 2.4.8 Creep data for two hot rolled aluminium bronzes

Fig 2.4.8

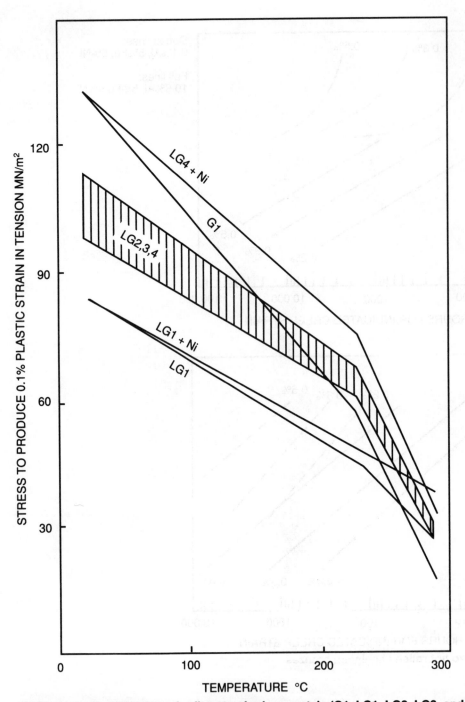

FIG 2.4.9 Permissible stress for five standard gunmetals (G1, LG1, LG2, LG3, and G4 of BS 1400) and two gunmetals with nickel additions. The data are based on typical values for the 0.1% proof stress at lower temperatures and the stresses to produce 0.1% strain in 10 000 h at higher temperatures

Fig 2.4.9

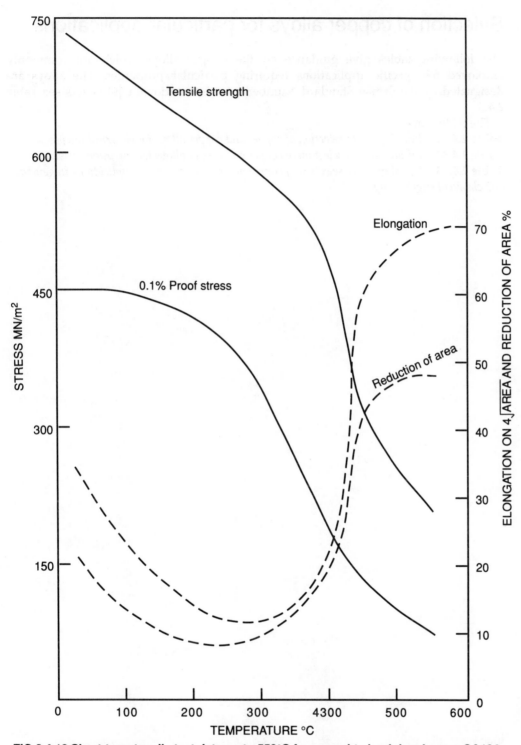

FIG 2.4.10 Short-term tensile test data up to 550°C for wrought aluminium bronze CA104 (9.6%Al, 5%Fe, 4.9% Ni)

Fig 2.4.10

2.4.9   Selection of copper alloys for particular applications

The following tables give guidance on the copper alloys which are commonly employed for specific applications requiring particular properties. The alloys are designated by the British Standard Numbers—for international equivalents see Table 2.4.3.

The tables cover:

Table 2.4.31   *Initial guide to selection of copper and copper alloys for electrical purposes*

Table 2.4.32   *Initial guide to selection of copper and copper alloys for corrosion resistance*

Table 2.4.33   *Initial guide to selection of copper and copper alloys for applications in general and chemical engineering*

TABLE 2.4.31 Initial guide to selection of copper and copper alloys for electrical purposes

Applications	Wrought	Cast
Power cables, overhead lines, industrial and domestic wiring.	C101, C102	
Telecommunication and coaxial cables.	C101, C102	
Overhead lines and electric-traction catenaries.	C101, C102 C108	
Bus bars.	C101, C102	
Flexible cables.	C101, C102, C108	
Generator windings and transformers at low and normal temperatures.	C101, C102	
Generator windings at raised temperatures.	CC101, CC102 Cu + 0.1 Ag	
Commutators.	C101, C102, A2/1 Cu + 0.1 Ag	
Slip rings.	CZ108	PB4, HTB1, SCB4, SCB6
Electronic components in vacuum, switch blades, etc.	C103, C110 C101, C102, CB101 Cu + 0.1 Ag	HCC1, CC1
Machined components for fittings (terminals, etc.)	C111, C109	
Cast components for conductors, spring contacts, etc.	PB101, PB102, PB103, C108, CB101, A3/2 NS103, NS105, NS106, NS107	HCC1

Table 2.4.31

TABLE 2.4.32 Initial guide to selection of copper and copper alloys for corrosion resistance

Environment and applications	Wrought	Cast
Rural and industrial		
Roofing sheet and cladding, flashing and gutters.	C101, C102, C104	
External decorative items, formed strip and sections.	C102, C104, CZ107, CZ121, CZ125, NS101, NS102, NS111, NS112	
Lightning conductors.	C101, C102, C104	
Wall ties and masonry fittings.	PB102, CA103	PB1, AB1, AB2
Window frames.	CZ121	
Statuary.	C101, C102, C104, C106	Special tin bronzes.
Solar heating panels.	C106, C107	
Damp proof course and weather strip.	C104, C107, CZ101, CZ102	
Tubes for water and gas.	C106	
Water cylinders, calorifiers.	C106, CS101	
Tubes for soil and waste systems	C106	
Hinges and butts.	CZ108, CZ121	
Water fittings, taps, etc.		SCB1, SCB3, SCB6, DCB1, DCB3, PCB1
Valves for water.		SCB1, SCB3, SCB6, DCB3, LG2, LG4, G1
Nails and screws.	C102, CS101, CZ106, CZ108, CZ121, CZ124	
Marine		
Condenser tubes.	CN101, CN107, CZ109, CZ110, CZ112, CA105	
Tubes for sea-water.	CN101, CN107, CZ110, CS101, CA104, CA105	
Oil tank heaters	CZ110	
Fittings for sea-water.	CZ112, HTB1, CN107, CS101, CA104, CA105	PB1, PB4, G1, LG2, LG4
Valves and pump components.	CA104, CA105	G1, LG2, LG4, PB1, PB4, AB2
Boiler feed water fittings.	CT1, PB1, PB3	AB2, G1, LG2, LG4, PB1, PB4
Propellers.		CMA1, (CMA2), AB2, HTB1
Portholes, deadlights, windows.	CA104	G1, AB2
Non-magnetic fittings.	CZ112, CA103, CA104	G1, LG2, LG4, AB1, AB2
Small fittings and cleats.	CZ110, CZ111, CZ112, CS101, CA104	HTB1, LG2, LG4
Nails, clouts and screws.	CS101, CZ112, CA103, CA105	

Table 2.4.32

TABLE 2.4.33 Initial guide to selection of copper and copper alloys for applications in general and chemical engineering

Applications	Wrought	Cast
Bellows and diaphragms.	PB101, PB102	
Valve bodies for high duties.	CA104, CA105	AB1, AB2
Pressure vessels.	C106, C107, CS101	
Wear resistant cams, guides, etc.	CB101, CA105	AB1, AB2
Pump components.	CA106, PB103, PB104	AB1, AB2, CMA1, G1, LG2, LG4
Springs.	CB101, PB102, PB103, NS104, NS106, NS107 CZ106, CZ107, CZ108, C101	
Bearings — heavy-duty, rolling mills. — marine. — non-critical, low-loads. — light duties. — average duties, good lubrication. — average duties, poor lubrication. — for hard shafts. — for soft shafts, low loads. — plates for bridges. — ball and roller cages.	CA106, CA105, Copper lead CZ120, CZ121, CZ122 PB104, CZ124 PB104, CZ124 PB103, PB104, CZ121, CZ124 CA104, PB104 A/3/2	AB2 G1 LG2, LG4, LB4 DCB1, DCB2, DCB3 LG2, LG3 LB2, LB3, LB4, LPB1 PB1, PB2, PB4, G1 LB1, LB4, LB5 PB1, PB2, PB4
Automobile radiators — tubes. — strip. — tanks.	C101, C102, C106, CZ105, CZ106 Copper + 0.1 Ag CZ105, CZ106, CZ107	
Gears — light duty. Gears — moderate duty. — very heavy duty, low speed. — high loading. — high abrasive loading. — pinions. — clocks and similar. — instruments (high precision).	 PB103, PB104, CA103, CA104, CA105 CA103, CA104 CA104 PB103, PB104 CZ118, CZ120, CZ122 CZ120, CZ122	G1, LG2, LG3, LG4, DCB1 DCB2, DCB3, AB1 PB1, PB2, PB4 PB2, AB2 PB1, PB2, PB4, AB2 AB1 PB1
Bushings for sleeves.	PB103, PB104, CZ120, CZ121, CZ122	
Deep drawn and pressed items.	CZ105, CZ106, CN104, NS104, NS105, C104	
Repetition machined items.	C109, C111, CZ118, CZ119, CZ120 CZ121, CZ122	
Brazed assemblies.	C106, C107, CZ103	
Non-sparking tools, etc.	CB101, CA103, CA104, CA105	AB1, AB2
Chains.	CB101, CA104	AB1
Hot forgings and stampings.	CZ109, CZ120, CZ122, CZ123	

Table 2.4.33

Aluminium and its alloys

Contents

List of tables

List of figures

2.5.1 Introduction and properties

2.5.1.1 Introduction

From the rare metal of 100 years ago, aluminium has become the most widely used non-ferrous metal on a volume basis. Whilst more expensive on a tonnage basis, it is the least expensive of metals other than steel on the basis of volume or area.

The advantages and limitations of aluminium alloys are listed in:
Table 2.5.1 *Characteristics of aluminium alloys*

The word 'aluminium' is often used loosely as the generic term for the metal itself in several grades of purity and for the family of aluminium base alloys in which the soft pure metal is strengthened by the addition of one or more of the elements as detailed in Section 2.5.1.2.

2.5.1.2 Basis and range of alloy types

Pure aluminium can be strengthened by alloying with small amounts of Mn (up to 1.25%) and Mg (up to 3.5%). The addition of larger percentages of Mg produces still higher strengths, but precautions are needed for satisfactory performance. These alloys and pure aluminium can be further hardened by cold work up to tensile strengths of 200 or even 300 MPa.

Higher strengths are achieved in alloys which are heat-treatable. These alloys are quenched from elevated temperatures and then aged by reheating at lower temperatures. They fall into three main groups:

Alloys containing small amounts of Mg + Si + Mn (or Cr) which give strengths up to 330 MPa.

Alloys, with additions of Cu + Mg + Si + Mn, some of which age at room temperature after quenching and often require precipitation treatment to give strengths up to 450 MPa.

Alloys containing Zn + Mg (+ Cr) some of which give strengths in the same range as the Mg + Si + Mn series while others range up to 600 MPa. Care is required when resistance to impact fracture, toughness and stress corrosion are required from alloyed material heat-treated to the higher strength levels.

A new alloy series containing essentially small additions of lithium, with copper, are available. The alloys are heat-treatable and as they are of current interest for aircraft components they are covered by DTD specifications.

The relationship between the properties, and applications of aluminium alloys is given in:
Fig 2.5.1 *Variation of some properties of aluminium alloys with tensile strength*

2.5.1.3 Classification of aluminium-based materials

Aluminium and its alloys are divided into two broad classes, castings and wrought (mechanically worked) products. The latter class is subdivided into non-heat-treatable and heat-treatable alloys, and into the various forms produced by mechanical working. A complete list of equivalent or near-equivalent national designations is given in:
Table 2.5.2 *British alloy designations and international equivalents*

2.5.1.4 Available forms

The forms available to the designer are sand, permanent mould and pressure die-castings; rolled flat products (plate down to 6 mm and sheet, strip and foil); extruded sections and tubes; drawn tubes, bars and wire; forgings and impact extruded components and powders.

2.5.1.5   Resistance to environmental attack

Aluminium and its alloys have good resistance to attack by the atmosphere, most waters and a wide range of chemicals due to the film of aluminium oxide which is always present on the surface when the metal is exposed to air.

GALVANIC CORROSION

The severity of such attack is indicated by the electrolytic solution potentials of a range of alloys against commercial purity aluminium (99.3% pure) listed in:

Table 2.5.3   *Bimetallic (galvanic) corrosion of aluminium alloys in contact with other metals*

Painting is recommended for protection, particularly of the heat-treated alloys exposed to aggressive conditions. Both site painting and factory-applied paint coatings require pre-painting treatment of the surface to provide a key. Chemical commercial coatings may also be applied to large areas such as roof cladding.

More details on the suitability of aluminium alloys for particular applications requiring corrosion resistance are given in:

Table 2.5.4   *Guide to exposure of aluminium alloys to rural, marine and industrial environments,* and

Table 2.5.5   *Initial guide to corrosion resistance of aluminium alloys at normal temperatures*

Designations are given according to British Standards—for international equivalents see Table 2.5.2.

More information on the comparative corrosion resistance of aluminium alloys is given in Vol. 1, Chapter 3.

TABLE 2.5.1 Characteristics of aluminium alloys

Advantages	Limitations
Low density—at 2.69, only magnesium is less dense among the common metals.	Relatively low modulus of elasticity about ⅓ that of steel and ½ that of copper.
Corrosion resistance—high in normal atmospheres and conditions though the Al–Cu and Al–Zn–Mg alloys need protection.	Preferentially attacked when joined most other metals in corrosive media (except Mg and Zn).
High electrical and thermal conductivity—61% IACS on a volume basis but higher conductivity than copper weight for weight.	Some alloys in certain conditions of heat treatment are prone to intercrystalline attack and stress corrosion.
High specific strength.	Fatigue endurance low relative to tensile strength.
Tensile properties improve at cryogenic temperatures with no embrittlement.	Limited high temperature capability (120–200°C),
Available in all wrought forms—plate, sheet and strip, foil, extruded rod and sections, drawn tubes and wires.	Some heat-treated alloys are difficult to weld.
Readily fabricated by normal working processes (rolling, extrusion, forging, drawing).	Not suitable for use in certain environments, e.g. coal mines, due to sparking possibility.
Castability generally good by sand and chill casting methods, die-cast and pressure processes.	
Most alloys are easily welded by appropriate processes.	
Range of surface finishes available to improve appearance and wear resistance.	
Non-magnetic.	

Table 2.5.1

TABLE 2.5.2 British alloy designations and international equivalents[a]

CASTINGS

UK BS/DTD	ISO	AICMA	USA AA	USA SAE	USA ASTM	Canada	France AFNOR	West Germany DIN alloy No.	Italy UNI Spec.	USSR	Sweden SIS Specs.	Japan	Switzerland VSM Specs.
LM0	Al 99.5						A5-Y; A5-Y4						10845
LM2	Al Si10 Cu2 Fe							G-Al Si9 (Cu)		A125		ADC12	
LM4 LM22	Al Si5 Cu3		319	326	SC64D	SC53	A-S5 U3	GD-Al Si6 Cu3		A16			
LM5	Al Mg5		314	320	G4A		A-G3T	G-Al Mg5			144163		
LM6	Al Si12				S12C	S12P S12N	A-S 13-Y	G-Al Si12 / GD-AlSi12	4514	A12	144261	AC3A	10855
LM9	Al Si12 Mg		A360	309	SG 100A		A-S10G	GD-Al Si10 (Cu) / G-Al Si12	3049		144253		10856
LM10	Al 10Mg		520	324	G10A	G10	A-G10; A-G11	G-AlMg10	3056	A18, A27		AC78	
LM12	Al Cu10 Si2 Mg		222	34	CG100A	CG100	A-U10G-Y						
LM13	Al Si11 Mg Cu		A332	321	SN122A		A-S12 UN-Y / A-S12 N2 G			A130		AC8A	
LM16	Al Si5 Cu1 Mg	AL-C21	355	322	SC51A	SC51		G-Al Si5 Cu1	3600			AC4D	
LM18	Al Si5		443	35	S5A	S5		G-Al Si5 Mg	5077				
LM20	Al Si12 Cu Fe		413	305	S12A/B		A-S13-Y4	G-Al Si12 (Cu)	3048		144260		10855
LM21	Al Si6 Cu4 Zn		319	329	SC64C			G-Al Si6 Cu4	3052			AC2B	
LM22	Al Si5 Cu3 Mn												
LM24	Al Si8 Cu3 Fe	AL-C25	A380	306	SC84A	SC84		{G-Al Si8.5 Cu / G-Al Si7 Cu3}	5075	A132	144252	ADC10	
LM31													
LM25	Al Si7 Mg	AL-C22	356	323	SG70A	SG70	A-S7G-Y	G-Al Si7 Mg	3599	A19	144244	AC4C	
LM26	Al Si9 Cu3 Mg		F332	332	SC103A								
LM30	Al Si7 Cu4 Mg		390										
L35	Al Cu4 Ni2 Mg2	AL-C13		39	CN42A	CN42	A-U4T-Y	AlCu Ni Mg	3045				
L51	Al Si2 Cu Fe Ni						A-S2U-Y						
L52	Al Cu2 Si1 Ni Mn Fe												
L91; L92	Al Cu4 Ti	AL-C11		310	ZG61A	ZG61	A-U5GT	G-Al Cu4 Ti					
DTD 5008B	Al Zn5 Mg						A-Z5G-Y		3502		144438		

Table 2.5.2

TABLE 2.5.2 British alloy designations and international equivalents[a]—continued

WROUGHT MATERIALS

UK BS/DTD	ISO	AICMA	USA AA	Canada CSA	France AFNOR alloy No.	West Germany DIN alloy No.	Italy UNI Spec.	USSR	Sweden SIS Specs.	Japan	Switzerland VSM Specs.
1	Al 99.99		1099; 1199		A9	Al 99.98R					10842
1060	Al 99.8		1060		A8	Al 99.8	4509	A6	144020		
1050	Al 99.5	Al 99.5	1050	995	A5	Al 99.5	4507	A5	144017, 144007	A1 X 1	10842
1E	Al 99.5		EC		A5/L	EAl 99.5			144008		
1200	Al 99.0		1200	990	A4	Al 99.0	3567	A0	144022, 144010	A1 X 3	10842
	Al Si 12		4047	S12		S-Al Si12					
	Al Si5		4043/4543	S5		S-A Si5					
3103	Al Mn1		3103/3003	M1	A-M1	Al Mn	3568	1400	144054		10848
3105	Al Mn Mg		3105		A-MG 05						
5052	Al Mg2	AL-P31	5052	GR20	A-G2	Al Mg2	3574		144120	A2 X 1	10849
5005	Al Mg1		5005		A-G06	Al Mg1	4510	1510	144106	A2 X 8	10849
5154	Al Mg3.5		5154A	GR40	A-G3	Al Mg3	3575	1530 (Si 0.6)	144133		
5454	Al Mg3 Mn		5454	GM31	A-G3						
5056	Al Mg5	AL-P32	5056A	GM50N	A-G5	Al Mg5	3576	1550	144146	A2 X 2	10849
5083	Al Mg4.5 Mn		5083	GM50R	A-G4MC	Al Mg4.5 Mn	5462				
6463	Al Mg Si		6463			E-Al Mg Si	3570		144102		10851
6063	Al Mg Si		6063	GS10	A-GS	Al Mg Si0.5	3569		144104		
	Al Cu2 Ni1 Mg Fe Si				A-U2N						
2014	Al Cu4 Si Mg	AL-P12	2014	CS41N	A-U4SG	Al Cu Si Mn	3581	1380, 1185	144338	A3 X 1	10859
2618	Al Cu2 Mg1.5 Fe1 Ni1	AL-P11	2618		A-U2GN			1141		A4 X 1	
7005	Al Zn4.5 Mg1		7005			Al Zn Mg1	6170		144425		
6061	Al Mg1 Si Cu		6061	GS11N	A-SGM	Al Mg Si1	3571				
6082	Al Si1 Mg Mn	AL-P21	6082	GS11R					144212		10850
			2218		A-U4N						
L86	Al Cu2 Mg	AL-P14	2117	CG30	A-U2G	Al Cu Mg0.5	3577				10853
DTD5074A		AL-P42	7075	ZG62	A-Z5GU	Al Zn Mg Cu1.5	3735				
L95; L96							3736				10858
L97; L98	Al Cu4 Mg1	AL-P13	2024	CG42	A-U4G1	Al Cu Mg2	3583	1160			
DTD324B		AL-P22	4032	SG121	A-S12UN		3572				
DTD5004A			2219		A-U6MT						
FC1	Al Cu5.5 Pb Bi		2011	CB60		Al Cu Pb Si					

(Data courtesy of The Aluminium Federation)

[a] In many cases there are no direct equivalents, and in some cases more than one equivalent is indicated. It is suggested that these alloys might be used in similar circumstances, but the suggestion is only a guide, and actual specifications should always be referred to before ordering material.

Table 2.5.2—continued

TABLE 2.5.3 Bimetallic (galvanic) corrosion of aluminium alloys in contact with other metals

Metal in contact	Environment				
	Atmospheric			Immersed	
	Rural	Urban	Marine	Freshwater	Seawater
Aluminium Bronze	1	3	3	3	3
Brasses—all types	1	3	3	3	3
Bronze (tin)	1	3	3	3	3
Cadmium	0	0	0	0	0
Carbon	1	2	3	3	3
Cast iron	0	1	2	1	3
Chromium	1	1	2	2	3
Copper	1	3	3	3	3
Cupronickel	1	3	3	3	3
Gunmetal	1	3	3	3	3
Lead	0	0	0/2	2	3
Magnesium and alloys	0	1	2	1	2
'Monel'	1	3	3	3	3
Nickel	1	2	3	2	2
Nickel–copper alloys	1	3	3	3	3
Nickel–chromium alloys	1	2	3	2	3
Nickel silver	1	3	3	3	3
Phosphur bronze	1	3	3	3	3
Silicon bronze	1	3	3	3	3
Solders — lead/tin	1	2	3	2	3
— silver	1	1/3	3	3	2/3
Stainless steels	0	1	2	2	3
Steels (carbon and low alloy)	1	1	3	2	3
Tin	1	2	3	2	3
Titanium and alloys	0	1	2	2	3
Zinc and its alloys	0	0	0	0	0

Code: 0 = Little or no additional attack.
       1 = Slight corrosion, possibly tolerable.
       2 = Fairly severe additional corrosion necessitating protection (insulation)
       3 = Corrosion severe, therefore avoid contact.

The figures provide only a general guide to additional corrosion arising from differences in potential and, in some cases, are based on limited data. Variations in environment and in composition of the materials give rise to differing results. Protective measures such as insulation of components or painting or metal spraying of mating surfaces are recommended when doubts arise.

Table 2.5.3

TABLE 2.5.4 Guide to exposure of aluminium alloys to rural, marine and industrial environments

Castings (BS 1490)	Wrought (BS 1470–75)	Notes
Group 1 LM5–M LM10–TB LM6⅞M, LM9–TE and TF LM18–M, LM20–M	180, 1050, 1200 3103, 3105 5005, 5251, 5154, 5454, 5083 6060–T4, 6061–T4, 6082–T4, 6463–T4. 2014C–T4, 2024C–T4.	Weathers to pleasant grey colour, deepened to black in industrial atmospheres. Superficial pitting occurs initially and gradually ceases. Seldom needs painting except for decoration. May be anodised for appearance but some alloys (e.g. LM6) give dark films.
Group 2 LM4–M, LM2–M LM16–TF, LM13–M LM21–M, LM22–TB, LM24–M LM27–M, LM31–M LM20–M	6060–T6, 6061–T6, 6065–T6 6082–T6, 6463–T6	Weathers as above. Is normally painted in severe industrial environments and for marine service. May be anodised as above.
Group 3 LM12–TF	2014–T4 2024–T4 and T6	Painting needed in marine and industrial atmospheres, but coatings need only infrequent renewals. Sprayed aluminium coatings, or cladding, give excellent protection. Seldom anodised.

Aluminium retains its initial appearance if it is washed periodically—3–12-month intervals, depending on the severity of the pollution in the air. For many applications the weathered surface is satisfactory but anodising is often undertaken, particularly for architectural items, to preserve a smooth appearance but periodic washing is desirable, especially when the anodic film is coloured.

Table 2.5.4

TABLE 2.5.5 Initial guide to corrosion resistance of aluminium alloys at normal temperatures

Exposure media/applications	Wrought products (BS 1470–77, 4300)	Castings (BS 1490)
Inland atmospheres/ building components, roofs.	**1st choice:** 1050, 1090, 1200, 3103, 3105, 5005, 5251. **2nd choice:** 5083, 5154, 5454, 6061, 6063, 6082, 7020.	**1st choice** LM0, 5, 6, 9, 10, 18, 25. **2nd choice:** LM2, 41, 13, 16, 20, 21, 22, 24, 26, 27, 28, 29, 30.
Marine—boats and ships, fittings.	**1st choice:** 5005, 5083, 51540, 5251, 54540, 6061TB, 6063TB, 6082TB. **2nd choice:** 5154H, 5454H, 6061TF, 6062TF, 6082TF, 7020TB and TF.	**1st choice:** LM5, 10. **2nd choice:** LM0, 9, 18, 25.
Chemical and food plant.	**1st choice:** 1080, 1050, 3103, 3105, 51040.	**1st choice:** LM0, 5, 10.
Structural items.	**1st choice:** 6061TB, 6063TB, 6082TB. **2nd choice:** 6061TF, 6063TF, 7020, 6082TF, 2014A (clad).	**1st choice:** LM5, 10. **2nd choice:** LM6, 9.

Table 2.5.5

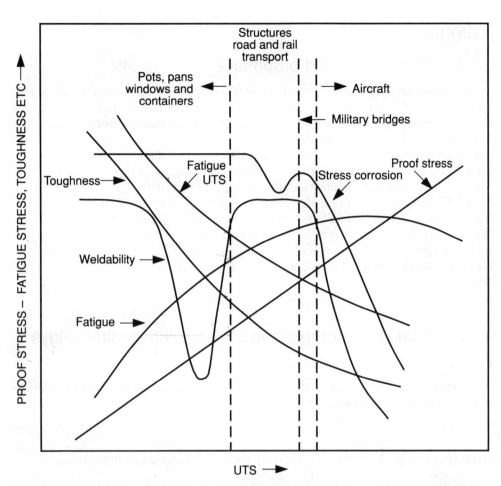

The diagram is not drawn to any scale; it is intended to show the compromise that has to be made in choosing a suitable alloy for a particular application

FIG 2.5.1 Variation of some properties of aluminium alloys with tensile strength

Fig 2.5.1

2.5.2   Castings

2.5.2.1   Introduction and properties

Three processes, sand, permanent mould and die-casting are commonly used for aluminium alloys. As a general rule, heat-treatable alloys are either sand or permanent mould cast, but die-castings are not generally suitable for heat treatment unless they have been made by special processes.

The compositions of the aluminium casting alloys are listed in:
Table 2.5.6   *Compositions of aluminium alloy ingots and castings*
They are rated according to their specific characteristics in:
Table 2.5.7   *Comparative rating of aluminium casting alloys*
Their physical and mechanical properties are listed in:
Table 2.5.8   *Typical properties of aluminium casting alloys*
and their elevated temperature mechanical properties in:
Table 2.5.9   *Mechanical properties of some casting alloys at elevated temperatures* *(short-term tests)*

2.5.2.2   Qualitative comparison of aluminium casting alloys

PURE ALUMINIUM

Pure aluminium is rarely used in the cast condition, but special castings may be required for electrical applications.

Al–Si

Silicon is the principal element for conferring good castability on aluminium alloys.

Al–Si–Mg

Good casting alloys, which can be solution-treated and age-hardened to give good mechanical properties. The alloys do age-harden at room temperature to some extent over a period of about 5 days following solution treatment.

Al–Si–Cu

A mixed class of good casting alloys, many of which have properties which are useful in the as-cast condition, although they may be further improved by solution treatment and precipitation. In some cases the solution treatment is dispensed with. In the case of the hypereutectic alloy LM30 only stress relieving is often carried out. Shock resistance is generally low.

Al–Mg

The 10% Mg alloy has the best combination of strength and toughness of all the aluminium casting alloys. However, it is unfortunately one of the most difficult alloys to cast, being subject to porosity, high drossing and poor fluidity. For this reason only relatively few foundries attempt to cast it, since it requires specialised techniques. The 5% Mg alloy is a little easier to cast.

Al–Cu

Heat-treatable alloys of moderately high strength, medium or poor impact resistance, and fast casting properties. Poorest corrosion resistance of Al alloys.

Al–Zn–Mg

A heat-treatable alloy requiring special skill in overcoming its poor castability.

Al–Mn

A cheap, non-heat-treatable alloy of poor mechanical properties, fair castability, but exceptional for non-load-bearing application at temperatures of up to 500°C (for instance the gas burners of domestic cookers).

Al–Ag

Aluminium–silver alloys, which were developed for sand and permanent mould castings, can be strengthened by heat treatment to give properties rivalling those of forgings, and show promise for use at elevated temperatures. A typical composition is given and the properties of the alloy summarised in:

Table 2.5.10   *Composition and typical properties of an aluminium–silver casting alloy*

2.5.2.3   Low and elevated temperature properties of aluminium casting alloys

LOW TEMPERATURE PROPERTIES

In general the tensile strength and elongation of the pure metal and its alloys are both greater at sub-zero temperatures than at room temperature. None of the alloys suffers from brittleness at low temperatures, and there is no transition point below which brittle fracture occurs.

ELEVATED TEMPERATURES

Details of the elevated temperature properties of aluminium casting alloys are given in Table 2.5.9. Guidance on selection is given in:

Table 2.5.11   *Initial guide to aluminium alloys for use at elevated temperatures*

TABLE 2.5.6 Compositions of aluminium alloy ingots and castings

Alloy type	Nominal composition (ISO)	BS alloy no.	Al	Cu	Mg	Si	Fe	Mn	Ni	Zn	Pb	Sn	Ti	Notes
							Composition (%) (Single values are maxima unless otherwise stated)							
Pure Al	Al	LM0	99.50 min.	0.03	0.03	0.30	0.40	0.03	0.03	0.07	0.03	0.03	—	
Al–Si	Al-5 Si	LM 18	rem	0.1	0.10	4.5-6.0	0.6	0.5	0.1	0.1	0.1	0.05	0.2	
	Al-12 Si	LM6/L33	rem	0.1	0.10	10.0-13.0	0.6	0.5	0.1	0.1	0.1	0.05	0.2	
Al-Si-Mg	Al-5Si-Cu-Mg	LM16	rem	1.0-1.5	0.4-0.6	4.5-5.5	0.6	0.5	0.25	0.1	0.1	0.05	0.2	
		LM31												
	Al-5Si-Mg-Fe-Ni	DTD716/722/ 727/735	rem	0.1	0.3-0.8	3.5-6.0	0.6	0.5	0.1	0.1	0.05	0.05	0.25	
	Al-7Si-Mg	DTD 5028 LM25/L99	rem	0.1	0.20-0.6	6.5-7.5	0.5	0.3	0.1	0.1	0.1	0.05	0.2	
	Al-11Si-Mg-Cu	LM13	rem	0.7-1.5	0.8-1.5	10.0-13.0	1.0	0.5	1.5	0.5	0.1	0.1	0.2	
	Al-12Si-Mg	LM9	rem	0.1	0.2-0.6	10.0-13.0	0.6	0.3-0.7	0.1	0.1	0.1	0.05	0.2	
	Al-19Si-Mg-Cu-Ni	LM28	rem	1.3-1.8	0.8-1.5	17-20	0.7	0.6	0.8-1.5	0.2	0.1	0.1	0.2	Cr 0.5 Co 0.5
	Al-23Si-Mg-Cu-Ni	LM29	rem	0.8-1.3	0.8-1.3	22-25	0.7	0.6	0.8-1.3	0.2	0.1	0.1	0.2	Cr 0.6 Co 0.5
Al-Si-Cu	Al-2Si-1Cu-Fe-Ni-Mg	L51/LM23	rem	0.8-2.0	0.05-0.20	1.5-2.8	0.8-1.4	0.1	0.8-1.7	0.1	0.05	0.05	0.25	
	Al-2Si-10Cu	LM12	rem	9.0-11.0	0.2-0.4	2.5	1.0	0.6	0.5	0.8	0.1	0.1	0.2	
	Al-5Si-1Cu	L78	rem	1.0-1.5	0.4-0.6	4.5-5.5	0.6	0.5	0.25	0.10	0.05	0.05	0.25	
	Al-5Si-3Cu	LM4	rem	2.0-4.0	0.15	4.0-6.0	0.8	0.2-0.6	0.3	0.5	0.1	0.1	0.2	
	Al-5Si-3Cu-Mn	LM22	rem	2.8-3.8	0.05	4.0-6.0	0.6	0.2-0.6	0.15	0.15	0.1	0.05	0.2	
	Al-6Si-4Cu-Zn	LM21	rem	3.0-5.0	0.1-0.3	5.0-7.0	1.0	0.2-0.6	0.3	2.0	0.2	0.1	0.2	
	Al-7Si-2Cu	LM27	rem	1.5-2.5	0.3	6.0-8.0	0.8	0.2-0.6	0.3	1.0	0.2	0.1	0.2	
	Al-8Si-3Cu-Fe	LM24	rem	3.0-4.0	0.30	7.5-9.5	1.3	0.5	0.5	3.0	0.3	0.2	0.2	
	Al-9Si-3Cu-Mg	LM26	rem	2.0-4.0	0.5-1.5	8.5-10.5	1.2	0.5	1.0	1.0	0.2	0.1	0.2	
	Al-10Si-2Cu-Fe	LM2	rem	0.7-2.5	0.30	9.0-11.5	1.0	0.5	0.5	2.0	0.3	0.2	0.2	
	Al-11Si-Cu-Mg	LM13	rem	0.7-1.5	0.8-1.5	10.0-12.0	1.0	0.5	1.5	0.5	0.1	0.1	0.2	
	Al-12Si-Cu-Fe	LM20	rem	0.4	0.2	10.0-13.0	1.0	0.5	0.1	0.2	0.1	0.1	0.2	
	Al-17Si-4Cu	LM30	rem	4.0-5.0	0.4-0.7	16-18	1.1	0.3	0.1	0.2	0.1	0.1	0.2	
Al-Mg	Al-5Mg	LM5	rem	0.1	3.0-6.0	0.3	0.6	0.3-0.7	0.1	0.1	0.05	0.05	0.2	
	Al-10Mg	LM10/L53	rem	0.1	9.5-11.0	0.25	0.35	0.10	0.10	0.10	0.05	0.05	0.2	
Al-Cu-X	Al-2Cu-Mg-Si-Fe-Ni	L52	rem	1.3-3.0	0.5-1.7	0.6-2.0	0.8-1.4	0.1	0.5-2.0	0.1	0.05	0.05	0.25	
	Al-4Cu-Si	L154/155	rem	3.8-4.5	0.10	1.0-1.5	0.25	0.1	0.1	0.1	0.05	0.05	0.5-0.25	
	Al-4Cu-Ni	L119	rem	4.5-5.5	0.10	0.30	0.50	0.20-0.30	1.3-1.7	0.10	0.05	0.05	+Zr 0.50	Zr 0.10-0.30 Co 0.10-0.30 Sb 0.10-0.30
	Al-4Cu-Ti	L91/L92	rem	4.0-5.0	0.10	0.25	0.25	0.10	0.10	0.10	0.05	0.05	0.25	
	Al-4Cu-2Mg-2Ni	L35	rem	3.5-4.5	1.2-1.7	0.6*	0.6*	0.6	1.8-2.3	0.1	0.1	0.05	0.25	*Si+Fe 1.0
	Al-10Cu-2Si	LM12	rem	9.0-11.0	0.2-0.4	2.5	1.0	0.6	0.5	0.8	0.1	0.1	0.2	
Al-Zn-Mg	Al-5Zn-Mg	DTD5008	rem	0.1	0.5-0.75	0.25	0.5	0.1	0.1	4.8-5.7	0.05	0.05	0.25	Cr 0.5-0.6
	Al-7Mg-Zn	DTD5018	rem	0.2	7.4-7.9	0.25	0.35	0.1-0.3	0.10	0.9-1.4	0.05	0.05	0.25	

Table 2.5.6

TABLE 2.5.7 Comparative rating of aluminium casting alloys

Alloy type	Nominal Composition	BS alloy No.	Tensile strength		Ductility		Casting characteristics						Machinability	Corrosion resistance	Finishing Suitability for				
							Suitability for										Anodising		
			AC	HT	AC	HT	Sand	Perm. mould	Die	Fluidity	Resistance to hot tearing	Pressure tightness			Plating	Vitreous enamelling	Protective	Colour	Bright
Pure Al	Al	LM0	2	—	5	—	2	1	1	1	1	2	2	5	5	5	5	5	5
Al–Si	Al–5Si	LM18	3	—	4	—	4	4	1a	3	5	5	2	5	2	5	3	2b	0
	Al–12Si	LM6/L33	3	—	4	—	5	5	5	5	5	5	2	5	2	5	2	0	0
Al–Si–Mg	Al–5Si–Cu–Mg	LM16	—	5	—	2	4	3	1a	3	3	3	3	3	2	—	3	2b	0
	Al–7Si–Mg	DTD 5028 LM25/L99	3	5	3	2	3	5	1a	3	3	3	2	5	2	—	3	2b	0
	Al–11Si–Mg–Cu	LM13	—	5	—	1	3	3	1a	3	5	2	2	3	2	—	2	0	0
	Al–12Si–Mg	LM9	3	5	2	1	3	5	1a	3	5	3	2	5	2	—	2	0	0
	Al–19Si–Mg–Cu–Ni	LM28	—	3	—	1	1	2	1a	2	3	2	1	3	2	—	0	0	0
	Al–23Si–Mg–Cu–Ni	LM29	—	3	—	1	1	2	1a	2	3	2	1	3	2	—	0	0	0
Al–Si–Cu	Al–2Si–1Cu–Fe–Ni–Mg	L51	—	3	—	2	1	—	—	—	—	—	—	—	—	—	—	—	—
	Al–2Si–10Cu	LM12	3	—	—	—	2	3	0	2	3	3	5	1	3	—	2	2	0
	Al–5Si–1Cu	L78	—	5	—	—	—	—	—	—	—	—	—	—	—	—	—	—	—
	Al–5Si–3Cu	LM4	3	5	2	2	3	3	4	3	3	3	3	3	3	—	3	2b	0
	Al–5Si–3Cu–Mn	LM22	—	4	—	—	2a	3	1a	3	3	3	3	3	3	—	3	2	0
	Al–6Si–4Cu–Zn	LM21	3	—	2	—	3	3	2a	3	3	3	3	3	2	—	2	0	0
	Al–7Si–2Cu	LM27	3	—	2	—	3	5	2a	3	3	3	3	3	3	—	3	2b	0
	Al–8Si–3Cu–Fe	LM24	3	—	2	—	2a	2a	5	3	3	3	2	3	2	—	2	2b	0
	Al–9Si–3Cu–Mg	LM26	—	4	—	—	3	3	2a	3	3	2	2	3	2	—	2	0	0
	Al–10Si–2Cu–Fe	LM2	3	—	2	—	2a	2a	5	3	5	4	2	3	2	—	2	0	0
	Al–12Si–Cu–Fe	LM20	4	—	2	—	5a	5	3	5	5	5	2	3	3	5	3	2b	0
	Al–17Si–4Cu	LM30	3	—	—	—	—	2	3	3	3	2	1	3	2	—	0	0	0
		LM31																	
Al–Mg	Al–5Mg	LM5	4	—	5	—	3	3	1a	2	2	1	3	5	2	0	5	5	3
	Al–10Mg	LM10/L53	—	6	—	5	2	2	1a	2	3	1	3	5	0	0	5	2	0

Strength (MPa)	14–70	71–140	141–210	211–280	281–350	351 +
Rating	1	2	3	4	5	6
Ductility % El.	0–1.0	1.1–3.0	3.1–5.0	5.1–10.0	10.1 +	
Rating	1	2	3	4	5	

AC – as cast.
HT – heat treated.
a Not usually cast by this method.
b Dark colours only.

Other ratings

5 — Excellent.

4 — Very good.

3 — Good.

2 — Fair.

1 — Poor.

0 — Unsuitable.

— — No data.

Table 2.5.7

TABLE 2.5.8 Typical properties of aluminium casting alloys

Alloy type	BS	Alloy type	Density	Melting range[a] (°C)	Coefficient of linear expansion (10⁻⁶ K⁻¹)	Thermal Conductivity at 25°C W/m per K	%IACS	Electrical resistivity at 20°C (μΩcm)[b]	conductivity at 20°C (% IACS)
Pure Al	LM0 –M	Al 99.5	2.70	643–657	24	209	53.1	3.02	57
Al–Si	LM18–M	Al Si5	2.69	565–625	22	142	36.1	4.65	37
	LM6 –M	Al Si12	2.65	565–575	20	142	36.1	4.65	37
Al–Si–Mg	LM16–TB –TF	Al Si15 Cu1 Mg	2.70	560–625	23	142	36.1	4.78	36
	LM25–M –TE –TB7 –TF	Al Si7 Mg	2.68	550–615	22	151	38.4	4.4	39
	LM13–TE –TF –TF7	Al Si11 Mg Cu	2.70	525–560	19	117	29.7	5.95	29
	LM9 –M –TE –TF	Al Si12 Mg	2.68	550–575	22	147	37.3	4.55	38
	LM28–TE –TF	Al Si19 Cu Mg Ni	2.68	520–675	18	134	34.0	—	—
	LM29–TE –TF	Al Si23 Cu Mg Ni	2.65	520–770	16	126	32.5	—	—
Al–Si–Cu	LM12–M	Al Cu10 Si2 Mg	2.94	525–625	22	130	33.0	5.23	33
	L78 –TF	Al 5Si 1Cu	2.74	525–625	21	121	30.7	5.56	31
	LM4 –M –TF	Al Si5 Cu3	2.75	525–625	21	121	30.7	5.39	32
	LM22–TB	Al S5 Cu3 Mn	2.77	520–620	21	121	30.7	5.39	32
	LM21–M	Al Si6 Cu4 Zn	2.81	520–615	21	121	30.7	5.39	32
	LM27–M	Al Si7 Cu2	2.75	525–605	21	155	39.4	6.39	27
	LM24–M	Al Si8 Cu3 Fe	2.79	520–590	21	96	24.5	7.18	24
	LM26–TE	Al Si9 Cu3 Mg	2.76	520–580	21	105	26.7	6.63	26
	LM2 –M	Al Si10 Cu2 Fe	2.74	525–570	20	100	25.4	6.63	26
	LM20–M	Al Si12 Cu Fe	2.68	565–575	20	155	39.4	4.65	37
	LM30–M –TS	Al Si17 Cu4 Mg	2.73	505–650	18	134	34.0	8.62	20
Al–Mg	LM5 –M	Al Mg5	2.65	580–642	23	138	35.1	5.56	31
	LM10–TB	Al Mg10	2.57	450–620	25	88	22.3	8.62	20
Al–Cu–X	L51 –TE	Al Cu2 Si1.3 Ni1.3 Mg1 Fe1	2.75	600–645	22.5			4.3	40
	L52 –TF	Al Cu2.2 Si1.3 Ni1.3 Mg1 Fe1	2.75	600–645	22.5	155	39	4.3	40
	L91 –TE	Al Cu4	2.80	545–640	23	138	35	4.9	35
	L92 –TF	Al Cu4	2.80	545–640	23	138	35	4.9	35
	L119 –TF	Al Cu4.5 Ni1.5 Mn Fe Zn CoSb	2.84	545–650	22.5	155		4.7	36
	L35 –TF (Y alloy)	Al Cu4 Ni2 Mg2	2.82	530–640	22.5	133	33	5.2	33
Al–Zn–Mg	LM31	Al Zn5 Mg	2.75		24	103		6.0	25
	DTD 5008 (age hardened)	Al Zn5 Mg	2.75		24	135	34.5	6.0	25
	DTD 5018 TB	Al Mg7.7 Zn1.2	2.64		24	100	25.4	7.2	24
Al–Mn	M	Al Mn1.5	2.7	650–658	23				

[a] Values are approximate, since actual temperatures depend on rate of heating.
[b] μ Ω cm = Ω m × 10⁻⁸.

Table 2.5.8

TABLE 2.5.8 Typical properties of aluminium casting alloys—continued

0.2% Proof stress		Tensile strength		Elongation (%)		Fatigue strength		Hardness Brinell		Max. service Tempera-ture (°C)	Modulus of elasticity (GPa)
Sand cast (MPa)	Chill cast (MPa)	Sand cast (MPa)	Chill cast (MPa)	Sand cast	Chill cast 5.66 $\sqrt{S_o}$	50×10^7 cycles Sand cast (MPa)	Chill cast (MPa)	Sand cast	Chill cast		
30	30	80	80	30	40	—	—	25	25		69
70	80	120	150	5	6	45	60	40	50		71
70	80	170	200	8	13	51	68	55	60	150	71
130	140	210	250	3	6	70	85	80	85		71
240	270	280	310	1	2	70	93	100	110		
90	90	140	180	2.5	5	—	—	60	60		
130	150	170	220	1.5	2	55	—	70	80	150	71
100	100	170	230	3	8	—	75	65	65		
220	240	250	310	1	3	60	95	105	105		
—	—	—	—	—	—	—		—	—		
200	280	200	290	—	1	85	100	115	125		71
140	200	150	210	1	1	—	—	75	75		
—	—	—	200	—	3	—	—	—	—		
120	150	180	250	2	2.5	55	75	70	80		71
220	280	250	310	—	1	70	90	100	100		
—	170	—	190	—	0.5	—	—	—	120		82
120	170	130	200	0.5	0.5	—	—	120	120		
120	170	130	210	0.3	0.3	—	—	120	120		88
120	170	130	210	0.3	0.3	—	—	120	120		
—	150	—	180	—	—	—	60	—	95		71
240	270	280	310	~1	~2	70	93	100	110		71
100		150		2		75	85	70		150	71
250	250	280	310	1	3	—	—	105	110		
—	120	—	260	—	9	—	—	—	75		71
130	130	180	200	1	2	—	—	85	90		71
90	100	150	180	2	3	—	—	75	80		71
—	110	—	200	—	2	—	—	—	85		71
—	180	—	230	—	1	—	—	—	105		71
—	90	—	180	—	2	—	—	—	80	150	71
—	80	—	220	—	7	—	80	—	60		71
—	150	—	150	—	—	—	—	—	110		82
—	150	—	160	—	—	—	—	—	110		
90	90	170	230	5	10	54	100	60	65		71
180	180	310	360	15	20	60	—	85	95		71
125	140	160–180	200–225	2–3	3–4			70	75	200	71
245–280	295–325	280–295	325–355	1	1	76	—	100–150	120–150	250	71
165–200	165–200	220–250	265–310	7–10	13–15	70	93	80	90	200	71
200–240	200–240	280–310	310–350	4–6	9–12	70	93	100	105	250	71
190–200	—	215–225	—	1	—	100	—	80–100	—	250	68
210–240	230–260	220–250	280–310	1	1–3	80	110	100–130	100–130	270	71
170	—	215	—	4	—			75		150	
170	—	215	—	4	—			75		150	70
170	170	275	305	10		—	—				69
117	108	140	180	1–3	—					500	

Data courtesy of The Aluminium Federation and other sources.

Table 2.5.8—continued

TABLE 2.5.9 Mechanical properties of some casting alloys at elevated temperatures (short-term tests)

Alloy type	Material designation		Property	Temperature (°C)							
	ISO	BS		20	100	150	200	250	300	350	400
Al–Si	Al Si12	LM6–M	PS	70	67	57	52	43	30	24	16
			TS	183	155	130	105	76	59	42	31
			EL	12.5	12.0	12.0	17.5	17.5	18.0	21.5	27.0
			BHN	55.8	51.2	43.1	36.8	28.1	20.2	12.8[a]	7.5
			FS	52	44		33		25		
Al–Si–Mg	Al Si7 Mg	LM25–TF[b]	PS	206	193	154	59	34	20	14	—
			TS	234	207	160	83	50	28	17	—
			EL	2	2	6	18	35	60	80	—
Al–Si–Cu	Al Si5 Cu 3	LM4–M	PS	117	105	88	114	98	70	40	25
			TS	176	164	147	155	125	88	52	35
			EL	1.4	1.8	1.9	2.0	2.4	3.0	6.0	9.0
			BHN	68.5	57.1	56.4	50.6	37.2	20.3	13.5[a]	9.6
			FS	63	54		60		41		
	Al Si11 Cu 3	LM13–TF Chill cast	TS	332	325	308	270	155	92	54	37
Al–Mg	Al Mg5	LM5–M	PS	83	83	84	95	97	65	44	30
			TS	147	152	162	150	132	94	64	37
			EL	4.5	5.0	8.3	5.5	7.5	8.0	8.0	8.5
			BHN	63.9	57.2	53.8	52.8	41.5	28.7	17.0[a]	11.5
Al–Cu–X	Al Cu4 Ni2 Mg	L35 Fully heat-treated	PS	238	236	225	233	220	154	65	33
			TS	255	260	235	249	220	174	88	50
			EL	0.3	0.6	0.5	0.4	0.3	1.1	6.3	15.0
			BHN	113	94.8	100	99.3	64.2	32.8	24.2[a]	9.7
	Al Cu4.5	2L92 Fully heat-treated	PS	202	196	185	185	125	76	54	16
			TS	322	303	262	235	170	122	70	25
			EL	4.8	6.0	4.0	3.4	5.0	10.5	13.0	17.0
			BHN	94.5	87.7	82.8	62.5	43.9	28.1	8.7[a]	5.9
	Al Cu10 Si2 Mg	LM12–TF	TS	364	355	340	308	232	138	77	

PS = 0.1% proof stress, MPa.
TS = Tensile strength, MPa.
FS = Fatigue strength in MPa at 10×10^6 cycles.
EL = Percentage elongation on $4\sqrt{S_o}$ test piece.
BHN = Brinell hardness, measured by applying 125 kg load for 30 s with 5 mm ball; where marked [a] load was reduced to 25 kg. With the two exceptions marked, determinations were made with sand cast test bars.
      [b] The figures for LM25 are based on a soaking of 10 000 h, whereas the other figures are for 1000 h.
M   = As cast.
TF = Solution heat-treated and precipitation-treated.

Table 2.5.9

TABLE 2.5.10 Composition and typical properties of an aluminium–silver casting alloy

Property	Value	
Melting range, °C Specific gravity Thermal expansion coefficient, K^{-1} Electrical conductivity, % IACS Thermal conductivity, W/m per K	570–650 2.796 34×10^{-6} 27–34 121.4	
	As cast	*Heat-treated*
Tensile strength, MPa 0.2% proof stress, MPa Elongation, % Brinell hardness Shear strength, MPa	410–420 250–320 15–17 130 290	440–450 330–370 8–12

Composition: Al–4.5% Cu–0.7% Ag–0.34% Mg–0.10% Si–0.14%
                        Fe–0.4% Mn–0.4% Cr.

Table 2.5.10

TABLE 2.5.11 Initial guide to aluminium alloys for use at elevated temperatures

UTS (MPa)	Temperature (°C)	BS References	
		Castings	*Wrought forms*
100–155	100	LM4, LM5, LM6, LM10TB, LM12 TF, LM13 TF, LM25 TF, L35, 2L92 TF.	1200 H8, 2014 T6, 2618 T6. 3103 H4–H8, 5083 O, 5251 O H4–H8, 5454 O/H2–H4, 6063 T6, 6082 T6, 7075 T6.
	150	LM4, LM5, LM6, LM10 TB, LM12 TF, LM13 TF, LM25 TF, L35, 2L92.	1200 H8, 2014 T6, 2618 T6, 3103 H4–H8, 5083 O, 5251 O/H4–H8, 5454 O/H2–H4, 6063 T6, 6082 T6, 7075 T6.
	200	LM4, LM5, LM6, LM10 TB, LM12 TF, LM13 TF, L35, 2L92.	2014 T6, 2618 T6, 5083 O, 5251 O/H4–H8, 5454 O/H2–H4, 7075 T6.
	250	LM4, LM5, LM10 TB, LM12 TF, LM13 TF, L35, 2L92.	5083 O, 5454 O/H2–H4, 6082 T6, 7075 T6.
	300	LM10 TB, LM12 TF, L35, 2L92.	—
150–200	100	LM4, LM5, LM6, LM10 TB, LM12 TF, LM13 TF, L25 TF, L35, 2L92.	1200 H8, 2014 T6, 2618 T6, 3103 H8, 5083 O, 5251 O/H4–H8, 5454 O/H2–H8, 6063 T6, 6082 T6, 7075 T6.
	150	LM5, LM10 TB, LM12 TF, LM13 TF, LM25 TF, L35, 2L92.	2014 T6, 2618 T6, 3103 H8, 5083 O, 5251 O/H4–H8, 5454 O/H2–H4, 6082 T6, 7075 T6.
	200	LM10 TB, LM12 TF, LM13 TF, L35, 2L92.	2618 T6, 5083 O, 5251 H4–H8, 5454 O/H2–H4.
200–250	100	LM10 TB, LM12 TF, LM13 TF, LM25 TF, L35, 2L92.	2014 AFT, 2618 T6, 5083 O, 5251 H4–H8, 5454 O/H2–H4, 6063 T6, 6082 T6, 7075 T6.
	150	LM10 TB, LM12 TF, LM13 TF, L35, 2L92.	2014 T6, 2618 T6, 5083 O, 5251 H4–H8, 5454 H2–H8, 7075 T6.
	200	LM10 TB, LM12 TF, LM13 TF, L35, 2L92.	2618 T6.
	250	LM12 TF, L35.	—

Based, in general, on static tests undertaken at temperature after soaking at the same temperature.

Table 2.5.11

2.5.3   Wrought products

2.5.3.1   Qualitative comparison of wrought aluminium alloys

ALUMINIUM

Superpurity aluminium (99.99+%) is limited to certain chemical plant items, flashing for buildings and other applications requiring maximum resistance to corrosion and/or high ductility, justifying its high cost.

Commercial purity metal (99.00–99.80%) is available in three purities and a range of work-hardened grades, for a wide variety of general applications plus a special composition for electrical purposes.

Al–Mn(–Mg)

The addition of about 1.25% Mn increases strength without impairing ductility. The alternative alloy with small amounts of both Mn and Mg has slightly higher strength whilst retaining good ductility.

A special alloy containing up to 2.5% Mn and 2.5%Ni has found limited use for pressure die-castings or hot and cold formed items for use at raised temperatures.

Al–Mg

There are five standard compositions with Mg contents up to 4.9%, with Mn or Cr in small amounts. There are work-hardening alloys with high to moderate strength and ductility, and high resistance to sea-water corrosion, but alloys with more than 3.5% Mg require care as corrosion resistance may be impaired. They are readily welded using filler metal of slightly higher Mg content than the parent metal and anodise well.

Al–Mg–Si

These heat-treatable alloys, containing small additions of Mn or Cr, are very readily extruded and provide moderate strength with good ductility. They can be readily welded and anodised.

Al–Cu–Mg

These involve the first age-hardening alloys ('Duralumin') and now cover a range of compositions. They are high strength materials but their copper content reduces corrosion resistance. Rolled plate and sheet are often clad with a layer of pure aluminium approximately 5% of the sheet thickness on each side: 'Alclad' is the well known trade name.

Al–Zn–Mg

Additions of up to 5% Zn and 1.25% Mg give alloys of maximum strength but their lower corrosion resistance requires care in their production and use. A lower range of Zn/Mg additions provide reasonable levels of strength and good weldability. Rolled flat products may be clad with Al–1% Zn alloy.

Al–Li

A series of alloys of special interest for aircraft purposes, based upon lithium additions of 2–3%, plus Cu and Mg. Aluminium lithium alloys are lower in density, (2.53–2.59 of 2.80 for conventional alloys), 10% higher in elastic modulus and approximately 3 × price of conventional aluminium alloys.

Al–Sn

Non-standard alloys of aluminium with 6–7% Sn are used as bearing materials, particularly clad on to steel shells for automobile engines and similar applications.

2.5.3.2   Quantitative comparison of wrought aluminium alloys

The compositions of wrought alloys are listed in:

Table 2.5.12   *Composition of wrought aluminium alloys*

Temper designations, British and the international equivalents are given in:

Table 2.5.13   *British temper designation and international equivalents*

The physical and mechanical properties of wrought aluminium alloys are listed in:

Table 2.5.14   *Typical properties of wrought aluminium alloys*

A tabulated guide to the relative merits of each aluminium alloy according to a number of significant criteria will be found in:

Table 2.5.15   *Comparative characteristics of wrought aluminium alloys*

The effects of temperature on the tensile properties of wrought aluminium alloys are listed in:

Table 2.5.16   *Typical properties of wrought aluminium alloys at various temperatures*

TABLE 2.5.12 Composition of wrought aluminium alloys

	Alloy designation			Composition (%) (single values are maxima unless otherwise stated)			
	ISO (nominal composition)	BS		Al	Si	Fe	Cu
		1986	Former				
Pure aluminium	Al 99.99	—	1	99.99 min.	*	*	*
	Al 99.8	1080A	1A	99.8 min.	0.15	0.15	0.03
	Al 99.5	1050A	1B	99.5 min.	0.25	0.40	0.05
	Al 99.0	1200	1C	99.0 min.	Si + Fe = 1.0		0.05
	Al 99.95		1E	rem		—	0.05
Non-heat-treatable	Al–12 Si		N2	rem	10.0–13.0	0.6	0.10
	Al–5 Si		N21	rem	4.5–6.0	0.6	0.10
	Al–1 Mn	3103	N3	rem	0.50	0.7	0.10
	Al Mn Mg	3105	N31	rem	0.6	0.7	0.30
	Al Mg 1	5005	N41	rem	0.30	0.7	0.20
	Al Mg 4.5 Mn	5083	N8	rem	0.40	0.40	0.10
	Al Mg 3.5	5154	N5	rem	0.50	0.50	0.10
	Al Mg 2 Mn	5251	N4	rem	0.40	0.50	0.15
	Al Mg 3 Mn	5454	N51	rem	0.25	0.40	0.10

Table 2.5.12

TABLE 2.5.12 Composition of wrought aluminium alloys—*continued*

Composition (%) (single values are maxima unless otherwise stated)						Notes
Mn	Mg	Ni	Zn	Ti	Cr	Notes
	—	—	—			*Cu + Si + Fe = 0.01 max.
0.02	0.02	—	0.06	0.02	—	Ga 0.03. Others 0.02 each
0.05	0.05	—	0.07	0.05	—	Others 0.03 each
0.05	—	—	0.10	0.05	—	Others 0.05 each. Total 0.15
—	—	—	—	—	—	(For electrical uses)
0.5	0.2	—	0.2	—	—	
0.5	0.2	—	0.2	—	—	
0.9–1.5	0.30	—	0.20	—		Ti + Zr = 0.10
0.30–0.8	0.20–0.8	—	0.20	0.10	0.20	Others 0.05 each. Total 0.15
0.20	0.50–1.1	—	0.25	0.15	0.10	
0.40–1.0	4.0–4.9	—	0.25	—	—	
0.50	3.1–3.9	—	0.20	0.20	0.25	Mn + Cr = 0.10–0.50
0.10–0.50	1.7–2.4	—	0.15	0.15	0.15	
0.50–1.0	2.4–3.0	—	0.25	0.20	0.05–0.20	

Table 2.5.12—*continued*

TABLE 2.5.12 Composition of wrought aluminium alloys—*continued*

Alloy designation			Composition (%) (single values are maxima unless otherwise stated)			
ISO (nominal composition)	BS		Al	Si	Fe	Cu
	1986	Former				
Al Mg Si	6060	H9	rem	0.30–0.6	0.10–0.30	0.10
Al Mg Si	6061	H20	rem	0.40–0.8	0.7	0.15–0.40
Al Mg Si	6063	91E	rem	0.20–0.6	0.35	0.10
Al Si 1 Mg Mn	6082	H30	rem	0.7–1.3	0.50	0.10
Al Mg Si	6463	4300/4	rem	0.20–0.6	0.15	0.20
Al Cu 4 Si Mg	2014	H15	rem	0.50–0.9	0.50	3.9–5.0
Al Cu 4 Mg 1.5	2024	H14	rem	0.50	0.50	3.8–4.9
Al Cu 2 Mg 1.5 Ni 1	2031	H12	rem	0.50–1.3	0.6–1.2	1.8–2.8
Al Cu 2 Mg 1 Ni 1 Fe Si	2618A	H18	rem	0.15–0.25	0.9–1.4	1.8–2.7
Al Zn 4.5 Mg 1	7020	4300/14	rem	0.35	0.40	0.20
Al Li 2.5 Cu 1 Mg	DTD XXX	—	rem	0.20	0.30	1.0–1.40
—		BTR1	rem	0.15	0.3–0.8	0.10
—		BTR2	rem	0.15	0.7–1.2	0.10
—		BTR6	rem	0.20	0.4–0.8	0.2–0.5
Al Cu 5.5 Pb Bi		FC1	rem	5.0–6.0	0.1	0.40

Heat-treatable / Bright trim / Free-cutting

Table 2.5.12—*continued*

TABLE 2.5.12 Composition of wrought aluminium alloys—*continued*

Composition (%) (single values are maxima unless otherwise stated)						Notes
Mn	**Mg**	**Ni**	**Zn**	**Ti**	**Cr**	**Notes**
0.10	0.35–0.6	—	0.15	0.10	0.05	Others 0.05 each. Total 0.15
0.15	0.8–1.2	—	0.25	0.15	0.04–0.35	Others 0.05 each. Total 0.15
0.10	0.45–0.9	—	0.10	0.10	0.10	Others 0.05 each. Total 0.15
0.40–1.0	0.6–1.2	—	0.20	0.10	0.25	Others 0.05 each. Total 0.15
0.05	0.45–0.9	—	0.05			Others 0.05 each. Total 0.15
0.40–1.2	0.20–0.9	0.10	0.25	0.15	0.10	Ti + Zr = 0.20. Total other 0.15
0.30–0.9	1.2–1.8	—	0.25		0.10	Others 0.5 each. Total 0.15
0.50	0.6–1.2	0.6–1.4	0.20	0.20	—	Others 0.05 each. Total 0.15
0.25	1.2–1.8	0.6–1.4	0.20	0.20	—	Others 0.05 each. Total 0.15
0.05–0.05	1.0–1.4	—	4.0–5.0	Ti + Zr = 0.8–0.25	0.10–0.35	Zr 0.08–0.20. Others 0.05 each. Total 0.15
—	0.5–0.90	Li 2.30–2.60	—	Zr 0.10–0.14		Na 0.002. Others 0.10 each. Total 0.20
0.10	0.2	—	—	0.05	—	
0.10	0.3	—	—	0.05	—	
0.15	0.05	—	—	0.05	—	
0.7	0.2	0.2	0.2	0.2		Pb : 0.20–0.70 Sn : 0.05 Bi : 0.20–0.70 Sb : 0.05

Table 2.5.12—*continued*

TABLE 2.5.13 British temper designation and international equivalents

United Kingdom F As Fabricated. M As Manufactured (formerly).	Canada F As Fabricated	France F As Fabricated	Germany O As Fabricated	Italy O No Treatment	USA F As Fabricated
O Annealed	O Annealed Recrystallised	O Annealed Recrystallised	.1 Soft, annealed Recrystallised with or without cold worked by flattening	R Annealed	O Annealed Recrystallised
H.2 Strain Hardened ¼ Hard	H.12 Strain Hardened ¼ Hard	H.12 Strain Hardened ¼ Hard	.2 Cold Worked ¼ to ¾ Hard	H.25 Numbers after 'H' show the cold working in % of the reduction of the section.	H.12 Strain Hardened ¼ Hard
H.4 Strain Hardened ½ Hard	H.14 Strain Hardened ½ Hard	H.14 Strain Hardened ½ Hard	.2 Cold Worked ¼ to ¾ Hard	H.25 Numbers after 'H' show the cold working in % of the reduction of the section.	H.14 Strain Hardened ½ Hard
H.6 Strain Hardened ¾ Hard	H.16 Strain Hardened ¾ Hard	H.16 Strain Hardened ¾ Hard	.2 Cold Worked ¼ to ¾ Hard	—	H.16 Strain Hardened ¾ Hard
H.8 Strain Hardened Full Hard	H.18 Strain Hardened Full Hard	H.18 Strain Hardened Full Hard	.3 Cold Worked Hard	—	H.18 Strain Hardened Full Hard
—	H.19 Strain Hardened Extra Hard	—	—	—	—
H.22 Strain Hardened partially annealed ¼ Hard	H.22 Strain Hardened partially annealed ¼ Hard	H.22 Strain Hardened partially annealed ¼ Hard	.2 Cold Worked ¼ to ¾ Hard	—	H.22 Strain Hardened partially annealed ¼ Hard
H.24 Strain Hardened partially annealed ½ Hard	H.24 Strain Hardened partially annealed ½ Hard	H.24 Strain Hardened partially annealed ½ Hard	.2 Cold Worked ¼ to ¾ Hard	—	H.24 Strain Hardened partially annealed ½ Hard
	H.26 Strain Hardened partially annealed ¾ Hard	H.26 Strain Hardened partially annealed ¾ Hard	.2 Cold Worked ¼ to ¾ Hard	—	H.26 Strain Hardened partially annealed ¾ Hard
H.28 Strain Hardened partially annealed Full Hard	H.28 Strain Hardened partially annealed Full Hard	H.28 Strain Hardened partially annealed Full Hard	.3 Cold Worked Hard	—	H.28 Strain Hardened partially annealed Full Hard
H.12 Strain Hardened after annealing Stabilised ¼ Hard	H.32 Strain Hardened Stabilised ¼ Hard	H.32 Strain Hardened Stabilised ¼ Hard	.2 Cold Worked ¼ to ¾ Hard	—	H.32 Strain Hardened Stabilised ¼ Hard

Table 2.5.13

TABLE 2.5.13 British temper designation and international equivalents—*continued*

United Kingdom	Canada	France	Germany	Italy	USA
H.14 Strain Hardened after annealing Stabilised ½ Hard	H.34 Strain Hardened Stabilised ½ Hard	H.34 Strain Hardened Stabilised ½ Hard	.2 Cold Worked ¼ to ¾ Hard	—	H.34 Strain Hardened Stabilised ½ Hard
H.16 Strain Hardened after annealing Stabilised ¾ Hard	H.36 Strain Hardened Stabilised ¾ Hard	H.36 Strain Hardened Stabilised ¾ Hard	.2 Cold Worked ¼ to ¾ Hard	—	H.36 Strain Hardened Stabilised ¾ Hard
H.18 Strain Hardened after annealing Stabilised Full Hard	H.38 Strain Hardened Stabilised Full Hard	H.38 Strain Hardened Stabilised Full Hard	.3 Cold Worked Hard	—	H.38 Strain Hardened Stabilised Full Hard
	—	—	—	—	H.321/H.323 Strain Hardened ¼ Hard Stabilised
	—	—	—	—	H.343 Strain Hardened ½ Hard Stabilised
T.3 (formerly TD) Solution-treated cold worked, naturally aged	T.3 Solution heat-treated and then cold worked	T.3 Solution heat-treated. Cold worked by flattening	.5 Solution heat-treated. Cold worked by flattening	THN Solution heat-treated. Cold worked by flattening	T.3 Solution heat-treated. Cold worked by flattening
T.4 (formerly TB) Solution heat-treated naturally aged (no cold work after solution heat treatment, except as may be required for flattening or straightening).	T.4 Solution heat-treated and naturally aged to a substantially stable condition.	T.4 Solution heat-treated	.4 Solution heat-treated	TN Solution heat-treated	T.4 Solution heat-treated
TE Cooled from an elevated shaping process and precipitation-treated.	T.5 Artificially aged only		—	—	T.5 Artificially aged only
T.6 (formerly TF) Solution heat-treated and precipitation-treated	T.6 Solution heat-treated and then artificially aged	T.6 Solution heat-treated and artificially aged	.7 Solution heat-treated and artificially aged	TA Solution heat-treated and artificially aged	
T.8 (formerly TH) Solution heat-treated. Cold worked. Precipitation-treated	T.8 Solution heat-treated. Cold worked. Artificially aged	T.8 Solution heat-treated. Cold worked. Artificially aged	—	THA Solution heat-treated. Cold worked. Artificially aged	T.8 Solution heat-treated. Cold worked. Artificially aged

Table 2.5.13—*continued*

TABLE 2.5.14 Typical properties of wrought aluminium alloys

	Designation		Density (g/cm³)a	Meltingb range (°C)	Coefficient of linear expansion (20–100°C) 10^{-6}/°C	Thermal Conductivity 0–100° Ω		Electrical		
	BS	Alloy type				W/m°C	% IACS	Resistivity ($\mu\Omega$ cm)c	Conductivity (20°C) (% IACS)	
Pure aluminium	1–0	Al 99.99	2.70	(660)	23.5	244	61.9	2.7	63.8	
	1080–0 –HB	Al 99.8	2.70	(645)	24	230	58.4	2.8	61.6	
	1050A–0 –HB	Al 99.5	2.71	(635)	24	230	58.4	2.8	61.6	
	1200–0 –H4 –H8	Al 99.0	2.71	660 (630)	24	226	57.4	2.9	59.5	
Non-heat-treatable alloys	3103–0 –H4 –H8	Al Mn 1	2.73	645–655 (640)	23	172	43.7	4.0	43.1	
	3105–0 –H4 –H8	Al Mn Mg	2.70	625–650	24		(43)	4.0	43.1	
	5005–0 –H3 –H6	Al Mg 1	2.69	595–650 (600)	24	155	39.4	4.7	36.7	
	5251–0 –H4 –H8	Al Mg 2	2.69	595–640	24		(38)			
	5154–0 –H2 –H4	Al Mg 3.5	2.67	600–640 (590)	23	138	35.1	5.4	31.9	
	5454–0 –H2 –H4	Al Mg3 Mn	2.68	600–640	24	147	37.3	5.1	33.8	
	5083–0 –H2 –H4	Al Mg 4.5 Mn	2.67	580–645	24.5	109	27.7	6.1	28.3	

aNumerically identical with specific gravity. 1 g/cm³= 10^3 kg/m³.

bFigures are approximate, since actual temperature depends on rate of heating. Figures in parentheses show first sign of melting with very rapid heating.

c$\mu\Omega$ cm = Ωm × 10^{-8}.

Since the tensile properties, particularly, vary with form, shape and size, the figures given are an indication for the sake of comparison in choosing a suitable alloy.

(Data courtesy of The Aluminium Federation)

Table 2.5.14

TABLE 2.5.14 Typical properties of wrought aluminium alloys—*continued*

0.2% Proof stress (MPa)	Tensile strength (MPa)	Elongation (%)		Shear strength (MPa)	Fatigue strength at 50×10^6 cycles (MPa)	Hardness		Service temperature max. (°C)	Modulus of elasticity (GPa)
		on 5.65 $\sqrt{S_o}$	on 50 mm			Brinell	Vickers		
—	58	—	40					—	69
50 125	75 135	—	32 5	60		19		100	69
55 130	75 145	25 —	32 5			21		100	69
60 110 140	80–105 90–125 140–160	20	20–30 4–8 2–4	50 80 100	40 55 76	22 33 40		100	69
70 120 130	90–130 140–175 180–210		20–24 5–9 2–4	70 90 105	54 70 85	29 42 51		100	69
— 150 190	110–155 160–195 215–240		16–20 2–5 1–2	70 85 120		33 50 55		100	69
— 100 150	95–145 125–180 185–210	16	18–20 4–6 1–3	65 80 110	92 124	32 48 53		100	71
85 130 175	160–200 200–240 225–275		18–20 4–8 2–5	45 72 80		48 54 75		100	71
85 165 225	215–275 245–310 275–325	20	12–18 5–8 4–6	140 150 170	108 139 148	55 85 90		100	71
85–120 165–180 200–225	215–250 250–300 270–320	20	12–18 4–8 3–6	140 150 170		60 85 90	62 78 87	100	71
125 235 270	275–350 310–375 345–405	16	12–16 5–8 4–6	155 160 180	124	85 92 98		100	71

Table 2.5.14—*continued*

TABLE 2.5.14 Typical properties of wrought aluminium alloys—*continued*

	Designation		Density (g/cm³)ª	Meltingᵇ range (°C)	Coefficient of linear expansion (20–100°C) $10^{-6}/°C$	Thermal Conductivity 0–100° Ω		Electrical	
								Resistivity	Conductivity (20°C)
	BS	Alloy type				W/m°C	% IACS	$(\mu\Omega\,cm)^c$	(% IACS)
Heat-treatable alloys	6060–T4 –T6	Al Mg Si	2.69	580–660	23	197 201	50.0 51.1	3.5 3.3	49.3 52.2
	6061–T4 –T6	Al Mg 1 Si Cu	2.69	570–660	24	156	39.6	4.3 4.0	40.1 43.1
	6063–T6	Al Mg Si	2.69	570–640	22	168	42.7	4.1	42.1
	6082–T4 –T6	Al Mg Si Mn	2.69	580–640 (515)	22	142 149	36.1 37.8	5.3 4.5	32.5 38.3
	6463–T4 –T6	Al Mg Si	2.69	570–640		151	38.4	4.4	39.2
	2014–T4 –T6	Al Cu 4 Si Mg	2.80	530–610	22	134	34.0	4.6	37.5
	2024–T4 –T6	Al Cu 4 Si Mg	2.70	570–660 (590)	24	156	39.6	4.3 4.0	40.1 43.1
	2031–T4 –TF	Al Cu 2 Mg 1.5 Ni	2.70	570–660 (590)	23	172	43.7	4.1 3.7	42.1 46.6
	2618–T4 –T6	Al Cu 2 Mg 1 Ni 1 Fe Si	2.70						
	7020–T4 –T6	Al Zn 4.5 Mg 1	2.78		24	134	34.0	4.6	37.5
	DTD XXX	Al Li 2.5 Cu 1 Mg	2.52						
Free Mach.	FC1–TD –TF	Al Cu 5.5 Pb Bi	2.83		24	163	41.4	4.4	39.2

Table 2.5.14—*continued*

TABLE 2.5.14 Typical properties of wrought aluminium alloys—*continued*

0.2% Proof stress (MPa)	Tensile strength (MPa)	Elongation (%)		Shear strength (MPa)	Fatigue strength at 50×10^6 cycles (MPa)	Hardness		Service temperature max. (°C)	Modulus of elasticity (GPa)
		on $5.65\sqrt{S_o}$	on 50 mm			Brinell	Vickers		
60–90 100–180	145–155 190–210	14–16 7–8	10 7	80 105	79 85	48 75		120	69
115–125 240–260	190–215 280–305	16–18 8–11	14 7	110 160		60–70 90–100		120	69
60–180	180–210	7–8	7	120		75		200	
120–150 240–270	190–215 280–310	15–18 8–9	14 7–8	130 175	170 170	60–70 90–100		200	74
75–90 160–170	175–190 185–210	16 10	14 9	260 290	125	65 72		300	
255–290 370–420	400–420 450–480	10–12 8	8–10 6	260 290				100	71
				270 290				150	69
								150	69
270–290 340–380	320–340 430–450	8 5							
170–190 270–290	280–300 320–360	10 8	12 10						
370–420 420–480	440–500 490–540		6 5–6						78
280 255	340 340					90 100		100	71

Table 2.5.14—*continued*

TABLE 2.5.15 Comparative characteristics of wrought aluminium alloys

Alloy type	Nominal composition and designation	Condition	Tensile properties			Resistance to atmospheric attack	Suitability for	
			0.2% proof stress	Tensile strength[a]	Elong. %		Cold forming	Machining
Pure aluminium	Al 99.99	–0	—	1	5	5	5	1
	Al 99.8–	–0	1	2	5	5	5	1
	1080	HB	2	2	5	5	2	1
	Al 99.5–	–0	1	2	5	4	5	2
	1050	HB	2	3	4	4	2	3
	Al 99.0–	–0	1	2	5	4	5	2
	1200	H4	2	2	4	4	4	2
		H8	3	3	3	4	2	3
Non-heat-treatable	Al Mn 1–	–0	1–2	2	5	4	5	2
	3103	H4	2–3	3	4	4	4	2
		H8	3	3	3	4	2	3
	Al Mn Mg–	–0	—	2	5	4	4	—
	3105	H4	3	3	3	4	4	—
		H8	3	4	3	4	4	—
	Al Mg 2–	–0	2	3	5	4	4	3
	5251	H3	3	4	3	4	3	3
		H6	3	4	3	4	3	4
	Al Mg 1–	–0	—	2	5	4	4	2
	5005	H4	2	3	4	4	3	3
		H8	3	3	3	4	2	3
	Al Mg 3.5–	–0	2	4	5	4	4	3
	5154	H2	3	4	5	3	3	4
		H4	4	5	4	3	2	4
	Al Mg 3 Mn–	–0	2	4	5	4	4	3
	5404	H2	3	4	4	3	3	4
		H4	4	5	3	3	3	4
	Al Mg 4.5 Mn–	–0	2–3	5	5	4	3	3
	5083	H2	4	5	4/5	2	2	4
		H4	5	6	4	2	2	4

Table 2.5.15

TABLE 2.5.15 Comparative characteristics of wrought aluminium alloys—*continued*

| Suitability for | | | | | | | |
| Welding | | | Anodising | | | Plating | Vitreous enamelling |
Inert gas shielded arc	Oxy–Gas	Resistance spot, seam, etc.	Protective	Colour	Bright		
4	4	3	5	5	5	—	—
4	4	3	5	5	4–5	—	4
4	4	3	5	5	4–5	—	4
4	4	4	5	5	4	4	3
4	4	4	5	5	4	4	3
4	4	4	4	4	3	4	3
4	4	4	4	4	3	4	3
4	4	4	4	4	3	4	3
4	4	5	3	3	2	3	4
4	4	5	3	3	2	3	4
4	4	5	3	3	2	3	4
4	4	4	3	3	2	—	—
4	4	4	3	3	2	—	—
4	4	4	3	3	2	—	—
4	4	5	4	4	3–4	—	0
4	4	5	4	4	3–4	—	0
4	4	5	4	4	3–4	—	0
5	4	5	5	5	5	—	—
5	4	5	5	5	5	—	—
5	3	5	5	5	5	—	—
5	2	5	4	4	3	—	0
5	2	5	4	4	3	—	0
5	2	5	4	4	3	—	0
5	2	5	4	4	3	—	0
5	2	5	4	4	3	—	0
5	2	5	4	4	3	—	0
5	2	5	4	4	3	—	0
5	2	5	4	4	3	—	0
5	2	5	4	4	3	—	0
5	2	5	4	4	3	—	0

Table 2.5.15—*continued*

TABLE 2.5.15 Comparative characteristics of wrought aluminium alloys—*continued*

Alloy type	Nominal composition and designation	Condition	Tensile properties			Resistance to atmospheric attack	Suitability for	
			0.2% proof stress	Tensile strength[a]	Elong. %		Cold forming	Machining
Heat-treatable	Al Mg Si–	–T4	2	3	5	4	4	3
	6060	T6	3	3–4	4	3	3	4
	Al Cu 2 Ni1 Mg Fe Si	–T4	3	5	4	2	—	3
		T6	5	6	3	2	—	4
	Al Cu 4 Si Mg–	–T4	4	6	4	1	3	3
	2014	T6	6	6	3	1	2	4
	Al Cu 2 Mg 1.5 Fe 1 Ni 1	–T6	6	6	4	2	—	3
	Al Zn 4.5 Mg 1–	–T4	3	5	4	3	—	2
	7020	T6	5	5–6	3	3	—	3
	Al Mg 1 Si Cu–	–T4	2	4	4	4	—	3
	6061	T6	4	5	3	3		4
	Al Si Mg Mn–	–T4	2	4	5	4	3	3
	6082	T6	4	5	4	3	3	3
Free cutting	Al Cu 5.5 Pb Si	–T4	4–5	5	3	1	2	5
	—	T6	4	5	3	1	2	5

Strength (MPa)	14–70	71–140	141–210	211–280	281–350	351+
Rating	1	2	3	4	5	6
Ductility % El.	0–1.0	1.1–3.0	3.1–5.0	5.1–10.0	10.1+	
Rating	1	2	3	4	5	

Table 2.5.15—*continued*

TABLE 2.5.15 Comparative characteristics of wrought aluminium alloys—*continued*

Suitability for							
Welding			Anodising			Plating	Vitreous enamelling
Inert gas shielded arc	Oxy–Gas	Resistance spot, seam, etc.	Protective	Colour	Bright		
4	2	4	4	4	3–4	—	—
4	2	4	4	4	3–4	—	—
0	0	0	2	2[b]	0	4	0
0	0	0	2	2[b]	0	4	0
0	0	5	2	2[b]	0	4	0
0	0	5	2	2[b]	0	4	0
0	0	4	3	0	0	—	—
4	0	4	3	3	0	—	—
4	0	4	3	3	0	—	—
4	2	4	3	3	2	—	—
4	2	4	3	3	2	—	—
4	2	4	3	3	2	—	—
4	2	4	3	3	2	—	—
0	0	0	—	—	—	—	0
0	0	0	—	—	—	—	0

[a]Since actual figures depend on size and form of materials, these values should be used only as a rough guide in comparing alloys.
[b]Dark colours only.

Other ratings
5 — Excellent.
4 — Very good.
3 — Good.
2 — Fair.
1 — Poor.
0 — Unsuitable.
— — No data.

Table 2.5.15—*continued*

TABLE 2.5.16 Typical properties of wrought aluminium alloys at various temperatures[a]

Alloy type	Material designation BS	Alloy type	Property	Temperature (°C)									
				−196	−80	−30	25	100	150	200	250	300	350
Pure aluminium	1200–0	Al 99.0	TS	170	105	95	90	68	55	41	26	20	13
			PS	40	37	35	35	30	28	24	16	—	—
			El	50	45	40	40	45	55	65	75	85	85
	–H8		TS	235	180	170	165	145	125	41	26	20	13
			PS	180	160	155	150	130	95	24	16	—	—
			El	30	15	15	15	16	19	65	75	85	85
Non-heat-treatable	3103–0	Al Mn 1	TS	228	138	118	110	90	75	60	40	28	17
			PS	57	48	45	42	40	35	30	23	—	—
			El	45	41	41	40	45	48	60	65	69	71
	–H4		TS	243	165	150	148	145	125	95	50	28	17
			PS	172	150	145	143	130	110	60	25	—	—
			El	30	20	16	16	16	17	20	63	69	71
	–H8		TS	280	220	206	200	180	160	95	50	28	17
			PS	225	200	190	185	145	110	60	25	—	—
			El	25	12	11	10	10	12	20	63	69	71
	5251–0	Al Mg 2	TS	303	200	193	190	190	160	118	83	53	35
			PS	110	90	88	86	85	83	76	51	39	—
			El	45	33	32	32	35	49	60	78	108	128
	–H4		TS	380	275	262	260	260	206	165	83	53	35
			PS	246	220	214	215	215	185	105	52	39	—
			El	28	20	18	17	18	25	45	78	108	128
	–H8		TS	415	303	290	290	275	235	173	83	53	35
			PS	304	262	255	253	248	193	105	52	39	—
			El	25	20	15	15	16	25	44	78	108	128

Table 2.5.16

TABLE 2.5.16 Typical properties of wrought aluminium alloys at various temperatures[a]—continued

	Alloy											
Non-heat-treatable	5454–0 Al Mg 3 Mn	TS	372	255	248	246	246	200	150	118	75	40
		PS	131	118	116	116	116	110	105	75	50	—
		EI	40	30	28	25	28	50	60	80	110	130
	–H2	TS	407	290	282	275	270	220	172	118	75	40
		PS	248	214	205	205	200	180	130	75	50	—
		EI	31	22	20	18	20	40	45	80	110	130
	–H4	TS	435	318	303	304	295	235	180	118	75	40
		PS	283	248	240	240	235	193	130	75	50	—
		EI	30	20	18	15	18	30	45	80	110	130
	5083–0 Al Mg 4.5 Mn	TS	410	295	290	290	275	215	150	120	80	42
		PS	165	145	143	143	140	130	118	75	55	—
		EI	35	30	25	25	35	50	60	80	106	128
Heat-treatable	6060–T6 Al Mg Si	TS	325	263	249	242	215	145	63	30	22	15
		PS	248	228	220	214	193	137	45	—	—	—
		EI	25	20	20	19	15	20	40	75	75	100
	2014–T6 Al Cu 4 Si Mg	TS	580	510	495	485	435	275	110	65	45	30
		PS	495	450	425	415	395	240	90	50	35	25
		EI	15	14	14	15	15	20	40	50	65	70
	2031–T6 Al Cu 2 Mg 1.5 Fe 1 Ni 1	TS	540	462	442	455	425	350	220	90	52	35
		PS	420	380	372	375	365	305	180	62	31	25
		EI	11	11	10	10	10	15	15	50	80	120
	6061–T6 Al Mg 1 Si Cu	TS	415	340	320	310	290	235	130	50	32	20
		PS	325	290	282	275	263	214	105	35	20	—
		EI	20	18	16	16	19	20	30	60	80	100
	6082–T6 Al Si 1 Mg Mn	TS	525	420	—	375	334[b]	180[b]	105[c]	32[d]	—	—
		PS	430	360	—	305	316[b]	155[b]	76[c]	21[d]	—	—
		EI	17	13	—	14	—	—	—	40[d]	—	—

[a] Based on specimens tested at temperature after soaking for approximately 10 000 h.
[b] Extrapolated to 10 000 h from 1000 h.
[c] Extrapolated to 4000 h from 1000 h.
[d] 1000 h.
TS = Tensile strength in MPa. PS = 0.2% proof stress in MPa. EI = Elongation, as a percentage of a 50 mm gauge length.

(Data courtesy of The Aluminium Federation)

Table 2.5.16—continued

2.5.4   New aluminium alloys

In recent years a number of new aluminium alloys have been developed. These include:

(i)   Improvements to standard alloys for specific applications developed by specialist component manufacturers—these include improvements in composition and processing—squeeze casting is an example of the latter.

(ii)   Aluminium–lithium alloys.

(iii) Aluminium alloys produced by powder metallurgy methods.

(iv) Aluminium alloys reinforced with ceramic fibres.

(v)   Rapidly solidified materials and vapour deposited materials. These processes permit the production of aluminium alloys with compositions and microstructures which are not possible by the conventional cast or wrought methods.

2.5.4.1   Improvements to standard alloys

A good example of the improvements which are capable for most 'standard' alloys is the development of special low expansion alloys with increased fatigue strength for pistons. Traditional LM13-Lo-Ex (Low Expansion) alloys suffer premature failure due to fatigue initiated at micro-shrinkage pores and coarse intermetallic phases. Phosphorus treatment, a special composition, greatly reduces the scatter of fatigue results and improves the median. See:

Table 2.5.17   *Eutectic Lo-Ex piston alloy compositions*, and

Fig 2.5.2   *Effects of Lo-Ex alloy composition and heat treatment on fatigue life (after 100 h prior exposure at test temperature)*

Squeeze casting, a process in which solidification takes place under pressure, also improves tensile properties and fatigue strength. See:

Table 2.5.18   *Typical properties of commercial Lo-Ex piston alloys*

Reinforcing the squeeze casting with ceramic fibres (see Section 2.5.4.4) has a marked effect on the elastic modulus of aluminium alloys, especially at elevated temperatures. See:

Table 2.5.19   *Effect of fibre reinforcement on the elastic modulus of squeeze-cast W109 alloy*

2.5.4.2   Aluminium–lithium alloys

Aluminium–lithium alloys are now available commercially from several sources. These materials are intended for aerospace applications where the 10% reduction in density and 10% improvement in elastic modulus are attractive and the threefold increase in cost is not prohibitive. A summary of the properties of aluminium–lithium alloys is given in:

Table 2.5.20   *Aluminium–lithium alloys*

2.5.4.3   Aluminium alloys produced by powder metallurgy methods

Powder metallurgy (PM) offers the possibility of producing unique formulations which are not possible by conventional melting techniques. The PM route can also be a cost-effective method of manufacturing components with conventional aluminium alloys especially for small parts requiring close dimensional tolerances, eg connecting rods for refrigeration compressors.

A comparison of powder metallurgy aluminium alloys and aluminium lithium alloys with conventional aluminium alloys is given in:

Table 2.5.21   *Summarised comparative advantages and limitations of aluminium–lithium alloys and powder metallurgy alloys compared with standard aluminium alloys*

2.5.4.4   Aluminium alloys reinforced with ceramic fibres (metal-matrix composites)

The main benefits of reinforcing aluminium alloys with ceramic fibres are a large increase in elastic modulus (especially at elevated temperatures) and improved creep strength and heat erosion resistance. The disadvantages are decreased elongation to fracture and more difficult machining characteristics.

The properties of metal matrix composite aluminium alloys are compared with other new aluminium alloys in:

Table 2.5.22   *Comparison of new aluminium alloys*

An alternative method of making aluminium metal-matrix composites is to grow reinforcing aluminium oxide network in molten aluminium while it is in a mould of the desired component shape—the so-called LANXIDE† process.

2.5.4.5   Rapidly solidified aluminium alloys

Most processes under this heading produce either powder, ribbon or wire. The latter two have to be comminuted before being consolidated by powder metallurgy methods. Uniquely good fatigue, corrosion and high temperature properties are possible by these methods but material qualities are very limited.

A fundamentally different process is the vapour deposition process developed by RAE at Farnborough in the UK. Thick deposits (2.5cm) of alloys with unique compositions and properties can be built up very rapidly on a moving collector and subsequently worked or machined. A summary of the properties of the alloy designated RAE 72, produced by this method, is given in:

Fig 2.5.3   *Comparison of specific tensile strength (strength/density) of vapour deposited Al–Cr–Fe, Ti–6Al–4V and two aluminium aircraft alloys, Concorde material CM001-1C and 7075-T6, measured at different test temperatures,* and

Table 2.5.23   *Typical mechanical properties of a vapour deposited Al–Cr–Fe alloy compared with those of other alloys used in aircraft components*

The major advantages of aluminium alloys produced in this way are improved high temperature properties, corrosion, fracture toughness and fatigue strength.

The room temperature tensile strength of the vapour deposited Al–Cr–Fe alloy is very impressive as can be seen from Table 2.5.23. However, it is in the retention of its room temperature strength when used at elevated temperature that the Al–Cr–Fe alloy shows its advantage over conventional alloys. In particular the high temperature specific strength (UTS/density) superiority is most marked as shown in Fig 2.5.3.

The vapour deposited Al-Cr-Fe alloy retains its room temperature strength entirely after 1000 hours exposure at 250°C and testing at room temperature. After 1000 hours at 300°C and testing at room temperature, approximately three quarters of its strength is retained.

---

†Lanxide is the trade name and trading name of a company based in Delaware in the USA.

TABLE 2.5.17 Eutectic Lo-Ex piston alloy compositions

Element \ Alloy	Traditional Lo-Ex (LM 13)	Medium duty (W.132)	Premium grade (W.109)
	(%)	(%)	(%)
Silicon	10.0–13.0	10.0–12.5	11.0–12.5
Copper	0.7–1.5	0.7–1.5	0.7–1.5
Magnesium	0.8–1.5	0.8–1.5	0.7–1.0
Nickel	1.5 max.	0.7–1.3	0.7–1.3
Iron	1.0 max.	0.7 max.	0.5 max.
Manganese	0.5 max.	0.45 max.	0.25 max.
Zinc	0.5 max.	0.5 max.	0.1 max.
Phosphorus treatment	No	Yes	Yes

Source: Wellworthy Ltd, UK.

TABLE 2.5.18 Typical properties of commercial Lo-Ex piston alloys

Casting route	Alloy	Heat treatment	Tensile strength (MPa)		0.2% proof stress (MPa)	Fatigue strength[a] (MPa)
			20°C	200°C[b]	200°C[b]	200°C[b]
Gravity Die-cast	Medium duty (W.132)	TE	215	160	130	70
	Premium grade (W.109)	TE	220	165	140	75
		TXF	230	170	150	77
		AXP	245	180	160	80
Squeeze cast	Premium grade (W.109)	X2	260	180	160	82
		X5	280	200	175	85

NB based on 130 mm piston diameter. [a] 10^7 cycles [b] 100 h prior exposure.
Source: Wellworthy Ltd, UK.

TABLE 2.5.19 Effect of fibre reinforcement on the elastic modulus of squeeze-cast W109 alloy

Temperature (°C)	Squeeze-cast (GPa)	Fibre-reinforced squeeze-cast (GPa)
20	73.6	94.4
150	70.0	93.4
350	60.8	75.5
450	43.8	60.4

Source: Wellworthy Ltd, UK.

Table 2.5.17, Table 2.5.18 and Table 2.5.19

TABLE 2.5.20 Aluminium–lithium alloys.
(a) Requirements and properties for Lital A (8090) (density 2.54)

	Sheet (T6) XXXA		Plate (T651) YYYA			Extrusion (T651) ZZZA		
	L	T	L	T	ST	L	T	ST
0.2% PS (MPa)								
Specified minima	380	380	410	390	360	430	—	—
Provisional minima[b]	350	355	410	380	335	440	395	355
Typical	365	375	450	420	365	480	420	385
TS (MPa)								
Specified minima	440	440	460	450	420	480	—	—
Provisional minima	430	450	455	455	400	495	470	420
Typical	465	470	495	480	435	550	490	470
Elongation (%)								
Specified minima	6	6	5	5	2.5	6	—	—
Provisional minima	5	4	4	4	1	3	3	1
Typical	5.5	7.5	6	7	2	5	5	3.5
Fracture toughness (MPa $\sqrt{m}$)								
Specified minima	70[a]	—	30	25	18	30	25	—
Provisional minima			30	26	15	31	16	—
Typical	66	—	37	33	16	38	20	—
Elastic Modulus (GPa)								
Typical tension	78	78.5	79	79	—	79.5	79	—
Typical compression	—	—	81.5	82.5	—	—	—	—

[a] On 400 mm wide CCT panels. K_{app}.
[b] Based on statistically derived 'A' values wherever possible, otherwise worst results.
Data courtesy of C.J. Peel, B. Evans and D.M. McDarmaid, RAE, Farnborough, UK.
Lital A, Lital B and Lital C are trade names of British Alcan, Banbury, UK.

Table 2.5.20

TABLE 2.5.20 Aluminium–lithium alloys—*continued*
(b) Requirements and properties for Lital B (density 2.55)

	Sheet (T6) XXXB[a]		Plate (T651) YYYB			Extrusion (T651) ZZZB	
	L	T	L	T	ST	L	T
0.2% PS (MPa)							
Specified minima	420	420	455	455	420	560	510
Provisional minima[b]	385	385	510	470	415	—	—
Typical	410	400	530	500	430	610	510
TS (MPa)							
Specified minima	490	490	525	525	490	600	580
Provisional minima	470	460	540	535	465	—	—
Typical	505	490	565	550	490	630	540
Elongation (%)							
Specified minima	6	6	6	6	—	—	—
Provisional minima	5	8	—	—	—	—	—
Typical	8	9	5	5	1.5	4	4
Fracture toughness (MPa $\sqrt{m}$)							
Specified minima			—	—	—		
Provisional minima			16[b]	14			
Typical			24	21	10		
Elastic Modulus (GPa)							
Typical tension	78.5	78.5	80	80.5	—		
Typical compression	—	—	82	82.5	—		

[a] Extended ageing variant.

[b] Considerably improved toughness is obtained at slightly reduced strength levels.

Table 2.5.20—*continued*

1212   **Aluminium and its alloys**

TABLE 2.5.20 Aluminium–lithium alloys—*continued*
(c) Requirements and properties for Lital C (8090) (density 2.60)

	Sheet XXXC		Plate (T651) YYYC			Extrusion (T651) ZZZC		
	L	T	L	T	ST	L	T	ST
0.2% PS (MPa)								
Specified minima	290	290	300	300	260	320	—	—
Provisional minima	290	300	370	310	260	—	—	—
Typical	330	335	390	340	290	370	325	—
TS (MPa)								
Specified minima	420	420	430	430	380	440	—	—
Provisional minima	400	410	430	420	380	—	—	—
Typical	440	450	460	440	415	460	440	—
Elongation (%)								
Specified minima	6	6	7	7	3.5	6	—	—
Provisional minima	4	7	5	6	2	—	—	—
Typical	6	9	6	9	3.5	5	6.5	—
Fracture toughness (MPa $\sqrt{m}$)								
Specified minima	83[a]	—	35	28	25	—	—	—
Provisional minima			36	32	20	—	—	—
Typical	76	—	>40	>35	25	>45	>45	—

[a] 400 mm wide CCT panels. K_{app}.

Table 2.5.20—*continued*

TABLE 2.5.21 Summarised comparative advantages and limitations of aluminium–lithium alloys and powder metallurgy alloys compared with standard aluminium alloys

Material	Advantages	Limitations
Al–Li alloys	10% lower density and up to 10% higher modulus than conventional alloys. Straightforward fabrication of components with existing equipment. No lightning strike problems of cfrp composites. Superplastic forming possibilities.	Cost still more than three times the cost of conventional alloys. Special precautions needed in processing. Not available below 1mm thickness.
Powder metallurgy alloys	30% stronger than conventional Al alloys. Up to 100°C better temperature resistance. Double fatigue strengths.	3.5% higher density. Difficult to process by conventional routes. Expensive.

Table 2.5.21

TABLE 2.5.22 Comparison of new aluminium alloys

Property	Conventional aluminium alloys		Aluminium–lithium alloys			RSR[a]	Alloys	Metal-matrix composites (MMC)		
								SiC/Al	SiC/Al	Al_2O_3/Al
Alloy Ref.	2024 T3	7075 T6	CP271	CP276 Alcoa B	Alcan Lital A B C	RAE 72	PA West Germany	40% particulate	45–50% continuous	45–50% continuous
Density (s.g.)	2.80	2.80	2.53	2.59 2.52	2.54	2.89	2.95	27	27	29
Ult Tensile strength (MPa)	470–480	565	540		440	723	520	600–700	1200–1600	860
Yield strength (MPa)	360	480/510	490	570–580	360–390	709	400	450–700	—	—
GPa modulus	72	70–72	81	81.5–80.2	79–80	89	92	145–150	120–220	150
Elongation (%)	17	10–11	3-5-10	6–12	6	7.5	6	Around 1	Very low transverse	
Fatigue strength MPa at 10^7 cycles	120					250	250	Probably low		
Fracture toughness								20–30	High	High

[a] Rapid solidification route.

TABLE 2.5.23 Typical mechanical properties of a vapour deposited Al–Cr–Fe alloy compared with those of other alloys used in aircraft components

Alloy	Density (g/cm³)	Tensile strength (MPa)	Specific tensile strength (kNm/kg)	Proof stress 0.2% (MPa)	Elongation (%)
Al–7.5 Cr–1.2 Fe	2.89	723	250.2	709	7.5
2014–T6	2.80	410	146.4	255	8
7075–T651	2.79	570	204.3	505	10
Concorde CM001–1C	2.75	390	141.8	320	7
Ti–6Al–4V	4.51	1000	221.7	830	8
Al–Li 8090	2.53	440	173.9	380	6

Table 2.5.22 and Table 2.5.23

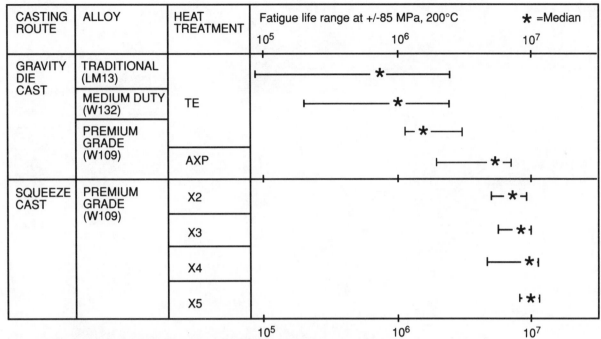

CASTING ROUTE	ALLOY	HEAT TREATMENT	Fatigue life range at +/-85 MPa, 200°C ★ =Median 10^5 10^6 10^7
GRAVITY DIE CAST	TRADITIONAL (LM13)	TE	
	MEDIUM DUTY (W132)		
	PREMIUM GRADE (W109)		
		AXP	
SQUEEZE CAST	PREMIUM GRADE (W109)	X2	
		X3	
		X4	
		X5	

FIG 2.5.2 Effects of Lo-Ex alloy composition and heat treatment on fatigue life (after 100 h prior exposure at test temperature). (Source: Wellworthy Ltd, UK)

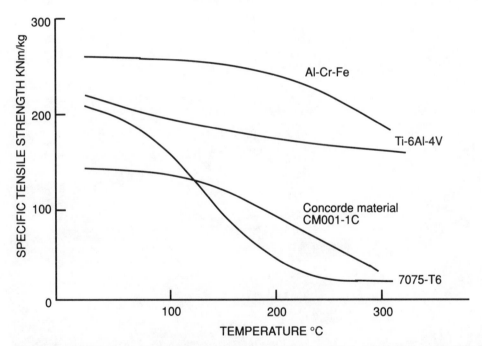

FIG 2.5.3 Comparison of specific tensile strength (strength/ density) of vapour deposited Al-Cr-Fe, Ti-6Al-4V and two aluminium aircraft alloys, Concorde material CM001-1C and 7075-T6, measured at different test temperatures

Fig 2.5.2 and Fig 2.5.3

Titanium and its alloys

Contents

List of tables

List of figures

2.6.1   Characteristics of titanium and its alloys

2.6.1.1   Introduction

The technical use of titanium depends on its low specific gravity combined with (in comparison with other light metals) its high strength at room, and moderately elevated, temperatures, and its very tenacious oxide film which resists corrosive environments. The characteristics of titanium and its alloys are listed in:

Table 2.6.1   *Advantages and limitations of titanium*

The applications of titanium and its alloys are listed in:

Table 2.6.2   *Applications of titanium and its alloys*

2.6.1.2   Different types of titanium alloys

The properties of titanium and its alloys depend on their basic structure and the way in which this is manipulated during their mechanical and thermal treatments during manufacture.

The atomic structure of titanium undergoes a transformation from a close packed hexagonal arrangement (so-called α-Titanium) to a body centred cubic arrangement (so-called β-Titanium) at 882°C. This transformation can be considerably modified by the addition of alloying elements to produce at room temperature alloys which have all α, all β or α + β structures.

Four main types of titanium alloy have been developed:

Alpha phase alloys—usually contain alpha stabilisers and have lowest strengths. However, they are formable and weldable. Some contain beta stabilisers to improve strength.

Alpha plus beta alloys—widely used for high strength applications and have moderate creep resistance.

Near alpha alloys—with medium strength but better creep resistance. They can be heat treated from the beta phase to optimise creep resistance and low cycle fatigue resistance. Some are weldable.

Beta phase alloys—are usually metastable, formable as quenched and can be aged to the highest strengths but then lack ductility. Fully stable beta alloys need large amounts of stabilisers and are therefore too dense. In addition, the modulus is low (<100 GPa) unless the beta phase structure is decomposed to precipitate alpha phase. They have poor stability at 200–300°C, have low creep resistance and are difficult to weld without embrittlement. Metastable beta alloys have some application as high strength fasteners.

The advantages and limitations of the four types of titanium alloy are summarised in:

Table 2.6.3   *Advantages and limitations of titanium alloy types.*

2.6.1.3   International specifications

In the tables which follow most materials are related to the British specification, for the international equivalents, see:

Table 2.6.4   *Equivalent international specifications of titanium and its alloys*

TABLE 2.6.1 Advantages and limitations of titanium

Advantages	Limitations
Low density (5.6 ≈ 4.51) and high strength to weight ratio, i.e. same strength but half density of mild and alloy steels.	Low elastic modulus (105–125 GPa) compared to 150–210 GPa for steels and 60–70 GPa for aluminium.
Low interstitial grades have excellent low temperate properties. Excellent retention of strength and creep resistance to 550–600°C. High fatigue and corrosion fatigue resistance (40–55% of UTS) Transmits ultrasonic energy with little attenuation.	Sensitive to fretting corrosion.
Low thermal expansion 8–10 $\times$ 10^{-6} K^{-1} between 0 and 500°C (half that of stainless steel, two thirds alloy steel, less than half aluminium). Non-magnetic.	Low thermal conductivity—approx. 5% of copper. Low electrical conductivity—approx. 3% of copper.
Exceptional corrosion resistance especially to oxidising solutions (including nitric acid up to boiling point), moist chlorine gas, aqueous chlorides and dilute alkalis (up to boiling point). Resistant to reducing solution if inhibited. Oxidation resistant to 550–600°C.	Attacked by reducing media (sulphuric hydrochloric, phosphoric, hydrofluoric, formic, oxalic acids) unless passivated by inhibitors. Attacked by aluminium chlorides and boiling concentrated NaOH, KOH. Can have pyrophoric reactions with fuming nitric acid with less than 2% water or more than 6% nitrogen dioxide; liquid oxygen on impact; anhydrous liquid or gaseous chlorine, liquid bromine, hot gaseous fluorine and oxygen-enriched atmospheres. Strong affinity for hydrogen (>130°C), nitrogen (>800°C) and oxygen (>700°C) which cause embrittlement.
Not normally susceptible to stress corrosion. Good resistance to cavitation erosion and crevice corrosion in sea-water.	Stress corrosion can occur in some alloys by chlorides on stressed parts at high temperature. Also in dry methanol (less than 2% water). Crevice corrosion can occur at higher temperatures and if there is no cathodic protection (e.g. from copper alloy tube plate).

C.P. Ti and most alloys can be welded but inert gas shielding essential. Contamination must be avoided.
 	
Some alloys can be formed superplastically at low stresses to complex shapes.
 	
Can be diffusion bonded simultaneously with superplastic forming.
	High cost of raw and semi-finished products:
	  (a) expensive extraction of metal from ore.
	  (b) need to melt under vacuum.
	  (c) care needed to avoid embrittlement from air during hot working.
	
	Machinability, comparable to austenite stainless steel.

Table 2.6.1

TABLE 2.6.2 Applications of titanium and its alloys

Industry	Components	Reasons for use
Aerospace	Airframe parts, sheet forgings, fasteners.	Strength/weight ratio. High temperature properties.
Gas turbine engines	Compressor blades, discs, casings, ducts, engine cowlings, exhaust shrouds, heat shields.	As above.
Metal finishing (anodising, electroplating, 'Tufftride')	Heat exchanger coils, mesh baskets, jigs, carriers, linings.	Corrosion resistance. Electrochemical properties.
Chemical process (chlorine manufacture, fibre processing, acids and fertiliser products)	Pumps, valves, heat exchangers.	Resistance to wet chlorine, acids, etc.
Metal refining (copper)	Cathodes.	Oxide film acts as parting agent. (Reduced labour costs.)
Oil refining	Condenser tubes.	Corrosion resistance.
Surgical implants	Hip, elbow joints, screws, plates.	Compatibility with human environment.
Mechanical engineering	Crankshafts, connecting rods, torsion bars, cycle chassis, high speed rotating or reciprocating components.	Strength/weight ratio.
Power generation	Integrally finned heat exchanger tube. Steam condensers.	Corrosion resistance to sea-water, brackish water and brine.
Desalination	Heat exchangers for multi-stage flush distillation units.	Good corrosion resistance to sea-water and brines.
Shoe manufacture	Ultrasonic hammers.	Low damping capacity.
Non-engineering	Jewellery, watch bezels, sculptures.	Decorative appearance of faceted selectively anodised and coloured surface.

Table 2.6.2

TABLE 2.6.3 Advantages and limitations of titanium alloy types

Alloy type	Advantages	Limitations
α	Weldable with properties comparable to parent metal.	Low to medium strength.
	Readily formable by sheet metal working processes.	Galls readily; take care with lubrication.
	Commercially pure titanium strengthened inexpensively by interstitial oxygen, nitrogen and carbon.	Interstitial elements impair low temperature properties, reduce ductility and are not effective strengtheners above 200–300°C.
	Copper additions allow heat treatment to raise strength at high temperatures.	Slightly reduces formability and corrosion resistance.
	Al and Sn raise strength and creep resistance.	Not cost-effective compared to other Ti alloys.
	Pd improves corrosive resistance in non-oxidising solutions. Mo + Ni cheaper, less effective, alternative.	Pd raises price compared to other alpha phase alloys.
	Heat treatable to higher strength and creep resistance.	Care needed with thermomechanical treatment and subsequent heat treatment to avoid deleterious microstructures.
α + β	Presence of β phase improves formability, forgeability, strength at low total alloy contents.	Higher alloy contents reduce formability but better hardenability.
	Al, Sn, Zr added to increase strength and creep resistance of α phases Mo, V, Zr strengthen β phase.	Increase in alloy additions decrease ductility, and forgeability.
	Characteristics of alloys controlled by heat treatment temperature to vary α and β proportions.	Need to forge or process in α + β range and ensure adequate cold work.
	Duplex alloys with fine structure are superplastic, enabling complex shapes to be formed.	Strain rates for superplastic forming very low; argon atmospheres (or vacuum) needed to avoid surface contamination or alteration.
	Reduction in β phase content by reducing β stabilisers increases creep resistance.	Lower strength at ambient temperature. Lower depth of hardening with less β stabilisers.
Near α	Improved creep resistance by β treatment.	Lower low cycle fatigue resistance by β treatment unless β grain size reduced, also lower strength and ductility.
	β treated alloys do not have to be forged in α + β range by large amounts to refine grain structure.	Less forging needed but thermomechanical processing needed to refine β grain size.
	β treated alloys readily weldable.	
	β treated alloys can be forged in β range.	
β phase	Highest strength of all titanium alloys for airframe applications.	Have amounts of β stabilisers increase density, introduce instability at 200–300°C and cause embrittlement. Lower elastic modulus.
	Readily forged at lower temperatures than α/β alloys. Formable in compression.	Less formable at ambient temperature unless strain-induced transformations.
	Weldability depends on alloy and the tendency to form embrittling phases on cooling. Ductile welds as quenched.	Alloys must be selected to avoid omega or other embrittling phase changes on cooling.

Table 2.6.3

TABLE 2.6.4 Equivalent international specifications of titanium and its alloys

Alloy composition	IMI desig-nation	UK BS	UK DTD	France	Germany	Japan JIS	AICMA	USA ASTM	USA AMS	MIL
Commercially pure with increasing interstitial content	110					Class 1 H4600, 4630 4650, 4670		GRADE 1 (B265, 337, 338, 348, 367, 381)		
	115	2TA 1	5073	T35	LW 3, 7024	Class 2 (ditto)	Ti-PO1			
	125	2TA 2, 3ˣ, 4ˣ, 5ˣ	5013B, 5033B 5183, 5293	T40	LW 3, 7034	Class 3 (ditto)	Ti-PO2	GRADE 2 (ditto)	4902, 4941 4942, 4951	T-9046
	130		5003B, 5023C 5193, 5283, 5293	T50				GRADE 3 (ditto)	4900	T-9046
	155	2TA 6	5063B	T60	LW 3, 7064		Ti-PO4	GRADE 4 (B265)	4901, 4921	T-9046 T-9047
	160	2TA 7,8,9,						GRADE 4 (B348, 367, 381)		
Ti-0.15%Pd	260							GRADE 11 (B265, 337, 338, 348, 381)		
	262							GRADE 7 (B265, 337, 338, 348, 367, 381)		
Ti-2.5%Cu	230	2TA 21, 22, 23, 24 TA, 52, 53, 54, 55, 58	5123, 5133 5233, 5243 5253, 5263	TU2			Ti-P11	GRADE 12 (B265, 337, 338, 348, 367, 381)		
Ti-0.3%Mo-0.8%Ni	Code 12									
Ti-5%Al-2.5%Sn	(317)		5083, 5093				Ti-P65	GRADE 6 (B265, 348, 367, 381)	4910, 4926 4966	T-9046
Ti-6%Al-4%V	318	2TA 10, 11, 12, 13, 28 TA 56, 59	5303,5313 5323	TA6V	LW 3, 7164		Ti-P63	GRADE 5 (B265, 348, 467, 381)	4906, 4907,4911, 4928, 4930, 4935, 4954, 4965, 4967	T-9046H T-9047F
Ti-4%Al-4%Mo-2%Sn-0.5%Si	550	TA 45, 46, 47, 48, 49 50, 51, 57	5103 5153, 5203	TA4DE	LW 3, 7184		Ti-P68			
Ti-4%Al-4%Mo-4%Sn-0.5% Si	551	TA 38, 39, 40, 41, 42	5223						4974	T-9047E T-009047E

Table 2.6.4

TABLE 2.6.4 Equivalent international specifications of titanium and its alloys—*continued*

Alloy composition	IMI desig-nation	UK BS	UK DTD	France	Germany	Japan JIS	AICMA	USA ASTM	USA AMS	USA MIL
Ti–2.25%Al–1%Mo–11%Sn–5%Zr–0.2%Si	679	TA 18, 19, 20, 25, 26 27ˣ	5113							
Ti–2.25%Al–4%Mo–11%Sn–0.2%Si	680		5213	TE11DA						
Ti–6%Al–0.5%Mo–5%Zr–0.25%Si	685	TA 43, 44		TAGZD	LW 3, 7154		Ti–P67			
Ti–8%Al–1%Mo–1%V							Ti–P66		4972A 4973A	
Ti–6%Al–6%V–2%Sn					LW 3, 7174		Ti–P64		4971A, 4979	T–9047 F–83142
Ti–6%Al–2%Sn–4%Zr–2%Mo	6242S								4976, 4975	
Ti–5%Al–4%Sn–3.8%Zr 0.7%Nb–0.5%Mo–0.35%Si–0.06%C	834									

Note: In US specification

ASTM B265	MIL T–9046	Plate, sheet, strip, material.
ASTM B337		Seamless and welded pipe.
ASTM B338	MIL T–9047	Seamless and welded tube for condensers.
ASTM B348	MIL T–9047	Bar and billet.
ASTM B367		Castings.
ASTM B381	MIL T–9047	Forgings.

Table 2.6.4—*continued*

2.6.2 Mechanical and physical properties

2.6.2.1 Typical properties

The conventionally specified mechanical properties of titanium alloys and their forming characteristics are listed respectively in:

Table 2.6.5 *Typical properties of wrought alpha phase titanium alloys*
Table 2.6.6 *Typical properties of wrought alpha-beta phase titanium alloys*
Table 2.6.7 *Typical properties of wrought near alpha phase titanium alloys*, and
Table 2.6.8 *Typical properties of wrought beta phase titanium alloys*

Their physical properties are listed in:

Table 2.6.9 *Physical properties of titanium alloys*

Variation of yield strength with temperature is given in:

Fig 2.6.1 *Variation of yield strength with temperature*

Variation of impact with temperature is given in:

Fig 2.6.2 *Variation of impact strength with temperature*

2.6.2.2 Low temperature properties of titanium alloys

Special grades of titanium alloys have been developed for low temperature application. These are subject to processing which ensures an extra low interstitial (ELI) impurity content. The impurity interstitial atoms, carbon, oxygen and nitrogen can cause embrittlement at normal temperatures if they occur in excessive concentrations, and are especially harmful at low temperatures. ELI grades of unalloyed titanium and Ti–6%Al–4%V alloy are commercially available, but are more expensive.

Their mechanical properties at low temperatures are listed in:

Table 2.6.10 *Low temperature mechanical properties of ELI grades of titanium alloys*

It can be seen that tensile strength is increased at low temperature, whilst useful ductility and impact strength is retained at minus 196°C.

2.6.2.3 Fracture toughness of titanium alloys

Fracture toughness data for different forms and heat treatments, where known, of commercially available alloys of titanium are listed in:

Table 2.6.11 *Typical fracture toughness values for titanium alloys*

There are considerable amounts of data for these and other alloys in different forms of heat treatment published in research papers. The data in Table 2.6.11 is a guide to what can be obtained in simple shapes.

Fracture toughness is very dependent on microstructure, which in turn depends critically on the thermomechanical processing of the alloy and its subsequent heat treatment. It is important to take into account these factors if the component design is critically dependent on fracture toughness. Included in the important factors is a fine beta grain size, but coarse alpha platelets in a basket weave structure (multi-orientation within one grain) are preferable to an aligned structure (one orientation across one grain). Low interstitial contents (O, N, H) favour high fracture toughness. The beta phase alloy Ti–10%V–2% Fe–3% Al provides the best fracture toughness in relation to yield stress and shows little susceptibility to embrittlement under corrosive conditions. These properties are maintained up to a 125 mm thick section.

The relationship between fracture toughness and yield stress (ys) for the data in Table 2.6.11 together with the data for Corona 5 and Beta III, which were developed as high fracture toughness alloys, is plotted in:

Fig 2.6.3 *Failure analysis diagram for data on fracture toughness of titanium alloys >25 mm thick*

2.6.2.4   Fatigue properties of titanium alloys

The fatigue properties of titanium alloys depend upon the heat treatment and strength of the alloy tested and the type of test used. Data are available both for fluctuating load tests with zero mean stress (rotating bend tests) and direct tensile fatigue tests with zero minimum load. Most alloys show a fatigue limit by 10^7 cycles of stress based on tests from test bars, but greater variability is obtained from samples cut from forgings.

In general, the fatigue limit is typically:

50% of UTS in commercial purity (CP) titanium;
60–65% of UTS in 230;
55–60% of UTS in 318;
50–60% of UTS in 550;
40–50% of UTS in 551;
55–60% of UTS in 679;
60% of UTS in 680;
50% of UTS in 685;
50% of UTS in 729;
50% of UTS in 834.

The fatigue strength at 10^7 cycles, the high cycle fatigue strength, is a measure of the resistance of the alloy to fatigue crack nucleation. Another important property is the fatigue crack propagation threshold (FCP), e.g. the stress concentration range ΔK to cause a rate of crack propagation of 10^{-9}m/cycle. This can be measured for low mean stress and high mean stress, defined by the ratio R of minimum to maximum stress (i.e. $R = 0.1$ for low mean stress and $R = 0.7$ for high).

These properties are affected differently by microstructure as indicated in:

Table 2.6.12   *Preferred microstructural features to optimise fatigue properties of titanium alloys*

Thus, high fatigue strength is favoured by a fine grain structure with a duplex α phase plus β phase decomposed to fine interlocking α plates. i.e. fast cooling from the recrystallised structure plus annealing to a fine α structure. If a beta treatment is given, the α annealed structure should comprise fine interlocking plates. High strength by ageing and high oxygen content (0.2%) improve fatigue strength at high cycles.

On the other hand, crack growth rates are lower in coarse-grained, aligned structures. Though crack growth rates in a fine basket weave structure appear higher, the crack follows a longer, more convoluted, path and so overall, the factors favouring better high cycle fatigue strength are preferable as a compromise.

Data on fatigue life from different sources are difficult to compare because test conditions vary. Some fatigue data which are comparable are listed in:

Table 2.6.13   *Fatigue data for titanium alloys*

2.6.2.5   Creep resistance of titanium alloys

Alloys with considerable increased creep resistance have been developed in the last 30 years. These and others are included in:

Table 2.6.14   *Typical creep properties of titanium alloys*

A wider comparison based on a Larson–Miller Plot produced by IMI Titanium Ltd is shown in:

Fig 2.6.4   *Larson–Miller plot of creep resistant titanium alloys*

The constant, 20, used in the Larson–Miller parameter is an estimate and is not correct for all alloys. The plot is useful only as a comparison of different alloys.

The data show the improvements achieved by near α alloys, especially when beta-treated. The latest alloy, 834, is fabricated to a fine β grain size stabilised by solution heat treatment in the high $\alpha + \beta$ range to optimise properties other than creep, e.g. high temperature, low cycle fatigue resistance.

TABLE 2.6.5 Typical properties of wrought alpha phase titanium alloys

Alloy type — Nominal composition	Ref. number	Forms available	Alloy condition and form	0.2% proof stress (MPa)	Ultimate tensile strength (MPa)	Impact energy (J)	Elastic modulus (GPa)	Elongation (min.) (%)	Reduction in area (%)	Specific gravity	Formability Minimum bend to thickness ratio for sheet	Forgeability 100 = best	Machinability 100 = best	Welding 100 = best	Remarks
Ti–0.05% O_2/N_2	110	R	Annealed	130–170	270–350		105–120	30	70	4.51	1t	100	100	100	Excellent corrosion resistance especially in oxidising solutions.
Ti–0.07% O_2/N_2	115	All	Annealed	170–210	330–420		105–120	25	70	4.51	1t	100	100	100	Excellent formability. Strength increases and ductility falls as interstitial content is increased.
Ti–0.13% O_2/N_2	125	All but W	Annealed sheet / bar	290–330 / 240–280	390–540	60	105–120	22 / 20	55	4.51	1.5t	100	100	98	
Ti–0.2% O_2/N_2	130	All	Annealed sheet / bar	350–400 / 310–380	460–620		105–120	20 / 16	50	4.51	2t	100	94	96	Lowest strength and creep resistance of titanium alloys.
Ti–0.28% O_2/N_2	155	P, S	Annealed	460–520	570–730	50	105–120	15	48	4.51	2.5t	100	92	95	
Ti–0.30% O_2/N_2	160	R, B, W, E, C[a]	Annealed	430–500	540–740		105–120	16	45	4.51	—	100	90	94	
Ti–0.15% Pd –0.07% O_2/N_2	260	S, C[a]	Annealed	170–210	330–420	60	105–120	25 / 35	70	4.52	1t	100	100	100	As above but better resistance to corrosion by non-oxidising acids.
Ti–0.15% Pd –0.13% O_2/N_2	262	All but W and E, C[a]	Annealed sheet / bar	290–330 / 240–280	390–540		105–120	20 / 16	50	4.52	1.5t	100	100	98	
Ti–2.5% Cu	230	All but St, T	Annealed sheet	462–540	620–647		105–120	18–24	—	4.56	1.5t	90	75	90	Ductile weldable alloy, can be strengthened by heat treatment and has creep resistance to moderate temperatures, e.g. 200–350°C.
			bar	400–480	590–620			16–27	45						
			Sol. treat. (805°C) rapid cool aged 18–24 h 400°C + 8 h 475°C sheet	615–675	770–832		105–120	10–20	—	4.56	2.0t	90	70	90	
			bar	580–640	740–795		105–120	10–22	41						
Ti–0.3% Mo –0.8% Ni	(Code 12)	B, P, S, T, W	Annealed	200–230	420–450		105–120	20	25	4.51	2.0t (sheet) 2.5t (plate) to 4.75 mm	100	100	98	Intermediate in corrosion resistance between C.P. Ti and Ti–Pd alloys. Resistant to crevice corrosion in chlorides.

[a] C Also available in the cast form.

B = bar/billet,   S = sheet,   St = strip,   P = plate,   T = tube,   W = wire,   E = extrusion,   R = rod.

Table 2.6.5

TABLE 2.6.6 Typical properties of alpha–beta phase titanium alloys

Nominal composition	Usual ref. number	Forms available	Alloy condition	0.2% proof stress (MPa)	Ultimate tensile strength (MPa)	Impact energy (J)	Elastic modulus (GPa)	Elongation (min.) (%)	Reduction in area (%)	Specific gravity	Formability Minimum bend to thickness ratio (sheet)	Forgeability 100 = best	Machinability 100 = best	Welding 100 = best	Remarks
Ti–6% Al–4% V	IMI318	All but St, T, C[a]	Annealed (700°C + AC)	900–970	960–1270	22	105–120	8–12	35	—	approx. 5t (3t at approx. 700°C)	75	60	60	General purpose, most widely used (and available) alloy in sheet and forgings and the base from which others have been developed. Creep resistant to about 300°C, good fatigue strength. Used widely in new engines and for fasteners, also main casting alloy.
			Sol. treat. (845°C) WQ and aged (540°C)	990 (925–1080)	1020 (1000–1155)	—		15	49	4.42					
			Sol. treat. (954°C) WQ and aged (540°C)	1070 (925–1080)	1130 (1000–1155)	—		14	55						
Ti–4% Al–4% Mo–2% Sn–0.5% Si	IMI550	B, P, F, E	Sol. treat. (900°C) AC and aged (500°C)	970–1110	1060–1280	14	120 (110–130)	12 (9–20) 12 (8–16)	35 25	4.60 4.62	—	70 50	40 30	20 10	High strength forging alloy creep resistant to approx. 400°C. Used for compressor discs, blades and air-frame components.
Ti–4% Al 4%–Mo–4% Sn–0.5% Si	IMI551	B, R, F, E	Sol. treat. (900°C) AC aged (500°C)	1150 (1065–1233)	1310 (1200–1420)	10	120 (110–130)	10 8 8	40 20 20	4.86	— — —	55 55 55	60 45 —	10 10 —	High strength version of 550 retaining good forging properties. Used mainly for air-frame components such as under carriage components, mounting brackets as well as engine components. Also used in steam turbines.
Ti–11% Sn–2.25% Al–4% Mo–0.2% Si	IMI680	B, R, F, E	Sol. treat. AC and aged	970–1000	1050–1110	11	105–120	8 (8–10) 8	15 (15–20) 20	4.54	3 to 4.5t	55	45	20	High strength forging alloy with better creep resistance than 550, now overtaken by other alloys; hardenable up to 200 mm.
			Sol. treat.	1080 1190	1260–1355	11	105–120								
			Sol. treat. FC and aged	930	1050		105–120	(8–13)	(20–32)		—	—	—	—	

Table 2.6.6

TABLE 2.6.6 Typical properties of alpha–beta phase titanium alloys—continued

Composition	Designation	Forms	Condition									Remarks
Ti–6% Al–6% V–2% Sn	IMI662	B, S, P, W, E	Annealed (730°C) / Sol. treat. (870°C) WQ and aged (560°C)	890–1030 / 1000–1155	950–1090 / 1078–1230	114 / 114	12 / 8–15	34 / 25–45	4.65 / 4.65	— / —	60	Improvement on 318 with higher strength and greater hardenability. Rocket cases, air-frame structures are main uses.
Ti–6% Al–2% Sn–4% Zr–6% Mo	246	B, S, P, F	Sol. treat. (870°C) WQ and aged (600°C)	1120–1170	1240–1275	114	6		4.54			High specific strength forging alloy with creep resistance to 400°C. Components for advanced jet engine e.g. disc and fan blade compressor forgings. Based on 6242 (see near α alloys) and used more in USA.
Ti–5% Al–2% Sn–2% Zr–4% Mo–4% Cr	Ti17	B, F	Sol. treat. (800°C) WQ and aged (635°C)	1031–1170	1100–1240	110–113	16		4.54		60	A near beta alloy rich in beta stabilisers, Mo and Cr, for thick section aerospace components e.g. fan and compressor discs. Combines strength and creep resistance of alpha/beta alloys with high hardenability of beta alloys (up to 15 mm). Good fatigue properties; creep resistant to 350°C.
Ti–4.5% Al–5% Mo–1.5% Cr–0.13% O	Corona 5	B, P, F	Sol. treat. (913°C) WQ aged 593°C / Sol. treat. quenched and aged / Annealed and aged	1172 / 850 / 890	1255 / 950 / 956	115	16		4.54			Alloy of high fracture toughness developed to replace 318 at higher strength and fatigue resistant levels.

[a] Also available in the cast form.

B = bar/billet, S = sheet, St = strip, P = plate, T = tube, W = wire, E = extrusion, R = rod, F = forging.

Table 2.6.6—*continued*

TABLE 2.6.7 Typical properties of wrought near alpha phase titanium alloys

Nominal composition	Usual ref. number	Forms available	Alloy condition	0.2% proof stress (MPa)	Ultimate tensile (MPa)	Impact energy (J)	Elastic modulus (GPa)	Elongation (%) (min.)	Reduction in area (%)	Specific gravity	Formability Minimum bend to thickness ratio (sheet)	Forgeability 100 = best	Machinability 100 = best	Welding 100 = best	Remarks
Ti–11% Sn–2.25% Al–2% Zr–1% Mo–0.25% Si	679	B, F	α + β sol. treat (900°C) AC aged (500°C) / α + β sol. treat (900°C) OQ aged (500°C)	890–1000 / 970–1080	1030–1150 / 1120–1340	15	103–110	12 / 10	35 / 30	4.84 / —	— / —	50 / 50	60 / 50	10 / 10	High strength creep resistant alloy for engine components, e.g. blades. Creep resistant to 450°C towards 500°C.
Ti–6% Al–2% Sn–4% Zr–2% Mo–0.1% Si	6242S	B, P, S, F, E	α + β sol. treat (980°C) AC aged (593°C) / α + β sol. treat (950°C) AC aged (595°C) / β sol. treat (1030°C) AC aged (595°C)	893–1020 / 938 / 917	1050–1140 / 1027 / 1048		116 / 116 / 114	15 / 8 / 12	35 / 38 / 23	4.54 / — / —	—	50	50	50	Widely used in USA for creep resistant applications up to 470°C in 400 engines including cases. Effect of Si on creep requires β treatment.
Ti–8% Al–1% Mo–1% V	811	B, S, F, W, E	Sol. treat. WQ aged	855–110	1050–1140	20–34	124–128	8	20	4.38	—	50	50	50	Low density, high modulus with good creep resistance and weldability. Less creep resistant than other alloys in this table but used in USA for aero-engine parts. Tendency to embrittlement at elevated temperature.
Ti–6% Al–5% Zr–0.5% Mo–0.25% Si	685	B, E, R, F	β/sol. treat. OQ aged 550°C	850–940	990–1080		125	6–10	15–22	4.45	—	90	55	90	A weldable β-treatable creep resistant alloy developed for RB211 aero-engine and used for blades and discs up to 520–550°C. Developed for long term creep resistance at 500°C plus. Heat treatment can be optimised for creep resistance or the fatigue properties according to section size etc.

Table 2.6.7

TABLE 2.6.7 Typical properties of wrought near alpha phase titanium alloys—*continued*

Composition	Alloy	Form	Heat treatment											Remarks
Ti–5.5% Al–3.5% Sn–3% Zr–1% Nb–0.25% Mo–0.3% Si	829	B, E, R F	β sol. treat (1050°C) AC aged (625°C) (up to 30 mm)	820–870	950–1020	120	9	18	4.54	—	85	55	90	A weldable β-treatable creep resistant alloy for use at 550°C capable of withstanding 25% higher creep stresses than 685, achieved partly by composition and partly by control of structure through alternative β and α/β working.
Ti–5.8% Al–4.5% Sn–4% Zr–0.7% Nb–0.5% Mo–0.4% Si–0.06% C	834	B, R, F	α + β sol. treat (1030°C) AC aged (625–650°C) for <35 mm thick	910	1030–1080		6	15–24	4.59	—	70	50	85	A weldable α/β-treatable creep resistant alloy developed from 829 to improve fatigue resistance by control of transformation characteristics and thermo-mechanical treatment. Applications in aero-engine application as discs, rings, blades and impellers.
			α + β sol. treat (1030°C) QC aged (625–650°C) for >35 mm thick to 80 mm	974–1050	1110–1140		8–13	16–28	—		—			

B = bar/billet,   S = sheet,   St = strip,   P = plate,   T = tube,   W = wire,   E = extrusion,   R = rod,   F = forgings.
Q = Several variants on 6242S have been developed in the USA but as none show sufficient advantage over 6242S they are not available commercially.
    They are known as Ti–11, 55265.

Table 2.6.7—*continued*

TABLE 2.6.8 Typical properties of wrought beta phase titanium alloys

Nominal composition	Ref. number	Forms available	Alloy condition	0.2% proof stress (MPa)	Ultimate tensile strength (MPa)	Elastic modulus (GPa)	Elongation (%) (min.)	Reduction in area (%)	Specific gravity	Formability Minimum bend to thickness ratio (sheet)	Forgeability 100 = best	Machinability 100 = best	Welding 100 = best	Remarks
Ti–15% V–3% Cr–3% Sn–3% Al	Ti–15–3	S, St	Annealed (790°C)	749–763	770–783	75–96	22–24		4.76	2 to 3t	75	60	60	This is an inexpensive cold forming alternative to hot forming. Its cold forming properties are comparable to unalloyed titanium. It can be aged to very high strengths and is applicable to castings and sheet structure.
			Sol. treat. (790°C) AC and aged (510°C)	1107–1244	1210–1334	92–113	10–8							
			Sol. treat. (790°C) AC and aged (540°C)	983–1038	1100–1141	92–113	14–11							
			Sol. treat. AC, cold rolled 20% aged 454–653°C	1430–1183	1568–1390		3–10							
			as above but 60% cold roll	1567–1292	1629–1375		1–8							
Ti–10% V–2% Fe–3% Al–<0.16% O	Ti 10 V, 2 Fe, 3 Al	B, P, F	Annealed/Beta treat. at 815°C plus average at 620°C	948	1017	86–104	17	46	4.65	—	75	45	—	High strength forging alloy capable of a range of strengths in sections up to 125 mm section with good transverse properties. Properties are stable up to 300°C and it demonstrates creep resistance 300°C. Forgeable at 760–815°C (lower than 318). Good fracture toughness especially overaged.
			Sol. treat. (760°C) WQ and aged (510°C)	1000–1230	1066–1280	111	8–12	15–30						
			Sol. treat. (760°C) AC averaged (580°C)	860–942	928–990	100	10–22	20–56						

B = bar/billet, S = sheet, St = strip, P = plate, T = tube, W = wire, E = extrusion, R = rod, F = forging.

Table 2.6.8

TABLE 2.6.9 Physical properties of titanium alloys

Alloy type		Structure	Specific density	Thermal conductivity 0–100°C (W/m per K)	Thermal expansion co-efficient 20–100°C ($\times 10^{-6}$ K^{-1})	Electrical resistivity 20°C (10^{-6} $\mu\Omega$ cm)	Specific heat at °C (J/kg per K)	Mass magnetic susceptibility (10^{-6})
Nominal composition	Ref. number							
C.P. titanium	110–160	α	4.51	17	7.6	48	528 (50°C)	3.4
Ti–0.15% Pd–0.07% O + N	260	α	4.52	17	7.6	48	528 (50°C)	3.4
Ti–2.5% Cu	230	α	4.56	13	9.0	68		
Ti–0.3% Mo–0.8% Ni	Code 12	α	4.51	17	9.54	52	544 (50°C)	
Ti–6% Al–4% V	318	α + β	4.42	6	7.9	168	595 (50°C)	3.3
Ti–4% Al–4% Mo–2% Sn–0.5% Si	550	α + β	4.60	8	8.8	159		
Ti–4% Al–4% Mo–4% Sn–0.5% Si	551	α + β	4.62	6	8.4	170	375 (50°C)	3.1
Ti–11% Sn–2.25% Al–4% Mo–0.2% Si	680	α + β	4.86	7.5	8.9	166		
Ti–6% Al–6% V–2% Sn	662	α + β	4.54	7	9.5	157	649 (50°C)	
Ti–6% Al–2% Sn–4% Zr–6% Mo	6246	α + β	4.65	7.5	9.36			
Ti–5% Al–2% Sn–2% Zr–4% Mo–4% Cr	Ti 17	α + β	4.65		9.72 (20–400°C)			
Ti–4.5% Al–5% Mo–1.5% Cr–0.13% O	Corona 5	α + β	4.54					
Ti–11% Sn–2.25% Al–5% Zr–1% Mo–0.25% Si	679	Near α	4.84	7.5	8.1	164		
Ti–6% Al–4% Sn–4% Zr–2% Mo–0.1% Si	6242S	Near α	4.54	7	8.1	190		
Ti–8% Al–1% Mo–1% V	811	Near α	4.38		10.1	199		
Ti–6% Al–5% Zr–0.5% Mo	685	Near α	4.45	4.8	9.8	167		
Ti–5.5% Al–3.5% Sn–3% Zr–1% Nb–0.25% Mo–0.3% Si	829	Near α	4.54	7.8	9.45		530 (50°C)	
Ti–5.8% Al–4.5% Sn–4% Zr–0.7% Mg–0.5% Mo–0.4% Si–0.06% C	834	Near α	4.59		10.6			
Ti–15% V–3% Cr–3% Sn–3% Al	15–3	β	4.76		9.72 (20–425°C)			
Ti–10% V–2% Fe–3% Al–<0.16% O	10V2Fe3Al	β	4.65		9.72 (20–425°C)			

Table 2.6.9

TABLE 2.6.10 Low temperature mechanical properties of ELI grades of titanium alloys

ELI grade of titanium	Temperature (°C)	UTS (MPa)	0.1% Proof stress (MPa)	Elongation (%)	Impact strength (J)
Unalloyed	−78	411	275	34	—
	−196	641	431	34	—
Ti–6% Al–4%V	−78	1205	1127	11	18
	−196	1688	1560	5	15

Table 2.6.10

TABLE 2.6.11 Typical fracture toughness values for titanium alloys

Alloy	Type	Heat treatment	0.2% proof stress (MPa)	Ultimate tensile strength (MPa)	El %	Fracture K_{1c} (MNm$^{-3/2}$)	Toughness K_{1scc} (MNm$^{-3/2}$)
318	α+β	2H 700°C AC	1054	1060	15.3	—	20
		2H 700°C AC	956	1041	18	47	—
		Sol. treat. 960°C, WQ aged 700°C (2H)	970	1059	22	51	—
		700°C (2H) AC	—	—	—	66 to 71	—
		Sol. treat. 900°C, WQ	—	—	—	55 to 66	—
550	α+β	Sol. treat. 900°C AC, aged 500°C (24H)	1045	1205	16	50	—
		—	998	1101	14	65–69	
		—	1002	1093	13	57–70	—
		Sol. treat. 945°C AC aged 500°C (24H)	996	1152	18	70	
551	α+β	Sol. treat. 900°C AC aged 500°C (24H)	1191	1354	17	36	
		Sol. treat. 960°C AC aged 500°C (24H)	1133	1321	18	60	
680	α+β	Sol. treat. 810°C WQ aged 500°C (24H)	1230	1337	18	40	
		FC 875°C to 700°C aged 500°C (24H)	1077	1185	21	62	
662	α+β	1" plate annealed	1066	—	—	46.4 to 47.5	
		1" plate annealed	976	—	—	58.5 to 62.2	
		2" plate annealed	928	—	—	56	
		2" plate annealed	969	—	—	60.8	
		2" plate sol. treat.	1210	—	—	52.7	
		and aged	1224	—	—	50.1	
		4" plate annealed	928	—	—	66.2	
		4" plate annealed	969	—	—	60.8	
		0.3" panels annealed	1121	—	—	152.7[a]	
		0.3" panels annealed	1141	—	—	156.0[a]	
		0.3" panels sol. treat.	1203	—	—	131.9[a]	
		and aged	1224	—	—	120.9[a]	
Ti–17	α+β	Not specified	1155	—	—	28.6–34	—
			1107	—	—	58.2–71.4	
			1052	—	—	58.2–62.6	
679	Near α	Sol. treat. 900°C AC, aged 500°C (24H)	996	1170	17	42	—
		unspecified	—	—	—	53.8	—
811	Near α	Duplex anneal 790°C	938	1005	24	—	15.7
		as above + aged 525°C	1032	1084	14.2	—	12.3
		Sol. treat. 1050°C (β) AC	881	1012	10	—	30.6
		as above + aged 525°C (25H)	962	1063	14.8	—	37.0
685	Near α	Unspecified	910	1000	8	68.6	—
		Unspecified	925	1030	8	70.9	—
		Sol. treat. 1050°C OQ aged 500°C (24H)	914	1020	11	70	—
829	Near α	Treated and aged	820	920	9	78	—
		Treated and aged tested at 540°C	460	590	12	65	—
834	Near α	α/β treated and aged	—	—	—	40	—
10V 2Fe 3Al	β	—	963	—	—	110	—
		—	1100	—	—	81	—
		—	1238	—	—	54	—
		Forged sol. treat.	—	—	—	82	79
		and aged				56	50
		Sol. treat. 760°C				54	—
		WQ, aged 510°C Sol. treat. 732°C				102	—
		AC aged 580°C					

[a] Centre notched panels, plane stress KQ.

Table 2.6.11

TABLE 2.6.12 Preferred microstructural features to optimise fatigue properties of titanium alloys

Feature	High cycle fatigue (HCF)	Fatigue crack propagation threshold, ΔK for crack growth	
		R = 0.1	R = 0.7
Grain size	Fine	Coarse	Coarse
α + β structure	Fine equiaxed α + β with β transformed to fine basket weave	Coarser, aligned structure of α plates in β	Coarse aligned structure of α in β
Age hardening	Aged to high strength	Aged to high strength	Aged to low strength
Oxygen content	High	High	Low
Preferred orientation	High	High	—

TABLE 2.6.13 Fatigue data for titanium alloys

(a) Alternating stress fatigue data in rotating bend

Alloy No.	Stress (MPa) for 2 x 10⁷ cycles at:				
	20°C	200°C	400°C	450°C	520°C
550	±650	±550	±500		
551					
685	±440			±330	±260

(b) Direct stress fatigue data in tension

Alloy No.	UTS (MN/m²)	Stress range (MPa) at 25°C for fatigue lives in excess of:				
		10³	10⁴	10⁵	10⁶	10⁷ cycles
318	1020					0–460, 0–560[a]
662	1068					0–586
	1129					0–724
685	1035		0–940	0–660	0–650	0–650
829	985		0–1000	0–860	0–730	
834	1125		17–1000	17–880	0–650	0–570

[a] Two data sources.

Table 2.6.12 and Table 2.6.13

TABLE 2.6.14 Typical creep properties of titanium alloys

Alloy designation	Type	Heat treatment	Temperature (%)	Stress (MPa) to produce 0.1% total plastic strain in				
				100 h	300 h	1000 h	10 000 h	100 000 h
115	C.P. Ti	Annealed	20 50 100 150 200 250 300			309 252 188 145 116 102 93	278 232 170 131 108 97 90	260 213 157 122 104 94 86
230	Alpha precipitation hardened	Annealed Aged	20 100 200 300 400 200 300 400 450	360 279 235 202 125 435 375 220 109				
318	α/β	Annealed	20 100 200 300 400 500	832 704 638 576 287 32	818 680 636 568 144 18	788 676 635 — 102 —		
550	α/β	Solution treated and aged	300 400 450 500	724 551 254 82	718 519 174 51	710 471 101 31		
551	α/β	Solution treated and aged	400 450	621 307	575 217	501		
680	α/β	Solution treated and aged	200 300 400 450 500	862 804 555 298 88	856 788 540 209 51			
6246	α/β	Solution treated (870°C) WQ, aged 593°C (8H)	400 450	558 241	454 162			
Ti 17	α/β	Solution treated and aged	300 400	744 241	723 —			
679	Near α	Air-cooled and aged	300 400 450 500 540	674 600 448 131 150Q	664 579 386 51 —	517 — — — —		
685	Near α	Solution treated 1050°C (β OQ, aged 500°C)	300 400 450 500 520 540	551 510 465 390 — 280Q	541x 480x 431x 340x 300x —	535 462 426 — — —		
829	Near α	Solution treated 1050°C (β) WQ, aged 625°C	300 400 450 500 540 550 600	 495 478 420 340 300 130	 490 440 345 — 200 83	 480 405 295 — 140 59	 440 330 200 — 74 —	500 400 250 102 — <50 —
834	Near α	Solution treated 1030°C (α+β) OQ, aged 650°C	500 550 600	600Q 385Q 200Q	510Q 295Q —	435Q 235Q —	295 — —	— — —

x = 500 h.   Q = 0.2% total plastic strain.

Table 2.6.14

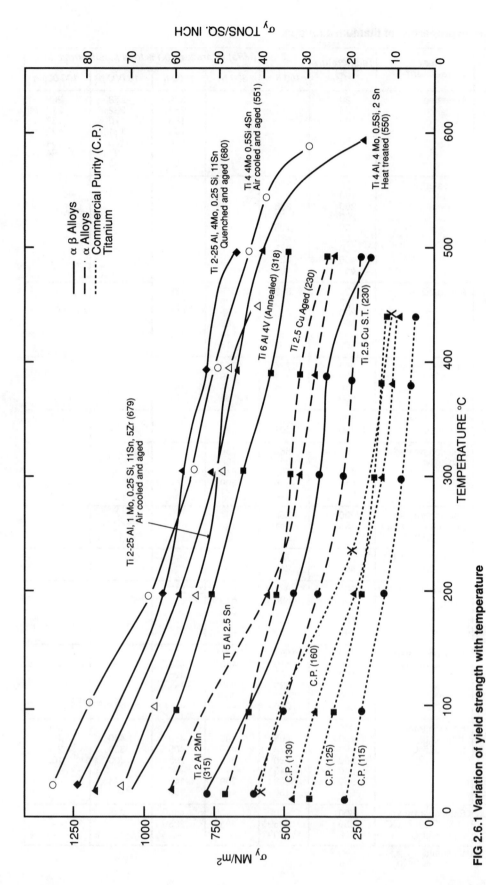

FIG 2.6.1 Variation of yield strength with temperature

Fig 2.6.1

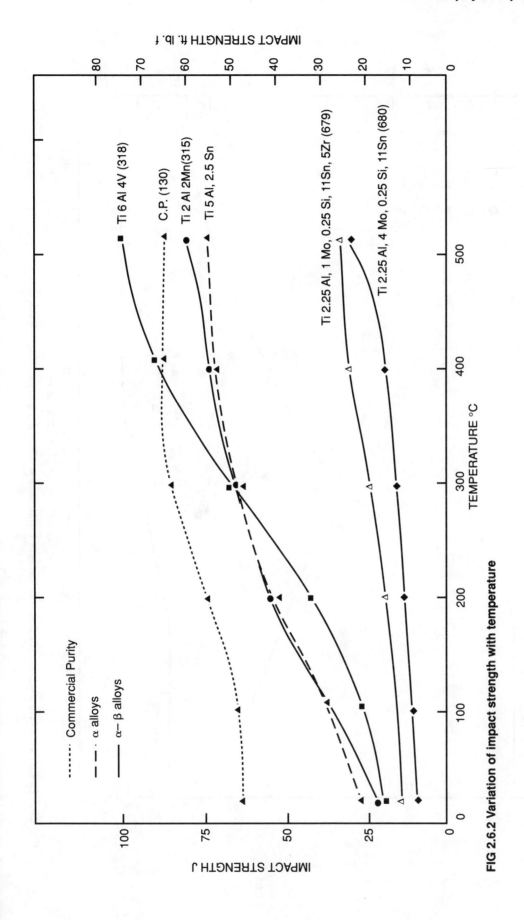

FIG 2.6.2 Variation of impact strength with temperature

Fig 2.6.2

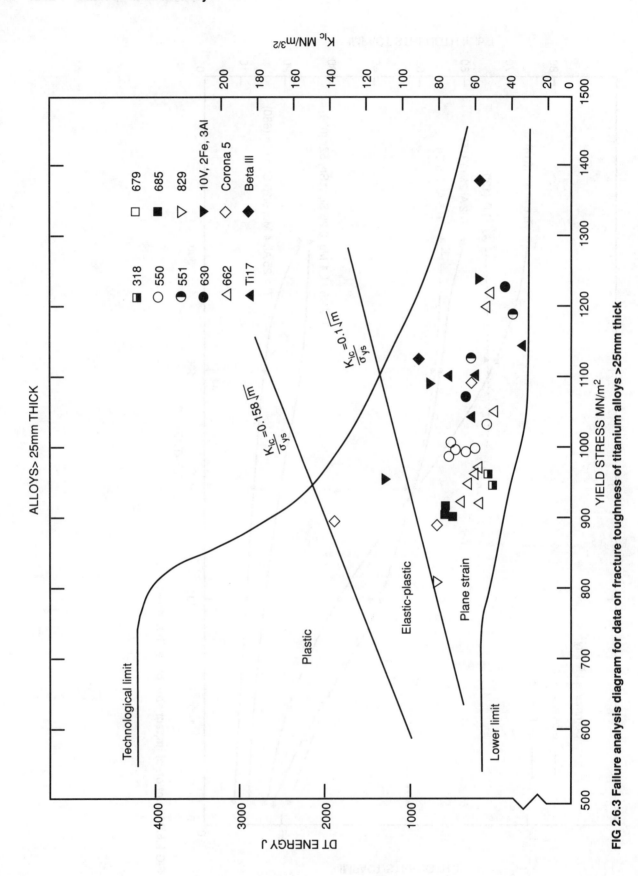

FIG 2.6.3 Failure analysis diagram for data on fracture toughness of titanium alloys >25mm thick

Fig 2.6.3

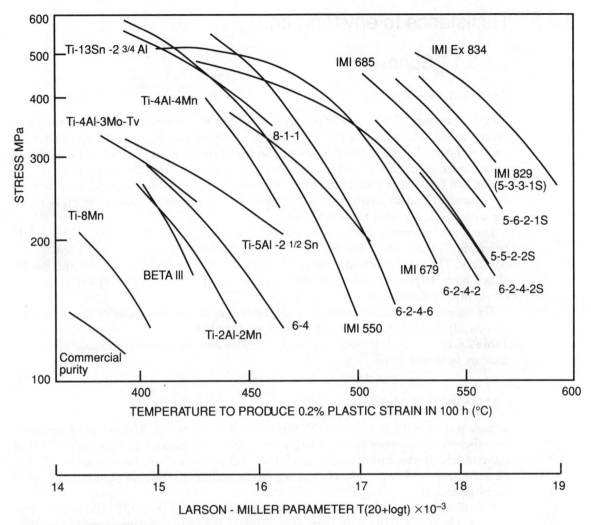

FIG 2.6.4 Larson–Miller plot of creep resistant titanium alloys

Fig 2.6.4

2.6.3   Resistance to environment

2.6.3.1   Corrosion

GENERAL CORROSION

Titanium is intrinsically a very reactive metal but is protected by a very strong and tenacious oxide film. Even if this is damaged, it reforms rapidly in oxidising or aqueous environments. Titanium has therefore exceptional resistance to corrosion by oxidising environments but less resistance in reducing conditions such that high rates of corrosion can occur. It is resistant to nitric acid solutions, sea-water, wet chlorine (but not dry chlorine) and to hydrogen-producing reducing solutions containing an oxidant, e.g. nitric acid, cupric or ferric ions, etc.

The alloys normally used under corrosive conditions are the α phase alloys, C.P. titanium, Ti–0.3% Mo–0.8% Ni, Ti–0.15% Pd, Ti–2.5% Cu and the α/β alloy Ti–6% Al–4% V. The most resistant alloy is usually C.P. titanium, but the Mo–Ni and the Pd alloys have improved corrosion resistance to crevice attack. The Ti–Pd alloys are especially resistant to non-oxidising acids.

The rates of attack of a number of aggressive media on the alloys of titanium used specifically for their resistance to corrosion are listed in:
Table 2.6.15   *General corrosion of commercially pure and other corrosion resistant alloys of titanium in various media*

GALVANIC CORROSION

Titanium is cathodic in contact with other metals in sea-water. The extent of attack on the other metal depends on the relative cathode/anode areas. For high ratios (10:1) of cathode/anode areas mild steel, aluminium and gun-metal are heavily attacked, aluminium brass and cupro-nickels less so, whilst stainless steel and aluminium bronze are not attacked.

A similar situation arises with non-oxidising acids, except that with mild steel hydrogen is evolved at the titanium surface and breaks down the protective oxide film. The titanium then corrodes, and is embrittled by the hydrogen.

The galvanic potential of titanium relative to other metals is listed in:
Table 2.6.16   *Galvanic series of metals based on potential measurements in flowing sea-water at 25°C*

STRESS CORROSION

Commercial purity titanium is resistant to stress corrosion at stress levels up to 80% of its UTS except in red fuming nitric acid with less than 2% water, methanol unless 2% water present and 10% hydrochloric acid.

CORROSION FATIGUE

The fatigue limit of C.P. titanium (10^7–10^8 cycles) is about 50% of the UTS. This is not reduced by corrosive environments except those which normally attack titanium in the absence of stress. If hydrogen is generated this will be absorbed by titanium and embrittle it. However there are some reports of a 15% fall in fatigue strength in 3.5% NaCl at 10^7 cycles but not at 10^8 cycles.

Alloys 318, 550, 551, 811 and Ti–6%Al–6%V–2%Sn all show little susceptibility to corrosion fatigue in 3.5% NaCl solution.

2.6.3.2 Friction and wear of titanium alloys

ADHESIVE WEAR

Titanium and its alloys are very prone to adhesive wear when in sliding contact with other metals and especially itself. This tendency to gall and smear on to mating surfaces is partly the cause of machining difficulties with titanium. At all but very light contact loads, the protective surface oxide film is ruptured during sliding and this allows bonding under pressure at the relatively high local temperatures so generated.

Some effective form of surface protection must be used to harden the surface, e.g. by cold working, by solid solution hardening or by applying hard layers.

Examples are:

Anodising to increase the thickness of the natural oxide layer.
Diffusion treatments such as carburising, nitriding, boronising, sulphurising, to produce a hard layer. This may result in a surface layer of TiN or TiB_2.
Electrodeposited coatings such as nickel, chromium or zinc.

The effect of these treatments on the fatigue properties must be taken into account. The combination of a compressive stress and high nitrogen content in the surface achieved by nitriding is preferred, but little data on this system is available.

A successful alternative technique is to plate a 15 μm thick chromium layer on a shot-peened surface, then to diffuse this in at 700°C and to restore the hard chromium surface by a further plating. A simpler version of this process is to electroplate to the thickness required and then to heat treat at 200°C for 2 h. This obviates the need for a second electroplate.

FRETTING

Titanium and its alloys are susceptible to fretting. The danger is not so much the removal of material and loss of dimensional control but the reduction in fatigue life caused by fretting. Titanium fretting on itself can reduce fatigue strength by a factor of 2.5 and fretting on harder metals such as steel or brass can cause reduction by a factor of 3–5.

Fretting can be reduced by shot peening or by using the coating treatments mentioned under Adhesive Wear. Coatings of molybdenum disulphide, or polymers, may be used to avoid the metal–metal contact which is needed for fretting. The best solution is to eliminate if possible the cause of the small relative movements which result in fretting.

TABLE 2.6.15 General corrosion of commercially pure and other corrosion resistant alloys of titanium in various media (data based on IMI Titanium source)

Medium	Concentration	Temperature	Rate of attack
Sea-water		Up to 130°C	No attack, pitting or crevice corrosion.
Natural waters		>130°C Ambient Ambient	Some crevice corrosion but not in Ti–Pd below 18 m/s in tubes or 8 m/s in rotating disc tests. No stress corrosion up to 80% UTS.
Nitric acid	Up to 100% 40–50% 20–70%	Up to 100% 100°–190°C 190°–240°C	Little or no attack (maybe weld attack if Fe>0.05%). Up to 3 mm/year. Up to 10 mm/year.
Red fuming nitric acid	<2% water >6% N_2O_4		Explosive hazard.
Chlorine (wet)	>0.7% water >0.95% water >15% water	25°C 140°C 200°C	No reaction. No reaction. No reaction.
Chlorine (dry)	<0.005% water	25°C	Ignition.
Hypochlorites	Up to 18% Up to 6%	Room 100°C	Little or no attack (<0.002 mm/year). <0.001 mm/year.
Bromine	Bromine water Moist gas Liquid or dry gas	25°C 25°C 25°C	No attack. <0.003 mm/year. Rapid solution.
Iodine chlorides	Gas (moist/dry)	25°C 130°C	<0.1 mm/year. Rapid (approx. 1.7 m/year.
$AlCl_3$	Up to 10% 10% 25% 25% 40%	<100°C 150°C Up to 60°C 100°C–BP 122°C	<0.002 m/year. 0.03 mm/year. <0.001 mm/year. 6.6 to 50 mm/year. 100 mm/year.
NH_4Cl	Saturated	100°C	No attack.
$BaCl_2$	Up to 25%	Up to boiling	No attack.
$CuCl_2$	Up to 55% 62% 73%	105°C 155°C 175°C	<0.001 mm/year. Attack starts. 0.75 mm/year.
$CuCl_2$	Up to 55%	118°C	<0.005 mm/year.
CuCl	50%	90°C	0.0025 mm/year.
$FeCl_3$	Up to 50%	110°C	<0.018 mm/year.
$LiCl_2$	50%	150°C	No attack.
$MgCl_2$	Up to 50%	<200°C	<0.01 mm/year.
$MnCl_2$	5, 20%	100°C	No attack.
$HgCl_2$	Saturated	100°C	0.001 mm/year.
$NiCl_2$	5, 20%	100°C	0.004 mm/year.

Table 2.6.15

TABLE 2.6.15 General corrosion of commercially pure and other corrosion resistant alloys of titanium in various media (data based on IMI Titanium source)—*continued*

Medium	Concentration	Temperature	Rate of attack
KC1	30% saturated	110°C 60°C	0.013 mm/year. No attack.
$SnCl_4$	24%	Boiling	0.05 mm/year.
NaCl	Saturated	Boiling	0.001 mm/year.
$ZnCl_2$	To 75% 80–90%	150°C 173–250°C	<0.06 mm/year. 2.1–30 mm/year.
NaCl	10% (all pH) 10% pH 5.5–9 pH 9–10.3	<140°C 150°C 150°C	Crevice corrosion not observed. Some pitting and crevice corrosion. No attack.
H_2SO_4	Up to 20% Up to 5% Higher concentration	0°C } 25°C } Higher temperatures	Little attack. Increasing attack as concentration and temperature rise. Corrosion at these higher concentrations and temperatures can be reduced to less than 0.3 mm/year by up to 1% copper sulphate, 0.5% CrO_3 or 8% ferric sulphate Ti–0.5Pd is resistant to 4% H_2SO_4 at BP, 1.0% H_2SO_4 at 70°C and 25% H_2SO_4 at 25°C.
HCl	Up to 7.5% Up to 3% 0.5%	25°C } 60°C } 100°C }	<0.1 mm/year. Higher concentration and temperatures cause accelerating corrosion. This is reduced significantly by oxidising ions.
H_3PO_4	Up to 30% 10% 2%	35°C } 60°C } 100°C }	<0.09 mm/year. Small amounts <0.1% of ferric, cupric, mercuric ions reduce corrosion in 50% H_3PO_4 to approx. 0.1 mm/year. Ti–Pd corrodes at 0.15 mm/year in boiling 10% H_3PO_4.
H_3BO_3	Saturated 10%	25°C } Boiling }	No attack.
H_2CrO_4	Up to 50%	Up to 80°C	<0.05 mm/year.
HBr	40%	25°C	No attack.
HI	10% 57%	Boiling 25°C	No attack. 0.15 mm/year.
HF fluoboric acid Fluosilicic acid			Rapid attack in acids with fluoride radical.
Aqua Regia (3:1 $HCl:HNO_3$)		25°C 80°C	No attack. 0.86 mm/year.
Alkalis (except NaOH, KOH)			No attack by barium, calcium and magnesium hydroxides; <0.003 mm/year in 28% NH_4OH at 25°C.

Table 2.6.15—*continued*

TABLE 2.6.15 General corrosion of commercially pure and other corrosion resistant alloys of titanium in various media (data based on IMI Titanium source)—*continued*

Medium	*Concentration*	*Temperature*	*Rate of attack*
KOH	Up to 50% 10–5%	25°C Boiling	0.01 mm/year. 0.13 to 2.7 mm/year.
NaOH	Saturated Up to 73%	25°C >60°C	<0.003 mm/year. 0.01–1 mm/year depending on temperature. Not suitable for boiling NaOH/KOH.
Inorganic salts of Al, ammonia, Ba, Ca, Cu, Fe, Mg, Hg, Ni, K, Ag, Na, Zn	Up to saturated		No or little attack unless mentioned elsewhere in this table.
Sodium bisulphate	10%	65°C Boiling	1.83 mm/year but resistant to sodium bisulphite.
Organic acids, except formic, oxalic, citric, propionic and chlorinated acetic acids	Up to 100%	Up to boiling	Resistant = no attack or less than 0.002 mm/year.
Citric acid (Aerated)	50% 50% 62%	100°C Boiling 150°C	<0.001 mm/year. 0.13–1.3 mm/year. Corrodes.
Formic acid	25% 50%	Boiling Boiling	2.4 mm/year only } if aerated; 7.6 mm/year } no attack if aerated.
Oxalic acid	0.5%–10% 0.5–10%	35°C 60–100°C	Approx. 0.1–0.2 mm/year. 2–21 mm/year depending on concentration and temperature.
Propionic acid	Vapour	190°C	Rapid attack.
Trichloracetic acid	100%	Boiling	14.55 mm/year. Little attack from mono- and dichloroacetic acids.
Hydrocarbons and substituted hydrocarbons except Benzyl Bromide	100% 100%	Boiling Boiling	Little or no attack. Rapid attack.
Other organic compounds except Phenol Methanol	Saturated	25°C	Little or no attack. 0.1 mm/year. Stress corrodes if >50 ppm chlorides/bromides present.
Ammonia			No attack except anhydrous at 40°C (<0.13 mm/year).
H_2S, SO_2 wet or dry			Little or no attack.
Molten metals			Only Al, Ga and Zn of low MP metals attack at >1.3 mm/year.

Table 2.6.15—*continued*

TABLE 2.6.16 Galvanic series of metals based on potential measurements in flowing sea-water at 25 °C

Metal	Steady state potential negative to saturated calomel half cell (V)
18/8 stainless steel (passive)	0.08
Hastelloy C	0.08
Monel	0.08
Titanium	0.08
Silver	0.13
Inconel	0.17
Nickel	0.20
70/30 cupro nickel	0.25
80/20 cupro nickel	0.27
90/10 cupro nickel	0.28
Admiralty Brass	0.29
Gun metal	0.31
Copper	0.36
37% zinc brass	0.36
Naval brass	0.40
18/8 stainless steel active	0.53
Steels and cast iron	0.61
Aluminium	0.79
Zinc	1.03

Source: IMI Titanium Ltd, data.

Table 2.6.16

2.6.4   Fabrication

2.6.4.1   Casting

The high reactivity of molten titanium alloys with crucible and mould materials, combined with their high melting points, have caused considerable difficulties in the production of sound precision castings with uncontaminated surfaces.

However, recent improvements have made it possible to make both conventional and investment castings for parts which were formerly welded constructions or forgings and for new components where the freedom to produce the complex forms allowed by casting permits the cost-effective use of titanium. Typical examples are impellers and pump housings and valves for the chemical industry as well as prostheses for the medical profession and engine component castings for the aircraft industry. The alloys normally cast are:

C.P. titanium (99.2–99.4% Ti)

Ti–Pd alloys (260 and 262)

Ti–6% Al–4%V (318)

Ti–5%Al–2.5% Sn (317) was formerly cast but is normally replaced by 318 now. Other alloys should be castable if the demand warrants it.

Two castings routes are available depending on the dimensional tolerances required. Conventional casting into specially prepared graphite moulds provides tolerances which are comparable to those for aluminium castings. The rammed graphite moulds are made from high purity electrographite and organic binders which, after firing and graphitising for several days, produces moulds capable of withstanding attack from molten titanium.

Precision castings are made by melting titanium in a consumable electrode vacuum skull melting furnace and casting centrifugally at up to 600 rpm into investment moulds made by the lost wax process, using special moulding materials fired in reducing conditions.

Design data for titanium castings are listed in:

Table 2.6.17   *Geometrical limitations, surface finish and mechanical properties of conventional (rammed graphite) and investment cast titanium*

For aircraft applications, castings can be hot isostatically pressed in argon to seal any internal shrinkage porosity. Surface defects are ground out and repaired by welding where necessary. If 318 alloy is Hipped after investment casting, the fatigue properties of castings are comparable with those of the wrought alloy in the same state of final heat treatment.

2.6.4.2   Forging

Titanium alloys can be readily hot worked at temperatures generally somewhat lower than those used for steels. To minimise surface contamination, titanium alloys should be held at high temperatures for only a short time before forging. The rate of contamination, relatively low up to 700°C, increases rapidly with increase in temperature.

All forging atmospheres contain free or combined oxygen, and some absorption of this element inevitably occurs. In addition to visible scaling, diffusion of oxygen results in hardening and embrittlement of a relatively shallow underlying layer. The effect of nitrogen is not usually significant at pre-heating temperature. Hydrogen, however, diffuses more rapidly than oxygen and may penetrate the full section of the work which can have a serious effect on properties. Such material can only be recovered by prolonged vacuum annealing. Hydrogen is absorbed from both reducing and oxidising gas- and oil-fired furnaces, but at a tolerably slow rate under strongly oxidising conditions.

The order of preference of preheating atmospheres is therefore: dried air (electric heating), undried air (electric heating), oxidising oil- or gas-fired furnaces. Direct flame impingement must be avoided.

Techniques for press and hammer forging are essentially the same as for low alloy steels. A final stress relief anneal is recommended. A comparison of the forgeability of commercially available alloys is given in Tables 2.6.5–2.6.8.

2.6.4.3   Superplastic forming and diffusion bonding

The barrier to the more widespread use of titanium alloys, especially in aerospace applications, is the high cost-in-position. To avoid wasteful machining to size and the effects of heat affected zones in fabrications by welding, more economic forming methods have been developed. These include superplastic forming, diffusion bonding (which can be carried out simultaneously with superplastic forming) and isothermal forging.

SUPERPLASTIC FORMING

Duplex alpha plus beta alloys with a grain size of 1–10 μm have high strain-rate sensitivity at 900–925°C when deformed at low strain rates in the range 10^{-4}–10^{-3} s^{-1}. Under these conditions, the alloys can be deformed to strains of over 1000% without local necking or variations in thickness. Alloys 318, 550 and 6–2–4–2 are suitable for this process.

Sheet alloys can be free blown to form a bubble by inert gas pressure or blown into a tool to produce more complicated shapes. The metal does not remain in contact with the die for long at high temperature so that die materials can include ceramics or austenite steel treated with boron nitride as a release agent. Two sheets can be edge welded and gas introduced through a pipe to internally pressurise the combination to expand the blank into a tool, forming a hollow component.

DIFFUSION BONDING

For diffusion bonding, smooth mating surfaces are brought into intimate contact under low pressure at temperatures of 900–925°C. Interdiffusion is aided by the absorption of the oxide film by titanium at bonding temperatures. The bond can be effected by either mechanical pressure or inert gas pressure.

Advantages	*Limitations*
No heat-affected zone.	If surfaces not smooth, pressure too low or bonding times too short, array of cavities may be left at interface.
Interface not detectable in good bond.	
Bonding can be produced by mechanical or gas pressure.	Cavities at interface reduce fatigue strength.
Bonding can be prevented locally by screen-printing boron nitride or yttria films, or spraying carbonaceous films.	Non-destructive examination of bonds difficult to carry out effectively.

COMBINED SUPERPLASTIC FORMING AND DIFFUSION BONDING

As both processes take place at 900–925°C, they can be used simultaneously to make complex multi-sheet components without the use of fasteners. Multi-layer sandwich components can be made by using stop-off materials printed on the sheets where no bonding is required.

These processes allow considerable freedom in design and enable ribbed structures to be made. High fatigue strengths are achieved in the absence of fasteners or heat-affected zones from welding. In airframe structures, savings of 10–45% in weight and 20–50% in cost have been achieved compared to titanium alloy fabrications.

Structures made by this process include:

Hollow fan blades for aero-engines.
Hot air ducts for engines.
Missile cones.
Engine frame supports.
Control surfaces in wings and tail planes.

ISOTHERMAL FORGING

More solid components can be shaped at 900–950°C between metal dies heated to the same temperature. This enables near net shape forgings to be made at temperatures at which no die chilling and no strain hardening of the forging occurs. This provides better utilisation of material, a reduction in the number of forging operations and a substantial saving in machining operations.

Normally, use is made of the superplastic properties of the alloy but where large strains are not involved (e.g. 100–500%), it is possible to deform at higher rates, e.g. 10^{-1}–1 s^{-1}. The forging dies must be made of heat resisting alloys capable of withstanding 900°C.

2.6.4.4   Sheet forming

Annealed and solution treated sheets can be pressed, stretch-formed, spun and dimpled but maximum deformation depends on slow application of the load. The edge condition of blanks is important and all burrs must be removed.

Simple shapes can be formed at room temperature, deformation being limited by the strength and springiness of the material. For more complicated designs, the workpiece and dies should be heated to facilitate forming. Heating increases ductility, permitting lower minimum bend radii, reduces loads and subsequent springback, and hence improves accuracy. Furthermore, at elevated temperatures the spread between yield and ultimate strengths is increased, which also aids forming. Prolonged heating is undesirable due to the possibility of contamination, but it is important that heating should be uniform if the workpiece is not to deform unevenly.

2.6.4.5   Heat treatment

The mechanical, fracture, fatigue and creep properties of titanium alloys are critically dependent on thermomechanical processing and the development of microstructure by subsequent heat treatment. A complex technology has now been developed to tailor the microstructure to optimise one or several of these properties.

2.6.4.6   Joining

WELDING

Titanium and many titanium alloys can be joined by fusion, resistance, flash-butt and pressure welding and, with adequate control of welding techniques, are among the easiest metals to join. Techniques are governed by the metal's affinity for atmospheric gases and consequent embrittlement. Appreciable contamination by oxygen, nitrogen and hydrogen can occur above 650°C and at its melting point, the metal rapidly dissolves these gases and carbon. Oxy-acetylene, metal-arc, carbon-arc, and atomic-hydrogen welding processes are therefore completely unsuitable.

Although the welding techniques employed for commercially pure titanium are also applicable to many titanium alloys, extra attention to ensure adequate shielding is necessary because these materials are less tolerant of contamination.

Fusion Welding

(a) Commercially pure titanium

Plasma arc, inert-gas metal-arc and electron-beam processes may all be used for welding titanium in appropriate circumstances. Protection from atmospheric gases is essential and is normally achieved by supplying argon to the heated surfaces, or by operating in a totally enclosed argon-filled welding cabinet.

The properties of the resultant welds are listed in:

Table 2.6.18   *Typical mechanical properties of argon-arc welds in commercially pure titanium*

Cracking in welds of commercially pure titanium occurs only after exceptionally severe contamination, which can be avoided if the recommended welding and shielding procedures are carefully followed. Cracked areas should never be rewelded. Repairs to such areas are best effected by removing the defective material and welding-in a patch. Titanium cannot be directly joined to mild steel because brittle intermetallic constituents form during welding.

(b) Titanium alloys

In general, although titanium alloys are less tolerant of contamination, their fusion welding characteristics are essentially similar to those of the commercially pure material. Properties obtainable after welding, however, depend to a large extent on the type of alloy under consideration. Thus, with single-phase alloys and those containing only small amounts of alloying additions, e.g. Ti–2.5%Cu (230), mechanical properties approach those of the parent material. With some of the more highly alloyed materials, e.g. the alpha-beta alloy Ti–6%Al–4%V (318), structure and properties are modified by the thermal cycle. Considerable research has been devoted to the evolution of new alloys and the improvement in properties of fusion welds in commercially available materials. Ti–6%Al–5%Zr–0.5%Mo–0.25%Si (685) and the other near alpha alloys (829 and 834) were developed with these aims in mind.

The high proof/tensile strength ratio of certain stronger alloys may lead to high residual stresses in welding, causing delayed cracking, stress corrosion and possible failure. A stress-relief anneal after welding is usually sufficient to prevent this.

The properties that may be anticipated in fusion welds in titanium alloys are listed in:

Table 2.6.19   *Typical mechanical properties of argon-arc welds in titanium alloys*

Resistance Welding

(a) Commercially pure titanium

Titanium, with an electrical resistance similar to that of stainless steel, is ideal for spot and seam welding. With sheet up to 1.5 mm thick sound joints can be obtained with conventional equipment over a wide range of machine settings.

(b) Titanium alloys

The ductility of spot welds in duplex alloys is lower than that of similar joints in commercially pure titanium.

Flash-butt welding

Satisfactory flash-butt welds can be produced in commercially pure titanium and most titanium alloys in sheet form and in sections up to about 1300 mm^2 cross-sectional area. In general, the process gives a more uniform structure than fusion welding because of the forging action and resultant grain refinement. Welding conditions must, however, be closely controlled.

Pressure Welding

Because of the high temperature involved, satisfactory fusion welds in titanium depend upon precautions taken to avoid contamination of the weld zone. Pressure welding, however, is possible at temperatures within the forging range, where oxidation and the diffusion of hardening elements are not excessive.

Pressure welding is ideal for butt-welding bar and tubular stock and joints in commercially pure titanium are as strong and ductile as the parent metal.

A comparison of the weldability of commercially available titanium alloys is given in Tables 2.6.5–2.6.8.

BRAZING

Several conventional brazing techniques are applicable to titanium and its alloys; most experience has been gained with argon-arc, oxyacetylene, furnace and induction brazing (the last two techniques utilising either inert gas atmosphere or a vacuum).

Certain limitations are imposed by the metal's chemical and metallurgical properties. Filler alloys tend to alloy with and attack the base metal, forming brittle intermetallic compounds. For commercially pure titanium, choice of filler alloy largely depends on the process and the application. When brazing titanium alloys, extra care must be taken to avoid impairing the properties of the parent metal. The brazing temperature, and hence filler alloy selection, must take account of the alpha-beta transition temperature of the titanium alloy and of the heat treatment normally applied.

2.6.4.7   Machining

Titanium and its alloys have two characteristics which make machining difficult—a low elastic modulus which results in spring-back, and a tendency to gall and smear on to other metals. To overcome these difficulties rigidity of the machine tool and workpiece, and use of sharp tools with good surface finish, is essential. In general, cutting speeds should be low and feeds as coarse as possible.

A good surface finish can be obtained using a large nose radius tool and coarse feeds but this will also increase the load on the workpiece with consequent deflection. If a good surface finish and dimensional accuracy are required, as is often the case, a compromise is necessary.

Titanium and its alloys can be turned, threaded, planed, milled, drilled, sawn and ground. For details of the recommended parameters of speeds, feeds, etc., reference to suppliers should be made.

N.B. Fine dry titanium swarf is inflammable and special care should be exercised. Fires should be extinguished with a mixture of dry mineral fibre and chalk powder; water must not be used.

2.6.4.8   Powder metallurgy

The high cost of machining and the low yield of titanium in machined titanium parts makes the near-set shaping of complex components an attractive route. Considerable efforts have been applied to devising methods of manufacture of components by powder metallurgy. This takes three forms.

Titanium sponge can be consolidated by conventional powder compacting techniques and also by hot isostatic pressing to remove all porosity. There are problems in avoiding high oxygen contents by absorption of surface oxides. Both pure titanium and 318 alloy have been made from elemental powders. Though acceptable strengths have been achieved, the fatigue strength has been inadequate especially for the aircraft industry.

An alternative route is to pre-alloy and atomise the powder. Techniques for control of quality have been developed and it is possible to manufacture components. The most cost-effective route to produce powder is by the Rotating Electrode Process which involves local melting only. Good quality products have been obtained but the costs have been unacceptable in comparison to those of cast and wrought products.

Using rapid solidification technology, compositions not attainable by other melting routes can be produced and this provides the potential for unusual alloys with very attractive properties.

TABLE 2.6.17 Geometrical limitations, surface finish and mechanical properties of conventional (rammed graphite) and investment cast titanium

	Investment cast		Rammed graphite	
Weight	50 kg up to 125 kg		80–90 kg up to 450 kg.	
Size (envelope)	500 mm×500 mm×500 mm		900 mm×900 mm×750 mm	
Wall sections Thickness	Minimum 2–3 mm in general; 1.5 mm locally		Minimum about 3 mm	
Tolerances	Up to 6 mm ± 0.1 mm 6–10 mm ± 0.12 mm 10–18 mm ± 0.20 mm		Up to 6mm ± 0.06 mm 6–10 mm ± 1.2 mm 10–18 mm ± 1.8 mm	
Tolerance on linear dimensions	Up to 50 mm ± 0.44 mm 50–80 mm ± 0.60 mm 80–120 mm ± 0.26 mm 120–180 mm ± 1.02 mm 180–250 mm ± 1.44 mm 250–315 mm ± 1.84 mm 315–400 mm ± 2.4 mm 400–500 mm ± 3.0 mm		Up to 50 mm ± 1.0 mm 50–80 mm± 1.2 mm 80–120 mm ± 1.3 mm 120–180 mm ± 1.6 mm 180–250 mm ± 1.8 mm 250–315 mm ± 2.0 mm 315–400 mm ± 2.2 mm 400–500 mm ± 2.4 mm	
Draft angles	1°–3°		2° recommended 1° possible for standard patterns 0°45' minimum in special cases	
Surface finish	better than 125 RMS as cast		250 RMS as cast 125 RMS hand finished	
Typical properties	C.P. Ti	318	C.P. Ti	318
0.2% proof stress (MPa)	370	815	350	800
Ultimate tensile strength (MPa)	470	880	450	900
Elongation (%)	25	5	10	5
Reduction in area (%)	35	10	15	10

(C.P. Ti as cast; 318 annealed)

Table 2.6.17

TABLE 2.6.18 Typical mechanical properties of argon-arc welds in commercially pure titanium

Material	Ultimate tensile strength (MPa)	Elongation on 50 mm (2 in) (%)	Bend radius	Hardness (HV 10)		
				Sheet	Heat-affected zone	Weld
Titanium 115	343	35^P	1t	154	148	163
Titanium 130	545	10–29$^{W/P}$	2t	182	193	201
Titanium 155	695–	14–15^P	3t	232	230	240–250

W, Fracture in weld metal.   P, Fracture in parent metal.

TABLE 2.6.19 Typical mechanical properties of argon-arc welds in titanium alloys

Material	Ultimate tensile strength (MPa)	Elongation on 50 mm (2 in) (%)	Bend radius
Ti–6Al–4V	1100–1120	7.10	6–7t
Ti–2.5Cu (Titanium 230) Annealed sheet Aged sheet	480–560 770	16–19 14	2t 2.5t

Table 2.6.18 and Table 2.6.19

2–7

Nickel alloys

Contents

List of tables

List of figures

2.7.1   Introduction

2.7.1.1   Introduction and general characteristics

Nickel base alloys are characterised by having a face-centred-cubic crystal structure, giving them, in general, high ductility and toughness over a wide temperature range. Except for the high-strength, high-temperature alloys, they are therefore reasonably easy to form to the required shape, even though a high rate of work hardening may require inter-stage annealing. The properties of nickel alloys which make them of industrial importance are varied, and include good corrosion resistance to many gaseous or liquid environments at normal temperatures, resistance to oxidation at elevated temperatures, good mechanical strength at high temperatures, and special physical properties–electrical, magnetic and dimensional. Detailed consideration of the commercially available alloys is therefore most appropriately carried out under separate headings according to the principal property requirement.

Most of the alloys listed as wrought products are available in all the common semi-finished forms; bar, rod, wire, sheet, plate, strip and tube. Many may also be produced as castings. The alloys listed as casting materials are normally only available in that form. Nickel and its more ductile alloys are readily weldable and suitable consumables such as electrodes and filler wires are available for most materials. The more highly alloyed materials, particularly the cast high temperature alloys, present greater welding problems. Nickel powder is available and may be used for alloy production when very close control of composition is necessary or for fully-dense or porous powder compacts which exploit the corrosion resistance of the metal. Pure nickel and many nickel-rich alloys are susceptible to serious attack at temperatures above about 600°C by sulphur compounds and special care is required to guard against this during heat treatment. Alloys containing chromium or manganese are much less seriously affected, and some of the high-chromium alloys are specifically recommended for resistance to hot corrosion by sulphur compounds.

2.7.1.2   Available types and typical applications

Commercial nickel base alloys are often developed and marketed with a particular application in mind, and in such cases quite minor modifications of composition may be introduced to satisfy the requirements. However, for a first consideration of the types of alloy available and their general characteristics, it is convenient to categorise the alloys into a few broad groups.The type of alloy under each main heading, its approximate range of composition, the common trade names and some typical fields of application are listed in:

Table 2.7.1   *Characteristics of nickel base alloys*
Table 2.7.2   *Main groupings of nickel alloys*

TABLE 2.7.1   Characteristics of nickel-base alloys

Advantages	Limitations
Wide range of physical properties possible within a single alloy system, e.g. high electrical resistance and low temperature coefficient of resistance.	Expensive materials because of costly alloying element additions which are necessary to achieve the desired physical properties, e.g. chromium, rare earth elements, etc.
Face-centred-cubic crystal lattice, therefore there is no transition from ductile fracture to brittle fracture as temperature drops. This applies down to the cryogenic range.	All the alloys work harden very rapidly and are hence difficult to work without intermediate annealing. Heat-treatment procedures are necessary to strengthen alloys by precipitation hardening.
Outstanding strength retention to within 0.7 Tm (melting point), a level surpassed only by special materials containing oxide phases, e.g. Al_2O_3 in Al.	Tendency to form brittle sigma phases after long time exposure in the temperature range of 650–870°C. Careful composition control is essential.
Pure nickel has good corrosion resistance to non-oxidising acid media and alkalis.	Very poor resistance to acids containing oxidising impurities and hot gases.
Excellent corrosion resistance with Ni–Cu, Ni–Mo, Ni–Mo–Cr, Ni–Mo–Cu–Cr alloys. Excellent oxidation resistance to 1100°C in Ni–Cr alloys with improved oxide adherence contributed by rare earth elements. All these alloys are tough, strong down to cryogenic temperatures.	Work hardening rates are high and sulphur has to be added to Ni–Cu alloys to improve machinability. Careful composition control of S+C is essential if weld sensitivity is to be avoided.
	Sulphidising atmospheres corrode these alloys very rapidly, especially at temperatures in the range of 750–1000°C. Silicon has to be added to improve castability of these alloys with a consequent drop in ductility.
Powder products can be made in many superalloy compositions, and these can be formed super-plastically because of their fine grain size, at low deformation stresses.	The high strength levels of superalloys at elevated temperatures leads to difficulties in hot deformation and thus causes problems in formability. Therefore, alloys with a high proportion of additional elements have to be used as cast.
Alloys of very low expansion coefficient and constant elastic modulus can be made (e.g. Invar has a coefficient equal to 0.1 of the coefficient of carbon steel at 200°C) and alloys of very high coefficient in the same basic Ni–Fe system.	Very sensitive to impurities and degree of cold work with marked loss in ductility which is only partially regained by heat treatment.
Magnetically soft and magnetically hard alloys can be prepared. Soft alloys are typically Ni(80)–Fe–Mo(40), and have excellent properties at low flux densities. Hard alloys are based on the Fe–Ni–Al system.	Magnetic saturation values of Ni–Fe are lower than those of iron, particularly at high permeability. Thus these alloys are not suitable for power plant where high flux densities are required (Fe–Si alloys are used). Magnetic properties deteriorate with section thickness.

Other characteristics:
Nickel is strongly magnetostrictive, i.e. contracts in all magnetic fields and shows a very large change (3×10^5) at a saturation field of 6500 G.

Table 2.7.1

TABLE 2.7.2 Main groupings of nickel alloys

Alloy type	Common or trade name	Typical applications
Corrosion-resistant alloys Ni 98/100 Ni–30 Cu Ni–20/30 Mo – (15/20 Cr) Ni–5/10 Mo – 20/30 Cr Ni–8/10 Si Ni–16 Mo–16 Cr	Nickel Monel Hastelloy Illium — Nirolium	Chemical and process plant, plating coinage Chemical and food processing, water treatment Chemical plant Chemical plant Chemical plant Chemical plant
High temperature alloys Ni–20/30 Cr Ni–50/60 Cr – (Nb) Ni–5/10 Fe – 15/20 Cr Ni–30/50 Fe – 18/20 Cr Ni–10/30 Cr – 1/4 Ti – 1/5 Al (Co, Mo)	— — Inconel Incoloy Nimonic, Nimocast	Furnace components Furnace and boiler components Furnace components, heat-treatment equipment Power generation, heat exchangers Gas turbines
Electrical alloys Ni–98/100 Ni–18/20 Cr Ni–20/40 Fe – 16/20 Cr Ni–40/45 Fe – 18/20 Cr – Si Ni–55 Cu Ni–2/5 Mn – 1/2 Si	Nickel Brightray, Nichrome- — Constantan, Ferry —	Electronics Heating elements Industrial furnace elements Resistors, thermocouples Sparking plugs
Magnetic alloys Ni–98/100 Ni–5/20 Co Ni–20/25 Fe – (Mo, Cu) Ni–45/50 Fe – (Mo, Cu, Si) Ni–30 Cu	Nickel — Permalloy, Mumetal — —	Magnetostrictive devices Magnetic shields, HF transformers, loading coils Transformer and rotor laminations Temperature compensators
Controlled-expansion and constant modulus alloys Ni–48/64 Fe Ni–15/20 Co – 50/65 Fe Ni–45/52 Fe – 5/12 Cr	Invar, Nilo Ni-Span, Elinvar	Bimetals, thermostats, glass-sealing Weighing machines
Miscellaneous alloys Ni–55/60 Ti Ni–Ca – rare earth metals Ni–3/5 Si – 2/4 B – (Cr)	Nitinol Hy-stor —	Shape-memory devices Hydrogen storage Brazing alloys

Table 2.7.2

Nickel alloys

2.7.2   Corrosion resistant nickel alloys

2.7.2.1   Introduction

Pure nickel itself is resistant to attack by a wide range of mildly corrosive media, particularly to the normal atmospheric environment. The extensive use of electroplated nickel, either alone or as an undercoat to chromium, exploits this property. The metal can readily be worked to all normal product forms and is suitable for the construction of many items of plant for chemical and food processing. Two grades of wrought nickel are normally considered for such applications and the details of these are given below. Nickel is not widely used in the cast form, but sand casting is feasible if appropriate deoxidation procedures are employed.

Nickel–copper alloys containing about 30% copper are characterised by good strength and high ductility and are resistant to corrosion by a wide range of media—particularly water of all types including sea-water, many non-oxidising acids and alkalis and many salts or organic materials. They are not fully resistant to oxidising acids. An age-hardening version enables high strength to be developed without significant change in corrosion resistance.

Nickel–molybdenum–iron alloys, often with additions of chromium, have exceptional resistance to high concentration acids at temperatures even up to the boiling point. The alloys have reasonably high strength and many of them retain their strength well at elevated temperatures so that they have also found structural applications in the high-temperature field.

Nickel–chromium–molybdenum–copper alloys are characterised by high resistance to strong mineral acids, to many fluorine compounds and to sea-water, and are mainly used in the cast state.

The nickel–iron–chromium alloys of the Inconel and Incoloy series are now principally regarded as alloys for service at elevated temperatures, hence the compositions and properties of most of these alloys are given in Section 2.7.3. However, two alloys, Inconel 625 and Incoloy 825, are primarily corrosion resistant alloys.

These and the other corrosion resistant nickel alloys are listed in:

Table 2.7.3   *Compositions and basic properties of corrosion resistant nickel alloys*

2.7.2.2   Corrosion resistance and special characteristics

Details of the environments for which corrosion resistant nickel alloys are suitable, along with the other characteristics of these alloys, are given in:

Table 2.7.4   *Appropriate environments for and special characteristics of corrosion resistant nickel alloys*

The forerunners of the Inconel and Incoloy alloy series, Inconel 600 and Incoloy 800 have good corrosion resistance at normal temperature, and comments on this aspect of their behaviour are given in Table 2.7.4. Their high temperature corrosion behaviour is given in Section 2.7.3.2.

2.7.2.3   Common or trade mark names and suppliers

The alloys listed above are typical of the principal types used commercially for corrosion resisting service, but there are many variants of these available under different proprietary names. Some trade mark names and suppliers are:

Monel	Wiggin Alloys Ltd
Inconel	Wiggin Alloys Ltd
Incoloy	Wiggin Alloys Ltd
Corronel	Wiggin Alloys Ltd
Hastelloy	Cabot Corporation, Stellite Division
Nirolium	Langley Alloys Ltd

TABLE 2.7.3 Compositions and basic properties of corrosion resistant nickel alloys

Alloy	Composition (%)													Density 10³ (kg/m³)	Specific heat at 20°C (J/kgK)
	Ni	C	Mn	Fe	S	Si	Cu	Cr	Co	Mo	Al	Ti	Others		
Nickel 200	99.5	0.08	0.18	0.2	0.005	0.18	0.13	—	—	—	—	—	—	8.89	456
Nickel 201	99.5	0.01	0.18	0.2	0.005	0.18	0.13	—	—	—	—	—	—	8.89	456
Nickel cast	98	0.1	1.0	0.2	—	1.0	—	—	—	—	—	—	Mg 0.1	8.8	456
MONEL alloy 400	63.0 min.	0.15	1.0 max.	2.5 max.	0.024	0.5 max.	31.0	—	—	—	—	—	—	8.83	420
MONEL alloy 410	66.0	0.2	0.8	1.0	0.008	1.6	30.5	—	—	—	—	—	—	8.63	540
MONEL alloy K-500	63.0 min.	0.15	1.5 max.	2.0 max.	0.010	0.5 max.	30.0	—	—	—	2.9	0.6	—	8.46	419
HASTELLOY alloy B	bal.	0.09	1.0	5.0	0.03	1.0	—	1.0	2.5	28.0	—	—	V0.3	9.24	381
HASTELLOY alloy C276	bal.	0.10	1.0	5.5	0.03	1.0	—	16.0	2.5	16.5	—	—	V0.3	8.90	425
HASTELLOY alloy D	bal.	0.12	1.0	2.0	—	10.0	3.0	1.0	1.5	—	—	—	—	7.80	454
HASTELLOY alloy G	bal.	0.05 max.	1.5	19.5	—	1.0 max.	2.0	22.0	2.5 max.	6.5	—	—	W1 max Nb+Ta 2	8.17	428
HASTELLOY alloy N	bal.	0.06	0.8 max.	5.0	—	1.0 max.	0.35 max.	7.0	0.2 max.	16.5	0.5 max.	—	B0.01 max.	8.93	419
ILLIUM alloy B	bal.	0.05	1.0	1.5	—	3.5	5.5	28.0	—	8.0	—	—	—	8.58	—
ILLIUM alloy G	56.0	—	—	—	—	—	6.5	22.5	—	6.5	—	—	—	8.58	449
ILLIUM alloy R	68.0	—	—	—	—	—	3.0	21.0	—	5.0	—	—	—	8.58	458
ILLIUM alloy 98	bal.	0.07 max.	1.5 max.	1.5	—	1.25	5.0	28.0	—	8.5	—	—	—	8.29	—
INCONEL alloy 625	60.5	0.10 max.	0.25 max.	5.0 max.	0.015 max.	0.5	—	21.5	—	9.0	0.25	0.25	Nb+Ta3.65	8.44	410
INCOLOY alloy 825	42.0	0.05 max.	1.0 max.	bal.	0.03 max.	0.5 max.	2.25	21.5	—	3.0	0.20	0.9	—	8.14	441
Nirolium alloy 215	bal.	0.06	0.6	1	—	1	—	16	—	16					

Table 2.7.3

TABLE 2.7.3 Compositions and basic properties of corrosion resistant nickel alloys—*continued*

Thermal conductivity at 20°C (W/mK)	Thermal expansion (10⁻⁶/K 20–95°C)	Electrical resistivity at 20°C (μΩcm)	Form	Tensile strength (MPa)	0.2% proof stress (MPa)	Elongation (%)	Hardness (HV)	Young's Modulus (GPa)
74.9	13.3	9.5	annealed → cold worked	380→550	100→210	40	90→240	204
79.2	13.3	7.6	annealed → cold worked	340→410	70→170	50	75→240	210
75	13.3	—	Cast bar	360–420	—	30–10	100	—
21.7	14.1	51.0	annealed bar	480–620	170–350	40	110–150	179
26.8	13.6	53.0	Cast bar	450–580	—	50–25	100–145	158
17.4	13.7	61.4	heat-treated bar	620–760	280–410	30–20	—	179
11.3	10.0	135	solution-treated bar	880	390	52	200–220	212
10.3	11.2	130	solution-treated plate	780	360	70	200–220	206
21.0	11.0	113	solution-treated cast bar	790	~790	1	300–380	200
14.4	14.6	112	annealed sheet	710	386	48	—	200
11.5	11.6	120	solution-treated sheet	700	315	50	—	218
—	—	—	—	415–485	345–450	—	—	—
12.1	12.2	123	—	470	195	—	—	168
13.0	12.0	120	—	780	295	—	—	216
—	—	—	—	370–540	285–295	—	—	—
9.8	12.8	129	annealed→cold worked	830→1040	410→620	40	140→250	207
11.1	14.0	113	annealed bar	590–730	240–450	50	120–180	193
				465	290	10		

Table 2.7.3—*continued*

TABLE 2.7.4 Appropriate environments for and special characteristics of corrosion resistant nickel alloys

Alloy	Resistant to corrosion by	Special characteristics
Nickel 200	Atmosphere, fresh and saline waters, organic materials, food caustic alkalis.	General purpose alloy.
Nickel 201	As Nickel 200.	Low-carbon alloy which is easy to deform cold and is also resistant to stress corrosion cracking caused by graphitisation at temperatures in the region of 300°C.
Nickel cast	As Nickel 200.	For castings.
Monel alloy 400	Atmosphere, including polluted industrial; fresh, brackish and sea waters, alkalis, neutral and alkaline salts (except hypochlorites), organic acids and compounds, non-oxidising mineral acids.	
Monel alloy 401	As Monel alloy 400.	For castings.
Monel alloy K500	As Monel alloy 400.	Age-hardenable for high strength retained to 650°C.
Hastelloy alloy B	Hydrochloric acid, all concentrations up to boiling point, non-oxidising mineral acids and salts.	Retains good strength to 750°C.
Hastelloy alloy C	Oxidising solutions, chlorides and hypochlorites, mineral acids, including oxidising acids at moderate temperatures.	Weldable without grain-boundary degradation and retains good strength to 750°C.
Hastelloy alloy D	Sulphuric acid at all concentrations, non-oxidising acids and salts.	For castings only, weldable with special care.
Hastelloy alloy G	Acid and alkaline solutions both oxidising and reducing, particularly hot sulphuric and phosphoric acids.	
Hastelloy alloy N	Molten fluorides up to 870°C.	Readily weldable.
Illium alloy B	Sulphuric acid at all concentrations up to boiling point.	For castings only.
Illium alloy G	Sulphuric acid up to 60% to 80°C, 60–98% to 40°C.	For castings only.
Illium alloy R	Sulphuric acid up to 60% to boiling point.	Wrought alloy.
Illium alloy 98	Sulphuric acid up to 98% to 65°C.	For castings only.
Inconel alloy 600	Many inorganic and organic acid and alkaline solutions. Resistant to chloride ion stress corrosion cracking.	Readily weldable.
Inconel alloy 625	As Inconel alloy 600 but especially resistant to sea-water.	Higher strength than Inconel alloy 600, particularly when cold-drawn to wire.
Incoloy alloy 800	Similar to Inconel alloy 600, but more resistant to stress corrosion cracking in chlorides and hydroxides.	Readily weldable.
Incoloy alloy 825	Strong mineral acids, reducing or oxidising, sulphuric and phosphoric acids at all concentrations up to boiling point.	

Table 2.7.4

2.7.3   High temperature alloys of nickel

2.7.3.1   Introduction

Alloys for service at high temperature require two main characteristics—first resistance to corrosive attack by the surrounding medium, and second adequate strength to resist deformation or fracture under the stresses and temperatures imposed. These properties are not completely independent of each other, but commercial alloys generally fall into one of two groups—those developed primarily to resist hot corrosion, and those with high strength. The former are generally called heat resisting alloys and the latter creep resisting alloys or 'superalloys'.

2.7.3.2   Heat resistant alloys

INTRODUCTION

Almost all nickel base, heat resisting alloys are developed from the 80 Ni–20Cr alloy introduced over 70 years ago. This basic material is well known for its use in wire or strip form as an electrical heating element (see Section 2.7.6) but it is also used in other forms for general high-temperature corrosion resistance. Nimonic alloy 75 is the representative material of this class available in most semi-finished wrought forms and is included amongst creep resisting alloys in Section 2.7.3.3. Although the maximum resistance to high-temperature scaling by oxidation in air is given by the 20% chromium alloy, higher resistance to certain fuel ashes (particularly oil ashes containing vanadium and sulphur compounds) is given by higher chromium contents. A limit to forgeability occurs at about 30% chromium. Alloys containing up to 60% chromium have been used in the cast form, but they are very brittle and much better ductility is obtained at about 50% chromium, particularly with the modified alloys containing additions of niobium.

Nickel–iron–chromium alloys, usually with 15–25% chromium, are available for service at temperatures up to about 1100°C in atmospheres involving oxidation, carburisation, sulphidation and other types of chemical attack. The nickel content may range from approximately 90% with negligible iron content to about 10% in certain austenitic steels. Alloys with less than 30% nickel are dealt with in Chapter 2.1.

The compositions and basic properties of the nickel base, heat resistant alloys are given in:

Table 2.7.5   *Compositions and basic properties of heat resistant nickel alloys*

SPECIAL HIGH-TEMPERATURE CORROSION CHARACTERISTICS

The high-temperature oxidation resistance of all nickel and nickel–iron–chromium alloys containing 15–20% chromium is very good and the rate of attack is generally such that oxide penetration to a depth of only about 0.01 mm is to be expected after heating at 1000°C for 100 h in clean air. The scale formed is protective so that subsequent penetration is much slower, but the rate of attack may be accelerated by heating and cooling cycles which cause the scale to flake and crack. Attack by more complex environments may be much more rapid. Quantitative comparison of rates of attack under standard conditions for different alloys is not generally available, but qualitative comments are given below.

Alloy	Characteristics
80 Ni–20 Cr (Nimonic 75)	General purpose wrought alloy, readily weldable.
50 Ni–50 Cr	Castings and cladding for tubes; resistant to fuel-ash corrosion.
IN 657	Castings only; more resistant to oxidation than 50 Ni–50 Cr, resistance to fuel-ash corrosion similar to 50 Ni–50 Cr.
Inconel 600	General purpose wrought alloy, similar to Nimonic 75.
Inconel 601	Similar to Inconel 600, but higher chromium content improves hot corrosion resistance.
Incoloy 800	General purpose wrought alloy; more resistant to embrittlement in service than lower-nickel austenitic steels.
Incoloy DS	Similar to Incoloy 800, but more resistant to carburisation and cyclic conditions than other alloys.

COMMON OR TRADE MARK NAMES AND SUPPLIERS

The names Nimonic, Inconel and Incoloy are trade marks of Wiggin Alloys Ltd, but similar alloys are available from other suppliers, such as for example BSC and Sandwick, usually incorporating the same distinguishing number, e.g. Alloy 800.

2.7.3.3   Creep resisting alloys (superalloys)†

INTRODUCTION

Nickel superalloys are developed from the 80% nickel–20% chromium composition which, with minor alloying additions as Brightray C and Brightray S has been used for the past 50 years or more for electrical resistance applications. They are normally used in the temperature range 600–1100°C where their strength and/or corrosion resistance is required. Properties are dependent on composition and manufacturing route. Additions of aluminium, titanium and niobium produce precipitation hardening and cobalt, molybdenum, tungsten, tantalum and hafnium have subsequently been added to produce solid solution strengthening and other effects. Alloying increases strength and temperature capability but decreases forgeability. Thus only the lower alloyed materials are available as sheet and forgings.

There are many uses for nickel superalloys in addition to those in the gas turbine. For example, Nimonics 80 and 81 are used for exhaust valves in reciprocating engines, Inconel 600 is used in furnace and heat treatment equipment and for spark plug electrodes, Nimocast 80 is used for diesel engine pre-combustion chambers and Inco 718 is used in pump bodies, rocket motors and various applications in nuclear power generation. The predominant use of nickel superalloys, however, is in the gas turbine engine and for this reason the alloys are discussed in categories based on gas turbine usage—turbine blades, discs and sheet components. The chemical compositions and physical properties of the alloys discussed are given in:

Table 2.7.6   *Chemical compositions and physical properties of nickel based superalloys*
The proof stress ranges of these alloys are shown in:
Fig 2.7.1   *Proof stress ranges of nickel based superalloys*
and the creep rupture ranges in:
Fig 2.7.2   *1000-h creep rupture stress ranges of nickel based superalloys*

†'Superalloys' may be based on iron or cobalt as well as nickel. Descriptions of their compositions and properties are included in the appropriate metal section.

TURBINE BLADE ALLOYS

Early blade alloys were forged or machined from wrought bar material. The strongest alloys cannot be forged and are used in the form of castings which must be produced *in vacuo*. Casting has the additional advantage that it can produce the complex internal air cooling geometries required by the most demanding operating conditions. Wrought alloys have lower creep strengths than cast alloys but typically have higher ductilities, toughnesses and fatigue strengths.

Alloy selection is normally based primarily on creep and corrosion/oxidation requirements, but toughness and fatigue may be significant for particular designs or operating environments. The strongest blade alloys tend to have the highest densities. Castability, assessed as general level of microshrinkage, which adversely influences creep properties at the higher stress levels, varies with alloy composition. Alloys with the higher titanium contents tend to suffer most in this respect. Hot isostatic pressing at temperatures up to around 1200°C can be used to heal this microshrinkage. Oxidation and corrosion resistance are controlled by alloy composition, particularly chromium, aluminium and titanium content. Increase in strength is normally associated with decrease in chromium and increase in aluminium. This generally results in an increase in oxidation resistance but a decrease in the so-called 'hot corrosion' resistance in the temperature range 600–900°C in environments contaminated by substances such as sea salt. These environments are common in industrial and marine gas turbine operation. Alloys such as IN738, IN939, Nimonic 81 and Nimonic 91 have been developed for hot corrosion resistance at various strength levels. They have higher chromium contents than other alloys of similar strength but the improved hot corrosion resistance is attained at the expense of high temperature oxidation resistance.

It must be appreciated that there is no universally accepted hot corrosion test and that different hot corrosion tests rank alloys of widely different corrosion resistance in essentially the same order but may rank alloys of broadly similar corrosion resistance somewhat differently. Protective coatings of the aluminide or MCrAlY overlay types are commonly used for all operating environments.

Recent processing developments include directional solidification and single crystal casting. Both these techniques can produce grain orientations that have low moduli in pre-selected directions in a component. Both can increase creep strength so that it is equivalent, at a temperature higher by between 15 and 40°C, to that of the strongest conventional casting operated at the lower temperature. Not all casting alloys are suitable for these techniques which are normally used only for the highest duty components. Recently a dispersion strengthened alloy, MA6000, has become available. Its corrosion resistance is broadly similar to that of IN738 while its creep strength at high temperatures is very good because of the dispersion strengthening.

DISC ALLOYS

Inco 901 and Inco 718 which contain significant quantities of iron were developed to replace the earlier creep resistance steels. Their temperature capability is limited to 600/650°C depending on specific operating environment. Conventional nickel base superalloys such as Waspaloy and Astroloy have superior properties at the higher disc temperatures.

The properties attainable in disc forgings are very dependent on disc size and geometry and any properties quoted should be regarded as appropriate indications of the level of strength attainable.

Microstructural control is essential and in highly alloyed materials such as Astroloy this can only be achieved by forging billet produced by compacting pre-alloyed powder.

High cycle fatigue strength in general increases with static strength and all alloys have similar crack propagation rates and fracture toughness. The superior creep properties of Waspaloy and Astroloy result in improved low cycle fatigue strength under conditions where creep/fatigue interaction occurs.

SHEET ALLOYS

Most nickel superalloy sheet materials are solid solution strengthened, but some, notably INCO718, C263 and Rene 41 are precipitation strengthened and two, TD Nickel and MA956 (which, although an alloy of iron has properties appropriate to this section) are strengthened by dispersions of ThO_2 and Y_2O_3, respectively.

The precipitation hardened alloys have the highest low temperature strengths but their strengths fall off as temperature increases between about 600 and 900°C so that INCO718 cannot be used about 650/700°C. See:

Fig 2.7.3   *1000-h creep rupture stress as a function of temperature for dispersion and precipitation strengthened nickel superalloy sheet materials*

The solid solution strengthened alloys, are, depending on the amount of strengthening, mostly as strong or stronger at temperatures above 900°C as the precipitation strengthened alloys. See:

Fig 2.7.4   *1000-h creep rupture stress as a function of temperature for solid solution strengthened nickel superalloy sheet materials*

The dispersion strengthened alloys have relatively low strengths in the lower temperature range but their strengths fall off only slowly with temperature and they therefore have the highest creep strengths above about 900°C—see:

Fig 2.7.5   *10 000-h creep rupture stress as a function of temperature for nickel superalloy sheet materials*

The precipitation hardened alloys have inferior oxidation resistance at temperatures above about 800°C showing significant intergranular oxidation at high temperatures.

The solid solution strengthened alloys have superior oxidation resistance, with alloys such as Nimonic 86 and Hastelloy S containing small amounts of rare earth elements cerium and lanthanum being the best. See:

Fig 2.7.6   *Rate of material loss in air of nickel superalloy sheet materials as a function of temperature*

MA956 has particularly good oxidation resistance but TD Nickel is relatively poor in this respect, although better than might be expected in view of its lack of chromium and aluminium. Because of its lack of alloying elements it has relatively high thermal conductivity.

With the exception of Rene 41 and the two dispersion strengthened alloys, all alloys can be readily welded by techniques such as TIG and resistance welding. Rene 41 is prone to heat affected zone cracking and welding destroys the dispersion in TD Nickel and MA956. C263 was developed by Rolls Royce as a readily weldable precipitation strengthened superalloy and is marketed as Nimonic 263 and Haynes alloy 263. MA956 currently experiences loss of ductility on soaking at temperature in excess of 1100°C. This may be of little concern in relatively low stressed applications and the alloy has been used in burners in oil-fired boilers in power generation.

ACKNOWLEDGEMENTS

Acknowledgement is made to the following organisations for information supplied:
Cabot Corporation (trade marks Hastelloy, Haynes)
Huntington Alloy Products Division of International Nickel Co.
    (trade marks Inconel, Incoloy, Inco)
Inco Alloy Products Ltd
Martin Marietta Corporation (trade mark MarM)
Special Metals (trade mark Udimet)
Teledyne Allvac
Universal Cyclops Specialty Steel Division
Wiggin Alloys Ltd. (trade marks Nimonic, Ninocast)

TABLE 2.7.5 Compositions and basic properties of heat resistant nickel alloys

Alloy	Composition (%)													Density (10³ kg/m³)	Specific heat at 20°C (J/kgK)
	Ni	C	Mn	Fe	S	Si	Cu	Cr	Co	Mo	Al	Ti	Others		
80 Ni–20 Cr	See Nimonic 75, Table 2.7.3														
50 Ni–50 Cr	50	—	—	—	—	—	—	50.0	—	—	—	—	—	—	—
IN 657	Bal.	0.1 max.	0.3 max.	1.0 max.	—	0.5 max.	—	48.52	—	—	—	—	Nb 1.5	7.96	452
INCONEL alloy 600	Bal.	0.15 max.	1.0 max.	8.0	0.015 max.	0.5 max.	0.5 max.	15.5	—	—	—	—	—	8.42	444
INCONEL alloy 601	60.5	0.05	0.5	14.1	0.007	0.25	0.25	23.0	—	—	1.35	—	—	8.06	448
INCOLOY alloy 800	32.5	0.10 max.	1.5	Bal.	0.015 max.	1.0 max.	0.75	21.0	—	—	0.38	0.38	—	7.95	460
INCOLOY alloy DS	37.0	0.1 max.	1.2	Bal.	—	2.3	0.5 max.	18.0	—	—	—	—	—	7.92	450

Table 2.7.5

TABLE 2.7.5 Compositions and basic properties of heat resistant nickel alloys—*continued*

Thermal conductivity at 20°C (W/mK)	Thermal expansion (10⁻⁶/K 20–95°C)	Electrical resistivity at 20°C (μΩcm)	Form	Tensile strength (MPa)	0.2% proof stress (MPa)	Elongation (%)	Hardness (HV)	Young's Modulus (GPa)
—	—	—	Cast and extruded clad	~300	—	15	—	—
12.7	10.7	98.1	Cast	600	370	10–30	—	205
14.8	13.3	103	Annealed bar	550–690	210–340	55–35	115–170	214
11.2	13.68	119	Annealed bar	740	340	40	—	207
11.5	14.2	93.3	Annealed bar	590	290	44	140	196
12.0	14.1	108	Annealed bar	730	390	—	—	178

Table 2.7.5—*continued*

TABLE 2.7.6 Chemical compositions and physical properties of nickel based superalloys

Alloy	Composition (%)													Young's Modulus (GPa) at 20°C	Density (10^3 kg/m³)	Thermal conductivity at 20°C (W/mK)	Thermal expansion (10^{-6}/K 20–95°C)
	C	Cr	Ti	Al	Co	Fe	Mo	W	Nb	Ta	Hf	V	Ni				
Nimonic 80A	0.07	19.5	2.4	1.4	—	—	—	—	—	—	—	—	Bal.	222	8.19	11.2	12.7
Nimonic 81	0.05	30	1.8	1.0	—	—	—	—	—	—	—	—	Bal.	196	8.06	10.9	11.1
Nimonic 90	0.08	19.5	2.4	1.4	17	—	—	—	—	—	—	—	Bal.	227	8.18	11.5	12.7
Nimonic 91	0.05	28	2.4	1.2	20	—	—	—	0.7	—	—	—	Bal.	—	8.08	—	12.2
Nimonic 105	0.13	15	4.7	1.3	20	—	5	—	—	—	—	—	Bal.	223	8.01	10.9	12.2
Nimonic 115	0.15	14.5	3.8	5.0	13.3	—	3.3	—	—	—	—	—	Bal.	216	7.85	10.6	12
Udimet 500	0.08	18	3	3	19	—	4	—	—	—	—	—	Bal.	—	8.02	11.9	13.5
Udimet 520	0.05	19	3	2	12	—	6	1	—	—	—	—	Bal.	—	8.22	—	—
Udimet 700	0.08	15	3.5	4.3	18.5	—	5.2	—	—	—	—	—	Bal.	—	7.92	—	—
Udimet 720	0.04	18	5	2.5	15	—	3	1.25	—	—	—	—	Bal.	—	—	—	—
IN 738	0.1	16	3.5	3.5	8.5	1.8	2.5	0.7	—	—	—	—	Bal.	—	8.11	11.5	11.3
IN 939	0.15	22.5	3.7	1.9	19	—	—	2	1	1.4	—	—	Bal.	—	8.16	11.3	11.8
INCO 713C	0.12	12.5	0.8	6.1	—	—	4.2	—	2	1.6	—	—	Bal.	207	7.91	—	10.7
IN 100	0.18	10	4.7	5.5	15	—	3	—	—	—	—	1	Bal.	225	7.75	—	13
MAR M 002	0.15	9	1.5	5.5	10	—	—	10	—	2.5	1.5	—	Bal.	—	8.52	—	—
INCO 718	0.04	19	1	0.6	—	20	3	—	5.2	—	—	—	Bal.	206	8.19	11.2	13
INCO 901	0.04	13	3	0.3	—	36	5.7	—	—	—	—	—	Bal.	201	8.16	—	13.5
Waspaloy	0.08	19	3	1.3	13.5	—	4.3	—	—	—	—	—	Bal.	—	8.19	11.5	—
Astroloy	0.06	15	3.5	4	17	—	5.2	—	—	—	—	—	Bal.	220	7.91	—	—
Nimonic 75	0.1	20	0.4	—	—	—	—	—	—	—	—	—	Bal.	210	8.37	11.7	11
Nimonic 86	0.05	25	—	—	—	—	10	—	—	—	(+0.03Ce)	—	Bal.	—	8.54	—	12.7
INCONEL 600	0.04	16	—	—	—	7	—	—	—	—	—	—	Bal.	—	8.33	—	14.8
INCONEL 601	0.05	23	—	1.4	—	14	—	—	—	—	—	—	Bal.	—	—	—	—
INCONEL 617	0.07	22	—	1	12.5	—	9	—	—	—	—	—	Bal.	—	—	—	—
INCONEL 625	0.05	22	—	—	—	2.5	9	—	3.7	—	—	—	Bal.	—	8.44	11.2	14.2
Hastelloy S	0.02	16	—	—	—	—	15	—	—	—	(+0.02La)	—	Bal.	212	8.75	14	11.5
Hastelloy X	0.1	22	—	—	1.5	18.5	9	—	—	—	—	—	Bal.	196	8.22	9.7	13.8
C 263	0.06	20	2.2	0.4	20	—	6	—	—	—	—	—	Bal.	224	8.36	11.7	10.3
RENE 41	0.09	19	3.1	1.5	11	—	10	—	—	—	—	—	Bal.	—	8.25	—	13.2
TD Nickel	—	—	—	—	—	—	—	—	—	—	(+2ThO₂)	—	Bal.	152	8.9	82.6	10.8
MA 956	—	20	—	5	—	Bal.	—	—	—	—	(+0.5Y₂O₃)	—	—	—	—	—	—
MA 6000	0.05	15	2.5	4.5	—	—	2	4	—	2	(+1Y₂O₃)	—	Bal.	—	—	—	—

Table 2.7.6

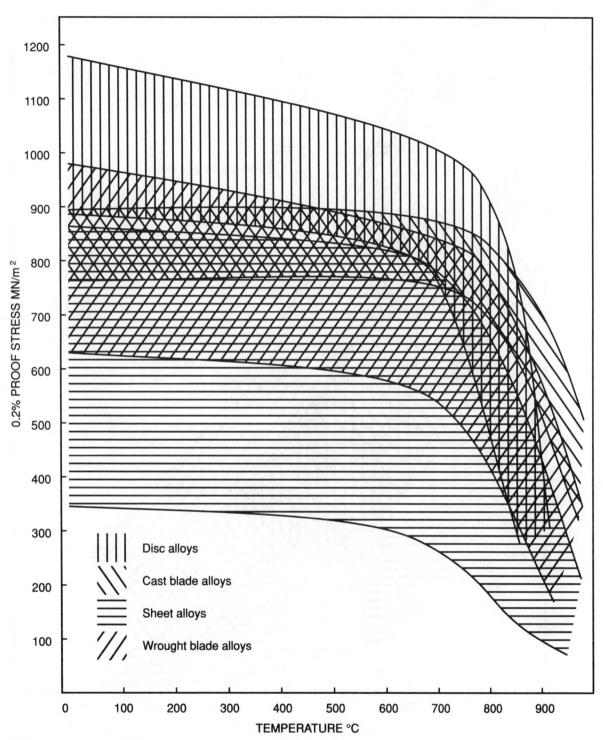

FIG 2.7.1 0.2% proof stress ranges of nickel based superalloys

Fig 2.7.1

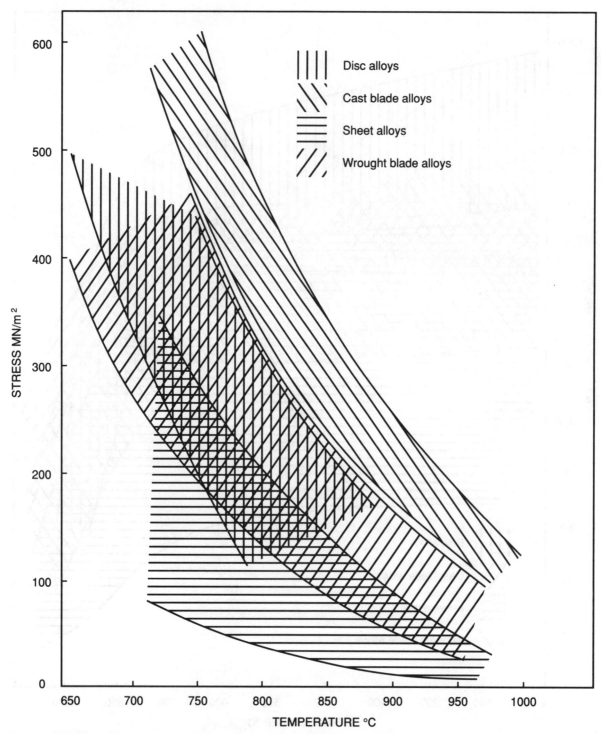

FIG 2.7.2 1000-h creep rupture stress ranges of nickel based superalloys

Fig 2.7.2

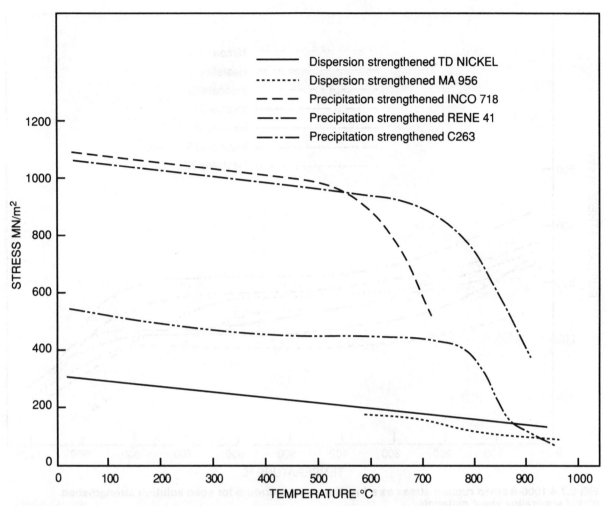

FIG 2.7.3 1000 h creep rupture stress as a function of temperature for dispersion and precipitation strengthened nickel superalloy sheet materials

Fig 2.7.3

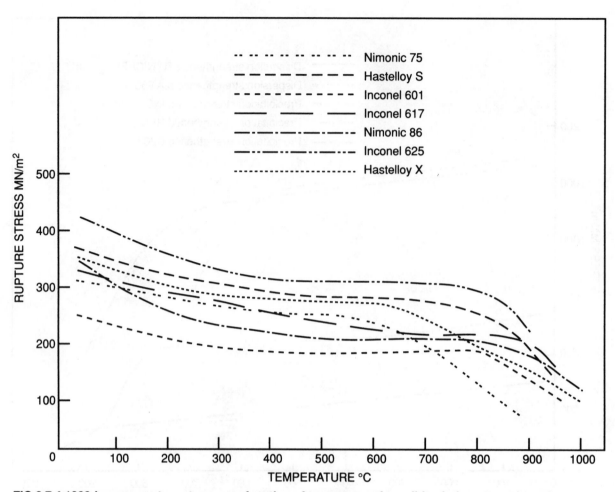

FIG 2.7.4 1000-h creep rupture stress as a function of temperature for solid solution strengthened nickel superalloy sheet materials

Fig 2.7.4

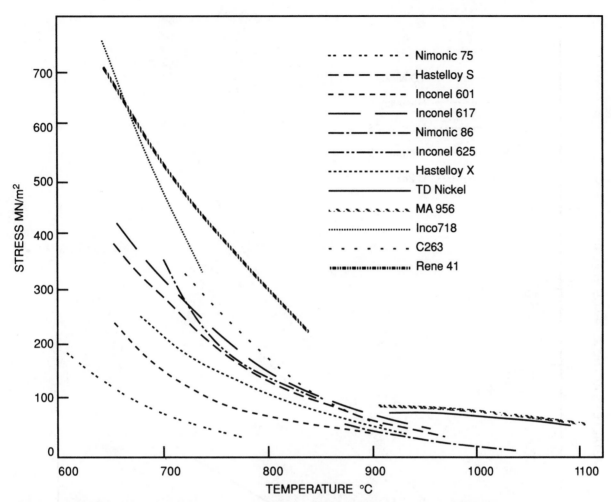

FIG 2.7.5 10 000 h creep rupture stress as a function of temperature for nickel superalloy sheet materials

Fig 2.7.5

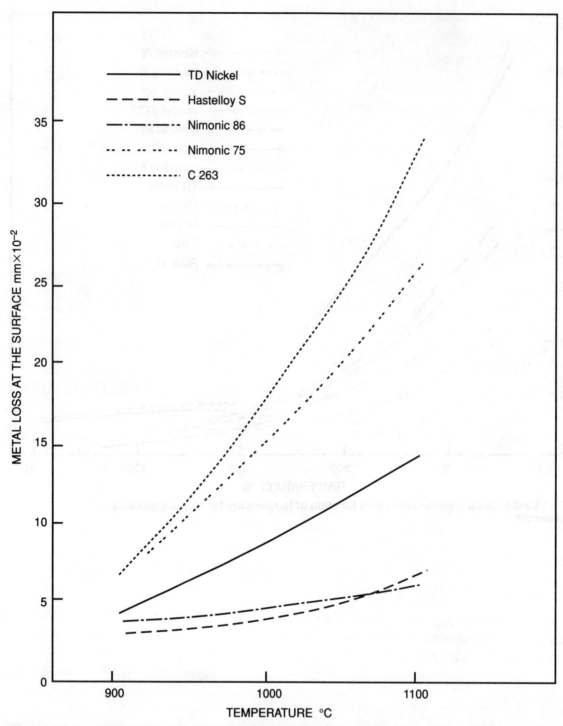

Fig 2.7.6 Rate of material loss in air of nickel superalloy sheet materials as a function of temperature

Fig 2.7.6

2.7.4   Electrical alloys of nickel

2.7.4.1   Introduction

Many nickel alloys find a number of applications in electronic or electrical devices, exploiting particular characteristics of resistivity or corrosion resistance. These are listed in:

Table 2.7.7   *Compositions and basic properties of electrical alloys of nickel*

A number of special grades of nickel are available for electronic vacuum devices in which the levels of other elements are carefully controlled to suit the requirements of strength, electron emission or aspects of the manufacturing process. Such alloys are usually produced as wire or strip.

Nickel–chromium or nickel–iron–chromium alloys are used for electrical heating elements because of their high resistance to oxidation and hot corrosion at temperatures up to about 1100°C, coupled with their high resistivity which is relatively insensitive to changes in temperatures. The nickel–copper alloy Constantan has a very low temperature coefficient of resistivity and is used for control resistors. The nickel–copper alloy Monel 400 is often selected for connecting leads to heating elements because of its low resistivity, high ductility, reasonable oxidation resistance at moderate temperatures, and because it may readily be soldered.

Nickel–chromium, nickel–aluminium and nickel–copper alloys form suitable elements for thermoelectric temperature measurements, and nickel–manganese–silicon alloys are standard materials for spark-plug electrodes because of a favourable combination of strength, ductility and resistance to spark erosion and hot corrosion.

2.7.4.2   Special electrical properties and applications

The special characteristics of the alloys listed in Table 2.7.2 (and some other nickel containing alloys) together with their most important applications are listed in:

Table 2.7.8   *Special electrical properties and applications of electrical alloys containing nickel*

2.7.4.3   Common or trade mark names and suppliers

Wires for electrical purposes are widely manufactured and the best-known trade names in the United Kingdom are as follows:

Brightray	— Wiggin Alloys Ltd
Monel	— Wiggin Alloys Ltd
Ferry	— Wiggin Alloys Ltd
Nichrome	— British Driver Harris
Tophet	— British Driver Harris
Chromel	— British Driver Harris
Alumel	— British Driver Harris
Pyromic	— Telcon Metals Ltd
Telconstan	— Telcon Metals Ltd
Constantan	— I.T.T. (Harlow) Ltd

TABLE 2.7.7 Compositions and basic properties of electrical alloys of nickel

Alloy	Composition (%)													Density 10³ (kg/m³)	Specific heat at 20°C (J/kgK)
	Ni	C	Mn	Fe	S	Si	Cu	Cr	Co	Mo	Al	Ti	Others		
Nickel 205	99.5	0.08	0.18	0.10	0.004	0.08	0.08	—	—	—	—	0.03	Mg 0.05	8.89	456
Nickel 212	97.7	0.10	2.0	0.05	0.005	0.05	0.03	—	—	—	—	—	—	8.72	430
Nickel 222	99.5	0.01	0.02	0.04	0.0025	0.01	0.01	0.01	0.06	—	0.01	0.01	Mg 0.08	8.89	456
Nickel 240	95.0	—	2.0	—	—	0.45	—	1.7	—	—	—	0.3	Zr 0.15	—	—
Nickel 270	99.98	0.01	0.003	<0.001	<0.001	<0.001	<0.001	<0.001	<0.001	—	—	<0.001	Mg 0.001	8.89	460
Constantan (Ferry)	45	—	—	—	—	—	55	—	—	—	—	—	—	8.92	395
Brightray B	59.0	0.1	1.0 max.	Bal.	—	0.35	—	16.0	—	—	—	0.17	—	8.22	461
Brightray C	Bal.	0.1 max.	0.25	1.0 max.	—	1.5	—	19.2	—	—	0.26	—	Rare earth metals 0.05	8.32	419
Brightray S	Bal.	0.1 max.	0.4	1.0 max.	—	1.0 max.	—	20.0	—	—	—	—		8.43	419
Brightway 35	37.7	0.1 max.	1.2	Bal.	—	2.2	—	18.0	—	—	—	—		7.92	410
Chromel alloy P	Bal.	—	—	0.2	—	0.4	—	10.0	—	—	—	—			
Alumel	Bal.	—	1,75	0.1	—	1.2	—	—	—	—	1.6	—			
Nicrosil	Bal.	—	—	—	—	1.5	—	14.3	—	—	—	—			
Nisil	Bal.	—	—	—	—	4.5	—	—	—	—	—	—	Mg 0.1		

Table 2.7.7

TABLE 2.7.7 Compositions and basic properties of electyrical alloys of nickel—*continued*

Thermal conductivity at 20°C (W/mK)	Thermal expansion (10⁻⁶/K 20–95°C)	Electrical resistivity at 20°C (μΩcm)	Form	Tensile strength (MPa)	0.2% proof stress (MPa)	Elongation (%)	Hardness (HV)	Young's Modulus (GPa)
74.9	13.3	9.5	Annealed rod	460	145	47	100	210
44.1	13.3	10.9	—	476	—	—	—	—
74.9	13.3	8.8	—	340	—	—	—	—
—	—	—	—	—	—	—	—	—
85.7	13.3	7.5	Annealed strip	340	110	45	—	210
21.6	14.7	49	Annealed bar	415	145	47	154	140
13.8	12.5	110	Annealed bar	686	285	44	181	200
13.4	12.5	110	Annealed bar	735	345	47	192	214
15.9	12.5	110	Annealed bar	734	345	47	192	214
13.8	14.2	110	Annealed bar	740	370	39	181	186

Table 2.7.7—*continued*

TABLE 2.7.8 Special electrical properties and applications of electrical alloys containing nickel

Alloy	Property				Application
Nickel 205	High purity gives high damping capacity to minimise vibration.				Electrodes in electronic valves.
Nickel 212	Manganese content resists sulphur embrittlement arising from sealing into glass envelope.				Support wires in valves and tungsten filaments.
Nickel 222	Manganese content promotes electron emission.				Sleeves for valve cathodes.
Nickel 240	Ductile and resistant to sulphur and lead salts.				Spark-plug electrodes.
Nickel 270	High purity powder metal product withstands heavy cold deformation.				Special electronic valves, resistance thermometers, etc.
	Resistivity ($\mu\Omega$ cm^{-1})			Temperature Coefficient	
	20°C	500°C	1000°C		
Monel alloy 400	51			0.0011 (0–100°C)	Connecting leads to heating elements.
Constantan	49			0.0003 (20–100°C) 0.0004 (20–500°C)	Control resistors.
Brightray B	110	117	119		Heating elements for domestic appliances up to 950°C.
Brightray C	108	112	110	Reactive element additions promote scale adhesion.	Heating elements for domestic appliances up to 1150°C.
Brightray S	110	113	113		Strip elements for industrial furnaces up to 1150°C.
Brightray 35	102	118	124	Resists carburisation.	Elements for industrial furnaces up to 1050°C.
Chromel/Alumel					Thermocouples up to 1100°C.
Nicrosil/Nisil	More stable than Chromel/Alumel			Thermoelectric potential	Thermocouples up to 1100°C.
Iron/constantan					Thermocouples up to 700°C.

Table 2.7.8

2.7.5 Magnetic alloys

2.7.5.1 Introduction

Although by far the major proportion of ferromagnetic material in industrial use consists of iron base alloys, some nickel base alloys have special properties of importance. The composition and properties of the nickel alloys are listed in:

Table 2.7.9 *Compositions and basic properties of magnetic alloys of nickel*

They are principally alloys having high magnetic permeability in low or moderate strength magnetising fields, or having special forms of magnetic hysteresis loop. Their applications are mainly in the fields of telecommunications or electronics for transformer cores, loading coils, magnetic screens, etc. They are available in the form of tapes or sheet, or as powder for powder cores, and their properties frequently depend on very careful control of melting and processing conditions, involving heat treatment in controlled atmospheres and sometimes in magnetic fields.

Pure nickel or certain nickel–cobalt alloys containing a high proportion of nickel have useful magnetostrictive properties and are used for ultrasonic generators and other transducers.

The nickel–30% copper alloy may be used as a magnetic temperature compensation alloy since its permeability varies rapidly with temperature around normal temperature. However, nickel–iron alloys containing about 30% nickel are more commonly used.

2.7.5.2 Special magnetic characteristics

Nickel 205	Magnetic field strength (kA/m)	Magnetostrictive effect ($\Delta 1/1 \times 10^6$)
	0.5	−3
	1.0	−8.5
	2.0	−14
	5.0	−20
	10.0	−27
	20.0	−34

	Relative initial permeability	Relative maximum permeability	Saturation induction (T)	Curie point (°C)
75 Ni–25 Fe	15 000–100 000	50 000–1 000 000	0.65–0.80	300–450
50 Ni–50 Fe	2 000–4500	20 000–70 000	0.40–0.80	350–500
36 Ni–74 Fe	c. 2000	c. 7000	0.90–1.30	180–270

	Flux density at 8 kA/m, (T)			Curie point (°C)
	0°C	20°C	40°C	
JAE metal	0.17	0.13	0.07	~50
30 Ni–70 Fe	0.35	0.24	0.14	~60

2.7.5.3 Common and trade mark names and suppliers

The magnetic alloys are marketed under a variety of trade mark names, the detailed magnetic properties depending critically upon composition and production procedure. The best known trade names in the United Kingdom are as follows:

Mumetal Radiometal	Telcon Metals Ltd
Permalloy	I.T.T. Harlow Ltd
Nilomag JAE metal	Wiggin Alloys Ltd

TABLE 2.7.9 Compositions and basic properties of magnetic alloys of nickel

Alloy	Composition (%)													Density 10^3 (kg/m³)	Specific heat at 20°C (J/kgK)
	Ni	C	Mn	Fe	S	Si	Cu	Cr	Co	Mo	Al	Ti	Others		
Nickel 205	See section 2.3.3.2.														
Nickel –4% Co	Bal.	—	—	—	—	—	—	—	4.0	—	—	—	—		
Nickel – 18% Co	Bal.	—	—	—	—	—	—	—	18.0	—	—	—	—		
75 Ni – 25 Fe	70–80	—	—	Bal.	—	x	x	—	—	x	—	—	Small	8.8	440
50 Ni – 50 Fe	50	—	—	Bal.	—	x	x	—	—	x	—	—	additions of	8.3	480
36 Ni – 74 Fe	36	—	—	Bal.	—	x	x	—	—	x	—	—	Si, Mo or Cu	8.1	470
JAE metal	70	—	—	—	—	30	—	—	—	—	—	—	—	8.94	—
30 Ni – 70 Fe	30	—	—	70	—	—	—	—	—	—	—	—	—	8.0	500

TABLE 2.7.9 Compositions and basic properties of magnetic alloys of nickel—*continued*

Thermal conductivity at 20°C (W/mK)	Thermal expansion (10^{-6}/K 20–95°C)	Electrical resistivity at 20°C ($\mu\Omega\,cm$)	Form	Tensile strength (MN/m²)	0.2% proof stress (MN/m²)	Elongation (%)	Hardness (HV)	Young's Modulus (GN/m²)
33	13	55–62	Annealed strip	540	—	—	110	—
13	10	40–55	Annealed strip	430	—	—	105	—
8	2	70–90	Annealed strip	530	—	—	115	—
—	13.7	40–46	Annealed bar	430	150	52	116	196
13	10	85	Annealed strip	430	—	—	110	—

Table 2.7.9

2.7.6   Controlled-expansion and constant-modulus alloys

2.7.6.1   Introduction

The thermal expansion coefficients of binary nickel–iron alloys depend critically on composition, falling to almost zero over a restricted range of temperature for the 36% nickel alloy. Other binary alloys have low, but not zero, coefficients over wider temperature ranges, while some ternary alloys containing cobalt or chromium have controlled coefficients suitable for specific applications. The alloys with minimal expansion coefficients find applications in metrology and chronometry, and as one member of thermostatic bimetals. The alloys with controlled coefficients are primarily used for sealing to glass or ceramics in various vacuum devices such as lamp bulbs, television tubes, etc. The alloys are chosen to match the expansion of the particular glass or ceramic involved. Certain of the low-expansion alloys can be strengthened by additions of titanium and aluminium followed by appropriate heat treatment, making them suitable for highly stressed applications.

The nickel–iron alloys, again in the region of 36% nickel, have positive temperature coefficients of elastic modulus, in contrast to the negative coefficient normal for most metallic materials. By careful control of composition, by the addition of small proportions of other elements and by special cold working and heat treatment procedures the coefficient can be brought very close to zero, and combined with mechanical properties suitable for use of the alloys as springs or vibrating devices insensitive to variations in temperature, e.g. watch hair springs, spring balances, tuning forks, etc.

Table 2.7.10 gives the composition and properties of the controlled-expansion and constant-modulus nickel base alloys.

Table 2.7.10   *Compositions and basic properties of controlled expansion and constant modulus nickel alloys*

2.7.6.2   Special expansion features

Alloy	Expansion coefficient ($\times 10^6$/K)	Temperature range (°C)	Forms and applications
Nilo 36	1.7	20-100	Most wrought forms for general use at ambient temperature.
Nilo 42	4.5–6.5	20-300	Wire and strip for thermostats and cores of copper-clad sealing wires.
Nilo 48	8.3–9.3	20-400	Bar, wire and strip for thermostats and sealing to soft glass.
Nilo K	5.95–6.45	20-500	Most wrought forms for sealing to borosilicate glasses.
Incoloy 903	8.0	20-400	Rod and bar for stressed components, e.g. gas turbine shafts.
Ni-Span C902	The temperature coefficient of the elastic modulus (for longitudinal vibrations) lies between 0 and $+50 \times 10^{-6}$/K depending on the mechanical and thermal treatment of the alloy, compared with about -200×10^{-6}/K for most metallic materials.		

2.7.6.3   Common or trade mark names and suppliers

Controlled-expansion alloys are produced under a number of different trade mark names and designations, with the properties varying with the detailed composition. Trade names of the materials produced in the United Kingdom include the following:

Nilo	Wiggin Alloys Ltd
Invar	
Telcoseal }	Telcon Metals Ltd
Therlo	British Driver Harris Ltd

Constant-modulus alloys are similarly produced as:

Ni-Span	Wiggin Alloys Ltd
Elinvar	Telcon Metals Ltd

TABLE 2.7.10 Compositions and basic properties of controlled expansion and constant modulus nickel alloys

Alloy	Composition (%)													Density 10³ (kg/m³)	Specific heat at 20°C (J/kgK)
	Ni	C	Mn	Fe	S	Si	Cu	Cr	Co	Mo	Al	Ti	Others		
NILO alloy 36	36.0	0.15 max.	0.5	Bal.	—	0.5 max.	0.5 max.	—	—	—	—	—	—	8.05	502
NILO alloy 42	42.0	0.15 max.	0.5	Bal.	—	0.5 max.	0.5 max.	—	—	—	—	—	—	8.12	502
NILO alloy 48	48.0	0.15 max.	0.5	Bal.	—	0.5 max.	0.5 max.	—	—	—	—	—	—	8.25	502
NILO alloy K	29.5	0.05 max.	0.3	Bal.	—	0.5 max.	0.5 max.	17.0	—	—	—	—	—	8.16	502
INCONEL alloy 903	38.0	—	—	Bal.	—	—	—	—	15.0	—	0.7	1.4	Nb 3.0	—	—
NI-SPAN alloy C–902	42.25	0.1 max.	0.5	Bal.	—	0.6	—	5.3	—	—	0.55	2.5	—	8.01	502

TABLE 2.7.10 Compositions and basic properties of controlled expansion and constant modulus nickel alloys—continued

Thermal conductivity at 20°C (W/mK)	Thermal expansion (10⁻⁶/K 20–95°C)	Electrical resistivity at 20°C ($\mu\Omega$cm)	Form	Tensile strength (MN/m²)	0.2% proof stress (MN/m²)	Elongation (%)	Hardness (HV)	Young's Modulus (GN/m²)
10.5	1.7	78	Annealed bar	460	265	42	135	—
10.5	5.3	57	Annealed bar	525	275	45	150	145
16.7	8.5	44	Annealed bar	494	270	—	—	152
16.7	6.0	46	Annealed bar	525	365	41	174	—
—	8.0ˣ	—	Warm-worked bar	—	1100	—	—	—
13.0	6.2	102	Heat-treated bar	1240	790	18	305	169

ˣ = 20–40°C

Table 2.7.10

2.7.7   Surfacing alloys

2.7.7.1   Introduction

Nickel alloys usually containing iron, chromium, silicon and boron with carbon contents varying between 0.1 and 1% are used for hard surfacing. These alloys have good fluidity and wetting ability which enable the formation of thin precise protective coatings on shafts, cylinders, etc.

The most wear resistant of these alloys are superior in this respect to other available materials at room temperature but cobalt based materials are superior both in resistance to wear at elevated temperatures and to corrosion in aqueous environments. There are a large number of commercial nickel base hardfacing alloys; typical properties are given in:

Table 2.7.11   *Typical properties of nickel based hardfacing alloys,* and

Fig 2.7.7   *Hot hardness of nickel hardfacing alloys (instantaneous values)*

and a guide to the effect on properties of increasing content of alloying elements is given in:

Table 2.7.12   *Effect on properties of increasing content of alloying elements*

Exceptional corrosion resistance combined with a good resistance to wear are shown by Hastelloy 'C' (see Table 2.7.3) and the intermetallic compound 'Triballoy T700' which contains 32% molybdenum and 3.4% silicon.

High nickel alloys are used also to overlay welds (hardpouring alloys). Most of these alloys were developed for and are exploited for their high temperature and corrosion resistance.

2.7.7.2   Common and trade mark names and suppliers

Deloro (–alloy No.) is the trade mark for nickel base, wear resistant alloys and Triballoy (T.No.) for intermetallic compounds supplied by the Wear Technology Division of the Cabot Corporation (Deloro Stellite) Stratton St Margaret, Swindon, Wiltshire, UK and Kokomo, Indiana 46901, USA.

Colmonoy (No.) is the trademark for nickel base wear resistant alloys supplied by Wall Colmonoy Ltd., Pontardawe, West Glamorgan, UK, Detroit, USA and Montreal, Canada.

TABLE 2.7.11 Typical properties of nickel based hard facing alloys

Specification			Deposit type	Typical properties											Deposition technique						Finishing technique	
AWS grade	ASM group	AWRA type		Chemical composition (%)								Vickers Hardness (kg/mm²)	Elastic modulus (GN/m²)	OA	MA	MIG	TIG	SA	SF	Turn	Grind
				C	Cr	W	Mo	Co	Si	B	Fe										
		5130		0.06	15	4	16				5	300	160	√	√	√	√			√	
	4B		Cr/B/Mo/W	0.4	11			1		2.5	3	400		√	√		√			√	
	4B			0.6	13			1		3	4	530		√	√		√			√	√
	4B			0.8	15			1		4	4	720		√	√					√	√
NiCr-A		5230		0.1	7				2.5	1.5	1.5	300		√						√	
NiCr-C			Cr/Si/B	0.6	14			1.5	3.25	3	3.25	450							√	√	√
				1.0	18			1	5.5	4.5	5.5	700							√	√	√

OA, Oxy-acetalene (includes powder weld).   MA, Manual arc.   MIG, Metal inert gas.

SA, Submerged arc.   TIG, Tungsten inert gas.

SF, Spray-fuse.

AWS, American Welding Society.
ASM, American Society for Metals.
AWRA, Australian Welding Research Association.

TABLE 2.7.12 Effect on properties of increasing content of alloying elements

Alloying elements	Hardness			Bulk toughness	Volume % carbides	Volume % borides	High temperature strength	Oxidation resistance	As-deposited thickness
	Bulk	Borides and carbides	Matrix						
C	↑			↓	↑				
Cr	↑	↑	↑		↑	↑		↑	
W	↑	↑	↑						
Mo	↑		↑				↑		
Cu	↑						↑		↑
B	↑					↑			↑

↑ = Increase.   ↓ = Decrease.

Table 2.7.11 and Table 2.7.12

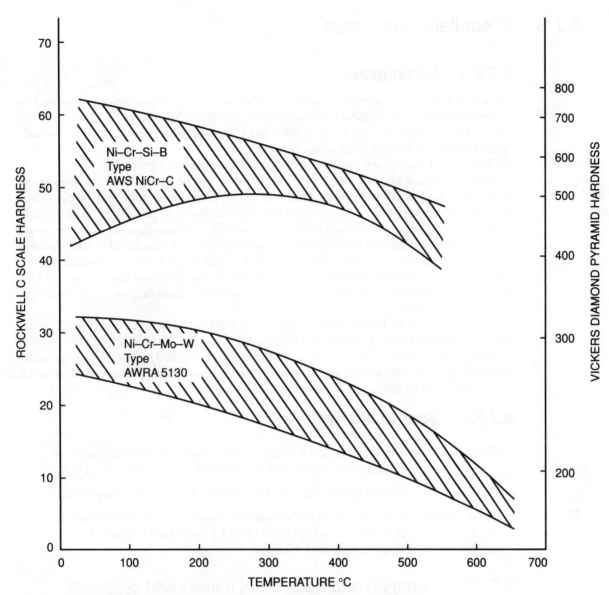

FIG 2.7.7 Hot hardness of nickel hardfacing alloys (instantaneous values)

Fig 2.7.7

2.7.8   Miscellaneous alloys

2.7.8.1   Introduction

The major types of industrially important nickel alloys have been described according to their principal applicational characteristics, but there are several special materials whose applications do not readily fall into classified groups and which are only approaching commercial production. See:

Table 2.7.13   *Compositions and basic properties of miscellaneous nickel alloys*

Nickel–titanium alloys close to the equiatomic proportion show shape-memory properties, i.e. after appropriate thermal setting to a given shape they may be plastically deformed at a lower temperature but regain the set shape on re-heating. Applications suggested for alloys of this class include thermostats, self-erecting antennae for space vehicles, tube fittings, various fixing and clamping devices in inaccessible locations and orthodontic appliances. Nickel–titanium alloys with higher nickel contents do not show the shape-memory effect but are susceptible to precipitation hardening and may find applications exploiting the combination of strength, corrosion resistance and non-magnetic properties.

Certain nickel base intermetallic compounds with calcium or rare earth metals are able to absorb hydrogen in large quantities and to release it by suitable changes of temperature and pressure. They are used in the form of crushed and activated granules enclosed in a suitable vessel and in some cases can store more than twice as much hydrogen as can be stored in the same volume of liquid.

2.7.8.2   Special characteristics

55 Nitinol	The shape-recovery temperature varies very critically with composition from a very low absolute temperature to ~115°C.
60 Nitinol	Non-magnetic high-strength alloy—no shape-memory effect.
Hy-Stor alloys	Different alloys of the available range have desorption curves with the pressure plateau at which hydrogen is evolved varying from about 0.5 atm. to over 20 atm.

2.7.8.3   Common and trade mark names and suppliers

Nitinol is the registered designation of the Naval Ordnance Laboratory, Department of Defence, Washington. A similar alloy is marketed by Raychem under the trade mark 'Tinel'.

Hy-Stor is the trade mark of MPD Technology Ltd.

TABLE 2.7.13 Compositions and basic properties of miscellaneous nickel alloys

Alloy	Composition (%)													Density 10^3 (kg/m³)	Specific heat at 20°C (J/kgK)
	Ni	C	Mn	Fe	S	Si	Cu	Cr	Co	Mo	Al	Ti	Others		
55 Nitinol	55	—	—	—	—	—	—	—	—	—	—	45	—	6.45	—
60 Nitinol	60	—	—	—	—	—	—	—	—	—	—	40	—	6.71	—
Hy-Stor	Ni₅ M where M is Ca and/or rare earth metals													—	—

TABLE 2.7.13 Compositions and basic properties of miscellaneous nickel alloys—continued

Thermal conductivity at 20°C (W/mK)	Thermal expansion (10^{-6}/K 20–95°C)	Electrical resistivity at 20°C (μΩcm)	Form	Tensile strength (MN/m²)	0.2% proof stress (MN/m²)	Elongation (%)	Hardness (HV)	Young's Modulus (GN/m²)
—	10.4	—	Annealed bar	860	—	60	190	700
—	—	—	Annealed bar	940	—	7	240–300	—
			Heat-treated bar	1070	—	—	700–750	1140
—	—	—	—	—	—	—	—	—

Table 2.7.13

Zinc and its alloys

Contents

List of tables

List of figures

2.8.1   Characteristics

The advantages and limitations of zinc and its alloys are listed in:

Table 2.8.1   *Characteristics of zinc and its alloys compared with other materials*

TABLE 2.8.1 Characteristics of zinc and its alloys compared with other materials

Advantages	Limitations
Available as castings, forgings, extrusions, continuously cast stock, plate, sheet, rod, wire, superplastic materials, surface coatings (hot dip, electroplate, spray, diffusion).	Specific stiffness (UTS/density) low for a metal. Corroded rapidly by acids and strong alkaline media. Saline atmospheres and solutions produce white powdery deposit.
Good corrosion resistance to natural atmospheres and neutral aqueous solutions.	
Position in the electromotive series makes it suitable for cathodically protecting other metals and for electrodes for batteries.	
Excellent castability—accurate reproduction of complicated shapes.	Poor creep resistance.
Compared with other die-casting materials zinc has the following advantages:	Compared with zinc other die-casting materials have the following advantages:

(i)   Lowest casting temperature giving:
    (a) longer die life;
    (b) closer tolerances;
    (c) lower fuel consumption;
    (d) faster production rates.
	Aluminium —higher service temperatures;
	    much better strength:weight ratio;
	    better corrosion resistance.

(ii) Thinner sections and finer detail can be cast.	Magnesium —much better strength:weight ratio.

(iii) Wider range of surface coatings.	Copper base —higher strength and service temperatures;
	    higher ductility;
	    much better corrosion resistance.

Compared with plastics, zinc die-castings are stronger in tension and compression, have higher impact strength, creep resistance and hardness, greater dimensional stability, are unaffected by moisture and conduct electricity.	Plastics are lighter, rarely require finishing, are dielectrics, have low thermal conductivity, are available in a wide range of colours and can be transparent, translucent or opaque.

Table 2.8.1

2.8.2   Pressure casting alloys and their properties

Unalloyed zinc has poor casting characteristics and low mechanical properties in the cast form. Alloying additions are made to produce grain refinement and improve quality and mechanical properties. Two principal alloys, A and B, are used for pressure diecasting. Their compositions are listed in:

Table 2.8.2   *Specifications for zinc die-casting alloy A and alloy B*

The relative advantages and limitations of alloy A and alloy B are outlined below:

Alloy A	*Alloy B*
No appreciable loss of impact strength on ageing at elevated temperatures.	Not recommended for application above room temperature.
Better dimensional stability.	Higher hardness and tensile strength.
Better ductility.	Slightly better castability.

Alloy A is more widely used.

The ZA family of zinc–aluminium high performance casting alloys, ZA8 (8%Al), ZA12 (11.5%Al) and ZA27 (27%Al), although developed as gravity casting alloys, are being increasingly used for the manufacture of pressure die-cast components. The ZA alloys are detailed in Section 2.8.3.

CREEP RESISTANCE

A severe limitation to the wider use of zinc die-casting alloys is their poor resistance to deformation under sustained load at elevated temperature.

THIN WALLED CASTING

It is now possible by correct design of sprues, runners, gates, etc., to obtain die-castings of wall thickness less than 1mm free from porosity in the thin sections. Larger overflows are necessary than with normal die-casting processes. However, the net metal yield is approximately the same due to the compensating effect of the thinner walls. The larger overflows do, of course, mean that more trimming is necessary after casting. A summary of the properties of die-casting alloys is given in:

Table 2.8.3   *General properties—die-casting alloys*

and the mechanical properties over a range of temperatures in:

Table 2.8.4   *Summarised mechanical properties of zinc die-casting alloys*

Typical applications are television instrument and car headlamp bezels.

TABLE 2.8.2 Specification for zinc die-casting alloy A and alloy B

(a) Chemical composition of zinc alloy ingot metal

Element			Alloy A (%)	Alloy B (%)
Aluminium			3.9–4.3	3.9–4.3
Copper			—	0.75–1.25
Magnesium			0.04–0.06	0.04–0.06
Impurities				
	Iron	not more than	0.05	0.05
	Copper	not more than	0.03	—
	Lead	not more than	0.003	0.003
	Cadmium	not more than	0.003	0.003
	Tin	not more than	0.001	0.001
	Nickel	not more than	0.001	0.001
	Thallium	not more than	0.001	0.001
	Indium	not more than	0.0005	0.0005
Zinc			Remainder	Remainder

(b) Chemical composition of zinc alloy die castings

Element			Alloy A (%)	Alloy B (%)
Aluminium			3.8–4.3	3.8–4.3
Copper			—	0.75–1.25
Magnesium			0.03–0.06	0.03–0.06
Impurities				
	Iron	not more than	0.10	0.10
	Copper	not more than	0.10	—
	Nickel	not more than	0.006	0.006
	Lead	not more than	0.005	0.005
	Cadmium	not more than	0.005	0.005
	Tin	not more than	0.002	0.002
	Thallium	not more than	0.001	0.001
	Indium	not more than	0.0005	0.0005
Zinc			Remainder	Remainder

Table 2.8.2

TABLE 2.8.3 General properties—die-casting alloys

		Alloy A	*Alloy B*
Specific gravity		6.7	6.7
Melting point	°C	387	388
Solidification point	°C	382	379
Specific heat	J/kg per K	4.18×10^2	4.18×10^2
Solidification shrinkage	in/ft mm/m	0.14 11.7	0.14 11.7
Thermal expansion coefficient	per °C	27×10^{-6}	27×10^{-6}
Electrical conductivity	Ω^{-1} m^{-1} at 20°C	157×10^5	153×10^5
Thermal conductivity	W/m per K at 18°C	113	109
Compression strength	tonf/in^2 MN/m^2	27 412	39 598
Modulus of rupture	tonf/in^2 MN/m^2	43 657	47 726
Shearing strength	tonf/in^2 MN/m^2	14 206	17 265
Fatigue strength (20×10^6 cycles)	tonf/in^2 MN/m^2	4.7 68.6	5.1 78.5

Table 2.8.3

TABLE 2.8.4 Summarised mechanical properties of zinc die-casting alloys

(a) Mechanical properties obtainable from pressure die-cast zinc alloys at 21°C

Alloy			Tensile strength (MPa)	Elongation % on 2 in (50 mm)	Impact strength (J)	Brinell hardness
Original value	A	Unstabilised	286	15	56.9	83
		Stabilised	275	17	60.8	69
	B	Unstabilised	335	9	57.9	92
		Stabilised	314	10	59.8	83
After 12 months ageing at room temperature	A	Unstabilised	265	25	57.9	67
		Stabilised	265	24	55.9	54
	B	Unstabilised	321	12	56.9	74
		Stabilised	291	14	60.8	72
After 8 years ageing at room temperature	A	Unstabilised	248	20	59.8	65
		Stabilised	243	19	56.9	61
	B	Unstabilised	293	14	55.9	74
		Stabilised	—	—	—	—
After 12 months dry ageing at 95°C	A		235	29	50.0	48
	B		255	23	13.5	64

(b) Mechanical properties at normal and sub-normal temperatures

Temperature °C	Tensile strength (MPa)		Elongation (% on 50 mm)		Impact strength (J)	
	Alloy A	Alloy B	Alloy A	Alloy B	Alloy A	Alloy B
20	282	349	11	8	56.9	59.8
10	—	—	—	—	42.2	55.9
0	298	406	9	8	10.2	53.9
−10	—	—	—	—	4.7	24.5
−20	—	—	—	—	3.4	5.2
−40	318	406	4.5	3	2.8	3.2

(c) Properties of zinc alloy die-castings at temperatures up to 100°C

Temperature °C	Tensile strength (MPa)		Elongation (% on 50 mm)		Impact strength (J)	
	Alloy A	Alloy B	Alloy A	Alloy B	Alloy A	Alloy B
20	282	349	11	8	56.9	59.8
40	247	294	16	13	56.9	62.3
95	197	240	30	23	57.9	—

Note: Brinell hardness values vary with temperature in accordance with the changes in tensile properties.

Table 2.8.4

2.8.3   Sand casting and gravity die-casting alloys

The mechanical properties of pressure die-casting alloys are dependent on the fine grain structure produced by rapid cooling, and the pressure applied to the die. Hence, if these alloys are sand or gravity die-cast the mechanical properties will be much inferior.

If prototype castings are required or if the numbers of components to be produced do not justify the cost of dies for pressure die-casting then zinc alloys, specially formulated ZA alloys for sand casting, or gravity die-casting should be employed.

ZA denotes zinc aluminium alloys, ZA8, ZA12 and ZA27 are commonly employed. Their compositions are:

ZA8:   Zn 8%Al 1.0%Cu
ZA12:   ZN 11%Al 0.75%Cu
ZA27:   ZN27%Al 2%Cu

Zinc–aluminium (ZA) alloys can be cast in sand, plaster, silicone rubber or graphite moulds.

ZA alloys are less dense than standard zinc pressure die-casting alloys and have a better creep resistance.

ZA8 is more fluid than either ZA12 or ZA27, is the densest, and casts at the lowest temperature.

ZA-27 is more difficult to cast and this is reflected in an inferior surface finish, but it offers 20% higher strength and 50% higher strength-to-weight ratio than ZA-12.

A very attractive feature of the ZA alloys is their bearing properties. Studies comparing ZA-12 and ZA-27 bearings to SAE 660 bronze bearing alloys show the ZA alloys to possess lower coefficient of friction (against steel), higher load-bearing capacities, greater wear resistance and superior lubrication retention characteristics. ZA alloys are also cheaper than bronze alloys. See:

Figure 2.8.1   *Comparison of operating limits for zinc base alloys and a leaded bronze*

The ZA alloys are compared to alloy A and alloy B in:

Table 2.8.5   *Comparison of the properties of zinc–aluminium alloys and common zinc die-casting alloys*

TABLE 2.8.5 Comparison of the properties of zinc–aluminium alloys and common zinc die-casting alloys

Properties		Units	Common die-casting alloys		Zinc–aluminium alloys		
			Alloy A	Alloy B	ZA–12	ZA–27	ZA–8
Chemical	Zn	%	rem	rem	rem	rem	rem
Composition	Al		3.9–4.3	3.9–4.3	11.0	27.0	8.0
	Cu		—	0.75–1.25	0.75	2.0–2.2	0.8–1.3
	Mg		0.03–0.06	0.03–0.06	0.20	0.15	0.015–0.03
Tensile strength		MN/m^2	275	314	345	410	374
Yield strength		MN/m^2	—	—	210	345	290
Elongation		%	17	10	2–5	1–3	6–10
Hardness		HB	69	83	110–125	110–120	100–106
Specific gravity		kg/dm^3	6.7	6.7	6.1	5.0	6.3
Melting point		°C	387	388	375–430	375–490	375–404

Table 2.8.5

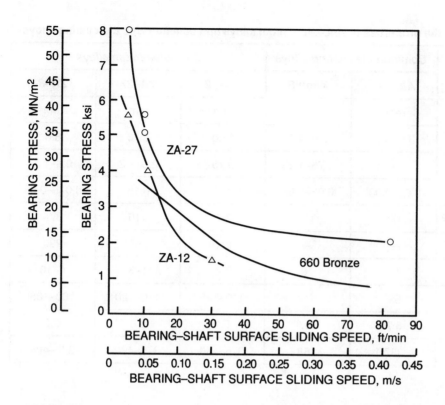

FIG 2.8.1 Comparison of operating limits for zinc base alloys and a leaded bronze

Fig 2.8.1

2.8.4   Dimensional stability of die-casting alloys

The ageing which takes place in zinc alloy die-casting causes dimensional changes which must be allowed for in design and should be minimised by stabilising heat treatment.

The shrinkages which may be anticipated for castings given a variety of heat treatments are listed in:

Table 2.8.6   *Shrinkage of zinc base die-castings as a function of their thermal history*

TABLE 2.8.6 Shrinkage of zinc base die-castings as a function of their thermal history

(a) Shrinkage during normal ageing

Time	Shrinkage (mm/m)		
	Alloy A		Alloy B Air-cooled
	Air-cooled	Quenched	
5 weeks	0.32	0.6	0.69
6 months	0.56	1.2	1.03
5 years	0.73	—	1.36
8 years	0.79	—	1.41

(b) Shrinkage after stabilising

Time	Shrinkage (mm/m)	
	Alloy A	Alloy B
5 weeks	0.20	0.22
3 months	0.30	0.26
2 years	0.40	0.37

Table 2.8.6

2.8.5 Finishes for zinc alloy die-castings

Zinc alloy die-castings have good natural corrosion resistance and where appearance is unimportant may be used unfinished. However, under conditions of high humidity, or more severe conditions such as marine exposure or where a decorative appearance is necessary, the following finishes are possible.

Surface preparation and finishing can typically contribute over 50% of the cost-in-position of a zinc die-casting. Surface preparation costs can be dramatically reduced by adopting vibratory finishing in place of buffing. Further information can be obtained from the sources referenced in Section 2.8.6.

Finish	*Characteristics*
CHROMATE	Gives dull green to yellow colour finish. Also used as pre-treatment to improve adhesion or corrosion resistance of other finishes. Not recommended as a decorative treatment.
ORGANIC COATINGS	
(a) Paint	Applied to degreased and pretreated (phosphated, chromated or etch primed) die cast surface.
(b) Clear lacquers	Useful as a finish for freshly polished zinc alloy which looks very like chromium plating. Still under development.
(c) Electroplating	Dense, uniform coating on complex shapes. One coat application only. Limited range of paints—colour matching difficult.
(d) Plastics Coatings	Thick, solvent-free coatings applied by spray or fluidised bed followed by oven stoving.
ELECTROPLATING	Decorative and (if thick enough) corrosion resistant coatings. Most common system consists of preliminary copper deposit (8 μm thick) followed by a corrosion resistant layer of nickel and then a decorative chromium layer.
VACUUM METALLISING	Bright finish for use where mechanical damage is unlikely. Colours possible by dyeing.
ANODISING	Recommended for salt water resistance. Abrasion resistant. Colours coated to light green, grey and brown.
PHOSPHATING	Used as temporary protection prior to paint bonding.

2.8.6   Acknowledgements

Much of the data in this chapter is taken from publications of the Zinc Development Association (ZDA) and the Zinc Alloy Die Casters Association (ZADCA), to whom grateful acknowledgement is made. These organisations provide an exemplary information service on both technical and economic aspects of the applications of zinc and its alloys.

Magnesium alloys

Contents

List of tables

2.9.1   Introduction

The advantages and limitations of magnesium alloys are listed in:
Table 2.9.1   *Characteristics of magnesium and its alloys*

TABLE 2.9.1 Characteristics of magnesium and its alloys

Advantages	*Limitations*
Lightest of all commercially available metals.	Not suitable for engineering applications in unalloyed form.
Competitive with aluminium alloys on a strength/weight basis.	All materials have poor corrosion resistance (inferior to Ti and Al alloys) and are suitable for uncontaminated atmospheres only. Epoxide coating commonly used for corrosion resistance.
Very high damping capacity.	
Excellent machinability in terms of metal removal rates (fire risk can be eliminated by proper precautions) and low tool wear.	Magnesium has a close packed hexagonal atomic structure which seriously limits its suitability for cold working, and there is only one alloy (containing 10% lithium) which is amenable to cold forming. This alloy has a body centred cubic structure which enables it to withstand cold deformation.
Certain alloys have very good casting properties and are particularly suited to die-casting at high production rates; magnesium alloys have the advantage over aluminium alloys in that they do not attack the die steel and so die wear and sticking are reduced.	
	Low modulus of elasticity—4.5 GN/m^2.
	Upper service temperature 300–350°C.
Available as casting (sand, gravity die, pressure die, shell moulded and precision), forgings, extrusions and sheet.	

Magnesium alloys are capable of deformation at elevated temperatures, usually above 300°C, but the ductility is not high compared with aluminium, for example. Extrusion is the most common method of hot working.

Table 2.9.1

2.9.2   Basis and range of alloy types

The major alloying additions made to commercial magnesium alloys are:

aluminium
manganese
zinc
zirconium
thorium
rare earth metals
silver

It is not usual to use alloy additions singly, but broadly each element has the following effect:

Aluminium (up to 10%)
Increases strength and gives precipitation hardening properties (causes casting porosity).

Manganese (1.0–2.0%)
Improves corrosion resistance with little effect on strength. Can cause reduction in fatigue strength.

Zinc (up to 6%)
Increases strength and improves workability of wrought alloys. When combined with Al or Mn causes precipitation hardening. Welding not satisfactory except with rare earth or thorium additions.

Thorium (up to 3%)
Increases creep and fatigue resistance, especially at elevated temperatures. Improves castability and reduces casting porosity. Generally used in conjunction with zinc and zirconium, where it restores weldability.

Silver (up to 3%)
Permits precipitation hardening of Mg–Zr–rare earth casting alloys to give room temperature strengths comparable with high strength aluminium alloys.

Zirconium
Refines grain structure with consequent increased strength. Improves hot workability of wrought products. Forms insoluble compounds with aluminium and manganese to give precipitation hardening. Up to 0.8% zirconium is generally used.

Yttrium (up to 5.5%)
Improves corrosion resistance.

Rare earths
Improves strength and creep properties, castability (reduces porosity—especially cerium), and reduces tendency to cracking in Mg–Zn–Zr alloys.

Most commercial alloy systems, whether cast or wrought, fall into two groups:

(i)   Mg–Al, Mg–Zn or Mg–Al–Zn systems, or
(ii)   Mg–Zr plus zinc, thorium, silver or rare earth.

In addition, mention should be made of hydride strengthened alloys. These are produced by allowing hydrogen to diffuse into a solidified Mg–Zn–rare earth type alloy by heat treating in a hydrogen atmosphere. Rare earth metal hydrides form from breakdown of the brittle Zn–rare earth metal phase, zinc enters solid solution and a structure free from microporosity and with improved strength and ductility is obtained.

The main types of magnesium alloys are compared in:

Table 2.9.2   *Comparison of magnesium alloys*

Table 2.9.3 gives the related specifications for magnesium casting alloys:

Table 2.9.3   *Related specifications for various magnesium casting alloys*

More details of the properties of the casting alloys are given in:

Table 2.9.4   *Properties of magnesium casting alloys*

Table 2.9.5 covers the related specifications of wrought magnesium alloys and Table 2.9.6 gives more details of the properties of these alloys:

Table 2.9.5   *Related specifications for various wrought magnesium alloys*

Table 2.9.6   *Properties of wrought magnesium alloys*

Physical property data for magnesium alloys is summarised in:

Table 2.9.7   *Physical properties of magnesium casting alloys*

ACKNOWLEDGEMENT

The information for Tables 2.9.3–2.9.7 was kindly supplied by Magnesium Elektron Ltd.

TABLE 2.9.2 Comparison of magnesium alloys

Alloy	Characteristics		Typical applications
	Advantages	**Limitations**	
8 Al–0.3 Mn–0.5 Zn	Improved strength. Heat treatable. Shock resistant.	Casting porosity.	General purpose. Sand and die-casting.
0.1 Al–2 Mn–0.1 Zn	Improved corrosion resistance.	Low strength. High casting shrinkage.	Where light weight and corrosion resistance is of prime importance.
6 Al–3 Zn	Increased strength. Heat treatable.	Casting porosity. Weld cracking.	General purpose. Sand casting.
6 Zn–0.8 Zr	High strength. Heat treatable.	Not weldable.	High strength sand casting— non-weldable.
6 Zn–0.8 Zr–2 Th	High strength. Heat treatable. Weldable. Improved castability.		High strength sand casting— weldable.
2.5 Ag–0.7 Zr–2.2 RE	High strength. Heat treatable. Weldable. Good castability.		High strength sand and investment casting.
2.2 Zn–0.6 Zr–2.7 RE	Medium strength. Heat treatable. Weldable.		Usable up to 250°C. Pressure tight.
2.2 Zn–0.7 Zr–3 Th	Medium strength. Heat treatable. Weldable.		Usable to 350°C. Pressure tight.
5.5 Al–0.2 Mn–0.5 Zn	Improved strength. Heat treatable.		General purpose extrusions and forgings.
8.5 Al–0.2 Mn–0.5 Zn	Improved strength. Heat treatable.		Highly stressed forgings.
0.1 Al–2 Mn–0.1 Zn	Improved corrosion resistance. Weldable.	Low strength.	Sheet, bar and tube where light weight and corrosion resistance are of prime importance.
3 Zn–0.7 Zr	Improved strength. Heat treatable.	Limited weldability.	General purpose. High strength. Extrusion and forging.
0.5 Zn–0.6 Zr–0.7 Th	High strength. Weldable. Heat treatable.		High strength. Extrusions. Usable up to 300°C.
Sy 2% RE ion 0.4 Zr	High strength. Heat treatable. Recommended for prolonged use at 250–300°C Weldable. Good castability.		Racing cars. Aerospace, e.g. helicopter transmissions.

Table 2.9.2

TABLE 2.9.3 Related specifications for various magnesium casting alloys

Alloy designation and condition	British specifications			American specifications[a]				German specifications			French specifications		
	Min. of Def. Procurement Executive (DTD Series)	BS series Aircraft	BS series General engineering	ASTM alloy designation and temper	ASTM specification	Federal or Military specification	AMS specification	Aircraft number	DIN 1729 number	Commercial designation	Air 3380	AFNOR	Standard AECMA
WE54 Solution and precipitation treated	—	—	—	—	—	—	—	—	—	—	—	—	—
ZRE1 Precipitation treated	—	2 L.126	2970 MAG6-TE	EZ33A-T5	B80-76	QQ-M-56B	4442B	3.6204	3.5103	ZRE1	ZRE1	G-TR3Z2	MG-C-91
RZ5 Precipitation treated	—	2 L.128	2970 MAG5-TE	ZE41A-T5	B80-76	—	4439A	3.6104	3.5101	RZ5	RZ5	G-Z4TR	MG-C-43
ZE63 Solution and precipitation treated	5045	—	—	ZE63A-T6	—	MIL-M-46062B	4425	—	—	—	—	—	—
ZT1 Precipitation treated	5005A	—	2970 MAG8-TE	HZ32A-T5	B80-76	QQ-M-56B	4447B	3.6254	3.5105	ZT1	—	G-Th3Z2	MG-C-81
TZ6 Precipitation treated	5015A	—	2970 MAG9-TE	ZH62A-T5	B80-76	QQ-M-56B	4438B	3.5114	3.5102	TZ6	TZ6	—	MG-C-41
EQ21A Solution and Precipitation treated	—	—	—	—	—	—	—	—	—	—	—	—	—
MSR-B Solution and precipitation treated	5035A	—	—	—	—	—	—	—	—	MSR-B	—	G-Ag2 5TR	MG-C-51
QE22A (MSR) Solution and precipitation treated	5055	—	—	QE22A-T6	B80-76	QQ-M-56B MIL-M-46062A	4418C	3.5164	3.5106	—	—	—	—
A8 As cast	—	—	2970 MAG1-M	AZ81A-F	B80-76	QQ-M-56B	—	—	3.5812	FT	G-A9	G-A9	MG-C-61
A8 Solution treated	—	3 L.122	2970 MAG1-TB	AZ81A-T4	B80-76	QQ-M-56B	—	—	3.5812	FT	G-A9	G-A9	MG-C-61
AZ91 As cast	—	—	2970 MAG3-M	AZ91C-F	B80-76	QQ-M-56B	—	—	—	F10	G-A9Z1	G-A9Z1	—
AZ91 Solution treated	—	3 L.124	2970 MAG3-TB	AZ91C-T4	B80-76	QQ-M-56B	—	3.5194	—	F10	G-A9Z1	G-A9Z1	—
AZ81 plus Be Die-cast	—	—	—	AZ81A	B93-76	—	—	—	—	—	—	—	—
AZ91 Solution and precipitation treated	—	3 L.125	2970 MAG3-TF	AZ91C-T6	B80-76	QQ-M-56B MIL-M-46062B	4437A	3.5194	—	F10	G-A9Z1	G-A9Z1	—
AZ91 plus Be Die-cast	—	—	—	AZ91B-F	B94-76	QQ-M-38B	4490E	—	—	—	—	—	—
C As cast	—	—	2970 MAG7-M	—	—	—	—	—	3.5912	—	—	—	—
C Solution treated	—	—	2970 MAG7-TB	—	—	—	—	—	3.5912	—	—	—	—
C Solution and precipitation treated	—	—	2970 MAG7-TF	—	—	—	—	—	3.5912	—	—	—	—

a The American specifications refer to sand castings except where otherwise stated; other specifications may cover the alloys in other forms.

Table 2.9.3

TABLE 2.9.4 Properties of magnesium casting alloys

Typical chemical composition – major alloying elements (%)	Elektron alloy	Tensile properties[a] 0.2% proof stress (N/mm²)	Tensile strength (N/mm²)	Elonga-tion[b] (%)	Comp. properties 0.2% proof stress (N/mm²)	Ultimate strength (N/mm²)	Fatigue endurance values[c] Unnotched (N/mm²)	Notched (N/mm²)	Hard-ness Brinell	Description	Specifications Min of Def. Procurement Executive (DTD series)	British Standards Aircraft	General Engineering
Y 5.25, Nd and other heavy rare earth metals 3.5, Zr 0.5	WE54 Solution and precipitation treated Sand cast / Chill cast	185 / 185	255 / 255	2 / 2	— / —	— / —	95–100 / —	— / —	80–90 / —	Excellent retention of strength after long exposure at 250°C. Good castability, weldable.	— / —	— / —	— / —
Rare earth metals 3.0, Zn 2.5, Zr 0.6	ZRE1 Precipitation treated Sand cast / Chill cast	95 / 100	140 / 155	3 / 3	85–120	275–340	65–75	50–55	50–60	Creep resistant up to 250°C. Excellent castability. Pressure tight and weldable.	— / —	2 L.126 / 2 L126	2970 MAG6-TE / 2970 MAG6-TE
Zn 4.2, rare earth metals 1.3, Zr 0.7	RZ5 Precipitation treated Sand cast / Chill cast	135 / 135	200 / 215	3 / 4	130–150	330–365	90–105	75–90	55–70	Easily cast, weldable, pressure tight, with useful strength at elevated temperatures.	— / —	2 L.128 / 2 L128	2970 MAG5-TE / 2970 MAG5-TE
Zn 5.8, rare earth metals 2.5, Zr 0.7	ZE63 Solution and precipitation treated Sand cast	170	275	5	190–200	430–465	115–125	70–75	60–85	Excellent castability, pressure tight and weldable with high developed properties in thin wall castings.	5045	—	—
Th 3.0, Zn 2.2, Zr 0.7	ZT1 Precipitation treated Sand cast / Chill cast	(85) / (85)	185 / 185	5 / 5	85–100	310–325	65–75	55–70	50–60	Creep resistant up to 350°C. Pressure tight and weldable.	5005A / 5005A	— / —	2970 MAG8-TE / 2970 MAG8-TE
Zn 5.5, Th 1.8, Zr 0.7	TZ6 Precipitation treated Sand cast / Chill cast	155 / 155	255 / 255	5 / 5	150–180	325–370	75–80	70–80	65–75	Stronger than, but as castable as RZ5, weldable, pressure tight.	5015A / 5015A	— / —	2970 MAG9-TE / 2970 MAG9-TE
Ag 1.5, Nd rich rare earth metals 2.0, Zr 0.6, Cu 0.07	EQ21A Solution and precipitation treated Sand cast / Chill cast	175 / 175	240 / 240	2 / 2	165–200	310–385	100–110	60–70	70–90	Heat-treated alloys with high yield strength up to 200°C, Pressure tight and weldable.	— / —	— / —	— / —
Ag 2.5, Nd rich rare earth metals, 2.5, Zr 0.6	MSR-B Solution and precipitation treated Sand cast / Chill cast	185 / 185	240 / 240	2 / 2	165–200	310–385	100–110	60–70	70–90		5035A / 5035A	— / —	— / —
Ag 2.5, Nd rich rare earth metals 2.0, Zr 0.6	QE22 (MSR) Solution and precipitation treated Sand cast / Chill cast	175 / 175	240 / 240	2 / 2	165–200	310–385	100–110	60–70	70–90		5055 / 5055	— / —	— / —

Table 2.9.4

TABLE 2.9.4 Properties of magnesium casting alloys—continued

Composition	Alloy	Condition	(a)	(b)	El. % [b]						Remarks		Spec.	BS 2970
Al 8.0, Zn 0.5, Mn 0.3	A8	As cast — Sand cast	(85)	140	2	75–90	280–340	75–85	58–65	50–60	General purpose alloy. Good founding properties. Good ductility, strength and shock resistance. Also available as a high purity grade.	—	—	2970 MAG1-M
		As cast — Chill cast	(85)	185	4							—	—	2970 MAG1-M
		Solution treated — Sand cast	80	200	7	75–90	325–415	75–90	60–70	50–60		—	3 L.122	2970 MAG1-TB
		Solution treated — Chill cast	80	230	10							—	3 L.122	2970 MAG1-TB
Al 9.0, Zn 0.5, Mn 0.3, Be 0.0015	AZ91	plus Be Die-cast	(150)	(200)	(1)						General purpose pressure die-casting alloy. Draft ISO specification.	—	—	—
Al 9.5, Zn 0.5, Mn 0.3	AZ91	As cast — Sand cast	(95)	125	—	85–110	280–340	77–85	58–65	55–65	General purpose alloy. Good founding properties. Suitable for pressure castings.	—	—	2970 MAG3-TB
		As cast — Chill cast	(100)	170	2							—	—	2970 MAG3-TB
		Solution treated — Sand cast	80	200	4	75–110	185–432	77–92	65–77	55–65		—	3 L.124	2970 MAG3-TB
		Solution treated — Chill cast	80	215	5							—	3 L.124	2970 MAG3-TB
		Solution and precipitation treated — Sand cast	120	200	—	110–140	385–465	70–77	58–62	75–85		—	3 L.125	2970 MAG3-TF
		Solution and precipitation treated — Chill cast	120	215	2							—	3 L.125	2970 MAG3-TF
Al 7.5–9.5, Zn 0.3–1.5, Mn 0.15 min.	C	As cast — Sand cast	(85)	125	—	65–90	278–340	73–80	58–65	50–60	General purpose alloy with good average properties.	—	—	2970 MAG7-M
		As cast — Chill cast	(85)	170	2							—	—	2970 MAG7-M
		Solution treated — Sand cast	(80)	185	4	75–90	330–415	77–85	62–73	50–60		—	—	2970 MAG7-TB
		Solution treated — Chill cast	(80)	215	5							—	—	2970 MAG7-TB
		Solution and precipitation treated — Sand cast	(110)	185	—	90–115	340–432	62–73	58–62	70–80		—	—	2970 MAG7-TF
		Solution and precipitation treated — Chill cast	(110)	215	2							—	—	2970 MAG7-TF

Approximate conversion factors 1 N/mm² = 0.065 TSI = 0.145 KSI.

The tensile properties quoted are the specification minima for the first specification listed for that alloy and condition. The ranges given are the specified minima; bracketed values are for information only.

a The values quoted are for separately cast test bars and may not be realised in certain portions of castings.

b Elongation values are based on a gauge length of 5.65√A except in the case of thin material where a gauge length of 50mm may be used (see BS 2 L.500, 3370 and 3373). With the latter gauge length, elongation requirements for sheet and plate depend on thickness and a range of minima is quoted.

c Endurance values for 50 × 10^6 reversals in rotating bending-type tests; semi-circular notch, radius 1.2 mm; SCF approx. 2.0. Reversed bending for sheet.

Table 2.9.4—*continued*

TABLE 2.9.5 Related specifications for various wrought magnesium alloys

Typical chemical composition – major alloying elements (%)	Alloy designation and condition	SH specifications			American specifications				German specifications		French specifications			Stand. AECMA
		Min. of Def. procurement executive (DTD series)	BS series Aircraft	BS series General engineering	ASTM designation	ASTM	Federal	AMS	Aircraft number	DIN 9715 number	Commercial designation	Air 9052	AFNOR	
ZCM711 Zn 6.5, Cu 1.2, Mn 0.7	Extruded bars & sections	—	—	—	—	—	—	—	—	—	—	—	—	
	Forgings	—	—	—	—	—	—	—	—	—	—	—	—	
ZW3 Zn 3.0, Zr 0.6	Extruded bars & sections & forging stock	—	2 L.505 & L.154	3373 MAG-E-151M	—	—	—	—	—	—		—	—	MG-P-43
	Forgings	—	L.154	3373 MAG-F-151M	—	—	—	—	—	—		—	—	MG-P-43
AZM Al 6.0, Zn 1.0, Mn 0.3	Extruded bars & sections & forging stock	—	L.152 & L.513	3373 MAG-E-121M	AZ61A-F	B107-76	QQ-M-31B	4350H	W.3510	3.5612	M1	G-A6Z1	G-A6Z1	MG-P-63
	Extruded tube	—	2 L.503	3373 MAG-E-121M	AZ61A-F	B107-76	WW-T-825B	—	W.3510	3.5612	M1	G-A6Z1	G-A6Z1	Mg-P-63
	Forgings	—	L.513	3372 MAG-F-121M	AZ61A-F	B91-72	QQ-M-40B	4358A	—	3.5612	M1	G-A6Z1	G-A6Z1	MG-P-63
AZ80 Al 8.5, Zn 0.5, Mn 0.12 min.	Forgings	—	—	—	AZ80A	B91-72	QQ-M-40B	4360D	W.3515	3.5812	—	G-A7Z1	—	MG-P-61
AZ31 Al 3.0, Zn 1.0, Mn 0.3	Sheet – soft	—	—	3370 MAG-S-1110	AZ31B-O	B90-70	QQ-M-44B	4375F	W.3504	3.5312	F3	G-A3Z1	G-A3Z1	MG-P-62
	Extruded bars & sections	—	—	3373 MAG-E-111M	AZ31B-F	B107-76	QQ-M-31B	— —	3.5312	F3	G-A3Z1	G-A3Z1	MG-P-62	
ZTY Th 0.8, Zn 0.6, Zr 0.6	Extruded forging stock & forgings	5111	—	—	—	—	—	—	—	—	—	—	—	—

Table 2.9.5

TABLE 2.9.6 Properties of wrought magnesium alloys

Typical chemical composition—major alloying elements (%)	Elektron alloy	Tensile properties[a]			Comp. properties		Impact value		Hardness (vpn)	Description	Min of Def. Procurement Executive (DTD series)	British Standards Aircraft	British Standards General Engineering
		0.2% proof stress (N/mm²)	Tensile strength (N/mm²)	Elongation[b] (%)	0.2% proof stress (N/mm²)	Compressive strength (N/mm²)	Unnotched (J)	Notched (J)					
Zn 6.5, Cu 1.2, Mn 0.7	**ZCM711** Extruded bars and sections 0–13 mm diameter									The highest stength Magnesium wrought alloy when fully heat treated. Weldable	—	—	—
	As extruded	160	240	7	—	—	—	—	—				
	Precipitation treated	200	250	5	—	—	—	—	—		—	—	
	Fully heat treated	300	325	3	—	—	—	—	—		—	—	—
	Forgings[c]	—	—	—	—	—	—	—	—		—	—	—
Zn 3.0, Zr 0.6	**ZW3** Extruded bars and sections 0–10 mm	200	280	8	200–250	385–465	23–31	9.5–12	65–75	High strength extrusion, and forging alloy. Weldable under good conditions.	—	2 L.505	3373 MAG-E-151M
	10–100 mm	225	305	8					65–75		—	2 L.505	3373 MAG-E-151M
	Extruded forging stock 0–10 mm	195	280	8					65–75		—	L.514	3372 MAG-E-151M
	10–100 mm	205	290	8			—	—	65–75		—	L.514	3372 MAG-E-151M
	Forgings[c]	205	290	7	165–215	370–340	6–27	4.7–9	60–80		—	L.514	3372 MAG-F-151M
Al 6.0, Zn 1.0, Mn 0.3	**AZM** Extruded bars and sections and extruded forging stock 0–75 mm	180	270	8	130–180	370–420	34–43	6.7–9.5	60–70	General purpose alloy. Gas and arc weldable.	—	L.512&3	3373 MAG-E-121M
	75–150 mm	160	250	7	115–165	340–400	—	—	55–65		—	L.512&3	3373 MAG-E-121M
	Extruded tube	160	260	7	130–180	—	—	—	60–70		—	2 L.503	3373 MAG-E-121M
	Forgings[c]	160	275	7	130–165	340–400	16–23	3.4–4	60–70		—	L.513	3372 MAG-F-121M
Al 8.5, Zn 0.5, Mn 0.12 min.	**AZ80** Forgings—precipitation treated	200	290	6	—	—	—	—	60	High strength alloy for forgings of simple design	—	—	—
Al 3.0, Zn 1.0, Mn 0.3	**AZ31** Sheet—soft[c] 0.5–6.0 mm	(120)	220–265	10–12	—	—	—	—	50–65	Medium strength sheet and extrusion alloy. Good formability. Weldable.	—	—	3370 MAG-S-111O
	Extruded bars and sections 0–10 mm	150	230	8	—	—	—	—	50–65		—	—	3373 MAG-E-111M
	10–75 mm	160	245	10	—	—	—	—	50–60		—	—	3373 MAG-E-111M
Th 0.8, Zn 0.6	**ZTY** Extruded forging stock 0–25 mm	130	230	6	—	—	—	—	50–60	Creep resistant up to 350°C. Fully weldable.	5111	—	—
	25–50 mm	110	200	8	—	—	—	—	50–60		5111	—	—
	above 50 mm	95	200	8	—	—	—	—	50–60		5111	—	—
	Forgings[c]	130	230	6	—	—	—	—	50–65		5111	—	—

Approximate conversion factors 1N/mm² = 0.065 TSI = 0.145 KSI.

Larger sizes than those shown above are available: when required property levels will be by agreement.

[a] The tensile properties quoted are the specification minima for the first specification listed for that alloy and condition. Where a range is quoted the specification requirements depend on sheet thickness. Bracketed values are for information only.

[b] Elongation values are based on a gauge length of $5.65\sqrt{A}$ except in the case of thin material where a gauge length of 50mm may be used (see BS 2 L.500, 3370 and 3373). With the latter gauge length, elongation requirements for sheet and plate depend on thickness and a range of minima is quoted.

[c] Forging properties quoted are those in the most favourable direction of flow; the manufacturer should be consulted on directionality.

Table 2.9.6

TABLE 2.9.7 Physical properties of magnesium casting alloys

Alloy	Specific gravity (20°C)	Coefficient of thermal expansion ($10^{-6}K^{-1}$) (20–200°C)	Thermal conductivity (W/m per K) (20°C)	Electrical resistivity (nΩm) (20°C)	Specific heat (J/kg per K) (20–100°C)
WE54	1.9	27	107	70	1000
ZRE1	1.80	26.8	100	73	1050
RZ5	1.84	27.1	109	68	960
ZE63	1.87	27.1	113	56	960
ZT1	1.83	26.7	105	72	960
TZ6	1.87	27.1	113	66	960
EQ21A	1.81	26.7	113	68.5	1000
MSR–B	1.82	26.7	113	68.5	1000
QE22	1.82	26.7	113	68.5	1000
A8	1.81	27.2	84	134	1000
AZ91	1.83	27.0	84	141	1000
C	1.81	27.2	84	134	1000

Table 2.9.7

2.9.3 Corrosion properties

Corrosion of magnesium alloys is considered in detail in Volume 1, Section 1.4.1.

2.9.4   Processing

Magnesium alloys can be fabricated by all the usual forming methods except cold forming. Alloys are available in bar, sheet, tube and extruded forms. Welding is possible using oxy-acetylene and argon-arc processes but alloys containing zinc are not weldable due to their propensity to weld cracking. Alloys containing zirconium cannot be oxy-acetylene welded. Soldering is not recommended. Other joining methods include riveting and adhesive bonding.

Magnesium alloys are readily machinable and very high cutting speeds are permissible. Dry machining is preferred, using compressed air-cooling if necessary. There is a fire risk associated with magnesium and so swarf should be removed frequently. Provided proper precautions are taken there is not a great risk of fire during machining or welding.

2–10

Lead alloys

Contents

List of tables

2.10.1   Introduction

The advantages and limitations of lead compared with other metals are listed in:
Table 2.10.1   *Characteristics of lead compared with other materials*

The relatively low cost, easy workability and good environmental resistance of lead have made it in the past an obvious choice for sheathing purposes. However, in more recent years its high toxicity, high density and its low creep strength have led to its substitution by plastics and metals such as aluminium, and its use is becoming restricted to applications where its special properties offer unique advantages to the end-product.

TABLE 2.10.1 Characteristics of lead compared with other materials

Advantages	Limitations
Excellent corrosion resistance to sulphuric, sulphurous, phosphoric and chromic acid responsible for use in piping, castings and lining material in the chemical industry.	Low melting point.
Resistant to atmospheric, fresh- and salt water attack.	Very low creep resistance in unalloyed state. (Maximum stress for unsupported material to resist creep at room temperature is about 17.5 kg/mm^2).
Good sound and vibration damping properties.	High density (although this can be an advantage when needed for counterweight applications and is responsible for radiation shielding applications).
High absorption of atomic radiation responsible for nuclear and X-ray shielding applications.	Toxic—cannot be used in contact with foodstuffs or handled without protective clothing.
Excellent ductility and absence of work hardening tendency.	Attacked by formic, acetic and nitric acids.
Hardness, fatigue and creep strength can be improved by alloy additions.	Low elastic limit.
Embeddability and lubricating properties responsible for bearing alloy applications.	Most alloy additions adversely affect corrosion resistance.
Very easy to fabricate and weld.	Low electrical conductivity and high coefficient of thermal expansion.
Available as castings, sheet, foil extrusions wire and surface coatings (cladding, hot dipping, electroplating, spraying) clad plywood, shot and wool.	

Table 2.10.1

2.10.2   Lead alloys and their properties

2.10.2.1   Alloy types

High purity lead (99.99%), sometimes known as chemical lead (Type A), is widely used in the chemical industry, its chemical resistance stemming from the ability to rapidly form protective films. Use of high purity lead in a lining supported by steel structure compensates for the lack of strength and creep resistance. High purity lead is available as ingot, sheet and pipe.

Alloy additions are designed to improve the creep and fatigue properties, generally to the detriment of corrosion resistance. The effect of commonly used alloying additions is summarised below.

Antimony, in proportion to the amount present, increases strength and hardness and improves castability without significantly affecting the corrosion resistance. However, it does make the material more difficult to weld or solder and lowers the melting point, consequently reducing the upper service temperature limit. Lead with antimony additions is known variously as Antimonial Lead, Lead Antimony Alloy, Regulus Metal or Hard Lead. Alloys containing less than 10% antimony are sufficiently malleable to be rolled or extruded and can be obtained in sheet and pipe form. Alloys containing more than 12% antimony are not ductile and are available only as castings.

Copper, silver and tellurium improve mechanical properties, largely through grain size control and stability, giving better creep and fatigue strength than chemical lead type A but inferior corrosion resistance. Alloys of lead with these elements are known as chemical lead type B.

Tin increases hardness and strength and improves castability. In solder, and lead alloy coatings, tin imparts the ability to bond with copper and steel. Lead–tin alloys are used mainly as solders and bearing alloys.

Dispersion strengthening of lead with its own oxide provides a means of improving stiffness, fatigue strength and creep resistance of pure lead and dilute alloys, without impairing significantly corrosion resistance. Mechanical properties are dependent on the production methods, degree of deformation etc. Forming and fabrication require special techniques, but machining is easier than with other lead alloys.

2.10.2.2   Applications

The applications of lead alloys are listed in:
Table 2.10.2   *Applications of lead alloys*

2.10.2.3   Properties

The short-term, room temperature, mechanical properties of lead alloys are listed in:
Table 2.10.3   *Typical mechanical properties of lead alloys*
and their fatigue resistance in:
Table 2.10.4   *Fatigue resistance of lead alloys*

TABLE 2.10.2 Applications of lead alloys

Alloy	Applications
High purity lead (99.99%) (Chemical lead Type A)	Linings, pipe, sheet, etc., used in the manufacture, handling and storage of corrosive chemicals. Preferred where corrosion resistance is more important than strength.
Copper, tellurium, silver leads (Chemical lead Type B)	As above, but where mechanical properties rather than corrosion resistance are likely to be life limiting factors.
Antimonial lead (Regulus metals)	Accumulator batteries (special purity grades), linings, pipes, pumps and impellers in the chemical industry where extra strength and stiffness are required. Linings for chromium plating tanks. Castings, bricks and sheets for nuclear shielding purposes.
Dispersion strengthened lead	This material must be considered to be still under development. The properties offered are corrosion resistance or high purity lead with the strength typical of wrought antimonial lead.

TABLE 2.10.3 Typical mechanical properties of lead alloys

Material		Condition	Tensile strength (MPa)	% El.	Hardness (VPN)
Chemical copper/ lead Type B	0.055–0.065% Cu	—	15.0	35	4.5–5.0
DSL	Dispersion strengthened lead	—	35	12	8.0–10.5
Antimonial lead	1% Sb	Chill cast	21.4	34	6.7
		Extruded	21.2	38	6.9
		Cold rolled	29.8	20	10.0
	4% Sb	Chill cast	40.9	26	12.5
		Extruded	28.1	41	8.5
		Cold rolled	29.8	47	8.3
	8% Sb	Chill cast	53.0	23	15.5
		Extruded	35.2	32	8.9
		Cold rolled	32.9	53	9.1
	12% Sb	Chill cast	—	—	16.0
		Extruded	38.3	49	10.0

Table 2.10.2 and Table 2.10.3

TABLE 2.10.4 Fatigue resistance of lead alloys

Material	20°C	80°C
	±MN/m²	±MN/m²
99.99% Lead	3.2	2.1
Copper chemical lead 0.055–0.065% Cu	4.0	3.0
Silver/copper chemical lead (0.003–0.005% Ag; 0.003–0.005% Cu)	4.2	3.0
Copper/tellurium chemical lead 0.06% Cu; 0.04% Te	7.7	5.1
8% Antimonial lead	10.3	6.0
Dispersion strengthened lead	13.8	12.5

20×10^6 cycles endurance limit.

Table 2.10.4

Cobalt alloys

Contents

List of tables

List of figures

2.11.1 General characteristics

2.11.1.1 Introduction

Cobalt is a silvery metal, intermediate in the periodic classification between iron and nickel, both of which it resembles. The unalloyed metal is soft and readily corroded by aggressive environments, but certain specific properties which are developed or enhanced by alloying give it technological importance. These properties include a high coercive force (capacity for magnetisation) exceptional hardness and wear resistance, very high hot strength, and very high resistance to corroding and oxidising environments both at ambient and elevated temperatures.

Properties of some typical cobalt alloys are listed in:
Table 2.11.1 *The properties of some typical alloys of cobalt*
The influence of individual elements on the properties of cobalt alloys is indicated in:
Table 2.11.2 *Effect on properties of increasing content of alloying elements in cobalt*

Other significant properties of cobalt are: its high melting point, its capacity to form alloys with useful dimensional properties and the capacity of the most abundant isotope, Co^{59} to be transmuted by nuclear irradiation to Co^{60} which emits 7×10^6 eV γ rays.

2.11.1.2 Applications

The major technological applications of cobalt, which between them account for two thirds of its use, are as a base for hard magnet alloys and, to an important but declining extent, as the major constituent of the cobalt base 'Superalloys' used principally in high temperature components for gas turbines. Other applications are: for soft magnet alloys, for metallurgical and glass making furnaces, for cutting and wear resistant alloys, for hardfacing alloy consumables, for dental and bone surgery materials, for alloys with special thermal expansion characteristics and as a material for stainless, constant rate springs. Cobalt also forms a base for permanent magnet alloys, sometimes alloyed with iron.

Cobalt alloys and their analyses, grouped according to application are listed later. There is a great family resemblance between the alloys in the separate groups and in many cases alloys designated for one application may be (and have been) used for other purposes.

1336   **Cobalt alloys**

TABLE 2.11.1 The properties of some typical alloys of cobalt

Alloy designation	Density (g/ml)	Tensile properties At 21°C UTS (MPa)	At 21°C 0.2% PS (MPa)	At 21°C Elongation (%)	At 871°C UTS (MPa)	At 871°C 0.2% PS (MPa)	At 871°C Elongation (%)	Dynamic modulus of Elasticity at 21°C MPa × 10⁻⁶	at 871°C MPa × 10⁻⁶	Mean coefficient of thermal expansion 21–982°C × 10⁶/°C
Superalloys and furnace hardware										
Cast										
X 40	8.60	745	524	9	324		16			
FSX 414	8.30	739	441	11	310	166	23			
MAR–M 509	8.85	786	573	4	352	290	20	225,000	137,000	16.5
Wrought										
L 605	9.13	1007	462	64	324	242	30			
HS 188	9.13	960	483	56	421	262	73			
UM Co 50	8.06	925	610	10	180	160	12	215,000	—	16.8
Spring material (Wire)										
Elgiloy	8.3	2208	1518	1.5	1700[a]	1214[a]	—	203,500	—	12.7[b]
Dental and prosthetic										
Cast										
BS3531 Pt. 2, 1981		700	500	7				230,000		12.7[b]
Wrought										
BS3531 Pt. 2, 1981		1000	840	52						
Wear resistant										
Wrought										
Stellite 6B	8.39	1007	632	11	385	270	18	210,000		15.4
Stellite 6K	8.39	1217	708	4	380	180	24	214,000		15.4

[a] At 538°C.
[b] 0–50°C.

Table 2.11.1

TABLE 2.11.2 Effect on properties of increasing content of alloying elements in cobalt

Alloying element	Hardness			Bulk toughness	Volume % carbides	Volume % borides	High temperature strength	Oxidation resistance
	Bulk	Borides and carbides	Matrix					
C	↑			↓	↑		↑	
Cr	↑	↑	↑				↑	↑
W	↑	↑	↑		↑	↑	↑	↓
Mo	↑		↑		↑	↑	↑	↓
Nb	↑		↑		↑			
Ni	↑		↑	↓				
B	↑		↑			↑	↑	↑

↑ ≅ Increase.   ↓ ≅ Decrease.

Table 2.11.2

2.11.2   Cobalt base superalloys

2.11.2.1   Introduction

The compositions of selected cobalt base superalloys are shown in:
Table 2.11.3   *Composition of cobalt base superalloys*

Their melting points are usually at least 100°C higher than those of the nickel and iron base alloys, their stress rupture properties (of the more recently developed alloys) are comparable at the highest temperatures and longest times, their susceptibility to 'corrosion' by fuel impurities in some cases approaches zero and their higher thermal conductivities render them less susceptible to thermal shock.

On the other hand, because an efficient strengthening mechanism has not yet been developed, cobalt alloys have lower strengths at intermediate temperatures, their oxidation resistances are lower, their densities for roughly equivalent tungsten content and their costs are higher.

These properties render cobalt alloys unsuitable for rotating parts for modern highly rated gas turbines. Wrought alloys have been used both for rotating blades and discs, and the cast alloys are suitable as blade materials for low rated industrial turbines burning low grade fuel.

The mechanical properties of the alloys are, however, satisfactory for stationary blades which operate at stresses between 35 and 69 MPa and temperatures between 870 and 1040°C and for combustion cans which may operate at or above 1100°C.

The creep rupture strengths of the nickel and cobalt base superalloys are compared with each other and with the stresses imposed on rotating and stationary gas turbine blades in:
Fig 2.11.1   *Relative merit of cast nickel and cobalt base superalloys at fixed and rotating blade temperatures and stresses.*

2.11.2.2   Alloying for mechanical properties

Unalloyed cobalt possesses an hexagonal close packed (ε) structure below 417°C and an FCC (γ) structure between this temperature and the melting point.

Strengthening of the FCC phase is promoted by solid solution strengthening by the refractory metals tungsten, molybdenum, vanadium, niobium, titanium and tantalum and precipitation hardening by the carbides of these metals and chromium. Boron added in small quantities improves mechanical properties.

The less desirable ε phase is promoted by these additions and nickel is usually added to stabilise the more ductile γ phase and to reduce the number of stacking faults (which tend to impair ductility) by increasing stacking fault energy.

The high mechanical properties which result from the alloy additions are developed by a heat treatment process of solution treatment and ageing.

2.11.2.3   Alloying for oxidation resistance

Additions of 25% of chromium decrease the oxidation rate to a minimum with the establishment of a protective scale of Cr_2O_3. Alloys with 20–30% of chromium are further improved by additions of aluminium, boron, calcium, and more recently lanthanum. Iron, yttrium and nickel appear to be innocuous while molybdenum and niobium additions promote catastrophic attack.

2.11.2.4   Fabrication

The earlier cobalt superalloys were air cast, but the properties of the more recently developed complex alloys benefit substantially from vacuum induction melting (of material for casting) and vacuum induction melting followed by electroslag refining for wrought alloys.

Improvements in properties in a longitudinal direction could be achieved by the unidirectional solidification process used for nickel superalloys if the requirement for cobalt alloy blades justified this.

Eutectic alloys with well aligned fibres of TaC have been produced but cannot compete with the standard nickel alloys.

Dispersion strengthening by ThO_2 using powder metallurgy techniques has been examined but the dispersions so far achieved are not fine enough to compete strengthwise with nickel superalloys.

2.11.2.5 Specific alloys

The compositions of the cobalt superalloys are shown in Table 2.11.3. Stress rupture properties are compared in:

Fig 2.11.2 *1000-h rupture strengths for cast and wrought cobalt base superalloys*
and oxidation and hot corrosion resistance in:
Fig 2.11.3 *Oxidation and hot corrosion resistance of cobalt and nickel base superalloys*

2.11.2.6 Applications of superalloys

GAS TURBINES

Cost, and potential supply restrictions are reducing the utilisation of cobalt alloys in gas turbines, but of the cast alloys, X40 and its derivatives X45 and FSX414 continue to be used for stationary blades for the lower rated gas turbines which burn lower grade fuel or operate in a marine atmosphere, while MAR–M509 has a very favourable combination of properties. Of the wrought alloys, S816 has been used extensively both for rotating blades and discs.

Alloy L605 has been utilised in the sheet form for combustor cans, spray bars, flame holders and liners in jet engines. More recently Haynes 188, to which lanthanum is added, has been shown in laboratory tests to have the highest resistance to corrosion by fuel oxides contaminated by sodium chloride of all available sheet materials (Fig 2.11.3). It also has very favourable creep resistance, as shown in:
Fig 2.11.4 *Creep characteristics of HS 188*

It would therefore appear to be an ideal material for duct components operating with low grade fuel or in marine atmospheres within the temperature range 800–975°C at which hot corrosion occurs. (Protective coatings, see below, are not used for sheet metal components because of problems which would arise in repair.) Both alloy L605 and HS188 are strong, ductile and easily formed and fabricated.

The oxidation resistance of cobalt (and nickel) superalloys is inadequate for the highest temperature gas turbine blades and can be increased by a protective coating. Coatings, which may contain any of the oxidation resistant elements, aluminium, chromium and silicon, may be applied by a variety of processes which include pack cementation, slurry coating, electrophoresis followed by diffusion, hot dipping, chemical vapour deposition, fused salt metalliding and physical vapour deposition.

In practice pack cementation with aluminium powder is used. This produces an adherent surface intermetallic containing near 50% Al which on oxidation forms a self-healing surface layer of tenacious aluminium oxide.

These layers provide long-term environmental protection and resistance to thermal fatigue to a maximum temperature of 980°C. Above this temperature thermal fatigue produces oxide spalling, aluminium loss and thermal fatigue cracking which allows direct access of the environment to the substrate.

FURNACE HARDWARE

The deformation, corrosion and oxidation resistant characteristics which fit the cobalt superalloys for gas turbine service make them suitable for grates, trays and rolls, skids and rails, slag notch rings, tundishes and muffles for glass, ceramic and metallurgical furnaces. Alloys which have been used include, in the wrought form, HS188 because of an excellent combination of properties, L605, UMCo50 because of its relatively low cost, and the UMCo50 equivalent, Stellite 250 in the cast form.

The compositions of the cobalt alloys used for furnace hardware are listed in:

Table 2.11.4   *Composition of miscellaneous alloys of cobalt*

and a typical example of creep properties is shown in:

Fig 2.11.5   *Creep characteristics of wrought UMCo50*

TABLE 2.11.3 Composition of cobalt base superalloys

Alloy designation	Nominal compositions (weight %)															
	Co	Cr	C	Ni	Mo	W	Ta	Nb	Al	Ti	Fe	Mn	Si	B	Zr	Other
Superalloys																
Cast																
X40	54	25.5	0.5	10.5	—	7.5	—	—	—	—	—	0.75	0.75	—	—	—
X45	54	25.5	0.25	10.5	—	7.5	—	—	—	—	—	0.75	0.75	—	—	—
FSX414	52	29	0.25	10	—	7.5	—	—	—	—	1	—	—	0.01	—	—
MAR.M509	55	23.5	0.6	10	—	7	3.5	—	—	0.2	—	—	—	—	0.5	—
AR13	58	21	0.45	1	—	11	—	2	3.5	—	2.5	0.5	—	—	—	0.1Y
AR213	66	19	0.18	—	—	4.7	6.5	—	3.5	—	—	—	—	—	0.15	0.1Y
AR215	64	19	0.35	—	—	4.5	7.5	—	4.3	—	—	—	—	—	0.13	0.17Y
MAR.M302	58	21.5	0.85	—	—	10	9	—	—	—	—	—	—	0.005	0.2	—
MAR.M322	61	21.5	1	—	—	9	4.5	—	—	0.75	—	—	—	—	2.25	—
NASA Co WRe	68	3	0.4	—	—	25	—	—	—	—	—	—	—	—	—	—
W152	63	21	0.45	—	—	11	—	2	—	—	—	0.75	0.25	—	—	—
Superalloys																
Wrought																
L605	52.9	20	0.1	10	—	15	—	—	—	—	—	1.5	0.5	—	—	—
HS188	39.2	22	0.1	22	—	14	—	—	—	—	1.5	0.75	0.4	—	—	0.08La
S816	42	20	0.38	20	4	4	—	4	—	—	4	1.2	0.4	—	—	—
CM7	48	20	0.1	15	—	15	—	—	0.5	1.3	—	—	—	—	—	—
MAR.M918	52	20	0.05	20	—	—	7.5	—	—	—	—	—	—	—	—	—
TD Co	60	18	—	20	—	—	—	—	—	—	—	—	—	—	—	2ThO$_2$
UM Co-50	51	28	0.1	—	—	—	—	—	—	—	21	—	—	—	—	—

Table 2.11.3

TABLE 2.11.4 Composition of miscellaneous alloys of cobalt

Alloy designation	Nominal composition (weight %)											
	Co	Cr	Ni	Mo	W	Al	Ti	Fe	Mn	Si	Be	C
Furnace hardwear alloys												
Stellite 250	52	28	—	—	—	—	—	20	—	—	—	0.1
Stellite 251	54	28	—	—	—	—	—	18	—	—	—	0.3
UM Co 50 L 605 HS 188	Listed under 'Superalloys'											
Low expansion alloy												
Stainless Invar	53.8	9.1	—	—	—	—	—	36.6	0.1	0.31	—	0.05
Spring alloy												
Elgiloy	40	20	15	7	—	—	—	16	2	—	0.04	0.15*

*Max.

Table 2.11.4

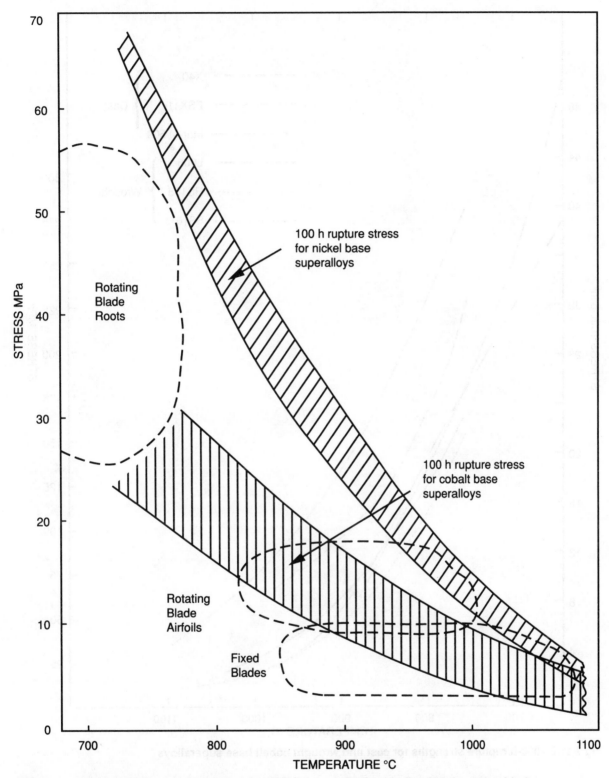

FIG 2.11.1 Relative merit of cast nickel and cobalt base superalloys at fixed and rotating blade temperatures and stresses

Fig 2.11.1

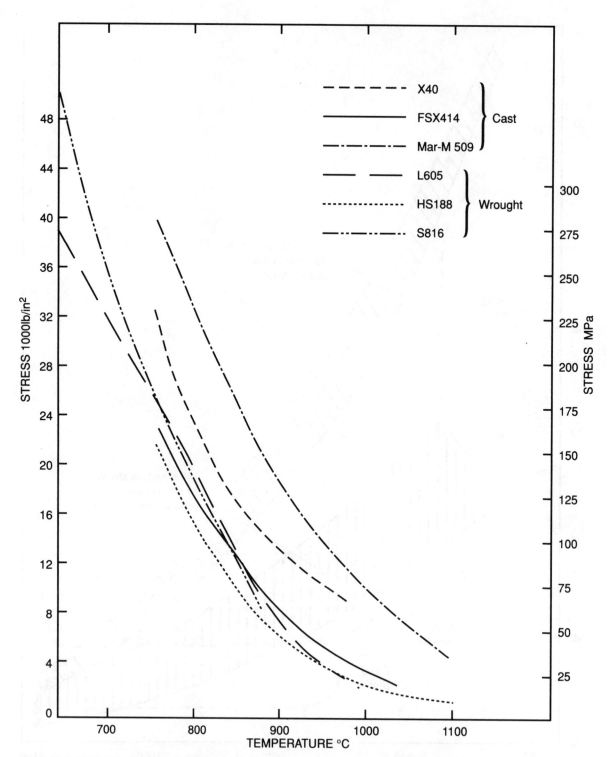

Fig 2.11.2 1000-h rupture strengths for cast and wrought cobalt base superalloys

Fig 2.11.2

(a) Rig tests on selected cobalt base alloys.

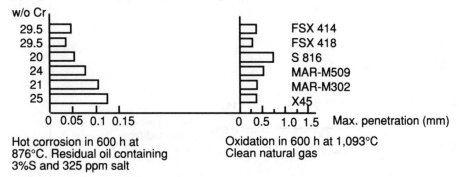

Hot corrosion in 600 h at 876°C. Residual oil containing 3%S and 325 ppm salt

Oxidation in 600 h at 1,093°C Clean natural gas

(b) Oxidation of HS 188 in 100 h intermittent exposure.

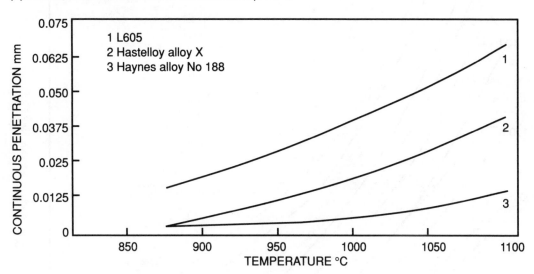

1 L605
2 Hastelloy alloy X
3 Haynes alloy No 188

(c) Hot corrosion, in 200 h at 900°C in flue gas from distillate fuel (0.3-0.45S) with added salt of HS 188 compared with Hastelloy X (the best nickel alloy) and other alloys.

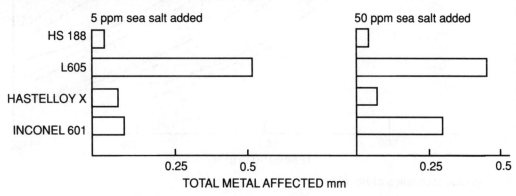

FIG 2.11.3 Oxidation and hot corrosion resistance of cobalt and nickel superalloys

Fig 2.11.3

1346   **Cobalt alloys**

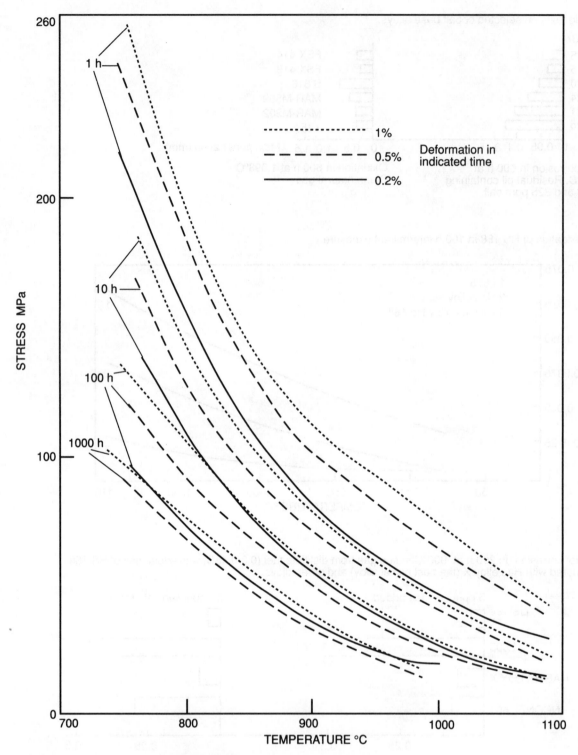

FIG 2.11.4 Creep characteristics of HS 188

Fig 2.11.4

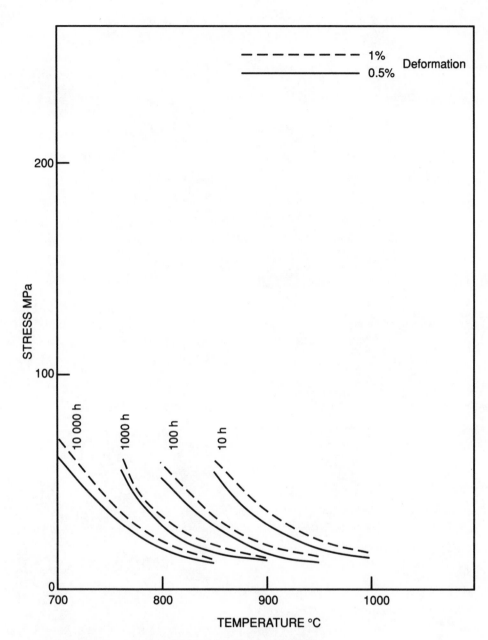

FIG 2.11.5 Creep characteristics of wrought UMCo50

Fig 2.11.5

2.11.3   Cobalt base wear resistant alloys

Wear resistant alloys may be used in the cast, powder compacted, wrought and spray deposited forms. One range often referred to as 'stellites' differs from the superalloys mainly in containing higher proportions of chromium, carbon and, in some cases, tungsten. The high chromium content ensures a very high resistance to most environments. The major objective in formulating these alloys is to increase the proportion of carbides in the cobalt matrix without making the alloy so brittle that it cracks or crumbles in. The effect of tungsten and carbon is shown in:

Fig 2.11.6   *Effect of tungsten and carbon content on high temperature (600°C) adhesive wear (metal to metal pressure 3.75 MPa) and abrasive wear resistance of Co–30% Cr hard facings*

An even higher proportion of carbide in a cobalt matrix is present in the sintered carbide materials. Similar wear and environmental resistance properties are shown by the 'triballoy' range of materials which are stable intermetallic powders generally containing cobalt, silicon and molybdenum. They can be fabricated (if necessary blended with other metals) in a variety of ways.

All the alloys have good corrosion resistance, good high temperature properties, low coefficient of friction and high resistance to seizing and galling. Many alloys have excellent oxidation or hot-gas corrosion resistance up to 1100°C and good short-term strength and creep resistance at elevated temperatures, see:

Fig 2.11.7   *Elevated temperature hardness of cobalt base hard facing alloys*

Compositions, available forms and comparison of relevant properties of the more commonly used alloys are listed in:

Table 2.11.5   *Cobalt based wear resistant alloys*

The physical and mechanical properties of two typical alloys are shown in Table 2.11.1 and corrosion data for a number of alloys in aqueous environments in:

Table 2.11.6   *Comparative corrosion data for certain cobalt, nickel and iron corrosion resistant alloys in aqueous environments*

Most of the alloys can be machined and welded and all can be brazed and ground.

ALLOYS USED IN MASSIVE FORM

These materials are used in applications involving severe conditions of wear or wear combined with corrosion and/or high temperatures. These include fans, mixers, knives, extrusion dies, feed screws and high temperature bearings. Stellite 100 which is produced by a carefully controlled chill casting process has the highest hot hardness and may be used for very severe cutting processes.

ALLOYS USED AS WELD OVERLAYS

The principal alloying elements are chromium and tungsten. Carbon, although present in lower proportions by weight, is an essential constituent. Alloys with nickel, silicon and boron have superior fluidity and may be used for the deposition of thin and precise coatings. A boron content of not less than 1.7% by weight confers a self fluxing property and alloys of this composition may be used for the 'spray fuse process'. Cobalt base hardfacing alloys are selected primarily for service at elevated temperatures, for high temperature hardness and for corrosion resistance. At lower temperatures iron or nickel base hardfacing alloys are generally more economic with equal or superior toughness and resistance to metal-to-metal wear, abrasive wear and erosion.

TABLE 2.11.5 Cobalt based wear resistant alloys

Alloy designation		Available form											Chemical composition (%)		
		Solid			Hardfacing consumable										
Stellite	Wallex	Cast	Wrought	Powder for compaction	Oxyacetylene inc. powder	Metal arc	Metal inert gas	Tungsten inert gas	Submerged arc	Plasma transferred arc	Plasma spray	Spray fuse	Cr	Ni	W
1	1	✓			✓	✓	✓	✓	✓	✓			30	3[a]	12.5
SF1												✓	19	13	13
3				✓									31	3[a]	12.5
	4				✓								30.5	3[a]	14
6	6	✓		✓	✓		✓	✓	✓	✓	✓		29	3[a]	4.5
SF6	42				✓	✓						✓	19	13.5	7.5
6B			✓										30	3[a]	4.5
6K			✓										30	3[a]	4.5
	7					✓	✓	✓	✓				30	3[a]	—
12	12	✓		✓	✓	✓	✓	✓	✓	✓			29	3[a]	8
SF12												✓	19	13	15
19		✓		✓									31	3[a]	10.5
20		✓											33	—	18
SF20												✓	19	13	15
21		✓				✓	✓	✓	✓	✓			27	2.8	—
31				✓							✓		25	10.5	7.5
	50	✓		✓								✓	19	17	10
100		✓											34	—	19
156				✓	✓					✓			28	3[a]	4
157						✓				✓		✓	22	2[a]	4.5
158								✓		✓			26	3[a]	5.5
F										✓			25	22	12
Intermetallic compounds (triballoy)															
T400				✓						✓	✓		8.5	—	—
T800				✓						✓	✓		17.5	—	—

[a] Max.
[b] Can vary from C20 to C48 according to metallurgical conditions.
[c] Merit order: most favourable properties '1'. Machinability '1': easily machined. '4': can only be ground.
✓ Available form of material.

Table 2.11.5

TABLE 2.11.5 Cobalt based wear resistant alloys—*continued*

Chemical composition (%)						Property merit order[c]						Specification		
B	Si	Fe	Mn	Mo	C	Hardness Rockwell C	Hot hardness	Adhesive wear	Abrasive wear	Room temp. impact.	Machinability	AWS grade	ASM group	AWRA type
—	1	3[a]	1[a]	1[a]	2.5	48	3	1	5	5	3	CoCr C	4A	4357
2.2	3	3	1[a]	—	1.3	55								
—	1	3[a]	1	—	2.4									
—	1.5[a]	3[a]	—	—	1	48					3			
—	1.25	3[a]	—	—	1	37	7	5	7	2	2	CoCr A	4A	4240
1.7	2.3	3	1[a]	—	0.7	43								
—	2[a]	3[a]	2[a]	—	1.1	38								
—	2[a]	3[a]	2[a]	—	1.6	45								
—	—	3	—	5.5	0.3	35					1			
—	1.4	3[a]	1[a]	1[a]	1.4	41	5	4	6	5	3	CoCr B	4A	4249
1.8	2.8	2[a]	1[a]	—	1	48								
—	1[a]	2[a]	1[a]	1.5[a]	3.1	48								
—	—	—	—	—	2.5	58								
3	2.8	2[a]	1[a]	—	1.3	60								
—	2[a]	2[a]	1[a]	5.5	0.25	+[b]	8	7	8	1				4130
—	1[a]	2[a]	1[a]	1[a]	0.5	28					1			
3.75	2.75	1	—	—	0.8	57					4			
—	—	—	—	—	2	62								
—	1.1	—	1[a]	1[a]	1.6	43	6	6	4	3				
2.4	1.6	2[a]	1[a]	1[a]	0.1	52	4		1	4				
0.7	1.2	2[a]	1[a]	1[a]	1.75	43								
—	1.1	3[a]	0.5	0.6[a]	1.75	43								
—	2.8	—	—	28.5	0.08[a]	54	2	2	3	7	4			
—	3.4	—	—	25.5	0.08[a]	53	1	3	2		4			

Table 2.11.5—*continued*

TABLE 2.11.6 Comparative corrosion data for certain cobalt, nickel and iron corrosion resistant alloys in aqueous environments

Media	Concentration and temperature	Triballoy			Stellite No. 6	Deloro No. 60[a]	Stainless steel type 316
		T–400	T–700[a]	T–800			
Acetic acid	50% Boiling	E	E	E	E	—	E
Ferric chloride	10% Room temperature	U	U	G	S	U	U
Formic acid	30% 150°F (66°C)	—	E	—	S	G	E
	45% Boiling	E	—	E	S	—	U
Hydrochloric acid	5% 150°F (66°C)	U	S	E	U	—	U
Nitric acid	65% 150°F (66°C)	U	G	S	U	U	E
	65% Boiling	—	U	—	U	U	G
Phosphoric acid	85% 150°F (66°C)	E	E	E	E	U	E
Sodium chloride	10% + 5% FeCl Room temperature	U	U	G	S	U	U
Sulphuric acid	5% 150°F (66°C)	—	G	—	E	U	G
	5% Boiling	—	G	—	U	U	U
	10% Boiling	U	—	S	U	—	U

Determined in laboratory tests. It is recommended that samples be tested under actual plant conditions.
[a]Nickel base alloys.

Code:
E – Less than 0.05 mm/year
G – Less than 0.51 mm/year
S – Less than 1.27 mm/year
U – More than 1.27 mm/year

Data courtesy of Cabot Wear Technology Division.

Table 2.11.6

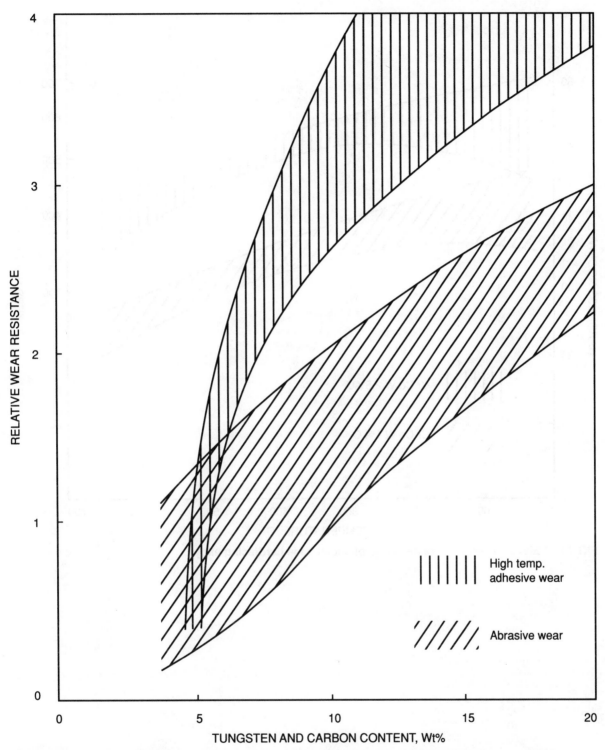

FIG 2.11.6 Effect of tungsten and carbon content on high temperature (600°C) adhesive wear (metal to metal pressure 3.75 MPa) and abrasive wear resistance of Co–30% CR hardfacings

Fig 2.11.6

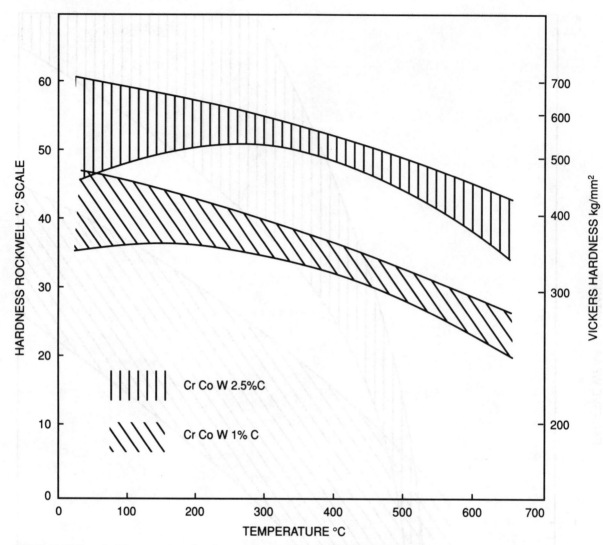

FIG 2.11.7 Elevated temperature hardness of cobalt base hardfacing alloys

Fig 2.11.7

2.11.4   Other uses of cobalt

2.11.4.1   Spring alloys

The heat-treatable high cobalt alloy 'Elgiloy' or 'Cobenium' (see Table 2.11.4), which was originally developed as a corrosion resistant, watch mainspring material, has extremely high strength at normal and elevated temperatures. (An alternative material for this purpose is a cold worked austenitic steel Sandwik 11R70Hv). Elgiloy is completely non-magnetic and cannot be magnetised. It responds to heat treatment from the solution heated condition only after cold work, as is shown in:

Fig 2.11.8 *Effects of cold reduction on strength and hardness of Elgiloy heat treated for 5 h at 586°C*

The composition of Elgiloy (together with that of Stainless Invar and the alloys used for furnace hardwear) is listed in Table 2.11.4. Besides its use for constant rate springs, Elgiloy has been used for instrument pivots and for ball bearings.

When the cobalt based spring alloys were developed they had the best resistance of any material to relaxation at high temperatures. For example, Elgiloy stressed to 345 MPa relaxes only 18% in 240 h at 450°C. However, the more recently developed nickel alloy, Nimonic 90 only relaxes 6% of this stress in 1008 h at the same temperature while Inconel X 750 is superior to both at 550°C.

The superalloy L605 also has good high temperature spring properties. It may also be used for diaphragms, bearings and rupture discs.

An alloy of cobalt with chromium and iron (Stainless Invar), see Table 2.11.4, combines the property of high resistance to corrosion with a zero coefficient of expansion between –60 and 20°C.

There are a number of cobalt base alloys with very low temperature coefficients of elasticity modulus. These include Co-Elinvar with compositions varying between 51.5 and 60% Co, 25 and 38.5% Fe and 8 and 15% Cr, and Velinvar 56–63% Co, 29–34.5% Fe and 7–10.2% V. These alloys have important applications for instrument springs and are listed in:

Table 2.11.7 *Composition and thermal properties of some constant modulus alloys containing cobalt*

2.11.4.2   Dental and prosthetic alloys

A number of alloys of approximate composition 65% Co, 30% Cr, 5% Mo are used in dentistry and as prosthetic devices for bone surgery. The dental alloys are fabricated by precision casting, and are available in a range of compositions with chromium varying between 26 and 30% and carbon varying between 0.2 and 0.5% to allow for melting in an induction furnace or an oxyacetylene flame and to provide variations in wear resistance. They are superior to gold for dental and prosthetic applications because of their lower density and higher strength. Their use is, however, decreasing in comparison with that of stainless steel:

Table 2.11.8 *Cobalt alloys for surgical implant applications*

Evidently, the precise composition is not critical provided the mechanical properties are adequate and no element (e.g. beryllium) is present which is incompatible with the human body or which promotes corrosion. When conducting bone surgery, it is important to use only tools made of the same alloy as that used for the implant to avoid electrolytic effects.

TABLE 2.11.7 Composition and thermal properties of some constant modulus alloys containing cobalt

Alloy composition (%)[a]				Thermal coefficient			Moduli	
Co	Fe	Cr	V	Expansion (10–50°C) (10^{-6} K^{-1})	Elasticity (0–50°C) (10^{-4} MPa/K)	Rigidity (20–50°C) (10^{-4} MPa/K)	Elasticity at 20°C (GPa)	Rigidity at 20°C (GPa)
Co-Elinvar[b]								
57	35	8	—	3.1	+ 1.7	+ 17.9	160	72
60	32	8	—	6.6	− 12.4	− 2.5	—	65
51.5	38.5	10	—	8.7	− 1.0	− 31.5	192	77
57.5	32.5	10	—	3.5	+ 1.2	+ 10.1	167	69
60	30	10	—	5.1	− 15.5	− 0.2	—	70
58.5	29.5	12	—	6.0	+ 4.4	+ 1.0	—	77
60	28	12	—	7.6	− 2.4	− 4.3	—	74
60	25	15	—	14.0	− 19.6	− 0.9	—	81
Velinvar								
63	29.2	—	7.0	8.52	—	− 6.7	—	54
61	30.4	—	7.9	7.77	—	+ 0.5	—	57
63	28.4	—	7.9	8.56	—	− 1.6	—	67
62	28.5	—	8.7	8.74	—	− 4.3	—	64
60	29	—	10.2	8.07	—	− 0.03	—	66
56	34.5	—	8.7	3.94	—	− 31.5	—	63

[a] 0.5% Mn and 0.1% Al added for deoxidation.
[b] 0.18–0.19% Ni, 0.46–0.68 Si, 0.07 C, traces of Mn, Al, P as impurities.

Table 2.11.7

TABLE 2.11.8 Cobalt alloys for surgical implant applications

Specification	Manufacture	Chemical analysis (max. unless stated) (%)											Mechanical properties		
		Cr	Mo	Ni	Fe	C	Si	Mn	Al	Ti	W	Co	UTS (MPa)	YS (MPa)	Elong. (%)
ANSI/ASTM F75–76	Cast	27–30	5–7	2.5	0.75	0.35	1	1	0.14	0.14	—	Bal	655	450	8
BS 3531 Pt 2 1980	Cast	26.5–30	4.5–7	2.5	0.75	0.35	1	1	0.14	0.14	—	Bal	700	500	8
BS 3531 Pt 2 1980	Wrought	19–21	—	9–11	3	0.05–0.15	1	1	—	—	14–16	Bal	860	310	30
ANSI/ASTM F90–76	Wrought	19–21	—	9–11	3	0.05–0.15	1	2	—	—	14–16	Bal	860	310	10
ANSI/ASTM F563–78	Wrought	18–22	3–4	15–25	4–6	0.05	0.5	—	—	0.05–3.5	3–4	Bal	600–1586	276–1310	50
ANSI/ASTM F562–78	Wrought	19–21	9–10.5	33–37	1	0.025	0.15	0.15	—	1	—	Bal	795–1790	240–1585	50–8

Table 2.11.8

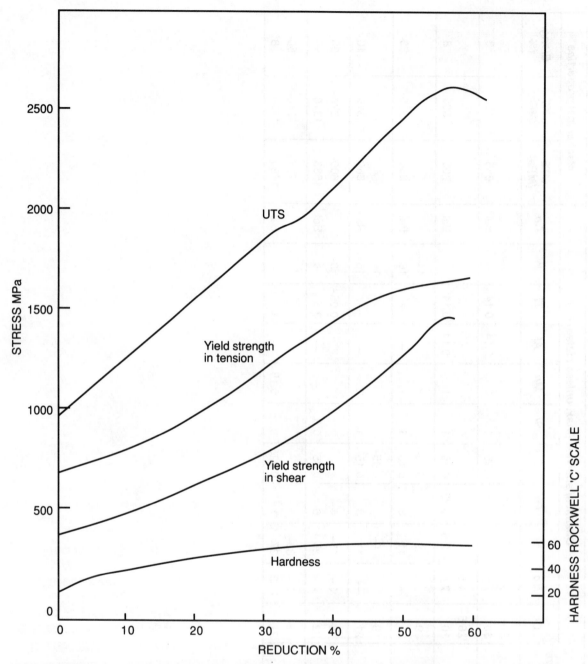

FIG 2.11.8 Effect of cold reduction on strength and hardness of Elgiloy heat treated for 5 h at 586°C

Fig 2.11.8

Precious metals

Contents

List of tables

2.12.1   General characteristics

2.12.1.1   Introduction

The precious metals are characterised by their resistance to corrosion and oxidation under normal conditions. Historically their precious nature was due to scarcity and their lasting qualities and thus the main applications were as currency standards.

Precious metals are frequently specified for applications which demand utmost reliability or complete freedom from corrosion, and for applications where their use is economic in relation to their high recovery value and the long, trouble-free service they provide.

2.12.1.2   Characteristics of precious metals

With the exception of gold, all the precious metals are silvery-white in appearance. The platinum group metals (platinum, palladium, iridium, rhodium, osmium, ruthenium), unlike gold and silver, have *high melting points*, are comparable in *strength at elevated temperatures* with the refractory metals (Table 2.12.1) and have excellent dimensional stability at high temperatures. See:

Table 2.12.1   *Tensile strengths at high temperatures of iridium and some refractory metals in the presence of carbon vapour*

In addition they can be used in *oxidising atmospheres* which would rapidly destroy high melting point base metals such as tungsten and molybdenum.

More information is given in Table 2.12.2 on the physical and mechanical properties of precious metals. Table 2.12.3 compares the advantages and limitations of precious metals and lists some typical applications.

Table 2.12.2   *Physical and mechanical properties of the precious metals*

Table 2.12.3   *Characteristics of precious metals*

TABLE 2.12.1 Tensile strengths at high temperatures of iridium and some refractory metals in the presence of carbon vapour

Metal	Tensile strength (MPa)					
	1000°C	1200°C	1400°C	1600°C	1800°C	2000°C
Iridium	313	186	92	68	49	39
Niobium	122	59	25	15	10	10
Tantalum	268	166	83	49	45	39
Molybdenum	259	122	79	49	34	20
Tungsten	322	249	196	157	108	68

Table 2.12.1

TABLE 2.12.2 Physical and mechanical properties of the precious metals

Property		Units	Platinum	Iridium	Osmium	Palladium	Rhodium	Ruthenium	Gold	Silver
Atomic number		—	78	77	76	46	45	44	79	47
Atomic weight		—	195.09	192.2	190.2	106.4	102.905	101.07	196.967	107.870
Tensile strength	Annealed	MPa	123	1100	—	150	700	—	100	140
	Hardened	MPa	250	—	—	300	—	—	220	160
Elongation	Annealed	%	—	—	—	45	—	—	45	50
	Hardened	%	—	—	—	4	—	—	4	4
Hardness	Annealed	HV	40	220	350	40	100	240	20	25
	Hardened	HV	92	—	—	100	—	—	75	100
Annealing temperature		°C	750	1400	—	800–900	1400	—	—	—
Modulus of elasticity in Tension		GN/m²	172	515	558	117	317	413	71	71
Specific density		—	21.45	22.65	22.61	12.02	12.41	12.45	19.3	10.5
Melting point		°C	1769	2443	3050	1552	1960	2310	1063	960.8
Thermal conductivity		W/m per K	305	619	364	317	627	439	293	418
Thermal expansion coefficient[a]		10^{-6} K^{-1}	9.1	6.8	6.1	11.1	8.3	9.1	14	19
Lattice structure		—	FCC	FCC	CPH	FCC	FCC	CPH	FCC	FCC
Resistivity at 0°C		$\mu\,\Omega$ cm	9.85	4.71	8.12	9.93	4.33	6.71	2.4	1.6
Temperature resistance coefficient[b]		K^{-1}	0.0039	0.0043	0.0042	0.0038	0.0046	0.0042	0.0034	0.0041
Thermal emf vs. Pt at 1000°C		mV	—	12.73[a]	—	-1.505	14.05[a]	—	—	—
Thermal neutron cross-section		Barns	9	440	15	8	156	2.6	—	—
Specific heat at 0°C		J/kg per K	132	129	129	244	247	230	—	—
Mass susceptibility (χ) (magnetic)		cm³/g x 10^{-6}	0.9712[a]	0.133[a]	0.052[a]	5.231[a]	0.9903[a]	0.427[a]	—	—
Thermionic function (A)		A/cm² per K²	64	170	—	60	100	—	—	—
Work function (∅)		V	5.27	5.40	5.40	—	4.99	4.90	—	—

[a] (0–100°C)   [b] (20–100°C)

Table 2.12.2

TABLE 2.12.3 Characteristics of precious metals

Metal	Advantages	Limitations	Applications
Silver	Highest electrical and thermal conductivity of all metals. Stable in alkaline solutions. Special grades of oxygen-free silver produced for premium applications. Soft and ductile, easily worked.	Soluble in oxidising acids, but not hydrochloric or aqua regia, because of protective film formation. Tarnishes in the presence of oxidising agents, or atmospheric sulphur. Rapidly attacked by cyanide solutions. Readily work hardened.	Electrical contacts (low current). Corrosion-resistant containers. Ornaments and jewellery. Manufacture of silver nitrate.
Gold	Very soft and ductile, very little work hardening. Completely free from oxide film formation. Pressure weldable. Unattacked by acids except certain oxidising and complexing mixtures such as aqua regia. Unattacked by alkaline solutions except in the presence of cyanide ions. Third highest electrical conductivity. Only coloured metal, other than copper.	Too soft and dense to be of practical industrial or decorative use except as lining or electrodeposit. . . . . . copper and silver higher.	Jewellery and coinage (often alloyed). Gold leaf—very thin fabricated sheet. Very fine wire. Electrical contacts. Instruments requiring corrosion resistance.
Platinum	Malleable, ductile, does not work-harden rapidly. Catalytic action. Resistant to aqueous media	Should not be heated in contact with refractory materials in reducing atmospheres (embrittlement). More expensive than silver. Low electrical conductivity. .except in the presence of complexing agents, (readily dissolved by aqua regia). Attacked at red heat by molten cyanides, sulphides, phosphides. Superficially attacked by fused alkali metal hydroxides, oxides, peroxides and nitrates.	Electrical contacts. Furnace windings. Laboratory crucibles and containers. Catalyst. Anodes. Standard resistance thermometers. Thermocouples. Jewellery. Apparatus for the manufacture of high quality optical glasses. Linings for fluoride process plant.
Palladium	Very similar to platinum. Passivating film protects against attack by nitric acid and sodium nitrate. Unaffected by alkaline solutions.	Oxidises in air below 600°C. Less corrosion resistant than platinum. Soluble in oxidising acids or acids containing complexing agents. Soluble in potassium hydrogen sulphate. Superficially attacked by molten alkalis.	Catalyst. Diffusion barrier for H_2 manufacture (see also platinum–silver alloy), high duty electrical contacts.
Iridium	Can be worked at elevated temperatures, or drawn into wire. Excellent high temperature characteristics. With rhodium, one of the most corrosion-resistant metals known. Insoluble in aqua regia.	Hard and difficult to fabricate. Forms volatile oxide above 900°C. Attacked by fused mixtures of potassium hydroxide and potassium nitrate.	Apparatus for the manufacture of high quality optical glasses.
Rhodium	Similar to iridium. Less volatile oxide than iridium, allowing its use at high temperatures under oxidising conditions. Most stable of metals because of oxide film (although lower electrode potential than gold). Extremely hard (800 HV) tarnish-free electroplate.	Dissolved, in finely divided form, by warm concentrated sulphuric acid or aqua regia. Superficially attacked by fused alkali metal hydroxides. Slowly dissolves in potassium hydrogen sulphate.	Alloying constituent, particularly with platinum.
Ruthenium	Stable in solutions of non-complexing acids. Low thermal neutron absorption cross-section.	Hard, brittle and difficult to fabricate, even at high temperatures. Highly volatile and toxic tetroxide formed when heated in air. Readily attacked by oxidising alkaline solutions (e.g. peroxides, hypochlorites).	Rarely used in wrought form. Useful as alloying additions to platinum for use under irradiation conditions.
Osmium	Highest melting point of all precious metals (3050°C)	Virtually unworkable. Highly volatile and poisonous tetroxide formed when heated in air.	Alloying addition. Osmic acid is used for staining biological specimens.

Table 2.12.3

2.12.2   Precious metal alloys

Although the precious metals are highly resistant to corrosion and have excellent high temperature characteristics, they are generally too soft to be used as constructional materials for many applications. Much harder materials with greater resistance to creep can be produced by alloying one precious metal with another without significantly affecting their chemical resistance. The advantages of lubrication of precious metal alloys are given in Table 2.12.4 together with typical applications. The variation in creep resistance of platinum and platinum alloys at various temperatures is shown in Table 2.12.5.

Table 2.12.4   *Characteristics of precious metal alloys*

Table 2.12.5   *Creep resistance of platinum and platinum alloys at temperatures up to 1250°C*

2.12.2.1   Available shapes and forms

A guide to the ductility of precious metals and alloys is provided by a comparison of the minimum sizes of wire that can be produced. Gold, silver, platinum and palladium can all be drawn down to 0.02 mm diameter wire, and in certain cases to below 0.01 mm diameter. Rhodium and iridium can be drawn down to 0.5 mm diameter wire, but osmium and ruthenium cannot be drawn to wire at all. However, rhodium and iridium can be rolled into sheets of 0.005 mm minimum thickness.

Nearly all the common precious metal alloys can be drawn down into wire 0.02 mm diameter among the exceptions being 40% rhodium–platinum alloy and 30% iridium–platinum alloy for which the minimum size is 0.5 mm diameter. See:

Table 2.12.6   *Typical physical properties of precious metal resistance wires*

2.12.2.2   Fabrication

All the noble metals, with the exception of osmium and ruthenium, have a face-centred cubic structure and, can therefore be expected to be workable. Gold, silver, platinum and palladium are easy to fabricate by conventional methods, being soft, malleable and ductile. Iridium and rhodium, however, can only be worked at high temperatures and careful temperature control is necessary. The reason for this has not been satisfactorily explained, but is more probably related to the rapid work hardening of the metals than to the presence of trace impurities. Osmium and ruthenium have a hexagonal close-packed structure and are practically unworkable.

Alloys of platinum and palladium with iridium, rhodium, osmium and ruthenium are harder and frequently more resistant to corrosion than pure platinum and palladium, and most of the alloys are workable by conventional methods.

All the precious metals except osmium can be cast and hot forged although the working temperatures depend upon the individual melting points which vary widely. Platinum, palladium, gold and silver can be cold worked by all the common methods such as rolling, spinning, swaging, drawing, stamping and deep drawing—for example, platinum wire can be drawn from 5.0 to 0.08 mm without annealing, a reduction of more than 99%.

All the precious metals can be fusion welded by arc or gas welding techniques, though silver, iridium and rhodium should be protected from air by a shield of argon. These metals should preferably be argon arc welded. Oxy-hydrogen torch welding can also be used and is preferable to acetylene welding because it is cleaner and therefore less likely to contaminate the metal. Because of the absence of oxide films on their surfaces, gold, platinum and palladium can be hammer welded at temperatures well below their melting points.

The more malleable precious metals and their alloys can be used economically as claddings for base metals to give greater strength or improved electrical conductivity.

The bimetals used in contact applications and the rolled golds used in jewellery are made by cladding with precious metal certain areas of a machined base metal ingot, and rolling the ingot to appropriate sizes of sheet or strip. Linings for the protection of chemical plant are made by shaping sheet metal to fit over the surface to be clad: linings in silver can be bonded to the base metal by soldering, while linings in other precious metals and alloys are made as tight-fitting sheaths whose edges are brazed to the base metal. Electrodeposition provides a useful means of producing thin, uniform precious metal coatings on articles of complex shape. Thin coatings of precious metals can also be produced by vacuum deposition.

TABLE 2.12.4 Characteristics of precious metal alloys

Alloy type	Description	Advantages	Limitations	Applications
Platinum alloys	Rhodium (up to 40%)–platinum	Only really ductile materials capable of taking tensile stresses in oxidising atmospheres above 1250°C. Volatilisation of pure platinum in oxygen at high temperatures reduced by rhodium additions.	More difficult to work as rhodium content increases.	Most widely employed of all precious metal alloys; cladding for refractory components of glass melting furnaces (10% alloy). Linings for autoclaves and pressure vessels. Furnace windings. Spinnerettes. Catalysts.
	Iridium (up to 30%)–platinum	Harder than rhodium–platinum alloys. Inert at room temperature.	30% iridium is the limit of easy workability. Restricted to non-oxidising atmospheres.	Standards of length and weight. Linings for autoclaves and pressure vessels.
	Ruthenium–platinum	Small additions of ruthenium greatly increase creep resistance. Ruthenium is a very effective hardener of platinum.	Not suitable for use in air at high temperatures.	
	Gold–platinum		Difficult to work.	Largely replaced by other platinum alloys. High melting point platinum solders. 70% gold alloy used in manufacture of spinnerettes.
	Copper–platinum			Jewellery and electrical contacts. Iridium–platinum generally preferred.
	Cobalt (23%)–platinum	Workable; can be shaped before heat treatment.		After special heat treatment, are one of the most powerful permanent magnet materials known.
Palladium alloys	Silver–palladium	Alloys of 30 to 40% silver have maximum hardness and tensile strength, coupled with the optimum electrical resistivity and temperature coefficient. 23% silver alloy is dimensionally stable when thermally cycled and is more permeable to hydrogen than pure palladium.		20% and 40% silver alloys widely used in electrical contact applications. 23% silver alloy is used as the diffusion membrane in hydrogen diffusion cells.
	Gold–palladium	Very ductile, easily worked alloys.		Dentistry. Cheap alternatives to platinum for chemical ware and thermocouples.
Other gold alloys		Great range of properties and colours available.		High malleability—deep medal stamping. High hardness—springs in brooches or clips. Brazing materials.
Other silver alloys	Copper–silver	Harder than fine silver.		Jewellery. 7½% copper alloy (sterling silver) and 10% alloy widely used.
	Gold–silver			Green carat gold.
	Silver solders			Brazing alloys.

Table 2.12.4

TABLE 2.12.5 Creep resistance of platinum and platinum alloys at temperatures up to 1250°C (MN/m^2)

Temp (°C)	Pure Pt		4% Pd–Pt		5% Rh–Pt		10% Rh–Pt	
	f_0	f_{100}	f_0	f_{100}	f_0	f_{100}	f_0	f_{100}
20	131.9	121.1	217.1	200.5	219.9	200.5	280.2	258.7
300	98.5	79.1	175.0	156.6	166.0	147.5	254.1	200.2
500	74.8	44.9	142.7	97.7	133.9	106.5	178.6	144.6
700	63.2	22.4	104.7	50.8	100.7	48.9	146.9	72.3
900	32.8	11.7	67.2	18.3	69.3	16.6	104.7	26.7
1100	17.5	4.9	23.1	4.9	35.1	7.5	58.2	11.7
1250	13.7	3.8	14.7	2.7	29.2	5.5	—	—
Temp (°C)	5% Ir–Pt		10% Ir–Pt		5% Au–Pt		4% Ru–Pt	
	f_0	f_{100}	f_0	f_{100}	f_0	f_{100}	f_0	f_{100}
20	238.7	213.2	352.0	327.2	344.9	312.6	431.2	391.1
300	214.0	163.2	264.1	249.4	301.8	—	301.0	—
500	155.5	87.7	210.9	142.6	293.3	125.5	287.9	220.2
700	103.6	43.1	149.6	57.7	204.8	55.7	194.0	102.5
900	65.4	12.4	87.0	19.5	114.2	21.5	123.9	32.3
1100	31.7	6.7	49.2	9.3	45.8	6.3	55.2	9.2
1250	20.9	4.6	36.6	4.9	28.3	3.6	—	—

f_0 = stress required to cause instantaneous failure.
f_{100} = stress required to cause failure in 100 h.

Table 2.12.5

TABLE 2.12.6 Typical physical properties of precious metal resistance wires

Alloy	Specific resistance ($\mu\Omega$ cm)	Temperature coefficient of resistance (0–100°C) per °C	Thermal emf at 100°C (mV) Against copper	Tensile strength		Minimum diameter (mm)
				Annealed (MN/m^2)	Hard drawn (MN/m^2)	
20% palladium–silver[a]	10.5	0.0005	−0.88	28	60	0.025
10% rhodium–platinum	19	0.0017	−0.10	47	117	0.013
10% iridium-platinum	24.5	0.0013	+0.55	55	125	0.013
5% ruthenium– 15% rhodium–platinum	31	0.0007	+0.03	101	173	0.013
20% iridium–platinum	32	0.00085	+0.61	70	165	0.013
40% silver–palladium	42	0.00003	−4.20	38	110	0.02[b]
10% ruthenium– platinum	42	0.00047	+0.14	79	140	0.013
8% tungsten–platinum	62	0.00028	+0.71	94	150	0.02[b]
5% molybdenum– platinum	64	0.00024	+0.77	94	140	0.02[b]
20% copper–platinum	82.5	0.000098	−0.67	60	140	0.02[b]
5% molybdenum– 40% palladium–gold	100	0.00012	−0.19	69	110	0.02[b]

[a] This alloy is not completely free from tarnish.
[b] Bare wire down to 0.0005 in diameter (0.013 mm) is supplied in limited lengths.

Table 2.12.6

Refractory metals and alloys

Contents

List of tables

List of figures

2.13.1   Introduction

The refractory metals are characterised by their high melting points, (MP), exceeding an arbitrary value of 2000°C, and low vapour pressures, two properties exploited by the electronics industry.

Only four of the refractory metals—molybdenum, MP 2620°C, niobium, MP 2468°C, tantalum, MP 2996°C, and tungsten, MP 3380°C—are available in quantities of industrial significance. Metals such as iridium, osmium and ruthenium, which also melt above 2000°C but which are chemically inert, are classified as precious metals (Vol. 2, Chapter 12).

Molybdenum, niobium, tantalum and tungsten have been produced commercially for many years, mainly as additives to steels, nickel and cobalt alloys and for certain electrical applications. However, with the advent of space exploration, nuclear engineering and the increasing temperature requirements of aircraft power units there has been a search for materials having superior high temperature properties to the conventional iron, nickel and cobalt base superalloys (see Vol. 2, Chapters 1, 7 and 11).

In addition to high temperature strength, the relatively low thermal expansions and high thermal conductivities of the refractory metals suggest good resistance to thermal shock. There are however two characteristics, ready oxidation at high temperatures and, in the case of molybdenum and tungsten, brittleness at low temperatures which limit their applications. Coatings have, however, been developed which confer limited protection against oxidation. The advantages, limitations and applications of refractory metals are listed in:

Table 2.13.1   *Characteristics of refractory metals*

TABLE 2.13.1 Characteristics of refractory metals

	Metal	Advantages	Limitations	Applications
Soft and ductile	Tantalum (structure BCC)	Anodic film has better dielectric properties than Al. Very low ductile–brittle transition temperature. Very versatile aqueous corrosion resistance; inert to HCl, HNO$_3$, resistant to aqua regia, perchloric and chromic acids, oxides of nitrogen, chlorine and bromine, organic acids, H$_2$O$_2$ and chlorides.	Combines with most gases above 500°C. Susceptible to H$_2$ embrittlement. Attacked by nascent hydrogen and F$_2$, HF, SO$_3$ and alkalis above 5% conc; 98% H$_2$SO$_4$ above 170°C, H$_3$PO$_4$ above 190°C. Even so, attack is uniform (no pitting).	Electrodes in thermionic valves. Capacitors. Corrosion resistant linings in chemical industry. Surgical implants. Corrosion resistant fabrications include thermocouple pockets, bayonet heaters, tube and sheet heat exchangers, etc.
	Niobium (structure BCC)	Fair oxidation resistance. Good corrosion resistance. Forms superconducting alloy with tin. Low neutron capture cross-section. Very low ductile–brittle transition temperature	Prone to H$_2$ embrittlement. Marginally less corrosion resistant than tantalum, but economics may lead to increased use at the latter's expense. .except when highly alloyed.	Gas turbine and rocket motors — high temperature parts. Linings and claddings (as tantalum). Has been used for cans for reactor fuel.
Strong, less malleable	Tungsten (structure BCC)	Good oxidation resistance below 500°C. Good corrosion resistance. Low thermal expansion. Worked forms (wire, sheet) ductile to low temperatures. As resistant as tantalum to mineral acids. High density. High Young's Modulus.	Ductile–brittle transition temperature may be 200°C for high strain rates (even as high as 1000°C (increasing amounts of O, C and N are bad)	Electric filaments. Welding electrodes. High temperature parts for rocket engines. Tungsten carbide manufacturer. Heavy duty electrical contacts. Glass-to-metal-seals. Radiation shield.
	Molybdenum (structure BCC)	High Young's Modulus. Worked forms (wire, sheet) ductile to low temperatures. Resistant to mineral acids . . . . Resistant to boiling HF.	Very low oxidation resistance above 450°C. Ductile–brittle transition temperature may be 200°C (increasing O, C, N are bad). unless oxidising agents are present.	High temperature furnace parts (must be protected from oxidation by atmosphere or coating), especially windings. Electrodes in glass melting furnaces. Metallising. Usage in components requiring rigidity, such as boring bars, tool shanks and grinding quills (improved work piece quality). Mould and corepin materials in diecasting (also TZM Mo alloy). Aerospace structural parts including leading edges and support vanes.

Table 2.13.1

2.13.2   Characteristics of refractory metals and alloys

2.13.2.1   Physical and mechanical properties

The physical and mechanical properties of the refractory metals are listed in:
Table 2.13.2   *Typical properties of refractory metals*
The elevated temperature properties of the refractory metals are shown in:
Fig 2.13.1   *Effect of temperature on the ductility of three refractory metals compared with that of nickel*, and
Fig 2.13.2   *Comparison of the elevated temperature strengths of tungsten, molybdenum, molybdenum alloys, tantalum,niobium and Nimonic 95 (tungsten, molybdenum and alloys in the stress relieved condition, tantalum and niobium in the annealed condition and Nimonic 95 in the fully heat-treated condition)*

2.13.2.2   High temperature gas reactions

While all the metals are relatively inactive at room temperature, they are very reactive at elevated temperatures. The oxide films formed by heating in air are protective at temperatures up to 500–600°C but are non-protective above those temperatures. The scale formed on tantalum and niobium is non-adherent and voluminous, and, if the oxidised material is heated above 1000°C, oxygen will diffuse into the body of the material and embrittle it. The oxide formed on tungsten is also non-adherent and at temperatures above 1000°C is volatile to an increasingly significant extent. The trioxide of molybdenum is volatile at much lower temperatures and the rate of oxidation at temperatures above 1000°C is quite disastrous—the metal almost literally 'going up in smoke'. Molybdenum and tungsten have low solubility for oxygen and consequently are not embrittled in the same way as tantalum and niobium when heated in oxygen-containing atmospheres.

The metals also react with, and are embrittled by, nitrogen-and carbon-containing atmospheres, such as town gas. These gases can, however, be used for protection of molybdenum and tungsten at temperatures up to about 1100°C, but not for niobium and tantalum.

Hydrogen is dissolved by tantalum and niobium at temperatures below 600°C but at higher temperatures this is driven off and little remains above 800°C. The embrittling effect of hydrogen in these two metals is strong and is used to enable process scrap to be ground before recovery. Molybdenum and tungsten are inert to hydrogen and the gas is commonly used as a protective atmosphere for these metals and their alloys. Acceptable atmospheres for all these metals, are the inert gases and vacuum.

2.13.2.3   Protective coatings

The inability of the refractory metals to withstand oxidising conditions, even at moderate temperatures, has led to the investigation of methods of reducing their oxidation rates. Alloying has been examined and in the main rejected since attainment of good oxidation resistance is at the expense of high temperature strength. Protective coatings seem to be the answer where oxidation is a problem.

A number of coatings have been developed for molybdenum. One of the most successful, designated W3 and based on siliconising, has been used for long periods of, say, 500 h at 1200°C, and even at 1700°C for short-term, 1-h applications. The coating, which is only about 0.002 in thick, is applied by a pack diffusion technique similar to chromising and coating procedures have been evolved for the protection of complex assemblies.

Coatings for tungsten have not been developed to the same extent as those for molybdenum. Siliconising can give protection for many hours at temperatures up to 1650°C and the maximum protective temperature for silicide coatings has been reported as 2000°C. Life at temperatures between 1650°C and 2000°C seems to be only a few hours.

The immediate potential of niobium for replacement material in gas turbine engines has given impetus to the development of a number of coating systems. Protective life varies, but up to 1000 h protection can be obtained at 1100°C and mostly over 100 h at 1250°C, but these values could well be altered when tested under the more stringent service conditions. The most successful coatings are based on the silicide process modified by the addition of other elements and in some cases the application of an overlay glaze coating to seal micro-cracks and provide protection at intermediate as well as high temperatures.

Coatings for tantalum have followed the patterns set down for niobium and the most advanced types of protective coatings are based on aluminides, beryllides and silicides.

2.13.2.4   Fabrication

SHEET METAL WORK

Tantalum and niobium may be worked at room temperature but it has been found advantageous to heat their high strength alloys. Both these metals may be readily formed by spinning, blanking, deep-drawing and pressing. Blanking sheet is straightforward, but steel dies and punches must be well lubricated to prevent galling. It is often advisable to use annealed sheet, anodised to aid lubrication, for deep-drawing and pressing operations and 'Narite' (a cast copper aluminium alloy) tools are preferred to steel in order to prevent galling.

The forming of molybdenum and tungsten is dominated by the ductile—brittle transition temperature. Despite the low transition temperature of molybdenum sheet, in practice most operations are carried out warm; it often being necessary to heat not only the workpiece but also the die. Notwithstanding this requirement, most forming operations are readily conducted on molybdenum including deep-drawing, spinning, flow turning, blanking and pressing. The forming of tungsten follows the same rules, but somewhat higher temperatures are needed.

MACHINING

Tantalum and niobium react to machining operations in a similar manner to stainless steels in that they tend to gall or seize to the tool. Satisfactory techniques for machining tungsten and molybdenum are well developed. Tungsten is the more difficult of the two as it is relatively hard (450–500 VPN for wrought material and about 380 VPN for annealed stock) but is only slightly more difficult to machine than the high strength superalloys based on nickel and cobalt. Molybdenum is a good deal easier to machine and may be likened to cast iron. Its hardness lies in the range 200–300 VPN.

JOINING

Because of the reactivity of all four metals at elevated temperature, welding procedures are restricted to those which prevent the weld metal or hot parent sheet from coming into contact with any substance which will embrittle them, particularly gases in the surrounding atmosphere. The practical methods are TIG and resistance welding and electron beam welding will probably become a third.

The TIG welding of tantalum and niobium is best conducted in a glove box, evacuated and back-filled with argon, using a normal shielded torch with a light flow of argon. With care welding can be conducted outside a glove box—high flow of argon must be used and the back of the joint must also be covered with argon. For thin sheet, resistance welding is preferred. Electron beam welding of these two metals clearly provides the ultimate in protection and has the advantages of very narrow weld and heat affected zones.

Owing to the brittleness of the as-cast structures the fusion welding of tungsten and molybdenum has not yet met with complete success and no one technique can be said to have given a consistently sound joint. The most favoured method of welding is the inert gas shielded arc and the operation should be preferably conducted in an argon-filled box. Molybdenum and in particular the powder metallurgical product suffers from hot shortness caused by oxygen and centre line cracking is often observed. This may be reduced by using an Mo–0.5 Ti alloy filler rod, the two major alloys are therefore somewhat easier to weld and strong joints are possible in these materials. Tungsten does not suffer from hot shortness and fusion welds are correspondingly easier to produce—even so, the welds are extremely fragile.

Again the electron beam process is well suited and welds ductile at room temperature can be made in molybdenum and molybdenum alloy sheet. Resistance welding of molybdenum and tungsten has been used for many years in the electronics industry but the major problems of electrode sticking, nugget porosity, weld cracking and brittleness are present when welding large-scale components.

All four metals can be brazed using the low temperature brazes based on copper and silver, but, whereas the embrittling recrystallisation is avoided in the case of tungsten and molybdenum, the high temperature strength of the bond is low. Brazes for use at temperatures above 1000°C have been developed.

Because of the difficulties involved in producing sound and ductile welds in molybdenum and tungsten, and of the limitation set by the physical and chemical properties of brazed joints, most commercial fabrication has been carried out with the aid of mechanical joints.

2.13.2.5   Refractory metal alloys

Tantalum and niobium alloys are undergoing development as potential gas turbine blade materials. The Nb–Zr, Nb–Ti and Nb_3 Sn alloys are used widely for their superconducting properties, whilst $W–ThO_2$ is used for incandescent filaments. The compositions and applications of the refractory metal alloys are listed in:

Table 2.13.3   *Commercially available alloys of refractory metals*

TABLE 2.13.2 Typical properties of refractory metals

Property	Units	Tantalum	Niobium	Tungsten	Molybdenum
Melting point	°C	2996	2468	3380	2620
Density	kg/dm^3	16.6	8.57	19.3	10.2
Softening temperature	°C	1500	1000	1200	900
Ultimate tensile strength (sheet)	MN/m^2	453(R) 755(C) 770(S) 1694(D)	462(S)	3126(T)	1540(T)
Elongation at break (sheet)	%	36(R) 40(S) 1(D)	33(S)	0–3(T)	15(T)
Yield stress (sheet)	MN/m^{-2}	405(R) 755(C)	—	3080(T)	1386(T)
Young's Modulus	GN/m^{-2}	186	110	411	324
Linear expansion coefficient[a]	10^{-6} K^{-1}	6.6	7.1	4.43	5.35
Specific heat capacity	J/kg per K	142	272	138	242
Thermal conductivity	W/m per K	54	52	130	145
Electrical conductivity	& IACS	13.9	13.3	31	36
Thermal neutron cross-section	Barns	21.3	1.1	19.2	2.5

[a]At 20°C

R — High purity sheet from EBM stock, recrystallised.
C — as R, but cold worked.
S — sheet from sintered stock, recrystallised.
D — as S, but cold worked.
T — stress relieved.

Table 2.13.2

TABLE 2.13.3 Commercially available alloys of refractory metals

Alloy base	Designation	Nominal composition	Description
Tantalum	Ta–10W	Ta–10%W	Alloying with tungsten increases high temperature strength whilst retaining workability and weldability. Ta–10W is possibly the most popular alloy but T111 and T222 are exciting much interest. S.G.S. was developed to combat embrittlement brought about by excessive grain growth at temperatures exceeding 2000°C. The technique has also been successfully applied to Ta–10W.
	T111	Ta–8%W, 2%Hf	
	T222	Ta–10%W, 2.5%Hf	
	S.G.S. Ta	Ta with extremely small additions of another element to give grain stabilisation.	
Niobium	AS30	Nb–20%W, 1%Zr, 0.1%C	Nb–1Zr alloy was proposed for nuclear canning applications requiring higher strength than pure Nb. Nb–25Zr and Nb–40Ti have been developed for their superconducting properties at low temperature, these alloys have higher critical temperatures than pure niobium (Nb_3Sn intermetallic has even better super-conducting properties, but fabrication is difficult, although techniques have been evolved for the production supported on a ductile metal base of wire and tape). The remaining alloys were developed for their high temperature properties. The highly alloyed, high strength alloys such as F48 and AS30 tend to be more difficult to fabricate and weld than the lower strength alloys. In some cases the lower strength alloys have improved oxidation resistance. On occasions, Cr, Al and Si are used to increase oxidation resistance. Ta and Ti promote solid solution strengthening whilst Zr promotes carbide dispersion strengthening.
	F48	Nb–15%W, 5%Mo, 1%Zr, 0.06%C	
	SV16	Nb–11%W, 3%Mo, 2%Hf, 0.08%C	
	D31	Nb–10%Mo, 10%Ti, 0.1%C	
	Nb–Zr	Nb–1%Zr	
	Nb–Ti	Nb–25%Zr	
		Nb–40%Ti	
Tungsten	W–ThO$_2$	W–1% ThO$_2$	Doped tungsten and thoriated tungsten are both deriva-tives of the electrical and electronics industry. Suitable controlled addition of thorium oxide enables tungsten to be used for incandescent electric filaments, but low temperature properties are only marginally better than those of unalloyed tungsten. The tungsten–rhenium alloys have better room temperature properties than unalloyed tungsten, without impairing high temperature strength, and the 26% Re alloy remains ductile to quite low temperatures. The rarity and high cost of Re tends to restrict the use of these alloys to specialised applications.
		W–2% ThO$_2$	
	Doped W	W containing very small amounts of alkali alumina silicates.	
	W–Re	W–3%Re	
		W–5% Re	
		W–26% Re	
Molybdenum	Mo–0.5Ti	Mo–0.5%Ti, 0.02%C	The two major alloys, Mo–0.5Ti and TZM, have improved high temperature properties, brought about by a precipita-tion hardening (carbide dispersion) mechanism.The precipitation mechanism in TZM may be affected by 'in process' thermal and mechanical treatments to give a wider range of properties. While the room temperature ductility of the wrought products is unlikely to be better than that of unalloyed Mo, they are less prone to embrittlement following exposure to high temperatures and consequently have improved weldability. Molybdenum may also be strengthened by W addition to form a solid solution.
	M.T.C.	As above	
	TZM	Mo–0.5%Ti, 0.1%Zr, 0.02%C	

Table 2.13.3

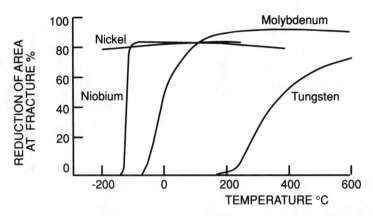

FIG 2.13.1 Effect of temperature on the ductility of three refractory metals compared with that of nickel

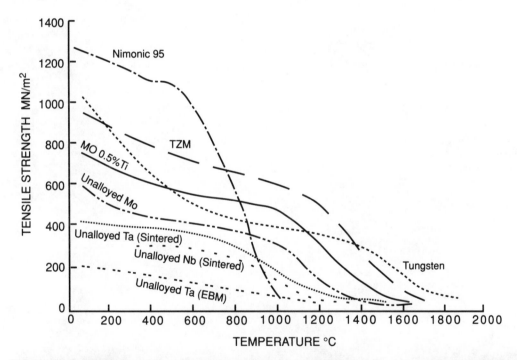

FIG 2.13.2 Comparison of the elevated temperature tensile strengths of tungsten, molybdenum, molybdenum alloys, tantalum, niobium and Nimonic 95 (tungsten, molybdenum and alloys in the stress relieved condition, tantalum and niobium in the annealed condition and Nimonic 95 in the fully heat-treated condition)

Fig 2.13.1 and Fig 2.13.2

Tin and bismuth alloys

Contents

List of tables

2.14.1   Terneplate

This section deals specifically with Terneplate and the following section (2.14.2) with the low-melting-point, or fusible, alloys.

2.14.1.1   General characteristics and properties

Terneplate is principally thought of as a steel product, consisting of mild steel strip or sheet, coated on both sides with a 'terne' alloy. Terne alloys are a series of lead–tin alloys, although some Terneplate coatings use antimony to replace some or all of the tin. The tin content of terne coatings is usually within the range 8–25%. Typical properties of Terneplate are listed in:

Table 2.14.1   *Typical properties of Terneplate*

The low melting point of terne alloys and their ability, when molten, to wet and alloy with clean mild steel, makes coating by hot dipping an ideal method of producing Terneplate and the overwhelming majority is produced this way. An alternative method, electroplating lead–tin coatings onto wide steel strip, is also feasible.

2.14.1.2   Applications

Terneplate is widely used due to its good corrosion resistance (particularly to atmosphere), solderability and improved drawability. Its applications include roofing, rainwater goods, fire-resisting doors, ducting, outdoor signs and components for roller blinds, oil heating furnace parts, office furniture, deed boxes and heavy-duty cleaning equipment.

However, the largest market for terneplate arises from its resistance to petroleum products and it is used as a material for fuel tanks (under competition from certain plastics). Excellent formability allows complex shapes to be drawn. The material is more extensively used in the US than in Europe.

TABLE 2.14.1 Typical properties of Terneplate

	Commercial quality	Drawing quality	Drawing quality Special skilled
Yield strength (MN/m²)	180/250	165/220	150/200
Tensile strength (MN/m²)	270/320	250/305	245/300
Elongation (min.%)	32/40	34/44	35/46
Hardness: Rockwell B VPN	40/60 90/110	40/55 90/105	36/50 80/100

Table 2.14.1

2.14.2   Fusible alloys

2.14.2.1   Characteristics

The term 'fusible alloy' is applied to alloys which melt at relatively low temperatures. Although it is often loosely applied to include the whole range of white metals, solders and amalgams, this section deals with the properties and applications of these materials as a primary engineering material.

Bismuth, tin, lead, cadmium and indium are all relatively low-melting point metals and when these are combined in various proportions, alloys with even lower melting points are obtained.

Most fusible alloys contain bismuth, a coarsely crystalline, silvery white metal which melts at 271°C. Tin is a constituent of very many fusible alloys, in amounts ranging from 1 to 60%. Alloys containing substantial amounts of tin will wet other metals.

PROPERTIES OF FUSIBLE ALLOYS

Melting points and compositions of some fusible alloys are listed in Table 2.14.2, and the mechanical properties of the more common compositions in Table 2.14.3:

Table 2.14.2   *Melting points and compositions of some fusible alloys*
Table 2.14.3   *Mechanical properties of some fusible alloys*

The mechanical properties are highly dependent on the elapsed time since casting, as well as on the actual casting and cooling conditions. The results are also affected by the testing conditions employed.

Most fusible alloys seem brittle when subjected to sudden shock, but exhibit fair ductility under slow rates of strain. Their load-bearing capacity is good, although under prolonged stress, deformation will occur. Mechanical properties improve slightly on ageing, due to minute structural changes which take place.

Typical applications of fusible alloys are listed in Table 2.14.4:

Table 2.14.4   *Applications of fusible alloys*

ADVANTAGES OF FUSIBLE ALLOYS

Some characteristics which are exploited in applications include:

> low vapour pressure;
> good thermal conductivity;
> ease of handling;
> high fluidity (for mould filling);
> controlled dimensional properties on solidification;
> ability to reproduce detail in castings, with a high degree of finish;
> re-usability.

DIMENSIONAL STABILITY

The dimensional stability of cast alloys depends upon the presence of bismuth. This metal *expands* on solidification (3.3% by volume) and this behaviour also influences the behaviour of its alloys. Generally speaking, alloys containing greater than 55% bismuth will expand on solidification, those containing less than 48% will contract and those containing 48–55% bismuth are dimensionally stable.

The presence of *lead* in bismuth-rich alloys tends to result in a gradual increase in volume over a period of days after casting.

TABLE 2.14.2 Melting points and compositions of some fusible alloys

Melting temperature °C	Typical composition (%)						Trade name/specification
	Bi	Sn	Pb	In	Cd	Other	
17		12.0				Ga82.0+	(Ternary Eutectic)
						Zn6.0	
20		8.0				Ga92.0	(Binary Eutectic)
47	44.7	8.3	22.6	19.1	5.3		117, NCP47
58	49.0	12.0	18.0	21.0			136, MCP58
61–65	48.0	12.7	25.63	4.0	9.6		Cerrolow 147
65–66	49.3	13.2	26.3		9.8	Ga1.4	
70	49.4	12.9	27.7		10		157, FA14, MCP70
70	49.5	13.1	27.3		10.1		(Quaternary Eutectic composition)
70–72	50.0	12.5	25.0		12.5		Wood's
70–73	50.0	13.3	26.7		10.0		158, Lipowitz', Cerrobend,. FA13
70–80	42.0	13.0	35.0		10.0		
70–85	40.0	13.0	37.0		10.0		
70–88	45.3	24.5	17.9		12.3		
70–90	42.5	11.3	37.7		8.5		160–190, Metspec 158–190, Cerrosafe
70–97	38.4	15.4	30.8		15.4		
92	51.7		40.2		8.1		(Ternary Eutectic)
96	52.0	15.5	32.0				(Ternary Eutectic) Indalloy 39, Ostalloy 205, FA11, MCP96
96–97	50.0	18.8	31.2				Newton's
96–98	50.0	25.0	25.0				D'Arcet's
96–100	50.0	20.0	30.0				Onion's or Lichtenberg's
96–110	50.0	22.0	28.0				Rose's
96–123	46.1	34.2	19.7				Malotte's
96–143	34.0	33.0	33.0				
103	53.9	25.9			20.2		(Ternary Eutectic)
103–113	50.0	25.0			25.0		
102–120	44.5	34.5			21.0		
103–227	48.0	14.5	28.5			Sb9	217–440, Indalloy 124, Cerromatrix
117		48.0		52.0			(Binary Eutectic)
117–120	55.0	1.0	44.0				
117–127		50.0		50.0			Cerroseal 35
124	55.5		44.5				(Binary Eutectic), Cewrrobase
128		46.0			17.0	Tl37.0	(Ternary Eutectic)
130	56.0	40.0				Zn4.0	(Ternary Eutectic)
134–135	57.4	41.6	1.0				
138	58.0	42.0					(Binary Eutectic) 281, Cerrotru, FA6, MCP137
138–170	40.0	60.0					FA5, Cerrocast, MCP150, Ostalloy 337, Indalloy 100, Asarcolo 281–338
142–166	10.2	48.8	41.0				
144	60.0				40.0		(Binary Eutectic)
145		51.2	30.6		18.2		(Ternary Eutectic)
145–160		40.0	42.0		18.0		
170		56.5				Tl43.5	(Binary Eutectics)
176		67.0			33.0		(Binary Eutectic)
183		63	37				Indalloy 106
188	47.5					Tl52.5	(Binary Eutectics)
199		92.0				Zn8.0	(Binary Eutectics)
203					17.0	Tl83.0	(Binary Eutectics)
221		96.5				Ag3.5	(Binary Eutectics)
227		99.25				Cu0.75	(Binary Eutectics)

Table 2.14.2

TABLE 2.14.3 Mechanical properties of some fusible alloys

Alloy	Tensile strength (MN/m²)	Elongation % in 50 mm (slow loading)	Hardness (Brinell No.)
Alloy 117	38	1.5	12
Alloy 136	45	50	14
Alloy 147	34	13.5	11
Alloy 158	41	200	9.2
Alloy 160–190	38	220	9
Alloy 217–440	90	1	19
Alloy Bi55.5/Pb44.5	45	60.70	10.2
Alloy 281	55	200	22

Table 2.14.3

TABLE 2.14.4 Applications of fusible alloys

Moulds:	Prototyping or short run mould. Cold-forming and pressing moulds. Hollow cavity moulds. Removable cores.
Casting and forging:	Patternmaking. Investment casting dies. Proof casting of cavities. Intricate cores. Electroforming patterns.
Safety devices and heat fuses:	Fire control systems. Pressure safety valves.
Magnetic devices:	Spacing material in permanent magnet chucks.
Processing and metalworking aids:	Joining and soldering. Jigs and supports. Tube bending. Tempering baths.
Composite matrices:	Metal matrix composites.
Medical uses:	Radiotherapy radiation shields. Lung material casts. Lens blocks for the ophthalmic industry. Denture moulds.
Decorative uses:	Costume jewellery. Small figures.
Test and measurement:	Flow measurements in molten metals. Temperature measurement. Electrolytic determination. Mounting metallographic specimens.
Fingerprinting	

Table 2.14.4

Engineering ceramics

Contents

List of tables

List of figures

2.15.1   General characteristics

2.15.1.1   Introduction

The advantages and limitations of ceramics are listed in:
Table 2.15.1   *General characteristics of ceramics*

2.15.1.2   Selection procedure

The selection of a ceramic material to fulfil an engineering task is often governed by the requirement for one or more of the following properties:

Hardness (scratch or erosion resistance).
Resistance to stress at high temperature.
Resistance to an oxidising or other corrosive environment.

In specific cases physical characteristics such as an electrical or optical property may be paramount.

2.15.1.3   Design considerations

Engineering design for the use of *metallic materials* relies on well-established principles which are backed up by easy access to information about the types of metal and alloys which are available (often specified in British Standards) and about the properties of these materials in the form of reliable data sheets. When designing for the use of ceramics it should be realised that manufacturers' data sheets for ceramics are often woefully inadequate. For instance, a manufacturer's data sheet for an engineering grade of silicon nitride states that the flexural strength of the material is 170 MPa. On testing 50 small specimens from one batch of this ceramic and assessing the results on a statistical basis it was found that the allowable flexural stress, for that batch of material, corresponding to a failure probability of 1 in 10^4 was 90 MPa. When allowance is made for the use of a sequence of batches and for the decrease in strength due to increase in the stressed volume (which must always be allowed for with ceramics where the components required are bigger than the specimens tested) the allowable stress falls to an even lower value, so that *a safety factor of about 3 to 1* becomes necessary. But a design engineer would have no means of fixing the value of the safety factor unless the necessary testing work had been done.

Prediction of lifetime under stress is also particularly important for ceramic materials, which can fail without warning. This can be achieved by statistical treatment of suitable test data, but cannot be forecast from manufacturers' information sheets.

The shape chosen by the designer of a ceramic component is almost always a critical factor. It should never be assumed that a shape suitable for a *metal* component will be suitable for a *ceramic* component which is being designed to supersede a metal one and improve on its performance. The reasons for this are to some extent due to the different shaping methods which will be used in making the ceramic part as against the metal one, but are also due to the difference in properties of the two types of material.

It will be useful for designers to consider some typical ceramic properties and the consequences which follow from these when designing for the use of ceramics.
This can be done with the help of:
Table 2.15.2   *The effect of ceramic properties on design and construction methods*
The need to adopt different design principles when designing for the use of ceramic materials, as compared with those principles which are used when designing for the use of metals, will now have become apparent.

DO'S AND DON'TS FOR DESIGNERS IN ENGINEERING CERAMICS

The following simple guidelines for design will be helpful. Although most of them will aid good design in metal components, they are even more important for engineering ceramics because of the brittleness problem.

> DON'T use sharp corners, notches, slots which can act as stress raisers.
>
> DO use generous radii.
>
> DO pay attention to surface finish. A good surface finish is essential.
>
> DO take advantage of the higher compressive strengths of ceramics by preferring compression joints and connections to threads and flanges.
>
> DO disperse stresses at assembly faces by using soft gaskets where possible to avoid high spot loadings.
>
> DO allow for thermal mismatch, particularly at ceramic/metal joints. Metals have expansion coefficients up to 10 times those of ceramics.
>
> DON'T incorporate abrupt changes of section which may lead to high thermal stresses in temperature gradients.
>
> DO make use of proof-testing procedures if possible.
>
> DO remember the batch-to-batch variability of ceramics and make trials on the grade and surface finish to be used in production.
>
> DO specify dimensions carefully in view of the greater difficulties in final machining than with metals.

In addition to these design guidelines, the following can be considered as the major steps involved in working with ceramics.

(From *Handbook of Properties of Technical and Engineering Ceramics*, Roger Morrell, NPL, 1989.)

1. Initially, choose materials in the low to medium cost bracket which would seem to fulfil technical requirements and which should do the job from the property point of view. It must be emphasized that in many cases the actual cost of the ceramic material is very small compared to the cost incurred in secondary machining operations. Some costs in the production of ceramic components may be reduced by using established design and technical experience in manufacturing organizations.

2. Consider the technical requirements, particularly those that apply only to a small part of the component. It may be better to use a small component in an assembly to meet such a local requirement than make the entire existing component in ceramic. Conversely, look at the original function of the design, because it may be possible to combine appropriate components or their functions into a single ceramic component. This is particularly advantageous if the component is to be made by one of the mass-production techniques.

3. Give serious consideration to the quantities of material or components required at each stage of product development, i.e. for material approval, application tests, prototypes, pre-production, and production.

4. Contact likely manufacturers, experts and research organizations. Be prepared to change designs according to the discussion. Do not resent change, because while you may have the greatest understanding and technical expertise in the design and function of your assembly, you may not be an expert in the design of the one ceramic component in the system. There is a wealth of ceramic-orientated knowledge in the organizations that deal with ceramics routinely. Do not ignore this valuable source of advice.

5. Ensure that the material and the manufacturer's process will give you properties and tolerances to do the job. Awareness of the influence of manufacturing methods on properties and sizes is useful here. Be as flexible as possible on tolerances.

6.   Establish clearly the ownership of the design that results from external help. This will clearly involve commercial, technical and patent appraisal of the design.

7.   Place your order according to a specification agreed with the manufacturer at all stages of design and development from the initial trials on materials through to the final production quantities of the finalized design. Do not omit to specify quality-control procedures to ensure acceptable dimensions and surface finish. Do not omit to specify special protective packaging if required.

2.15.1.4 Available types

The more important types of ceramic material, together with an indication of their more significant properties and applications, their basic structures and preparation techniques, and the section number under which their detailed properties and applications are to be found are listed alphabetically in:

Table 2.15.3 *Ceramics, their major characteristics, structures and references to property tables*

2.15.1.5 General properties

The properties of individual ceramics and their applications are listed in detail in Sections 2.15.2–2.15.9. As an initial guide to materials selection however some approximate property values for ceramics are listed in:

Table 2.15.4 *General properties of typical ceramics and glasses compared with those of cast iron*

TABLE 2.15.1 General characteristics of ceramics [a]

Advantages	Limitations
Low density compared to most metals.	
High Modulus of Elasticity.	
High strength retained at elevated temperatures.	
Good resistance to high temperature creep.	
Dimensional stability.	Cannot cold work. Local stresses cannot relax.
Good compressive strength.	Brittle, especially in tension and shear. The complete absence of any ductility prior to fracture means that impact strength is low.
Extreme hardness. Abrasion resistant. Wear resistant.	Difficult to machine after firing (diamond grinding is often the only possible method). Abrasive.
Corrosion and oxidation resistance.	Some products are not designed for resistance to corrosive media. Non-oxides may oxidise at high temperatures.
Wide range of thermal conductivity and expansion coefficient available.	Those ceramics with low thermal conductivity and relatively high coefficients of thermal expansion are very susceptible to thermal shock.
Good thermal insulation properties. Good electrical insulation properties.	All lose insulating properties at very high temperatures.
Electrical conductivity in special materials. Transparency or translucency.	
Cheap raw materials in many cases.	Expensive tooling and secondary shaping operations. Some raw materials are quite expensive.
Wide range of sizes can be made, 0.2 mm to 3 m possible.	Severe limits in some materials, determined by manufacturing techniques required. Small sizes are mechanically weak.
Non-toxic.	Beryllia is toxic as a dust.
Many specialist manufacturers of some types of materials.	Limited availability of some varieties of material, limited engineering expertise, limited understanding of performance.
	The brittleness of ceramics, the absence of ductility prior to failure, their consequent low impact strength and susceptibility to thermal shock failure and cracking at joints or fastenings between different materials (due to thermal expansion mismatch) will often rule out their use in applications where a ductile material can be used as an alternative.

[a] Adapted from *Handbook of Properties of Technical and Engineering Ceramics*, Part 1, Roger Morrell, NPL, 1989.

Table 2.15.1

TABLE 2.15.2 The effect of ceramic properties on design and construction methods

General property of ceramics	Consequence
Wear resistance and lack of ductility, malleability and toughness.	Form grinding of ceramic articles must be done, in most cases, with diamond wheels and is time-consuming and costly. Avoid as far as possible. Finish grinding should be cut to a minimum. Diamond wheel cut-off operations are, however, nearly always economically acceptable.
Rigidity or stiffness.	Susceptibility of ceramics to the stress concentrating effect of fixing holes, threads, sharp changes in section, etc. Designs should spread fixing or clamping stresses and avoid sharp corners by using suitable radii.
Generally low thermal conductivity; often (but not always) poor resistance to thermal stress or thermal shock.	Choose a low expansion ceramic (e.g. lithium aluminosilicate glass ceramic or magnesium aluminosilicate glass ceramic) in cases where good resistance to thermal stress or shock is required. Alternatively choose one of the higher thermal conductivity types (e.g. silicon carbide, cemented tungsten carbide or graphite). If neither of these courses is possible and only resistance to thermal *stress* is necessary, maintain uniform minimum cross-section.

Table 2.15.2

TABLE 2.15.3   Ceramics, their major characteristics, structures and references to property tables

Ceramic types	Major characteristics	Type of structure	Detailed properties listed in Section No.	Table No.
Aluminas	Characteristics depend to some extent on composition in the range 85–100 wt % Al_2O_3. Broadly they are strong, hard, wear and temperature resistant, good electrical insulators and have low dielectric loss especially when pure.	Sintered metal oxide	2.15.2	2.15.5 2.15.12
Aluminium titanate	Low expansion ceramic with a pronounced phase change on heating.	Sintered oxide compound	—	—
Alumino-silicates	A wide variety of material compositions which are primarily designed for refractoriness and thermal shock resistance. Open porosity may be up to 30%.	Sintered oxide compound	—	—
Barium titanates	Often contain small amounts of other (e.g. alkaline earth) titanates. Used when a very high dielectric constant is needed in electronic circuitry.	Sintered oxide compound	2.15.2	2.15.11
Beryllium oxides (beryllia)	Usually near to pure BeO which is toxic in the form of dust. The sintered materials are safe and have exceptionally high thermal conductivities (for ceramics) at low to moderate temperatures (approaching that of copper). Cost of raw material tends to limit their use to electrical or nuclear applications.	Sintered metal oxide	2.15.2	2.15.7
Boron carbides	Strong, very hard and resist high temperatures in inert atmospheres. Rather expensive.	Sintered (non-metal) carbide	2.15.3	2.15.16
Boron nitrides	Often almost pure BN. Less pure and somewhat cheaper materials are sold. Not wetted by many molten metals. Not outstandingly strong, but machinable and lubricious. Good thermal shock resistance.	Sintered (non-metal) nitride	2.15.3	2.15.15
Carbons	Not very strong except when produced as fibres, but creep resistant at high temperatures in non-oxidising conditions. Electrically conducting, lubricious.	Compacted with binder and fired or produced by condensation reaction	2.15.5	
Cemented carbides (hard metals)	Carbides of refractory metals, usually tungsten, used as wear resistant, cutting and metal making tool inserts.	Refractory metal carbide bonded with cobalt or nickel by liquid phase sintering	2.15.4	2.15.18
Cements (high temperature)	Often consist of aluminium oxide with small additions of e.g. sodium silicate to form a high temperature bond. Not unusually contain an organic bond to maintain cohesion at low temperatures.	Bonded by reaction with water	2.15.9	

Table 2.15.3

TABLE 2.15.3 Ceramics, their major characteristics, structures and references to property tables—*continued*

	Major characteristics	Structure			
Ceramic fibres	Usually oxides spun to fibre, bulked to felt and used for high temperature insulation including former applications of asbestos.	Spun from liquid phase	2.15.8		
Cordierites	Magnesium aluminium silicate ceramics of varying degrees of purity. Resistant to thermal shock because of low thermal expansion. Good electrical insulators. May be made by traditional ceramic route or as glass-ceramics.	Oxide compound sintered or crystallised from glass	2.15.2	2.15.9	2.15.12
Electrical porcelains	Not unlike continental tableware. Made to have high electrical resistivity and high electric strength. Can be made into very large pieces.	Sintered oxide compound	2.15.2	2.15.9	2.15.12
Forsterites	These are versions of steatites in which the magnesia level is raised by the addition of MgO or $MgCO_3$ so that the crystal phase forsterite ($2MgO. SiO_2$) is produced. Their expansion coefficients are higher than those of steatites, and are a good match to titanium and some nickel–iron alloys. Used for ceramic-to-metal assemblies in applications requiring low dielectric loss, such as high frequency devices and microwave tubes.	Sintered oxide compound	—	—	
Glasses	Oxide (silica), silicates and phosphates, and borosilicates comprise the bulk of glass compositions. Vitreous silica and a borate composition such as 'Pyrex' are resistant to thermal shock by virtue of their low thermal expansion coefficients. A range of special optical characteristics is available. Soda lime silica glass (window and container glass) can be made very strong by thermal or chemical strengthening.	Amorphous (very viscous liquid) oxide compound, produced from melt	2.15.6		
Glass-ceramics	Cover a wide range of compositions and can have low, medium or high thermal expansion depending on composition type. Many are good electrical insulators, some are transparent. A special composition can be machined with steel tools and rather resembles boron nitride in feel and texture.	Oxide compound crystallised from glass	2.15.7		
Graphites	Broadly as for carbons q.v. May be reinforced with carbon fibre or cloth to increase strength.	Laminar crystalline element	2.15.5		
Lead zirconate titanates	These materials are formulated to have strongly piezoelectric properties, and are not generally available in the form of large articles.	Crystalline oxide compound	—		
Magnesias	Less common as a fine grained engineering ceramic than as a coarse grained refractory material. Resist high temperatures, but liable to thermal stress cracking. Resists basic slags.	Sintered oxide compound	2.15.2		2.15.12
Magnesium titanates	Dielectric materials with a permittivity compensating property for electronic circuitry.	Sintered oxide compound	2.15.2		2.15.11

Table 2.15.3—*continued*

TABLE 2.15.3 Ceramics, their major characteristics, structures and references to property tables—continued

Ceramic types	Major characteristics	Type of structure	Detailed properties listed in	
			Section No.	Table No.
Molybdenum disilicide	Used almost exclusively for heating elements.	Sintered silicide	—	—
Porcelains	Well vitrified ceramics based on a mixture of clays, feldspars and a filler. The majority of technical porcelains have been designed to have high strength for use as insulators.	Sintered oxide compound	—	—
Sialons	Silicon *aluminium oxide nitride*. A new class of ceramic with exceptional strength and resistance to wear.	Sintered oxide/nitride compound	2.15.3	2.15.14
Silicon carbides	Versatile materials if electrical resistivity is not required. Varying purities and types of bonding are available. Fine grained materials are strong and resistant to chemical attack, high temperature and thermal stress.	Sintered (non-metal) carbide	2.15.3	2.15.13
Silicon nitrides	Often above 95% Si_3N_4 except in the case of the sialons which contain Al and O in compounded form. Silicon nitrides may be reaction bonded, in which case they are normally porous, or hot pressed to make them impermeable and stronger. They are resistant to high temperature, to thermal stress and shock, quite oxidation resistant and electrical insulators.	Reaction bonded or hot pressed (non-metal) nitride	2.15.3	2.15.14
Spinel	$MgAl_2O_4$ sometimes used as a fine grained ceramic for refractory purposes.	Sintered oxide compound	2.15.1	
Sprayed ceramic coatings	Ceramic coatings can be applied, usually to metals, by several spray methods and over a range of compositions. Three main hot spraying methods are used, the rod method and the powder method with conventional gas guns and the plasma spray method. Oxide, carbide and boride coatings can be applied, usually to provide hard facings on the base metal. Often an intermediate metallic layer is used. It is impractical to quote properties herein.	Spray deposit usually of molten particles	—	—
Steatites	The name usually used for ceramics composed mainly of enstatite ($MgO. SiO_2$) which have very low dielectric loss and are used, for example, as capacitors in electronic circuitry.	Sintered oxide compound	2.15.2	2.15.9
Titanium dioxides	Also used in electronic engineering, in this case because of high dielectric constants as well as low dielectric losses.	Sintered metal oxide	2.15.2	2.15.7
Titanium diboride		Sintered metal boride	-	—

Table 2.15.3—*continued*

TABLE 2.15.3   Ceramics, their major characteristics, structures and references to property tables—*continued*

Tungsten carbides	Usually bonded with cobalt and/or nickel, so that they could be called 'cermets'. But more usually they are called 'cemented carbides' or 'hardmetals'. A major use is as tips for metal cutting tools, and they have many other wear resistant applications.	Tungsten carbide usually bonded with cobalt or nickel but sometimes hot pressed dry	2.15.4 2.15.18
Vitreous carbon	A high density form of carbon which is impermeable and sometimes used where greater strength and corrosion resistance than is available with other carbons is needed. Rather expensive. Size limited.	Amorphous high density form of carbon	2.15.5
Vitreous silica	A glass of composition Si O_2. Transparent. Has very low thermal expansion and therefore resists thermal shock and thermal stress well. Resists higher temperatures than other glasses. There is a sintered form made from the powdered glass.	(Non-metal) oxide glass	2.15.7
Zirconia	Contains small amounts of other oxides to stabilise the ZrO_2 in the cubic form. Partially stabilised form has been found to be most resistant to thermal shock and stress. Electrically conducting at high temperatures upwards of 1000°C. A useful oxygen ion conductor at temperatures above about 500°C.	Sintered metal oxide	2.15.2 2.15.7
Other transparent materials	Often based on Li_2O–Al_2O_2 compositions. The range of commercially-available materials is small.	Sintered oxide compounds	— —

Table 2.15.3—*continued*

TABLE 2.15.4 General properties of typical ceramics and glasses compared with those of cast iron

Property	Units	Glasses			Glass-ceramics		Sintered alumina	Hot-pressed silicon nitride	Metal-bonded tungsten carbide	High density moulded graphite	Cast iron (Grade 10)
		Soda lime	Low expansion borosilicate	Vitreous silica	Low expansion	Machinable					
Composition	Weight %	13 Na$_2$O, 12(Ca, Mg) O, 72 SiO$_2$	12 B$_2$O$_3$, 4 Na$_2$O, 2 Al$_2$O$_3$, 80 SiO$_2$	c.100 SiO$_2$	2(Ti, Zr)O$_2$, 4 Li$_2$O, 20 Al$_2$O$_3$, 70 SiO$_2$	65 mica, 35 glass	c.100 Al$_2$O$_3$	96 Si$_3$N$_4$, 4 MgO	94 WC, 6 Co	c.100 C	3C 2 Si 0.5 Mn remainder Fe
Appearance		Clear transparent	Clear transparent	Clear transparent	White translucent or transparent	Off-white	White or transparent	Grey-black	Dark grey	Black	Typical grey metallic
Bulk density		2500	2200	2200	2500	2500	3900	3200	14900	1700	7200
Open porosity	% by volume	Zero	Zero	Zero	Zero	Zero	< 0.1	< 0.1	< 0.1	c.17	< 0.1
Hardness (Vickers)		c.550	c.650	c.750	c.700	c.255	c.2500	c.2200	c.1700	c.25 ^{f}	c.200
Wear resistance ^{a}		Good	Good	Good	Good	Fair	Excellent	Good	Very good	Fair	Fair
Bend strength ^{b}	MPa	45 ^{e}	50	60	130	100	250	600	2000	30	300
Weibull modulus (for bend strength values) ^{c}		c.6	c.6	c.6	c.12	c.12	c.12	c.15	c.25	c.12	c.25
Young's modulus of elasticity	GPa	73	66	71	87	64	350	310	580	9	150
Poisson's ratio		0.21	0.20	0.17	0.25	0.26	0.23	0.25	0.26	0.11	0.27
Impact resistance ^{a}		Poor	Poor	Poor	Fair	Fair	Fair	Fair-good	Good	Fair	Fair
Work of fracture or fracture energy	J/m^2	3.8	4.5	4.3	c.28	c.40	c.35	c.30	c.180	c.90	c.4000
Critical stress intensity factor	MN/m$^{-3/2}$	0.75	0.77	0.80	2.2	2.3	4.6	4.5	c.15	1.3	c.35
Thermal shock resistance ^{a}		Poor	Good	Very good	Very good	Good	Fair	Good	Fair-good	Good	Fair-good
Specific heat ^{c}	J/kg per K	770	750	710	850	800 est.	1070	710	210	850	500
Thermal conductivity ^{c}	W/m per K	c.1	c.1	c.1.5	c.2	c.2	c.30	18	c.75	c.150	c.50
Coefficient of linear thermal expansion (approximately mean values)	K^{-1}	8 × 10^{-6}	3 × 10^{-6}	5 × 10^{-7}	Zero	9 × 10^{-6}	7 × 10^{-6}	3 × 10^{-6}	5 × 10^{-6}	4 × 10^{-6}	12 × 10^{-6}
Approximate safe working temperature (no load) ^{d}	°C	500	525	1000	700	850	1600	1400	1000	3000 (inert atmosphere)	650
High temperature creep resistance ^{a}		Poor	Poor	Fair	Poor-fair	Fair	Good	Good	Fair-good	Excellent	Poor-fair
Volume resistivity	ohm metres	1 × 10^{12}	1 × 10^{15}	1 × 10^{19}	1 × 10^{15}	1 ×10^{17}	1 × 10^{16}	1 × 10^{13}	2 × 10^{-3}	11 × 10^{-2}	1 × 10^{-4}

Notes:

^{a} These qualitative statements of properties are self-consistent within this range of materials only.

^{b} In most cases these properties are particularly dependent on surface condition.

^{c} As with the other properties quoted, no attempt has been made to extend these values into the high temperature range. They are approximately correct for the range Room Temperature to 300°C (say 570°K).

^{d} There are a number of factors which may limit the safe working temperature. In individual cases the limitation may not be distortion or creep, but rather lack of chemical stability, e.g. oxidation of graphite (500°C in air).

^{e} Note that this value can be increased very considerably (to > 250 MPa) by thermal or chemical strengthening processes.

^{f} The hardness of graphite is rarely measured by a Vickers test, but the figure quoted gives a valid comparison with the other materials.

Table 2.15.4

2.15.2   Oxides

Oxides comprise the vast majority of ceramics. The most important numerically are grades of alumina because of the abundance and ease of fabrication of this material. Also included are other oxides such as magnesia, zirconia and beryllia, and compounds (and mixtures of oxides) porcelains, cordierites and steatites.

The properties (and Al_2O_3 content) of the aluminas are listed in:
Table 2.15.5   *General properties of non-absorbent and vacuum-tight aluminas*
and their applications in:
Table 2.15.6   *Some applications of alumina*

Other impervious oxides include beryllia, titania and zirconia. The properties (and base oxide content) of these materials are listed in:
Table 2.15.7   *General properties of non-absorbent and vacuum-tight oxide ceramics other than alumina*

The use of beryllia is restricted by hazard to health, but it has (for an oxide ceramic) an exceptionally high thermal conductivity and very low absorption for neutrons. Its applications are listed in:
Table 2.15.8   *Some applications of beryllia*

Oxide compounds are prepared, either by the addition of a more easily sintered oxide to a base oxide, or by the formulation of a compound such as cordierite or steatite. Many of these materials, some of which have specific favourable properties, are known as engineering porcelains. They can usually be sintered at lower temperatures than oxides but are much less refractory. The properties (and nominal compositions) of the more important oxide compounds are listed in:
Table 2.15.9   *General properties of non-absorbent oxide compounds, porcelains, etc.*
and the characteristics and applications of typical engineering porcelains in:
Table 2.15.10   *Characteristics and applications of engineering porcelains*

The impervious oxide ceramics also include a number of ceramic dielectric materials (presumably titanates) whose properties are listed in:
Table 2.15.11   *Properties of non-absorbent dielectric ceramics (compositions not disclosed, presumably titanates, etc.)*

Many of the compositions listed above as impervious ceramics and some others are available in a porous, water-absorbent form. These porous materials are listed in:
Table 2.15.12   *Properties of selected porous oxide ceramics (including low thermal expansion materials)*

TABLE 2.15.5 General properties of non-absorbent and vacuum-tight aluminas

Material		Nominal Al_2O_3 content (wt %)	Colour	Specific density	Flexural strength (MPa)	Young's Modulus (GPa)	Poisson's ratio	Hardness (GN/m^2)	Specific heat ($10^3 J$ $kg^{-1} K^{-1}$)	Thermal conductivity (W/m per K)		
		a	b	c	d	e	f	g (see note)	h (see note)	i (see note)		
										50°C	200°C	400°C
DERANOX 995	1	99+	White	3.87	332	382		15.27		25.6		12.4
DERANOX 995T	2	80		4.1	450	340				23		
DERANOX 975	3	97.5	White	3.78	366	338		15.05		23.7		13.1
DERANOX 970	4	97.0	White	3.73	234	328		12.35		23.7		13.1
DERANOX 934	5	93+	White	3.58	330	276						
HILOX 880/882	6	86		3.5	250	250				15.0		
CERALLOY 138	7	See note	Near white	3.89	>515	>345		>18.5				
AD–85	8	85	White	3.41	296					14.6		
AD–90	9	90	White	3.60	>300	270	0.22	10.6	0.920	16.7	13.4	7.9
AD–090	10	92		3.77	365	308				12.6		
AD–94	11	94	White	3.62	>310	277	0.21	10.9	0.880	18.0	14.2	7.9
AD–96	12	96	White	3.72	>315	297	0.21	10.9	0.880	24.7	18.8	10.0
AD–995	13	99.5	Ivory	3.89	370	365	0.22	14.4	0.880	35.6	25.9	12.1
AL23	14	99.7	White	>3.7	350	350		23	0.850	25		5
AF920	15	92	White	3.53	235	240		≥11	1.09	17		4
AF950R	16	95	Pink	3.6	235	260		≥11	1.09	21		4
AF995R	17	99.5	Brown	>3.85	240	300		≥11	1.09	25		0.63
AF9953B	18	99.5	Blue	>3.85	240	300		≥11	1.09	25		0.63
AF998P	19	99.8	Fawn	>3.90	240	310		≥11	1.09	25		0.63
A380	20	80		3.37	267					8.4		

Manufacturers/suppliers:
- 1–6   Morgan Matroc.
- 7   Ceradyne Inc. (All articles hot pressed. Sizes up to 250 mm dia. Al_2O_3 content 99+ wt%).
- 8–13   Coors Ceramics Co. (Other aluminas are made including a transparent one, Vistal). Coors is now owned by Vesuvius Zyalons.
- 14   Friedrichsfeld GmbH (Frialit-Degussit) and Bush Beach Engineering Ltd.
- 15–19   Desmarquest (twelve different alumina ceramics are made).
- 20   Narumi.

Notes:
- (g)   Materials 1, 3, 4, Vickers Microhardness 2.5 kg. Material 7, Knoop 100. Materials 9, 11, 12, 13, Knoop 1000. Material 14, Knoop 100. Materials 15–19, Vickers 3N.
- (h)   Materials 15–19, Values for 20–1000°C (mean).
- (i)   Materials 9, 11, 12, 13, first two columns are for 20°C and 100°C. Material 14, values are for 100 and 1000°C. Materials 15–19, values are for 20 and 800°C.
- (k)   Materials 9, 11, 12, second and third columns are for 500 and 700°C. Material 14, second and third columns are for 500 and 1000°C.
- (l) and (m)   Materials 1, 3, 4, at 9368 MHz. Materials 5, 9, 11, 12, 13 at 1 MHz. Material 14 at 10 MHz. Materials 15–19 at 1000 MHz.
- (n)   Depends on specimen thickness. Materials 9, 11, 12, 13, thickness 6.35 mm. Material 14, specimens clearly much thinner.

Table 2.15.5

TABLE 2.15.5 General properties of non-absorbent and vacuum-tight aluminas—*continued*

Coefficient of thermal expansion (10^{-6} K^{-1})			Volume resistivity 10^{-2} ohm m (ohm cm)			Loss tangent (power factor) 10^{-4}			Permittivity			Dielectric strength (kV/mm)	Safe working temp (°C) (no load)
j			k (see note)			l (see note)			m (see note)			n (see note)	o
20–200°C	20–500°C	20–1000°C	20°C	400°C	600°C	20°C	200°C	400°C	20°C	200°C	400°C		
		8.5	>1×10^{15}	>1×10^{10}	>1×10^{9}	9.70	9.85	10.02	5.0	4.2	4.6		1800
8.1													1500
		8.1	>1×10^{14}	>1×10^{9}	>1×10^{8}	9.49	9.68	9.97	4.3	4.4	6.3		1700
		8.2	>1×10^{15}	>1×10^{10}	>1×10^{8}	9.10	9.27	9.33	6.7	6.0	6.5		1700
		7.8	>1×10^{14}	>1×10^{9}	>1×10^{8}	8.90			10				1500
7.0			1×10^{14}			20							1200
		8.0	>1×10^{14}										
7.2			1×10^{14}			9.0							1400
6.1	7.0	8.1	>1×10^{14}	2.8×10^{8}	7.0×10^{6}	8.8			4			9.2	1500
8.4			1×10^{14}			200							1500
6.3	7.1	7.9	>1×10^{14}	2.5×10^{9}	5.0×10^{7}	8.9			1			8.7	1700
6.0	7.4	8.2	>1×10^{14}	4.0×10^{9}	1.0×10^{8}	9.0			1			8.3	1700
7.1	7.6	8.3	>1×10^{14}			9.7			3			8.7	1750
		8.1	>1×10^{14}	1.5×10^{11}	5.0×10^{6}	9.2			2			22	1950
		8.2	>1×10^{14}		1×10^{7}	8.9			8				1500
		8.5	>1×10^{14}		1×10^{7}	8.9			5				1650
		8.6	>1×10^{14}		1×10^{9}	9.6			1				1850
		8.6	>1×10^{14}		1×10^{9}	9.6			1				1850
		8.6	>1×10^{14}		1×10^{9}	9.6			1				1850
7.5			1×10^{14}			4.0							

Table 2.15.5—*continued*

TABLE 2.15.5—General properties of non-absorbent and vacuum-tight aluminas—*continued*

Material		Nominal Al$_2$O$_3$ content (wt %)	Colour	Specific density	Flexural strength (MPa)	Young's Modulus (GPa)	Poisson's ratio	Hardness (GN/m^2)	Specific heat (10^3 J kg^{-1} K^{-1})	Thermal conductivity (W/m per K)		
		a	b	c	d	e	f	g (see note)	h	i (see note)		
										50°C	200°C	400°C
P–3662	21	85	White	3.48	293	220				>12.5		
B–890–2	22	92	White	3.58	320	290				>12.5		
B–3142–1	23	95	Pink	3.72	340	317				>16.5		
P–3258	24	95	White	3.65	275	294				>16.5		
RUBALIT 702	25	85	White	3.5	300			73				
RUBALIT 705	26	92	White	3.7	340			77				
RUBALIT 935	27	94	White	3.6	320			—				
RUBALIT 708	28	96	White	3.75	350			78				
RUBALIT 717	29	>99.5	White	3.85	360			80				
A–459	30	90	Russet	3.6	278	250	0.30	12.5	0.83	16.7		
A–473	31	92	White	3.6	307	260	0.23	12.9	0.79	16.7		
A–476	32	96	White	3.8	268	308	0.23	14.4	0.79	20.3		
A–479	33	99	White	3.8	298	336	0.25	15.9	0.79	25.1		
A–490	34	99.5	White	3.8	268	384		15.4	0.79	25.1		
ALSINT 99.7	35	99.7	White	3.8	340					18.0		

Manufacturers/suppliers:
21–24   Diamonite Products Manufacturing Inc.
25–29   Rosenthal Technik AG. Ten different alumina ceramics are made and marketed altogether.
30–34   Kyocera Corporation. Make nineteen different impermeable alumina ceramics.
35   W. Haldenwanger Technische Keramik GmbH.

Notes:
(g)   Materials 21–24, only claim 9+ Mohs Scale. Materials 25–29, R45N, Materials 30–34, Vickers 500g. Material 35, only claim 9+ Mohs scale.
(i)   Materials 21–24, at 25°C. Material 35, mean 20–100°C.
(j)   Materials 21–24, third column to 700°C. Materials 25–29, first column to 100°C, second column to 600°C. Materials 30–34, second column to 400°C, third to 800°C.
(k)   Materials 21–24, second and third columns 250 and 500°C. Materials 25–29, first column 200°C, Materials 30–34, second column 300°C, third column 500°C.
(l)   Materials 21–24, at 10 GHz. Materials 25–29 at 1 MHz. Materials 30–34, at 1 MHz.
(m)   Materials 21–24 at 10GHz.
(n)   Depends on specimen thickness. Materials 21–24, 6.35 mm thick. Materials 25–29, thickness not stated, DIN standard? Materials 30–34, query thickness. Material 35, thin specimen.
(o)   Materials 21–24, firing temperature.

Table 2.15.5—*continued*

TABLE 2.15.5 General properties of non-absorbent and vacuum-tight aluminas—*continued*

Coefficient of thermal expansion (10^{-6} K^{-1})			Volume resistivity 10^{-2} ohm m (ohm cm)			Loss tangent (power factor) 10^{-4}			Permittivity			Dielectric strength (kV/mm)	Safe working temp (°C) (no load)
j (see note)			k (see note)			l (see note)			m (see note)			n (see note)	o (see note)
20–200°C	20–500°C	20–1000°C	20°C	400°C	600°C	20°C	200°C	400°C	20°C	200°C	400°C		
6.2	7.1	7.5		4.9×10^{10}	1.0×10^{8}	7.9			14			8.3	1520
6.2	7.9	7.8		6.0×10^{13}	6.0×10^{10}	9.15			10		13	8.6	1675
5.9	7.9	7.9		2.0×10^{14}	1.2×10^{10}	9.36			15		16	9.1	1675
6.2	7.4	7.8		9×10^{11}	3×10^{9}	8.9					14	8.9	1675
4.7	7.2	8.0	1×10^{11}	1×10^{10}	1×10^{8}	8			10			>10	1400
4.3	7.2	8.0	1×10^{12}	1×10^{10}	1×10^{8}	9			5			>10	1500
4.7	7.2	8.1	1×10^{12}	1×10^{10}	1×10^{8}	9			5			>10	1500
5.5	7.5	8.3	1×10^{12}	1×10^{11}	1×10^{8}	9.5			3			>10	1600
5.4	7.7	8.5	1×10^{15}	1×10^{13}	1×10^{11}	10			2			>10	1600
	7.0	7.8	1×10^{14}	1×10^{12}	1×10^{9}	8.5			3			10	1500
	6.9	7.7	>10^{14}	1×10^{13}	1×10^{10}	9.5			8			10	1500
	7.1	7.8	>10^{14}	1×10^{14}	1×10^{11}	10.2			2			10	1600
	7.1	7.9	>10^{14}	1×10^{14}	1×10^{11}	9.7			2			10	1600
	6.8	7.7	>10^{14}	1×10^{14}	1×10^{11}	9.7			2			10	—
	7.3	8.0	10^{14}									25	1900

Table 2.15.5—*continued*

TABLE 2.15.5—General properties of non-absorbent and vacuum-tight aluminas—*continued*

Material		Nominal Al_2O_3 content (wt %)	Colour	Specific density	Flexural strength (MPa)	Young's Modulus (GPa)	Poisson's ratio	Hardness (GN/m^2)*	Specific heat $(10^3 J$ $kg^{-1} K^{-1})$	Thermal conductivity (W/m per K)		
		a	b	c	d	e	f	g (see note)	h (see note)	i (see note)		
										50°C	200°C	400°C
DURAMIC HT-960	36	96	White	3.76	310	304			0.795	31		
DURAMIC HT-990	37	99	White	3.85	338	338			0.837	29		
DURAMIC HT-997	38	99.7	Ivory	3.94	338	371			0.837	32		
REGALOX	39	88	Deep pink	3.48	310	234	0.31	74	1.060	13.8	10.3	8.6
DURALOX	40	95	White	3.60	320	321	0.29	76	1.072	22.5	15.5	12.2
FIBRALOX	41	96	White	3.70	330	334	0.30	80	1.068	25.0	16.9	13.0
NOBALOX	42	99.7	Peach	3.88	220	336	0.25	82	1.072	32.0	20.9	14.8
SINTOX FA	43	95	Dark pink	3.76	320	330		82	0.733	22.6		
SINTOX FC	44	95	Light pink	3.69	300	310		85	0.733	22.6		
SINTOX DD	45	97	White	3.76	280	340		80		28.4		
SINTOX HK	46	92	Dark brown	3.83	330	330		83		18.8		
SINTOX AZ	47	88		4.1	500	280				18.5		
A85	48		White	3.70	350			15.7		18		
A110	49		Pink	3.80	350			16.6				
A112	50		Brown	3.70	320			16.6		15		
A114	51		Ruby	3.85	350			16.6				

*Except where stated in notes.

Manufacturers/suppliers:　36–38　Duramic Products Inc.
　　　　　　　　　　　　39–42　Coors Ceramics Company (now owned by Vesuvius Zyalons).
　　　　　　　　　　　　43–47　Lodge Ceramics Limited (two more aluminas FF and EH are made).
　　　　　　　　　　　　48–51　Unilator Technical Ceramics (two more aluminas A73 and A113 are made, no % Al_2O_3 contents stated).

Notes:　(g)　Materials 36–38, only claim 9+ Mohs Scale. Materials 39–42, Rockwell 45N. Materials 48–51, Vickers Pyramid.
　　　　(h)　Materials 39–42 at 25°C.
　　　　(i)　Materials 36–38, temp. not stated. Materials 43–46, 25–40°C. Materials 48 and 50, 20 to 200°C.
　　　　(j)　Materials 36–38, temp. not stated. Materials 39–42, columns first 100–300°C, second 100–500°C, third 20–280°C. Materials 43–46., second column to 600°C. Materials 48–51, 20–100°C.
　　　　(k)　Materials 39–42, first column 200°C. Materials 43–46, first colum 50°C.
　　　　(l)　Materials 36–38, at 1 MHz. Materials 39–42 at 1 MHz. Materials 43–46, at 1 MHz.
　　　　(m)　Materials 39–42 at 1MHz. Materials 43–46 at 1 MHz.
　　　　(n)　Materials 36–38, thickness not stated, ASTM test. Materials 39–42, 2.54 mm thick. Material 38, 20°C, 2–3 mm thick.

Table 2.15.5—*continued*

TABLE 2.15.5 General properties of non-absorbent and vacuum-tight aluminas—*continued*

Coefficient of thermal expansion (10⁻⁶ K⁻¹)			Volume resistivity 10^{-2} ohm m (ohm cm)			Loss tangent (power factor) 10^{-4}			Permittivity			Dielectric strength (kV/mm)	Safe working temp (°C) (no load)
j (see note)			k (see note)			l (see note)			m (see note)			n (see note)	o
20–200°C	20–500°C	20–1000°C	20°C	400°C	600°C	20°C	200°C	400°C	20°C	200°C	400°C		
6.4			$>1\times10^{15}$			9.3			18			8.9	1650
7.4			$>1\times10^{14}$			9.3			9			9.0	1700
6.4			$>1\times10^{14}$			9.5			9			9.0	1700
7.3	7.7	8.1	1.3×10^{11}	1.4×10^{8}	3.9×10^{6}	7.1			11.5			22.6	1250
7.3	7.8	8.1	1.3×10^{11}	2.4×10^{7}	1.4×10^{6}	7.6			4.7			20.6	1450
6.9	7.7	8.1	7.1×10^{13}	1.0×10^{12}	5.0×10^{9}	8.1			5.5			22.0	1450
7.2	7.6	8.3	1.0×10^{14}	5.6×10^{11}	5.0×10^{8}	8.2			1.9			22.2	1700
5.0	7.3	8.8	1×10^{16}	7.0×10^{12}	9.5×10^{9}	9.2			2			20	1450
5.0	7.7	9.0	4.5×10^{14}	$<1\times10^{8}$	$<1\times10^{8}$	8.9			4.5				1450
4.5	7.8	9.2	$>1\times10^{16}$	1.0×10^{12}	1.5×10^{10}	8.31			2.5				1650
5.0	7.6	8.9	8.4×10^{12}			9.7			5				1150
													1300
7			1×10^{14}										1500
7			1×10^{14}										1300
7			1×10^{14}										1150
7			1×10^{14}										1600

Table 2.15.5—*continued*

TABLE 2.15.5—General properties of non-absorbent and vacuum-tight aluminas—*continued*

Material		Nominal Al_2O_3 content (wt %) a	Colour b	Specific density c	Flexural strength (MPa) d	Young's Modulus (GPa) e	Poisson's ratio f	Hardness (GN/m^2)* g (see note)	Specific heat $(10^3 J kg^{-1} K^{-1})$ h	Thermal conductivity (W/m per K) i (see note) 50°C	200°C	400°C
UL-500	52	94.0	White	>3.64	350			78		20.5		
UL-600	53	96.0	White	>3.69	370			79		25.5		
UL-300	54	97.6	White	>3.72	300			75		26.8		
UL-995	55	99.5	White	>3.84	315			81		29.3		
AL-500	56	94.0	White	3.67	338			78		20.5		
AL-600	57	96.0	White	3.72	358			79		25.5		
AL-300	58	97.6	White	3.76	290			75		26.7		
AL-995	59	99.5	White	3.86	305			81		29.2		
998	60	99.8	White	3.85	345				1.05	33.2		8.6
AP35	61	99.0	White	3.7	345							
THERMALOX	62	99.7	White	3.8								
PUROX	63	99.7	White	3.78	165							5.0
DURAFRAX 80	64	87.0	White	3.46	170	170		93		33.2	16.7	9.4
VISTAL	65	99.9	Trans-lucent White	3.99	280	385	0.22	85	0.88	39.7	13.4	6.3
LUCALOX 45D9	66	99.9	Trans-lucent		275	390	0.23	85	0.75	43.2	14.4	

*Except where stated in notes.

Manufacturers/suppliers:
52–55   Wade (Advanced Ceramics) Ltd. ⎫ These materials appear to be identical.
56–59   Wesgo Division, GTE Products Corp. ⎭
60–61   McDanel Refractory Company.
62   Thermal Syndicate Ltd.
63   Morgan Refractories Ltd.
64   Carborundum Company.
65   Coors Ceramics Company. (Company now owned by Vesuvius Zyalons).
66   General Electric USA.

Notes:
(g)   Materials 52–55, Rockwell 45N. Materials 56–59, Rockwell 45N. Materials 64, 65 and 66, Rockwell 45N.
(i)   Material 60, third column 800°C. Material 63, at 1100°C. Material 65 at 20°C, 400°C, and 800°C. Material 66, at 5°C, and 295°C.
(j)   Materials 52–55, second column 400–600°C, third column 800–1000°C. Materials 56–59, second column 400–600°C, third column 800–1000°C
(k)   Materials 52–55, second column 600°C, third column 900°C. Materials 56–59, second column 300°C. Material 66, first column 240°C.
(l)   Materials 52–55, at 10 MHz, second column 300°C, third column 500°C. Materials 56–59 at 10 MHz, second column 300°C, third column 500°C. Material 65, 1kHz–100 MHz.
(m)   Materials 52–55, as for note (l). Material 65, all at 25°C at frequencies of 1 kHz, 1 MHz and 100 MHz.
(n)   Materials 52–55, 2.55 mm thick under oil. Material 60, greater than 230 V per mm quoted for ⅛ in thick specimen. Material 65, 1.27 mm thick.

Table 2.15.5—*continued*

TABLE 2.15.5 General properties of non-absorbent and vacuum-tight aluminas—*continued*

Coefficient of thermal expansion (10⁻⁶ K⁻¹)			Volume resistivity 10⁻² ohm m (ohm cm)			Loss tangent (power factor) 10⁻⁴			Permittivity			Dielectric strength (kV/mm)	Safe working temp (°C) (no load)
j (see note)			k (see note)			l (see note)			m (see note)			n (see note)	o
20–200°C	20–500°C	20–1000°C	20°C	400°C	600°C	20°C	200°C	400°C	20°C	200°C	400°C		
6.3	8.0	9.1	>1×10¹⁴	4.6×10⁸	3.5×10⁶	9.07	9.53	9.91			3.4	25	1600
6.4	8.2	9.0	>1×10¹⁴	5.2×10⁸	4.1×10⁶	9.30	9.65	10.10			3.3	26	1620
6.9	8.5	9.0	>1×10¹⁴	2.3×10¹⁰	5.0×10⁸	9.53	9.91	10.14	0.4	1.6	5.2	43	1650
6.9	8.3	9.4	>1×10¹⁴	6.0×10⁸	2.5×10⁶	9.58	9.92	10.20	0.3	0.9	4.0	31	1725
6.3	8.0	9.1	>1×10¹⁴	2.0×10¹²	4.6×10⁸	9.07	9.53	9.91	0.26	0.28	3.4	25	1600
6.4	8.2	9.0	>1×10¹⁴	2.0×10¹²	5.2×10⁸	9.30	9.65	10.10	0.30	0.61	3.0	26	1620
6.9	8.5	9.0	>1×10¹⁴	1.0×10¹¹	2.3×10¹⁰	9.53	9.91	10.14	0.40	1.6	5.2	43	1650
6.9	8.3	9.4	>1×10¹⁴	2.0×10¹¹	6.0×10⁸	9.58	9.92	10.20	0.30	0.90	4.0	31	1725
6.2	7.4	8.5	>1×10¹³									>9.2	1950
			>1×10¹³									9.5	1900
	8												1950
	8.4												1950
8.0													1540
6.5	7.4	8.0				10.1			5	0.4	0.6	20	
6.6	7.2	8.3	8×10¹³	7×10¹²	2×10¹⁰	9.9			0.25			1700 v/mil	1900

Table 2.15.5—*continued*

TABLE 2.15.6 Some applications of alumina

Industrial segment	Application	Properties responsible for use
Aerospace and aircraft	Fire seals and cable insulators in aircraft engines. Precision bearings in control equipment. Coil formers in hydraulic and pneumatic control equipment. Substrates in electronic engine control equipment.	Electrical insulation. Wear resistance.
Textile (man-made fibre production)	Thread guides for textile machinery; eyelets, pins, rollers and specially designed parts.	Wear resistance.
Chemical engineering	Sealing rings in mechanical seals for pumps. Floats in chemical flow measurement. Plungers in precision metering pumps. Mixer valves for fluid control. Laboratory ware.	Corrosion resistance. Wear resistance.
Gas industry	Insulators and spark electrode assemblies for ignition of Domestic Appliances and Industrial Burners using natural and other gases.	Electrical insulation. High temperature strength and resistance.
Mechanical engineering General	Sealing rings and busbar insulator assemblies in electrolytic plating operation. Nozzles in acid recovery plant. Thermocouple sheaths in pickling plants.	Electrical insulation. Corrosion resistance.
Cigarette-making	Abrasion resistant parts for cigarette forming machines.	Wear resistance.
Paper-making	Abrasion resistant parts for paper-making machinery.	Wear resistance. Abrasiveness, hardness.
Cement, chemical, paint and fertiliser production	High density grinding media for milling abrasive materials.	
Machine tools	Tool tips for high speed machining of metal parts. Abrasive elements.	
Wire industry Fine wire production	Drawing cones for drawing fine copper and alloy wires.	Wear resistance.
Insulated conductors	Nozzles for plastic coating fine wires for Electrical and Electronic Industries.	
Transcontinental telephone cables and power cables	Eyelets, pulleys and guides in the manufacture of the cable.	
Electrical and electronics engineering Radio and telecommunications	Aerial Insulators for radio, radar and telephone systems. Sealed terminals in crystal filter units for telephone equipment. Insulators for surge divertors.	High temperature, high strength electrical insulation.
Transformers and power factor correction equipment	Sealed terminals and bushings for oil filled transformers and capacitors.	
Heating elements	Bushes, sleeves, beads and bobbins for insulating industrial and domestic heating elements.	
Power generation Nuclear	Fuel element thermal and electrical insulation. Thermocouple insulators for instrumentation.	Electrical and thermal insulation.
Coal and oil-fired	High Tension Oil Igniters. Rappers bars in electrostatic dust precipitators. Ash ejector nozzles.	Hot abrasion resistance.
Oceanography	Instrument shield for deep water equipment. Rings for lead/zirconate transducers.	Resistance to sea-water.

Table 2.15.6

TABLE 2.15.7 General properties of non-absorbent and vacuum-tight oxide ceramics other than alumina

Material		Nominal oxide content (wt %)	Colour	Specific density	Flexural strength (MPa)	Young's Modulus (GPa)	Poisson's ratio	Hardness (GN/m²)	Specific heat (10³ J kg⁻¹ K⁻¹)	Thermal conductivity (W/m per K)		
		a	b	c	d	e	f	g (see note)	h	i (see note)		
										50°C	200°C	400°C
CERALLOY 418	1	>99.9 BeO	White	>2.93	>240	>140		>11		>85		
THERMALOX 995	2	99.5 BeO	White	>2.80	>205	345	0.26	ca. 64	1.09	270	165	
CERALLOY 908C	3	See note ThO₂		>9.75	>205	>85		>9		>10		
T-744	4	Ti O₂ (% not given)	Black	3.7	164			8.15				
T-792	5	Ti O₂ (% not given)	Greenish Yellow	4.3	144	172	0.25	7.65		3.3		
M306	6	Ti O₂ (% not given)	Buff	4.1	215			9.8		5		
M353	7	Ti O₂ (% not given)	Buff	4.1	215			9.8		5		
ZFE	8	97% Zr O₂ Ca O stabilised	White	5.6	580	140			0.67	0.92	1.7	4
ZFME	9	96% Zr O₂ Mg O stabilised	White	5.6	580	180			0.67	0.92	1.7	4
ZR23	10	94% Zr O₂ 5% Ca O	White	>5.0	190	160		15	0.75	2		2.1
Z140	11	94% Zr O₂ 6% Ca O	White	5.15	230							
Zirconia	12	Zr O₂ partially stabilised with Mg O	Ivory	5.4	405			68				

Manufacturers/suppliers:
1   Ceradyne Inc. (All articles hot pressed. Sizes up to 250 mm dia.).
2   Brush Wellman Inc.
3   Ceradyne Inc. (Hot pressed. Sizes up to 250 mm dia. Presumably 99+% ThO₂.).
4–5   Kyocera Corporation. (Make five titania ceramics in all).
6–7   Unilator Technical Ceramics.
8–9   Le Carbone
10   Friedrichfeld GmbH (Frialit-Degussit) and Bush Beach Engineering Ltd.
11   McDanel Refractory Company.
12   Coors Porcelain Company

Table 2.15.7

TABLE 2.15.7 General properties of non-absorbent and vacuum-tight oxide ceramics other than alumina—*continued*

Coefficient of thermal expansion (10^{-6} K^{-1})			Volume resistivity 10^{-2} ohm m (ohm cm)			Permittivity			Loss tangent (power factor) 10^{-4}			Dielectric strength (kV/mm)	Safe working temp (°C) (no load)
j (see note)			k (see note)			l (see note)			m (see note)			n (see note)	o
20–200°C	20–500°C	20–1000°C	20°C	400°C	600°C	20°C	200°C	400°C	20°C	200°C	400°C		
		9	$>1\times10^{17}$			6.4		6.9	1	1	4	24	
8	8	9											1200
		9.3	$>10^{10}$										Very high
	7.7		1×10^{12}										1000
	9.6		1×10^{4}			46							1000
8.5			1×10^{12}										1000
8.5			1×10^{12}										1000
		9.0	1×10^{15}	$3.5\times10^{\circ}$	$1\times10^{\circ}$								2400
		9.5	1×10^{15}	$3.5\times10^{\circ}$	$1\times10^{\circ}$								2400
		10.9	1×10^{11}	4×10^{1}	$3\times10^{\circ}$							9.5	2300
			$>1\times10^{13}$										2300
		4.9											Not stated

Notes: (g) Material 2, Rockwell 45N. Materials 4–5, Vickers 500g. Materials 6–7, Vickers Pyramid. Materials 8–9, reported 8 Mohs scale only. Material 10, Knoop 100.
 (i) Materials 8–9, at 1000, 1750 and 2400°C. Material 10, values are 100 and 1000°C
 (j) Materials 4–5, 40–400°C. Materials 6–7, 20–100°C. Material 12, note lower thermal expansion, hence more resistant to thermal shock.
 (k) Material 3, noted as being in air. Material 5, note low resistivity. Materials 8–9, at 20, 1000 and 1950°C. Material 10, values are for 20, 1000 and 1500°C. Material 11, will be electrically conducting at 1000°C+.
 (l) Material 2, at 1 MHz, third column 500°C. Material 5, at 1 MHz.
 (m) Material 2, at 1MHz, second column 300°C, third column 500°C.
 (n) Material 2, thickness 0.032 inches. Material 11, 250 volts per mil on ⅛ in thick specimen.

Table 2.15.7—*continued*

TABLE 2.15.7 General properties of non-absorbent and vacuum-tight oxide ceramics other than alumina—*continued*

Material		Nominal oxide content (wt %)	Colour	Specific density	Flexural strength (MPa)	Young's Modulus (GPa)	Poisson's ratio	Hardness (GN/m²)	Specific heat (10^3 J kg^{-1} K^{-1})	Thermal conductivity (W/m per K)		
		a	b	c	d	e	f	g (see note)	h	i		
										50°C	200°C	400°C
ZFCE	13	97%Zr O$_2$ partially stabilised with Ca O		5.6	180	140				1.9		
ZFMES	14	97% Zr O$_2$ partially stabilised with Mg O		5.4	180	140				1.9		
ZFYT	15	95% Zr O$_2$ Yttria modified		6.0	900	220				3.0		
ZFYE	16	92% Zr O$_2$ Yttria modified		5.9						1.7		
Zirconia Foam	17	94% Zr O$_2$		3.0	30	6				0.17		
SINTOX ZM	18	Zr O$_2$ (% not given) partially stabilised with Mg O		5.9	700	190				3.6		
CERAFINE	19	Zr O$_2$ (% not given) Yttria modified		6.05	1000	200				1.9		
CERAFINE HTZ	20	High toughness zirconia (% not given)		6.15	512	215				1.9		
Z800	21	95% Zr O$_2$ Yttria modified		6.00	800	205				1.9		
Z100/Z200	22	89% Zr O$_2$		5.7	480	160				2.5		
TTZ	23	Partially stabilised transformation toughened zirconia (% not given)		5.75	638	200						
THORIA	24	99.5% Th O$_2$		9.25	140	240				1.04		
NILGRA MS	25	Zr O$_2$ partially stabilised with Mg O	White	5.74	820	205	0.31		0.47	3.08 (at 20°C)		2.44
NILCRA TS	26	Zr O$_2$ partially stabilised with Mg O	White	5.73	716	205	0.31		0.47	3.08 (at 20°C)		2.47

Manufacturers/suppliers: 13–17   Le Carbone.
18   Smiths Industries Ltd. (now owned by Lodge Ceramics Ltd.).
19–20   Tenmat.
21–22   Morgan Matroc.
23   Vesuvius Zyalon
24   Le Carbone.
25–26   Nilcra Ceramics Ltd.

Yttria modified zirconias are subject to severe degradation of properties in the presence of moisture in the temperature range 100–400°C.

Table 2.15.7—*continued*

TABLE 2.15.7 General properties of non-absorbent and vacuum-tight oxide ceramics other than alumina—*continued*

Coefficient of thermal expansion (10^{-6} K^{-1})			Volume resistivity 10^{-2} ohm m (ohm cm)			Permittivity			Loss tangent (power factor) 10^{-4}			Dielectric strength (kV/mm)	Safe working temp (°C) (no load)
j			k			l			m			n	o
20–200°C	20–500°C	20–1000°C	20°C	400°C	600°C	20°C	200°C	400°C	20°C	200°C	400°C		
9.0			1×10^9											2000
7.0			1×10^9											1600
11.0														1000
11.5			1×10^9											2000
9.0			1×10^9											2000
8.0			1×10^{13}											
8.0														
9.1			1×10^9											1000
10.0														2000
10.1														
5.2			1×10^{10}											
10.2			$>10^{11}$		5×10^3									
9.9			$>10^{11}$		2.7×10^4									

Notes: (g) Material 13, 1200 HV. Material 14, 1000 HV. Material 15, 1300 HV. Material 16, 1100 HV. Material 25, 1120 HV. Material 26, 1020 HV.

Table 2.15.7—*continued*

TABLE 2.15.8 Some applications of beryllia

Industrial segment	Application	Properties responsible for use
Nuclear power	Moderator Reflector Fuel element cladding	Low neutron capture cross section. Excellent high temperature properties.
Electronics	Heat sinks	High thermal conductivity. Good electrical insulating properties.
Metal processing	Crucibles, thermocouples, etc.	High temperature properties and resistance to reactive metals.

Table 2.15.8

TABLE 2.15.9 General properties of non-absorbent oxide compounds, porcelains, etc.

Material		Nominal composition	Colour	Specific density	Flexural strength (MPa)	Young's Modulus (GPa)	Poisson's ratio	Hardness (GN/m²)	Specific heat (10³ J kg¹ K¹)	Thermal conductivity (W/m per K)		
		a	b	c	d	e	f	g (see note)	h	i (see note)		
										50°C	200°C	400°C
LOTEC-M Cordierite D	1	2MgO 2Al$_2$O$_3$ 5SiO$_2$	Near White	2.53	195	132				3.5	2.8	2.7
Mullite	2	3Al$_2$O$_3$ 2SiO$_2$	Light buff	2.82	180	152			0.848	4.1	3.9	
Forsterite F-1100	3	2MgO SiO$_2$	Light yellow	2.8	150	145		8.6		3.3		
Steatite S-210	4	MgO SiO$_2$	White	2.7	135			6.3		2.5		
Zircon Z-380	5	ZrO$_2$ SiO$_2$	White	3.5	165	155		8.1		5.0		
Spinet P-2101	6	MgO Al$_2$O$_3$	White	3.3	145					1.7		
Mullite K-635	7	3Al$_2$O$_3$ 2SiO$_2$	White	2.6	135							
Spinel SP23	8	MgO Al$_2$O$_3$	White	3.3	175			17.5				5.5
Pythagoras 1800Z	9	78% Al$_2$O$_3$ 20%SiO$_2$	White	3.1	190					3.5		
Mullite MV33	10	3Al$_2$O$_3$ 2SiO$_2$	Off white	2.9	200					6.2		3.9
MV30	11	Mullite+ 15% Glass	Buff	2.8	190					5.8		3.6
MV20	12	Mullite+ 20% Glass	Buff	2.4	140					5.0		
Triangle Mullite	13	3Al$_2$O$_3$ 2SiO$_2$	White	2.69	86							2.56
Triangle Porcelain	14	Alumino-Silicate	White	2.40	72							1.99

Manufacturers/suppliers:
1   Asahi Glass Co. Ltd. (Low expansion, thermal shock resistant).
2   Coors Porcelain Company.
3–7   Kyocera Corporation. (Make more than one material in each class except spinel).
8   Friedrichsfeld GmbH (Frialit-Degussit).
9   W. Haldenwanger Technische Keramik GmbH. (Make two other 'Pythagoras' ceramics).
10–12   McDanel Refractory Company.
13–14   Morgan Refractories Ltd.

Notes:
(g)   Materials 3–5, Vickers 500g. Material 8, Knoop 100. Material 9,Mohs scale 8–9 is all that is stated. Materials 10–12, Mohs values 8, 7.5 and 7 given.
(i)   Material 1, RT, 400 and 800°C. Material 2, 25°C and 100°C. Materials 3–6 at 20°C. Material 8 at 1000°C. Materials 10 and 11 at 24 and 800°C. Material 12 at 24°C. Materials 13 and 14, at 1100°C.
(j)   Material 1, % at 600, 1000 and 1200°C. Materials 3–7, second column 40–400°C, third 40–800°C. Material 9 for 20–300, 20–700 and 20–1000°C. Materials 10 and 11, first column to 250°C.
(k)   Material 1, second column 500°C. Material 2, second column 500°C, third 1000°C. Materials 3–6 at 20, 300 and 500°C.
(l)   Material 1 at 1 MHz. Materials 3–6 at 1 MHz. Material 11, at 1 MHz.
(m)   Material 1 at 1 MHz. Materials 3, 4 and 6 at 1 MHz. Material 11 at 1 MHz.

Table 2.15.9

TABLE 2.15.9 General properties of non-absorbent oxide compounds, porcelains, etc.—*continued*

Coefficient of thermal expansion (10⁻⁶ K⁻¹)			Volume resistivity 10^{-2} ohm m (ohm cm)			Permittivity			Loss tangent (power factor) 10^{-4}			Dielectric strength (kV/mm)	Safe working temp (°C) (no load)
j (see note)			k (see note)			l (see note)			m (see note)			n	o
20–200°C	20–500°C	20–1000°C	20°C	400°C	600°C	20°C	200°C	400°C	20°C	200°C	400°C		
0.10	0.23	0.28	$>1\times10^{14}$	1×10^{11}		10.6			60			14	Not stated
3.7	4.2	5.0	$>1\times10^{13}$	2×10^{9}	3×10^{5}							9.8	1700
	10.6		$>1\times10^{14}$	$>1\times10^{14}$	1×10^{11}	6.5			1			10	1000
	7.2	8.1	$>1\times10^{14}$	1×10^{11}	1×10^{8}	6.3			6			9	1000
	3.9	4.5	$>1\times10^{14}$	1×10^{11}	1×10^{8}	9.4						10	1200
	8.1	8.8	$>1\times10^{14}$	1×10^{14}	1×10^{11}	8			1			10	1100
	4.6	4.8											1100
		8.3											1950
4.8	5.6	6.0	$>1\times10^{13}$									25	1800
3.9	4–7	5.5											1550
3.3	4.0	5.0	$>1\times10^{13}$			5.8			35			10	1475
		5.5											1700
		5.0											1550

Table 2.15.9—*continued*

TABLE 2.15.9 General properties of non-absorbent oxide compounds, porcelains, etc.—*continued*

Material		Nominal composition	Colour	Specific density	Flexural strength (MPa)	Young's Modulus (GPa)	Poisson's ratio	Hardness (GN/m²)	Specific heat (10³ J kg¹ K¹)	Thermal conductivity (W/m per K)		
		a	b	c	d	e	f	g (see note)	h	i		
										50°C	200°C	400°C
Porcelain K81	15	Alumino-silicate	White	2.4	95			8–9		1.05		
Steatite SV 140	16	MgO SiO₂	White	2.6	130			7–8		2.9		
Cordex 203	17	2MgO 2Al₂O₃ 5SiO₂	Grey	2.4	110			7–8		1.26		
Porcelain 165	18	Alumino-silicate	White	2.4	100					1.5		
Steatite 221K	19	MgO SiO₂	White	2.9	180					2.5		
Thomit 600	20	Alumino-silicate	White	2.9	200					2		
Aluminous Porcelain	21	55% Al₂O₃	Brown	2.7								
Steatite S52A	22	MgO SiO₂	White	3.0	140					3		
Siliceous Porcelain	23	Alumino-silicate	White	>2.32	120ᵃ	69			1.10	1.59		
Aluminous Porcelain	24	Alumino-silicate	White	2.4	165ᵃ	124				2.93		
LHF20	25	Steatite	White	2.8	138	138						
25A2	26	Cordierite	Buff	2.4	97	97			1.08	2.09		
591	27	Alumino-silicate	White	2.4	90							
307	28	Aluminous	White	3.0	200							
104	29	Cordierite	Light buff	2.4	100							

ᵃGlazed.

Manufacturers/suppliers: 15–17   Park Royal Porcelain Company Ltd.
18–20   Rosenthal Technik AG.
21   Thermal Syndicate plc.
22   Unilator Technical Ceramics.
23–26   AI Ceramic Products Ltd.
27–29   AG Hackney & Co. Ltd.

Notes: (g) Materials 15, 16, 17 Moh 8–9, 7–8 and 7–8 claimed.
(j) Material 18–20, at 20–300, 600 and 1000°C. Material 22, 20–100°C. Materials 23–26, middle column 20–600°C.
(k) Materials 15–17, second column 300°C. Materials 18–20, first column 200°C. Materials 23–25, and 26 at 20, 500 and 700°C.
(l) Materials 15–17, believed at 1MHz. Materials 18–20 at 1 MHz. Material 22 at 1 MHz. Materials 23–26 from 50 to 10³ Hz.
(m) Materials 15 and 16 at 1 MHz dissipation %. Materials 18–20 at 1 MHz. Material 22 at 1 MHz. Materials 23, 25, 26 at 1 MHz as are 27–29.
(n) Materials 18–20. ASTM test. Material 23, depending on test piece.

Table 2.15.9—*continued*

TABLE 2.15.9 General properties of non-absorbent oxide compounds, porcelains, etc.—*continued*

Coefficient of thermal expansion (10^{-6} K^{-1})			Volume resistivity 10^{-2} ohm m (ohm cm)			Permittivity			Loss tangent (power factor) 10^{-4}			Dielectric strength (kV/mm)	Safe working temp (°C) (no load)
j (see note)			k (see note)			l (see note)			m (see note)			n (see note)	o
20–200°C	20–500°C	20–1000°C	20°C	400°C	600°C	20°C	200°C	400°C	20°C	200°C	400°C		
		6.5	1×10^{14}	6×10^{6}	3×10^{4}	5.5			0.55			26	
		8.3	1×10^{15}	3×10^{7}	1×10^{5}	6.1						24	
		2.5	1×10^{14}	5×10^{7}	7×10^{5}	5.6						24	
4.8	6.1	5.1	1×10^{9}			6			90			>10	1250
7.0	7.7	8.5	1×10^{12}	1×10^{9}	1×10^{7}	6.4			2			>10	
5.3	6.1	6.2	1×10^{12}	1×10^{9}		7.5			10			>10	
4.5													1500
8			1×10^{14}			6			5				1000
5.4	7.2	6.4	1×10^{14}	7×10^{4}	8×10^{3}	6			80			10–30	
5.0	6.8	6.1				7							
6.0	7.3	8.4	$>1\times10^{14}$	2×10^{9}	2×10^{7}	6.5			5				
2.4	3.5	4.4	$>1\times10^{14}$	3×10^{5}	2×10^{4}	6.5			220				
4.2		5.9	$>1\times10^{11}$		2×10^{4}	6.0			65			26	1150
4.8		7.5	$>1\times10^{11}$		2×10^{4}	6.4			65			26	1150
1.5		3.3	$>1\times10^{11}$		1.2×10^{5}	6.0			65			>20	1150

Table 2.15.9—*continued*

TABLE 2.15.10 Characteristics and applications of engineering porcelains

Porcelain type	Characteristics	Applications
Electrical	Typical composition 50% kaolin (china clay), 25% potash feldspar, 25% quartz (milled sand). High resistivity and dielectric strength. Large glass content. Poor thermal shock resistance—improves with clay content.	Line insulators and switchgear for electrical distribution at high voltages and low frequency. Suspension insulators. Pin insulators. Bushings, switch parts.
Chemical	Silica replaced by other refractory oxides—mullite, alumina or zirconia. Free undissolved silica must be eliminated. Low iron content. Chemically resistant except for alkalis and hydrofluoric acid. High strength and good thermal shock resistance.	Piping and valve parts and linings in chemical plant.
Mullite	Aluminous chemical porcelain withstands high temperatures ($\sim$1500°C). Good thermal shock, wear and chemical resistance and hot strength.	Ball mills, milling media, pump parts, crucibles, combustion tubes, burner tips, acid resistant components, sparking plugs. Resistor cores.
Zircon	Incorporation of zirconium silicate (zircon) produces high density porcelains with high strength, chemical inertness and low dielectric loss.	Mill linings, grinding media, blasting nozzles, extrusion dies, burner tips and nozzles.
Lithium silicate	Incorporation of lithium carbonate produces, by suitable formulation, zero or negative expansion coefficients. Can be obtained with consequently excellent thermal shock resistance and dimensional stability over wide range of temperature. Not highly refractory.	Specialist applications.

Table 2.15.10

TABLE 2.15.11 Properties of non-absorbent dielectric ceramics (compositions not disclosed, presumably titanates, etc.)

Material		BS category	Colour	Specific density	Flexural strength (MPa)	Young's Modulus (GPa)	Poisson's ratio	Hardness	Specific heat (10^3 J kg^{-1} K^{-1})	Thermal conductivity (W/m per K) (see note)		
		a	b	c	d	e	f	g	h	i		
										50°C	200°C	400°C
L103	1	C320	Off-white	3.44	90							
M98	2	C330	Off-white	4.45	105							
M303	3	C310	Buff	4.15	170					5		
K200	4	C340	Buff	4.40	170							
K1500	5	C350	Tan	5.6								
K1800	6	C350	Cream	5.65								
K4000	7	C350	Brown	5.50								
K5000	8	C350	Brown	5.50								

Manufacturer: 1–8 Unilator Technical Ceramics.

Notes: (i)   Material 3, 20–100°C.
        (j)   Materials 1–4, 20–100°C.
        (l)   Materials 1–4, at 1 MHz. Materials 5–8 at 1 kHz.
        (m)   Materials 1–4, at 1 MHz. Materials 5–8, at 1 kHz.

Table 2.15.11

TABLE 2.15.11 Properties of non-absorbent dielectric ceramics (compositions not disclosed, presumably titanates, etc.)—*continued*

Coefficient of thermal expansion (10^{-6} K^{-1})			Volume resistivity 10^{-2} ohm m (ohm cm)			Relative Permittivity			Loss tangent (power factor) 10^{-4}			Dielectric strength (kV/mm)		Safe working temp (°C) (no load)
j (see note)			k			l (see note)			m (see note)			n		o
20–200°C	20–500°C	20–1000°C	20°C	400°C	600°C	20°C	200°C	400°C	20°C	200°C	400°C	AC	DC	
10			1×10^{13}			13			3			10–13	20–25	
9.5			1×10^{13}			38			3			10–13	18–22	
8			1×10^{13}			95			5			4–5	6–9	
10			1×10^{13}			190			5			4–5	6–9	
			1×10^{11}			2000			50			6	8	
			1×10^{12}			1800			250					
			1×10^{11}			4000			200					
			1×10^{11}			5500			200			4	6	

Table 2.15.11—*continued*

TABLE 2.15.12 Properties of selected porous oxide ceramics (including low thermal expansion materials)

Material		Nominal composition	Colour	Specific density	Flexural strength (MPa)	Young's Modulus (GPa)	Poisson's ratio	Hardness	Specific heat (10^3 J kg^{-1} K^{-1})	Thermal conductivity (W/m per K)		
		a	b	c	d	e	f	g (see note)	h	i (see note)		
										50°C	200°C	400°C
LOTEC-M/C	1	2MgO 2Al$_2$O$_3$ 5SiO$_2$	White	2.02	17					1.8		1.5
LOTEC-T/F	2	Al$_2$O$_3$ TiO$_2$	Grey	2.99	15					1.04		1.04
LOTEC-T/M	3	Al$_2$O$_3$ TiO$_2$	Grey	3.26	16					1.16		1.16
LOTEC-Z/F	4	ZrO$_2$	Buff	4.37	65					3.7		2.7
SF	5	SiO$_2$	White	>1.8	20–35	2.4		6	1.09	0.33		0.92
M120	6	Alumino-silicate	Tan	2.3	60	54		6	1.047	1.3		
C-195	7	Alumino-silicate	Pink	2.4	45	54		4	1.047	1.3		
MG25	8	MgO	White	2.5	50							
Alsint	9	Al$_2$O$_3$	White	3.3	100							
60NG	10	Sillimanite	Near white	2.68	65							
KS	11	Sillimanite	Near white	2.3	35					1.4		
C-1100	12	2MgO 2Al$_2$O$_3$ 5SiO$_2$	White	1.9						0.8		
K-692	13	3Al$_2$O$_3$ 2SiO$_2$	Light cream	2.2	43					1.3		
TR72	14	Cordierite	Buff	2.0	31				1.110			
MISC.	15	Mullite-Cordierite	Buff	2.0	20–30							
Silicon carbide fibres in silicon matrix	16	98% SiO$_2$ 2% Sil		2.0	75	40				1.0		
SILICA	17	99.7% SiO$_2$	White	2.05	75	50				0.99		
MAGNESIA	18	98% MgO	White	8.4						1.14		

a Open porosity.

Manufacturers/suppliers:
  1–4   Asahi Glass Co. Ltd.
  5   Desmarquest.
  6–7   Duramic Products Inc.
  8   Friedrichsfeld GmbH (Frialit-Degussit).
  9–11 W. Haldenwanger Technische Keramik GmbH.
  12–13 Kyocera Corporation.
  14–15 Al Ceramic Products.
  16–18 Le Carbone.

Notes:
  (g) Material 5, Mho. Materials 6 and 7, Mho. Material 16, 500 HV.
  (i) Material 5, at 20°C and 1000°C. Materials 12–13, 20°C.
  (j) Materials 1–4, % expansion at 600, 1000 and 1200°C. Materials 9–11, 20–300°C, 20–700°C 20–1000°C.
       Materials 12–13, 40–400°C, 40–800°C. Materials 14 and 15 middle column 20–60°C.
  (k) Material 4, 500 and 1000°C. Material 5, second and third columns 100 and 1000°C. Material 14 at 500 and 700°C.
  (l) Material 5 at 10 GHz. Materials 6 and 7 at 1 MHz.
  (m) Material 5 at 10 GHz. Materials 6 and 7 at 1 MHz.

Table 2.15.12

TABLE 2.15.12 Properties of selected porous oxide ceramics (including low thermal expansion materials)—*continued*

Coefficient of thermal expansion (10^{-6} K^{-1})			Volume resistivity 10^{-2} ohm m (ohm cm)			Permittivity			Loss tangent (power factor) 10^{-4}			Dielectric strength (kV/mm)	Safe working temp (°C) (no load)	Water absorption (%)
j (see note)			k (see note)			l (see note)			m (see note)			n	o	p
20–200°C	20–500°C	20–1000°C	20°C	400°C	600°C	20°C	200°C	400°C	20°C	200°C	400°C			
0.05	0.14	0.18												8.7
−0.05	0.12													5.4
−0.05	0.05													1.4
0.32	0.32	0.40		1×10^6	2.5×10^6									4.2
		0.6	6×10^{14}	8×10^{13}	1×10^3	3.17			2				1100	8–15[a]
3.3			$>1\times10^{14}$			5.3			100			3.9	1100	2–3
3.1			$>1\times10^{14}$			5.3			100			3.9	1050	<4
		13											2300	40[a]
7.0	7.6	8.0											1900	5
4.6	5.2	5.7											1800	5
4.6	5.3	5.7											1650	12
	0.3	1.2											1200	25
	4.3	4.8											1600	15–25
2.7	3.2	3.5		8×10^6	9×10^3									
2.8	3.3	3.5												
0.5									250				1200	10
0.4				1×10^{12}					40					
5.8														

Table 2.15.12—*continued*

2.15.3   Carbides, nitrides and borides

Silicon carbides have exceptional hardness with good thermal and electrical conductivity. Silicon nitride has exceptional environmental resistance at high temperature and good friction and wear characteristics, sialon has quite exceptional strength and wear resistance, and boron carbide exceptional hardness with good insulating properties.

The properties of a number of manufacturers' grades of silicon carbide are listed in:

Table 2.15.13   *Properties of silicon carbides*

Those of silicon nitride in:

Table 2.15.14   *Properties of silicon nitride ceramics, including Syalon 101*

Those of boron nitride in:

Table 2.15.15   *Properties of boron nitride ceramics*

and those of titanium diboride, including one boron carbide, in:

Table 2.15.16   *Properties of titanium diborides and boron carbide*

Typical applications of silicon nitride and silicon carbide are listed in:

Table 2.15.17   *Applications of silicon nitride and silicon carbide*

TABLE 2.15.13 Properties of silicon carbides

Material		Nominal composition	Colour	Specific density	Flexural strength (MPa)	Young's Modulus (GPa)	Poisson's ratio	Hardness (GN/m²)	Specific heat (10³ J kg⁻¹ K⁻¹)	Thermal conductivity (W/m per K)		
		a	b	c	d (see note)	e	f	g (see note)	h	i (see note)		
					20°C	20°C				50°C	200°C	400°C
CERAROI C–600	1	Si C	Grey–black	>3.12	>540	>370	0.15	>23.5			44	
C–400	2	Si C	Grey–black	>3.08	>390	>380	0.15	>22.5			43	
REFEL	3	Si C	Grey–black	3.10	525	413	0.24	>24		200	83	39
REFEL 'S'	4	Si C	Grey–black	≥3.10	275	400	0.24	>24		160		
HEXALOY SA	5	Si C	Grey–black	3.10	540	400	0.14	28	0.67	125	102	77
CARBOFRAX	6	Si C	Grey–black	2.58	260							16
REFRAX 20	7	Si C	Grey–black	2.62	43							
CERALLOY 1461	8	Si C	Grey–black	3.21	>205	345		>24		>130		
CERALLOY 146A	9	Si C	Grey–black	3.21	>380	410		>27		>155		
SC–201	10	Si C	Black	3.1	480	385	0.16	94	0.58	67.0		
SC–5500	11	Si C	Cream	2.2	38					12.6		
Several	12	Si C	Grey–Black	>3.1	>490	>470		25		>63		
Silicon carbide/graphite composite	13	% not given		1.86	35	14				52		

Manufacturers/suppliers:
1–2   Asahi Glass Company Ltd. (also make C–900 special grade for small size products).
3–4   Tenmat.
5–7   The Carborundum Company.
8–9   Ceradyne Inc.
10–11 Kyocera Corporation.
12   Sumitomo Electric Industries Ltd.
13   Morgan Matroc

Notes:
(d)   Materials 1–2, see also Column I.
(g)   Materials 1–2, Vickers 500 g. Material 5, Knoop. Materials 8–9, Knoop. Material 10, Rockwell A (units?).
(i)   Materials 1–2 at 100°C. Materials 3–4 at 20°C, 500°C and 1200°C. Material 6 at 1200°C. Materials 10 and 11, temperature not stated.
(j)   Materials 1–2, temperature not stated. Material 5, to 700°C. Material 6, 20–1540°C. Materials 10 and 11, 40–400°C and 40–800°C.
(k)   Materials 1–2, second column 350°C. Material 5, * * indicates can be varied by doping.
(l)   Material 7, at 1350°C.

Table 2.15.13

Carbides, nitrides and borides

TABLE 2.15.13 Properties of silicon carbides—*continued*

Coefficient of thermal expansion (10^{-6} K^{-1})			Volume resistivity 10^{-2} ohm m (ohm cm)			Flexural strength at high temp. (MPa)			Weibull Modulus		K_{IC} value MN/m$^{3/2}$	Safe working temp (°C) (no load)	Apparent porosity (%)
j (see note)			k (see note)			l (see note)			m		n	o	p
20–200°C	20–500°C	20–1000°C	20°C	400°C	600°C	1000°C	1200°C	1400°C	20°C				
4.2			>1×10^5	>1×10^3			>510	>440			5.1	1400	Nil
4.2							>380	>370			4.5	1400	Nil
		4.3				>400	>400	200				1400	Nil
		4.3											Nil
	4.02		**	**	**				10		4.5	1650	Nil
		4.7										1650	14
								44				1750	15
		4.5	>1×10^3									1650	Nil
		5.1	>1×10^{-1}									1650	Nil
	3.6	4.2										1400	0.2 W.A.
	5.4											1400	10.15 W.A.
	~4.5										2.5 to 3.5		
4.0												1800	

W.A. = Water Absorption

Table 2.15.13—*continued*

TABLE 2.15.14 Properties of silicon nitride ceramics, including Syalon 101

Material		Nominal composition	Colour	Specific density	Flexural strength (MPa)	Young's Modulus (GPa)	Poisson's ratio	Hardness	Specific heat (10^3 J kg^{-1} K^{-1})	Thermal conductivity (W/m per K)		
		a	b	c	d	e	f	g (see note)	h	i (see note)		
					20°C	20°C				50°C	200°C	400°C
Reaction bonded silicon nitride	1	Si_3N_4	Grey	2.5	205	165			1.06			
Hot pressed silicon nitride	2	Si_3N_4	Black	3.19		310		1800			250	140
N600	3	Si_3N_4	Black	>3.1	590	285	0.24	1400		18		
N400	4	Si_3N_4	Black	>3.08	390	275	0.24	1350		17		
N600A	5	Si_3N_4	Black	>3.1	635	285	0.24	1400				
Ceralloy 147A	6	Si_3N_4	Black	>3.1	>480	285		>1700		31		
RB Silicon Nitride	7	Si_3N_4	Grey	<2.7	205	160			0.71	14		
HP Silicon Nitride	8	Si_3N_4	Black	3.19	845	305		1800		25		
SN–201	9	Si_3N_4	Grey	2.9	385	230	0.26	87		12.6		
SN–220	10	Si_3N_4	Black	3.2	580	290	0.28	91		16.7		
SN–501	11	Si_3N_4	Dark Grey	3.3	335	170	0.27	87		8.4		
SYALON 101	12	SiAlON	Grey	3.25	845	280	0.23	91	0.62	21.3		
SYALON 201	13	Yttria modified		3.24	825							
SYALON 501	14	Yttria modified electrically conducting		3.95	825					19.1		
KERSIA 201 (Hot pressed)	15	95% Si_3N_4		3.15	450					20		
KERSIT 301 (Sintered)	16	95% Si_3N_4		3.2	900	290				29		
NITRASIL (Sintered)	17	Si_3N_4		3.25	700					25		
RN 175	18	Si_3N_4		3.26	670	800				30		

Manufacturers/suppliers: 1–2   Advanced Materials Engineering.
                  3–5   Asahi Glass Co. Ltd.
                  6   Ceradyne Inc.
                  7–8   Duramic Products Inc.
                  9–11   Kyocera Corporation.
                  12–14   Vesuvius Zyalon.
                  15–16   Le Carbone.
                  17   Tenmat.
                  18   Fairey.

Notes: (g) Material 2, Vickers kg/mm² load not stated. Materials 3–5, Vickers kg/mm² at 500 g. Material 6, Knoop 100, kg/mm².
                Material 8, Vickers, load and units not stated. Materials 9–12, HRA.
      (i) Material 2, third column 1400°C. Materials 3–4 at 100°C. Materials 9–12 at 20°C..
      (j) Material 2, first column at R.T., third column at 1450°C. Materials 3–5, temperature not stated. Material 9, 40–400°C and 40–800°C.
                Materials 9–11, 40–800°C.
      (k) Material 3 at RT and 350°C. Materials 9–11 at 20, 300 and 500°C.

Table 2.15.14

TABLE 2.15.14 Properties of silicon nitride ceramics, including Syalon 101—*continued*

Coefficient of thermal expansion (10^{-6} K^{-1})			Volume resistivity 10^{-2} ohm m (ohm cm)			Flexural strength at high temp. (MPa)			Weibull Modulus			K_{IC} value (MN/m$^{3/2}$)	Safe working temp (°C) (no load)	Water absorption or porosity
j (see note)			k (see note)			l			m			n	o	p
20–200°C	20–500°C	20–1000°C	20°C	400°C	600°C	1000°C	1200°C	1400°C	20°C					
							270		15					20% P
2		5.2	1×10^{12}											Nil
3.3			>1×10^{16}	>10^{13}		390	295					5.3	1000	Nil
3.3						345	245					4.5	1000	Nil
3.3						440	345					5.3	1000	Nil
		3.0	>1×10^{13}											Nil
<3.2			6.6×10^{10}										1300	<15 W.A.
3.6			1×10^{12}										1300	Nil
	1.9	3.0	>1×10^{14}	1×10^{12}	1×10^{9}								1200	<1 P
		3.2	>1×10^{14}	1×10^{12}	1×10^{9}								1200	<0.1 P
		3.5	>1×10^{14}	1×10^{12}	1×10^{9}								1200	0.3 P
		3.04	1×10^{12}						10–15			7.7		
													1400	
5.6												5.7	1000	
3.2												4.5	1300	Nil
3.1												7.5	1300	Nil
3.0			1×10^{13}									8.0		
3.3												7.0	1200	Nil

Table 2.15.14—*continued*

TABLE 2.15.15 Properties of boron nitride ceramics

Material		Nominal com-position	Colour	Specific density	Flexural strength (MPa)	Young's Modulus (GPa)	Poisson's ratio	Hardness	Specific heat (10^3 J kg^{-1} K^{-1})	Thermal conductivity (W/m per K)		
		a	b	c	d	e	f	g	h	i (see note)		
										50°C	200°C	400°C
COMBAT Grade A	1	BN	White	2.08	81 WG 96 AG	65 WG 41 AG		385		23.1 44.0	18.1 36.3	16.3 31.6
COMBAT Grade M	2	BN	White	2.12	98 WG 106 AG	93 WG 105 AG		89.5		9.1 25.2	7.2 19.8	6.8 15.9
COMBAT Grade HP	3	BN	White	1.90	53 WG 44 AG	43 WG 76 AG		205		60.6 39.7	52.5 31.2	44.0 28.8
Boron Nitride	4	96% BN	White	2.08	79 WG 94 AG	63 WG 40 AG				23		

Manufacturers/suppliers: 1–3 The Carborundum Company
4 Morgan Matroc Duramic AG & WG (similar to material 1 except for very different values for thermal conductivity).

Notes: General—W.G. = with grain, A.G. = against grain.
(i) Materials 1–3, at 100, 350 and 700°C. Material 4, single value given thus, cf. Materials 1–3.
(j) Materials 1–3, from 75°C. Material 4, no temperature stated.
(k) Materials 1–3, second column 150°C.
(l) Materials 1–4 at 1 MHz.
(m) Materials 1–4 at 1 MHz.
(o) Materials 1–4, first temperature for inert or reducing atmosphere, second for oxidising atmosphere.

TABLE 2.15.16 Properties of titanium diborides and boron carbide

Material		Nominal com-position	Colour	Specific density	Flexural strength (MPa)	Young's Modulus (GPa)	Poisson's ratio	Hardness	Specific heat (10^3 J kg^{-1} K^{-1})	Thermal conductivity (W/m per K)		
		a	b	c	d	e	f	g (see note)	h	i (see note)		
										50°C	200°C	400°C
CERALLOY 225	1	Ti B$_2$	Grey	>4.45	>240			>2500		>69		
Titanium diboride	2	Ti B$_2$	Grey	4.52	235	420			0.963	25.3		
TIB/400	3	Ti B$_2$ composite	Grey	2.98	81					24.2		
CERALLOY 546	4	B$_4$ C		2.45	240	440		>2800		>31		

Manufacturers/suppliers: 1 and 4 Ceradyne Inc.
2–3 Duramic Products Inc.

Notes: (g) Material 1, kg/mm^2, Knoop 100.
(i) Materials 1 and 2, temperatures not stated.
(j) Material 2, temperature not stated..
(o) Material 1, Oxidation resistant to this temp. Oxidation limit.

Table 2.15.15 and Table 2.15.16

TABLE 2.15.15 Properties of boron nitride ceramics—*continued*

Coefficient of thermal expansion (10^{-6} K^{-1})			Volume resistivity 10^{-2} ohm m (ohm cm)			Permittivity			Loss tangent (power factor) 10^{-4}			Dielectric strength (kV/mm)	Safe working temp (°C) (no load)	% Porosity or water absorption
j (see note)			k (see note)			l (see note)			m (see note)			n	o (see note)	p
20–200°C	20–500°C	20–1000°C	20°C	400°C	600°C	20°C	200°C	400°C	20°C	200°C	400°C			
	2.3 1.1	2.8 0.9	>2×10^{14}	3×10^{13}		4.08			3.4			36	2775 985	1.1 W.A.
	1.8 0.2	2.5 0.4	>2.5× 10^{14}	3.2× 10^{13}		3.7			15			39	1400 1400	0.04 W.A.
	0.0 0.0	0.0 0.0	1.6× 10^{12}	4.7× 10^{12}		4.11			11			31	2775 1200	0.16 W.A.
4.1 2.0			>2×10^{14}			4.08			3.4			37	2775	1.1 W.A.

TABLE 2.15.16 Properties of titanium diborides and boron carbide—*continued*

Coefficient of thermal expansion (10^{-6} K^{-1})			Volume resistivity 10^{-2} ohm m (ohm cm)							Safe working temp (°C) (no load)	Water absorption or porosity (%)
j (see note)			k			l	m		n	o (see note)	p
20–200°C	20–500°C	20–1000°C	20°C	400°C	600°C						
		8.2	13–17×10^{-6}							1040°C	Nil
8.1			9–15×10^{-6}							1200°C	Nil
7			5–10×10^{-4}								Nil
		5.6	10^{-1} to 10^{1}							600°C	

Table 2.15.15 and Table 2.15.16—*continued*

TABLE 2.15.17 Applications of silicon nitride and silicon carbide

Material	Industrial segment	Possible applications	Relevant properties
Silicon nitride	Mechanical engineering	Plain bearings. Piston rings. Shaft seals in sea water pumps. Extrusion and wire drawing dies. Air bearings. Mechanical seals. Bearing plates.	Abrasion resistance. Anti-galling tendency. Low coefficient of friction. Low thermal expansion coefficient. Resistance to sea water.
	Chemical processing	Swirl chambers and paddle vanes. Electro-chemical machining, fitting and fixtures. Seals and valves. Nozzles.	Chemical resistance.
	Furnace refractories	Burner nozzles. Furnace runners and hearth rollers.	High temperature strength and stability.
	Aluminium and like industries	Thermocouple sheaths. Launders. Pump components. Immersion heating tubes. Plugs and valve seats. Metering valves. Filters.	High temperature strength and stability. Resistance to molten metals.
	Heat exchangers	Tube plates. Regenerator discs. Headers. Recuperator tubes. Diffusers.	High temperature strength and stability. Low coefficient of thermal expansion.
	Precious metal refining	Crucibles. Combustion boats.	Inert and resistant to molten metals.
	Industrial Heating	Furnace supports. Silver soldering. Vacuum brazing. Heating tubes. Jigs and fixtures. Copper brazing. Forging bushes. Saggers.	High temperature strength and stability.
Silicon carbide	High temperature technology	Gas turbine components. High temperature furnace tubes. Resistance heaters.	High temperature strength, oxidation resistance. Thermal shock resistance. Oxidation resistance. Electrical conductivity.
	Mechanical engineering	Abrasive elements. Abrasion-resistant components. Non-lubricated bearings.	Wear resistant properties. Hardness. Low friction, minimal wear.
	Electrical	Susceptors in HF induction fields.	Electrical properties.

Table 2.15.17

2.15.4 Cemented carbides ('hard metals')

Cemented carbides consist of refractory metal carbide compounds (chiefly tungsten and titanium carbides, but also chromium, tantalum and niobium carbides) incorporating a metallic binder material (usually cobalt, but sometimes, chromium, nickel or nickel–molybdenum binder are used).

They are formulated to maximise hardness, abrasion wear and corrosion resistance, to have good resistance to galling and good friction properties so that they can be used for cutting, working and crushing metals and minerals.

Cemented carbides are 'cermet' materials (i.e. ceramic metal composites) and thus exhibit a combination of ceramic-like and metallic properties. Like ceramics, they have high stiffness, low thermal expansion and fairly low specific heat, but they are electrically conducting and have better thermal conductivity than most ceramics. Most grades are also weakly magnetic although non-magnetic grades (employing Ni binders) are also available.

The properties of cemented carbide materials are affected by the composition and grain-size of the carbide constituents and by the composition and percentage of binder present. The higher the proportion of binder metal the less brittle the material.

In general, increasing the amount of cobalt binder in a tungsten carbide grade will decrease the hardness, density, elastic modulus and wear resistance of the material while increasing its impact resistance. There is thus a trade-off between hardness and toughness at increasing binder content.

Carbides containing titanium as well as tungsten have good friction properties and excellent resistance to galling. Corrosion of cemented carbides usually occurs by the selective attack of the binder constituent leading to pitting and progressive disintegration. Low-cobalt grades or grades utilising nickel or chromium–cobalt binders, may be preferred in corrosive environments.

Tungsten base carbides can be used in oxidising conditions up to about 550°C and to about 850°C in non-oxidising environments. Titanium base carbides can be used continuously at temperatures up to about 1100°C.

The lower binder content materials, and the mixed tungsten/titanium carbides are principally used for cutting tool bits. Where the requirement to resist abrasion is paramount, the cemented carbides are now increasingly augmented by ceramics, titanium carbide/alumina mixtures, aluminas and sialons. The higher binder content (9–12%) materials are used for heavy shock applications while materials with 16–20% cobalt are used for steel extrusion dies, rock cutting drills, crusher rolls and ball mill liners.

Mixtures of titanium and tungsten carbides are used for drawing dies and mandrels. Titanium carbides are used for wear resistance at high temperature.

The properties of a few of the many materials available are tabulated in:

Table 2.15.18 *Composition and properties of cemented carbides (hard metals) and other metal cutting materials.*

Attachment of cutting bit to Shank

Cutting bits, which are expensive and brittle are made to the minimum size that will perform satisfactorily. They are attached to a shank made from a cheaper and tougher material by methods listed and assessed in:

Table 2.15.19 *Attachment and assembly methods for cemented carbide inserts*

1438   Engineering ceramics

TABLE 2.15.18 Composition and properties of cemented carbides (hard metals) and other metal cutting materials.

Material		Chemical composition (%)				Grain size (μm)	Flexural strength (MPa)	Specific density
		WC	Ti C	Ta (Nb) C	Co/Ni			
		a	b	c	d	e	f	g
WHO5	1	96.9	0.1		3.0	0.8	2200	15.25
WH25	2	89.5		0.5	10.0	0.7	3000	14.50
WN50	3	85.0			15.0	1.8	3095	14.0
WT70	4	75.0			25.0	1.4	3200	13.20
SO5T	5	60.5	22	11	6.5	2.3	1200	10.0
U20T	6	77	4	10	9	3	1800	13.3
H40T	7	87.2	0.2	0.6	12	2	2300	14.3
B10T	8	94			6	2.3	1800	14.9
B50	9	85			15		2500	14.0
GO3	10	97			3		1100	15.1
MC2	11		100 TiC+ Al$_2$O$_3$			≤2	500	4.3
AC5	12		100 Al$_2$O$_3$			>2	480	3.99
TC30	13						1450	6.0
TC50	14						1850	7.7
NB905	15		100 TiC+Al$_2$O$_3$				930	>4.24
W80	16		100 Al$_2$O$_3$				780	>3.96
KO60	17		100 Al$_2$O$_3$				595	
KO90	18		100 TiC +Al$_2$O$_3$				890	
K68	19	Cemented carbide					2030	
KYON 2000	20	Sialon					735	

Manufacturers/suppliers: 1–4   Sandvik Ltd.
5–8   Metallwerk Plansee Gesellschaft GmbH.
9–12   GH-Metall Günther Hertel GmbH & Co. KG.
13–14   Kyocera Corporation.
15–16   Sumitomo Electrical Industries Ltd.
17–20   Kennametal Inc.

Note: (h)   Vickers 500g.

Table 2.15.18

TABLE 2.15.18 Composition and properties of cemented carbides (hard metals) and other metal cutting materials—continued

Hardness Vickers	Young's Modulus (GPa)	Poisson's Ratio	Coercive force (kA/m)	Resistivity micro (ohm cm)	Thermal conductivity (W/m per K)	Thermal expansion coefficient (10^{-6} K^{-1})	Remarks
h (see note)	i	j	k	l	m	n	o
				20°C	20°C	20–400°C	
>1900	675	0.22	>29.8	0.20	100	5.0	Leaflet HS-9950-ENG includes some figures for fracture toughness (relative)
1625	580	0.22	20.4	0.19	120	5.5	
1150	524	0.22	7.6	0.17	115	6.1	
950	462	0.24	7.8	0.16	95	6.7	
1650	490	0.22	11.9	91	25	7	
1450	550	0.22	10.5	28	60	5.5	
1300	600	0.22	10.7	17	90	5.5	
1450	650	0.22	11.9	19	100	5	
1070							
1670							
2000							Finishing cast iron etc.
1700							Roughing and finishing cast iron
1620	440	0.21			17	7.4	Cutting tool tips and other wear resistant parts
1470	410	0.23			13	7.8	
3000					21		Hot isostatically pressed
2400					16		
	430					8.2	* 5.8
	410					8.3	* 7.1
	600					6.0	*13.2
	295					3.2	* 7.3 (**)

*K_{IC} expressed as MPa m$^{1/2}$.
**Very high abrasion resistance claimed.

Table 2.15.18—continued

TABLE 2.15.19 Attachment and assembly methods for cemented carbide inserts

Classification	Design considerations and limitations

Mechanical fastening
—clamp mounting
  screw mounting
  wedge mounting
  shrink fit mounting, etc.
	Probably the most widely used attachment method. Thermal expansion mismatch problems are restricted to the range of temperatures encountered in service.
	
	Steel mounting parts can be heat-treated to desired hardness before insert assembly.
	
	Positive stops and shoulders should be provided as well as clamping forces. Clamping forces should operate in the same direction as the operating thrust for maximum support.
	
	Assemblies should be designed to take account of the high compressive strength of the brittle insert. Tensile stresses in the carbide are to be avoided.

Brazing	Cemented carbides can be brazed using conventional copper and silver alloy brazes. Thermal expansion mismatch may result in curvature of the assembly after brazing or in extreme cases, thermal strain may lead to fracture of the carbide. Brazing is therefore restricted to small or thin inserts. Larger inserts can be brazed by using a 'sandwich braze' containing a ductile copper shim capable of absorbing thermal strain by plastic flow.
Cementing	Carbide inserts can be assembled using epoxy-resin adhesives. These are limited by the fact that service temperatures must be below about 120°C and that bond strengths are only about 50 MPa (cf. 500 MPa for a brazed joint).

Table 2.15.19

2.15.5   Carbons and graphites

Both materials are electrically conducting, with good thermal conductivity and fair resistance to thermal shock. They are chemically inert to most corrosive gases and fluids, and, although they are limited to service temperatures of 500–600°C in air, they can be used at up to 3000°C in inert atmospheres. Carbons and graphites are relatively soft materials which can be machined to close tolerances using conventional machining techniques. They are self-lubricating and can operate wet or dry in frictional-contact conditions. These properties promote a very extensive field of applications which are summarised in:

Table 2.15.20   *Applications of carbon and graphite*

MATERIAL FORM

Carbon and graphite are available as solids, fibres or powder. The solid forms include vitreous carbon and pyrolytic graphite.

Vitreous carbon is a dense, non-porous form of carbon with the physical appearance of black glass. It is pure (99.98%C), has a good surface finish and is harder and stronger than graphite. It is non-porous and available only in small sizes.

Pyrolytic graphites are dense, impervious and, depending on the conditions of deposition, can be highly anisotropic.

Graphitisation

The properties of the hard carbon/graphites produced by the above process can be modified by a further heat treatment at 2500–3000°C to produce materials known as 'electrographites'. These have a more ordered crystal structure than ungraphitised materials and the properties are influenced by the degree of graphitisation which has occurred.

> thermal conductivity
> electrical conductivity
> lubricating properties   } increase with increasing graphitisation
> machinability
> anisotropy of properties

> hardness
> strength   } decrease with increasing graphitisation
> wear resistance

A selection of carbon and graphite materials and their more important properties are listed, according to material type in:

Table 2.15.21   *Typical properties of some industrial carbon and graphite materials*
and according to manufacturers' designation in:

Table 2.15.22   *Properties of carbons and graphites*

TABLE 2.15.20 Applications of carbon and graphite

Use	Requirement	
Brushes, contacts, current collectors for electrical equipment.	Good electrical conductivity with good lubrication properties.	Hard carbons of high resistance for fractional hp motors with difficult commutation. Resin-bonded graphites allow higher currents and faster rubbing speeds. Copper containing graphites are used for high current, low voltages. Electrographites are widely used.
Electrical components especially resistors and thermistors.	Electrical conductivity	
Electrodes for welding, brazing and electrodischarge machining. Spectrographic electrodes.	Electrical conductivity. Refractoriness.	
Heating elements for resistance and induction heating.	Electrical conductivity and susceptibility. Refractoriness. Mechanical strength.	Suitable grades of electrographite.
Jigs and formers in glass industry. Brazing fixtures. Dies for continuous casting and hot-pressing. Boats for sintering and zone-refining.	Refractoriness, chemical inertness. Not wetted by molten metal.	Carbon and electrographite grades.
Refractory brick furnace liners.	Refractoriness, good thermal shock resistance.	Carbon (or graphite) bonded with bitumen and fired.
Thermal insulation.	Low thermal conductivity.	Felt forms. Carbon bonded carbon fibre.
Bearings, pistons.	Good self-lubrication properties where conventional lubricants cannot be used because of: (1) high temperature (2) corrosive environment (3) risk of product contamination, etc.	
Seals, rings and gaskets.		Metal or resin impregnated grades for seals to avoid porosity. Carbon felt for gaskets.
Nuclear applications.	Low neutron scattering cross-section and resistance to radiation damage.	Specially developed graphite grades available for nuclear applications.
Rocket nozzles.	Low density, chemical inertness refractoriness.	May be tungsten coated. Vitreous carbon may be used.
Use in composite materials.	Reinforcement for polymer (and other matrix reinforcement). Filler for plastic–rubber materials.	High strength, high modulus carbon fibre. Carbon powder.

Table 2.15.20

TABLE 2.15.21 Typical properties of some industrial carbon and graphite materials

Material		Specific density	Apparent porosity (% vol.)	Flexural strength (MPa)	Compressive strength (MPa)	Young's Modulus (resonance method) (GPa)	Hardness (Shore Schleroscope)[a]
		1	2	3	4	5	6
Industrial carbons	A	1.3–1.8	10–15	35–50	80–160	6–15	55–90
Electrographites	B	1.5–1.9	6–25	6–60	15–120	5–15	30–70
Resin impregnated carbons	C	1.7–2	0.5–5	35–80	120–250	17–22	50–80
Metal impregnated carbons	D	2–3	<1	70–100	100–300	15–40	60–90
Vitreous carbon	E	1.5	0	60–100	150–300	26	120
Pyrolytic graphite	F[b]	2.2	0	240[b]	350[b]	—	—
Graphite foam	G[c]	0.07–0.18	80–95	—	Crushing strength ~3	—	—
Carbon and graphite felts	H[d]	0.07–0.15	—	Tensile[d] strengths 1–3 (a) 0.5–2 (b) Elongation[d] 5–15 (a) 25–30 (b)	—	—	—

[a] 6 The hardness of carbons and graphites is usually measured by the 'Shore Schleroscope' technique. The relative height of rebound of a metal ball from the surface of the material is taken as a measure of the hardness.

Indentation hardness methods are used to characterise graphite hardness although their use is not universal.

[b] F In pyrolytic graphite, the crystals are arranged in a highly ordered structure with the C-axis of the crystal cell parallel to the growth direction (i.e. the thickness of the deposit) and the hexagonal sheets (the 'a' planes of the crystal structure) almost parallel in the plane of the deposit.

Properties such as strength, conductivity and thermal expansion are therefore highly anisotropic and widely different values will be obtained for measurements parallel to the thickness and within the plane of the deposit.

F.3 Tensile stresses within the plane of the deposit.

F.4 Strength perpendicular to the plane of the deposit.

F.8 Parallel to the plane of the deposit.

F.9 Perpendicular to the plane of the deposit.

[c] G
[d] H } Properties dependent on grade of material.

H3 (a) along the length of the felt.
  (b) along the width of the felt.

H10 The range of figures quoted is the resistance in ohms of a sheet of felt material of square shape. Exact values depend on the thicknesses and structure of individual felt grades.

Table 2.15.21

TABLE 2.15.21 Typical properties of some industrial carbon and graphite materials—*continued*

Specific heat (10^3 J/kg per K)	Thermal conductivity (W/m per K)	Coefficient of thermal expansion (10^{-6} K^{-1})	Electrical resistivity (10^{-2} Ω m)	Comments
7	8	9	10	
0.8	5–30	1.3–5	$35–60\times10^{-4}$	
0.75–0.8	50–200	1.3–5	$6–40\times10^{-4}$	
0.8–1.1	5–30	1.3–5	$20–50\times10^{-4}$	Impregnations stable to 200–250°C. Acid and alkali-resistant impregnators available.Sections over 100 mm may be difficult to impregnate completely.
0.45–0.8	15–80	2–5	$1–20\times10^{-4}$	
0.8	4–8	3	50×10^{-4}	Tube and plate form in thicknesses up to 3 mm. Dish and crucible forms also available. Grades to withstand 1000°C or 2500°C in inert atmospheres are available.
0.8	375 (a) >10 (c)	0 (a) 23 (c) (o–300°C)	$2–4\times10^{-4}$ (a) 100–1000 $\times10^{-4}$ (c)	Available in thicknesses up to 8 mm.
0.8	0.07–0.2	3	$1–3\times10^{-1}$	
—	0.05–0.01	—	0.2–2^d ohms per square	Flexible materials which can be cut to shape with scissors.

Table 2.15.21—*continued*

TABLE 2.15.22 Properties of carbons and graphites

Material [b]		Graphite or carbon	Apparent specific density	Compressive strength (MPa)	Flexural strength (MPa)	Tensile strength (MPa)	Tensile strain to failure (%)	Poisson's Ratio
		a	b	c	d	e	f	g
ACF-10Q	1	Graphite	>1.78	193	110	69	0.62	0.16
AXF-5Q	2	Graphite	>1.80	131	83	55	0.82	0.15
AXZ-5Q	3	Graphite	>1.50	69	39	27		
D-555	4	Not stated	1.74	110	55	48		
D-657	5	Not stated	1.72	80	29	27		
MN	6	Graphite	>1.50	15	>4			
MN impregnated	7	Graphite resin	1.85	80	30			
R	8	Fine graphite	1.75	65	45			
MDS	9	Carbon	1.70	50	16			
CY10	10	Carbon	1.60	136	44			
CY9	11	Graphite (extruded)	1.65	45	23			
CY2C	12	Carbon/ resin	1.85	160	55			

Highest strength figure reported is quoted. For extruded grades strength is directional. For both materials there is great dependence on size of piece as formed, and on size of test piece.).

[a] Unitless   [b] Most manufacturers make a wide range of types. Selected ones only are mentioned here.

Manufacturers/suppliers: 1–3 Poco Graphite Inc.
  4–5 Duramic Products Inc.
  6–9 Sigri Engineering (property values for Material 6 are especially dependent on size). Also suppliers of graphite foams, felts, etc.
  10–12 Morganite Special Carbons Ltd..

Notes: (j) and (k) All materials—differences presumed due to different temperature ranges, not always stated.

Table 2.15.22

TABLE 2.15.22 Properties of carbons and graphites—*continued*

Young's Modulus (GPa)	Hardness— Shore Schleroscope[a]	Coefficient of thermal expansion (10^{-6} K^{-1})	Thermal conductivity (W/m per K)	Specific heat (10^3 J kg^{-1} K^{-1})	Electrical resistivity ohm cm $\times 10^{-4}$	Oxidation threshold (°C)	Porosity (%)
h	i	j (see note)	k (see note)	l	m	n	o
11	100	8.3	86.5		4.2	350	19
11	70	7.7	121		2.0	460	19
9	52	7.3	76		3.1	425	32
10	70	7.6	104	0.84	2.4		23
9	50	4.2	63	0.84	3.1		19
					<1.8	500	<25
					1.2	165	Zero
					0.9	500	8
					4.5	350	16
10	70	3.6	13				12
4	45	1.4	46				18
13	80	4.7	13				2

Table 2.15.22—*continued*

2.15.6   Glasses

Glasses are highly viscous, supercooled liquids produced from oxides, oxide mixtures or compounds high in silica. Their structure is amorphous and they are usually transparent, chemically resistant and non-porous. A selection of glasses, together with their more important properties, is listed, according to material type in:

Table 2.15.23   *Typical properties of commercial glasses*

APPLICATIONS

The engineering applications of glass usually make use of its light transmitting and refracting properties, or its material-containing capability. The applications of the principal glass types are listed in:

Table 2.15.24   *Applications of commercial glasses*

One major field of application is process plant. The reason for using glass in a number of processes, and the components available, are summarised in:

Table 2.15.25   *The use of glass in process plant*

TABLE 2.15.23 Typical properties of commercial glasses

Material	Specific density	Refract. index (sodium D line)	Thermal expansion coefficient 0–300°C $(10^{-7} K^{-1})$	Annealing temperature (°C)	Softening point (°C)	Log of volume resistivity 250°C (ohm–cm)	Dielectric constant at 20°C	Young's Modulus (GPa)	Poisson's Ratio	Vicker's hardness
Vitreous silica	2.20	1.458	5.4	1,140	1,667	12.0	3.8	75	0.16	700
Soda-lime silica	2.46	1.510	85	548	730	6.5	7.0	75	0.21	550
Borosilicate	2.23	1.474	32	565	820	8.1	4.6	60	0.22	600
Aluminosilicate	2.53	1.534	42	715	915	11.4	6.3	90	0.26	600
Glass fibre (soda-lime)	2.46	1.512	87	528	710	6.7	7.9	75	–	–
Glass fibre (E glass)	2.53	1.548	80	575	830	15	6.4	80	–	–

Table 2.15.23

TABLE 2.15.24 Applications of commercial glasses

Glass type	Applications	Limitations
Pure silica glass	Chiefly used for its low thermal expansion and transparency to wide range of wavelengths and its high service temperature for a glass.	Very high temperature needed for manufacture.
Vitreous silica	Lightweight mirrors for satellite telescopes, laser beam reflectors. Special crucibles for manufacture of single crystal silicon for transistors. Molecular sieve that lets through hydrogen and helium.	Expensive.
96% Silica glass (3% B_2O_3)	Made by forming an article, larger than required size, from a special borosilicate glass, leaching out the non-silicate components with acid, treating at high temperature to shrink the article and close the pores. Higher service temperature and lower expansion coefficient than all glasses except vitreous silica. Missile nose cones. Windows of space vehicles. Some laboratory glassware.	More expensive than borosilicate glass.
Soda-lime silica glasses	Commonest of all glasses used for plate and sheet (including windows), containers, lamp bulbs, etc.	Magnesia or alumina sometimes incorporated to improve corrosion resistance. Poor thermal shock resistance compared with borosilicate glasses.
Borosilicate glasses	Low thermal expansion and high dielectric strength. Good chemical resistance. Laboratory glassware. Industrial piping. High temperature thermometers. Large telescope mirrors. Household cooking ware (e.g. 'Pyrex'). Envelopes for hot lamps and high wattage electronic tubes.	High softening temperature makes it harder to work than lead glasses or soda-lime glasses.
Aluminosilicate glasses	Low expansion, chemically resistant glass with higher service temperature than borosilicate glass. High performance military power tubes, travelling wave tubes and similar applications to borosilicate. Particularly resistant to alkali if non-boron-containing compositions are used.	Harder to fabricate than borosilicate.
Alkali-lead silicate glasses	Lead reduces softening point and increases refractive index and dispersive power. Used as flint glass in optical applications and as 'crystal glass' for tableware. Thermometer tubes. Neon-sign tubes.	Lower service temperature than borosilicate or soda-lime glasses.
Others	Photochromic glasses—darken reversibly on exposure to UV light (Chance–Pilkington, Corning, Schott). Semi-conducting glasses for electronics applications, solder glasses and glasses for glass–metal seals, etc. (Corning). Opthalmic glasses (Chance–Pilkington).	

Table 2.15.24

TABLE 2.15.25 The use of glass in process plant

Examples of processes using glass	Advantages	Available components
Nitric acid concentration Bromine recovery–production Hydrochloric acid production and purification Sulphur trioxide absorption Phenol recovery Solvents recovery (Corning, Schott)	**Corrosive resistance**—affected only by hydrofluoric acid, phosphoric acid and hot, strong caustic solutions. **Non-contamination of products**—glass is an inert material and connections between various components are usually compression joints in which only PTFE gaskets (similarly inert) contact the product. **Transparency**—means that the process can be checked visually and blockages located easily. Amber-stained glass can be used if protection of product from light is required. **Wide operating temperature range**—from −200 to 300°C. Sudden temperature changes of more than 120°C may lead to thermal shock failure. **Integrity**—spherical vessels can be used under full vacuum.	Pipeline available in sizes from 15 to 600 mm bore and lengths to 3000 mm. Reaction vessels in a variety of shapes up to 40 litres capacity; spherical vessels can be used under full vacuum.

Table 2.15.25

2.15.7   Glass–ceramics

The properties of glass–ceramics depend on the composition of the starting glass and can be to some extent tailored to particular applications; in general, glass–ceramics are 70–100% crystalline with the remainder a glassy phase; there is no porosity. As a family, they are from three to four times stronger than glass and have unabraded strengths comparable to high density aluminas. Abraded strength is usually reported, however. They have good thermal shock resistance due to their high strength and low thermal expansion coefficients. They have good wear resistance and are resistant to corrosion.

A glass–ceramic, Corning's 'MACOR', can be machined to precise tolerances with conventional metalworking tools. It is supplied in standard rod and sheet. Some glass ceramics are photosensitive (their light transmission decreases with increase in intensity of incident light). The properties of typical glass ceramics are listed in:

Table 2.15.26   *Properties of glass–ceramics*

and compared with ceramics and glass reinforced polymers in:

Table 2.15.27   *Comparative properties of glass–ceramics*

APPLICATIONS

The applications for glass–ceramics include those dependent on the general properties (higher strength and heat resistance compared with glass) and those dependent on some special property (low thermal expansion or photosensitivity). Applications for glass ceramics and the properties which promote their use are listed in:

Table 2.15.28   *Applications for glass–ceramics*

TABLE 2.15.26 Properties of glass–ceramics

Material		Type of appli-cation	Colour	Specific density	Flexural strength (MPa)	Young's Modulus (GPa)	Poisson's ratio	Hardness (Knoop)	Specific heat (10^3 J kg^{-1} K^{-1})	Thermal conductivity (W/m per K)		
		a	b	c	d (see note)	e	f	g	h	i (see note)		
										50°C	200°C	400°C
MACOR	1	Machinable	White	2.52	100	62		250	0.75	1.30	1.32	1.35
0330	2	Flat	Grey	2.54		85	0.26	522				
9606	3	Radomes	White	2.6		120	0.24	657				
9608	4	Cookware	White	2.5		85	0.25	593				
9618	5	Cookware	Trans-parent slightly amber	2.5	69	97	0.24	650	0.83	1.67		
ZERODUR	5	Low expansion	Brown trans-parent	2.53	>75[a]	91	0.24	630	0.82	1.64		
CERAN	6	Low expansion		2.58		92	0.24	>575	0.825			
ROBAX	7	Low expansion		2.56		92	0.24					
FOTOCERAM	8	Photo-sensitive	Various	2.41	148	87	0.19	500	0.87	2.55	1.96	
NEOCERAM–O	9	E.G. lasers	Trans-parent	2.51	98	78		530	0.45	1.38		
NEOCERAM–11	10	E.G. cookware	White	2.50	147	78		570	0.48	1.51		
MICATHERM	11	Cookware		4.00	45					0.9		

All are non-absorbent and most are vacuum-tight. MACOR is vacuum compatible.

Manufacturers/suppliers:
- 1–4 and 8   Corning Glass Works. Also supplied by Duramic Products Inc. and McGeoch & Co. (Birmingham) Ltd.
- 5–7   Schott Glasswerke.
- 9–10   Nippon Electric Glass (UK—Southern Ceramic Supplies Ltd.)
- 11   Morgan Matroc.

Notes:
- (d)   [a]Data available on strength/time relationship.
- (i)   Materials 9 and 10, 25°C.
- (j)   Material 1, third column to 900°C. Materials 2, 3, 4 from 0 to 300°C. Material 4 see Remarks. Material 5, 20–300 and 600°C, Material 6, 20–300 and 700°C. Material 7, 20–500 and 700°C. Materials 9 and 10, 30–380°C and 30–800°C.
- (k)   Material 1, third column at 500°C. Materials 3 and 4, second column 250°C, third 350°C. Material 5 at 20, 500 and 700°C. Material 6 at 250 and 350°C. Materials 9 and 10, first column 50°C.
- (l)   Material 1, at 1 kHz, third column 250°C. Materials 3 and 4 at 1 MHz. Materials 5, 6 at 1 MHz. Material 8 at 100 kH. Materials 9 and 10, at 1 MHz, 25° and 150°C.
- (m)   Material 1, at 1 kHz, third column 250°C. Materials 3 and 4 at 1 MHz. Materials 5, 6 at 1 MHz. Material 8 at 100 kH. Materials 9 and 10, at 1 MHz, 25° and 150°C.
- (n)   Material 1, V/mm. Material 8, V/mm. Materials 9 and 10, kV/mm at 25°C.

Table 2.15.26

TABLE 2.15.26 Properties of glass–ceramics—*continued*

Coefficient of thermal expansion (10^{-6} K^{-1})			Volume resistivity 10^{-2} ohm m (ohm cm)			Permittivity			Loss tangent (power factor) 10^{-4}			Dielectric strength	Safe working temp (°C)	Remarks
j (see note)			k (see note)			l (see note)			m (see note)			n (see note)	o	p
20–200°C	20–500°C	20–1000°C	20°C	400°C	600°C	20°C	200°C	400°C	20°C	200°C	400°C			Machining data available
7.4	9.2	11.1	>1×10^{14}	1×10^{8}	1×10^{7}	6	6.7	10	35	160	3000	1000	1000	
9.7													535	
5.7			1×10^{16}	1×10^{10}	1×10^{5}	5.6			3.0				700	
0.4			1×10^{13}	1×10^{8}	>1×10^{7}	6.9			3.4				700	Range of expansion available
0.60													650	'PYROCERAM'
0.05	0.20		2×10^{13}	7×10^{3}	4×10^{2}	7.4			155					
−0.25	0.10			1×10^{7}	1×10^{5}	7.8			190				700	
0.05	0.10												700	
	10.4			1×10^{7}		5.6	6.0		80			4000	750	Also make Fotoform and Fotoform opal
	−4	−2	4.2×10^{11}	6.6×10^{4}	1.8×10^{3}	8.4	12.9		120	470		11	740	Graphs of prop. variation with temp. available
	9	12	5.0×10^{12}	6.0×10^{6}	7.0×10^{4}	6.6	7.1		29	380		11	1100	
11.6			1×10^{13}						20					Can be moulded around metal parts

Table 2.15.26—*continued*

TABLE 2.15.27 Comparative properties of glass–ceramics

Property		Units	Alumina (94% Al$_2$O$_3$)	Boron nitride (96% BN)	Macor machinable glass–ceramics (Code 9658)	Glass–ceramic (Code 9608)	Glass	Glass-bonded mica	PTFE	Thermoplastic polyester (PBT) (30% glass fibre)	Polyimide (30% glass fibre)
Dielectric strength (a.c.)		10^k Vm^{-2}	670	380	400	400	400^b	350	—	236	—
Dielectric constant (1 MHz, 25°C)		—	7.2	—	5.89	5.6	6.3^b	4.7	2.05	3.2–3.8	4.74
Volume resistivity	25°C	1 g (ohm m)	>14	>14	>14	—	—	—	>16	12.9	13.7
	300°C		7	—	9.7	9.0	13.0^b	12.5	—	—	—
Loss tangent (1 MHz, 25°C)		—	0.0001–0.0005	—	0.0024	0.0015	0.0016	0.0022	0.0002–0.0004	0.012–0.018	0.0055
Max. cont. service temperature		°C	1450	1650	800	800	100–500	—	260	180	>260
Thermal conductivity		W/m per K	17.5	13–27	1.46	3.3	1.7–15.1	—	0.21	0.24	0.48
Thermal expansion (0–400°C)		10^{-6} K^{-1}	7.1	4.1	9.4	5.7	0.5–9.0	—	44	4–8^c	15
Compressive strength		MN/m^2	2100	310	345	1000	—	—	—	105–125	225
Modulus of Rupture		MN/m^2	225	—	75	—	—	—	—	—	—
Tensile elastic (Young's) Modulus		GN/m^2	277	48	64	90–120	60–80	—	0.43	6–10	3.5
Hardness^a		—	2000K	<32K	250K	670–730K	470–730K	—	D50–55S	M80–90R	—
Porosity		%	0	1.1	0	0	0	—	0.01	—	—
Density		kgdm^{-3}	3.62	2.08	2.52	2.6	2.2–3.0	—	2.18	1.31	1.9

^{a} K = Knoop;   S = Shore (D);   R = Rockwell (M).
^{b} Data for aluminosilicate glass.
^{c} Higher figure across flow direction.

Table 2.15.27

TABLE 2.15.28 Applications for glass–ceramics

Type	Applications	Special properties
Non-machinable	Missile nose cones, etc.	High strength, good thermal shock resistance to rain corrosion.
	Domestic ovenware	Thermal shock resistance.
	Ceramic-to-metal seals	Ease of fabrication. Matching of thermal expansion coefficients.
	Bearings	High strength. Good abrasion resistance.
	High temperature heat exchangers	High strength—thermal shock resistance, chemical stability, high specific heat.
	Light sensitive spectacles	Photosensitivity.
Machinable	Manufacture of prototypes and special parts for limited production items where dielectric or electrical insulation properties are needed.	Ease of fabrication. Required electrical properties.
	Welding and brazing fixtures.	Ease of fabrication and good thermal insulation.
	Applications requiring tight tolerances on finished components.	Expensive grinding of conventional ceramics to fine tolerance is eliminated.

Table 2.15.28

2.15.8 Ceramic fibres

Ceramic fibres are fibres spun from the molten ceramic and are (usually) supplied as a flock, or bulked into a felt.

The use of these fibres has increased stepwise as a replacement for asbestos once the toxic nature of this mineral became appreciated. The available types and important applications of ceramic fibres are listed in:

Table 2.15.29 *Composition of ceramic fibres and applications*

TABLE 2.15.29 Composition of ceramic fibres and applications

Fibre types	Compositions (wt %)							
	Al_2O_3	SiO_2	$(Na, K)_2O$	B_2O_3	Fe_2O_3	TiO_2	CaO	MgO
FIBERFRAX—	51.7	47.6	0.3	0.15	0.02			
FIBERFRAX H—	62.3	37.2			Trace	Trace		
TRITON Kaowool (Standard)—	43–47	50–54	0.2–2.0	0.06–0.1	0.6–1.8	1.2–3.5	0.1–1.0	Trace
TRITON Kaowool (High Duty)—	52–56	43–46	0.05–0.4		0.05–0.08	0.02–0.05		Trace
SAFFIL—	95	5	875Na (ppm)				525Ca (ppm)	130Mg (ppm)

'FIBERFRAX'
The Carborundum Company PLC

'TRITON Kaowool'
Morganite Ceramic Fibres PLC

SAFFIL
ICI PLC (Manufacturers)
Morganite (Marketing agents)

Table 2.15.29

TABLE 2.15.29 Composition of ceramic fibres and applications—*continued*

Compositions (wt %)		Temperature limit for continuous use (°C)	Fibre products
Unspecified inorganics	Other metals (as oxides)		
0.2		1260	Bulk fibre, washed fibre, chopped and milled fibre, long staple fibre, blanket (several types), felt, dense felt, cloth, tape, sleeving, rope (low and high density), square section braid, paper. Coating cements, boards, blocks, slabs, castable and mouldable mixes, moist packs, sprayable and pumpable mixes. Shaped and die-cut parts. Also, for large jobs, felt or board with mineral wool backing. Ancillaries: rigidisers and adhesives.
0.5		1420	Bulk fibre, blanket, felt and paper.
0.2–0.3		1260	Bulk fibre, blanket, strip, paper, board, non-woven textile, wet felt, mastic and cement. Shapes (standard, special, super), cut shapes (board, paper, blanket, wet felt); foil backed, mesh enclosed and board in non-standard sizes. Ancillary material: hardener.
		1400	Bulk fibre, blanket.
	140Ni 60Cr (ppm)	1600	Bulk fibre, low density mat, blanket, board, shapes (standard or made to order), paper.

Table 2.15.29—*continued*

2.15.9 Heat resistant cements

Heat resistant cements are cements which develop reasonable strengths when heated to their service temperature. Slow, even heating is usually required. Some cements contain a low temperature bonding agent.

The manufacturers of most of the heat resistant adhesive cements are reluctant to disclose the compositions of their products of which there are a fair number. Satisfactory service can best be assured when the composition of a product is known and is judged to be suitable for a particular application. Wholly inorganic adhesive cements which contain sodium silicate are unsuitable for electrical use. By no means all applications of heat resistant adhesive cements require good electrical properties, however. Solid materials which are themselves heat resisting may need to be bonded together, or their surfaces may require coating. The choice of a cement then depends on the temperature resistance needed, on the adhesion properties and on the thermal expansion match or mismatch.

There are a very large number of specialised heat resistant cements varying in price from relatively cheap to very expensive. They are listed, classified by trade name and manufacturer in:

Table 2.15.30 *Heat resistant cements*

CEMENTING WIRE WOUND FURNACES

This is a most important application for heat resistant cement and one in which the choice of material is critical.

Especial care should be taken in selecting cements for the embedment of electrical resistance wires when furnace heating elements on ceramic formers are under construction. These cements must be of sufficiently high refractoriness to withstand the operating condition and at the same time be of suitable chemical composition to avoid the possibility of reaction between the cement and the element.

The ALCEM and TRIMOR ranges of alumina cements are recommended for this purpose as long as the lack of strength prior to heating is not a disadvantage. FORT-AFIX Heavy Grade Adhesive Cement develops strength at room temperature and is claimed to be suitable for electric furnace applications up to 1200°C. It would be unwise to use it in contact with a platinum alloy winding. The cements which are suitable for each type of winding are listed in:

Table 2.15.31 *Cements compatible with specific furnace winding materials*

TABLE 2.15.30 Heat resistant cements

Cement type and manufacture	Grade	Comments
ALCEM alumina cements Refractory Mouldings & Castings Ltd.	C60	For use with nickel–chromium wire wound resistors in the construction of furnace heating elements. Fine ground for maximum plasticity.
	C40	Also for Ni/Cr resistor applications. Coarser mix for coatings >5 mm thick.
	CC60	Similar to C60 but for use with platinum group precious metal resistors.
	CC40	Also for platinum resistors, where thick layers are required.
	101	Minimum impurities. Recommended for insulation of molybdenum.
		(All grades free from low temperature bonding material)
AUTOSTIC adhesive cement Carlton Brown & Partners Ltd.	Standard	For application by spray gun (low viscosity).
	FC4	Consistency for application by brush.
	FC8	Thick, for application by trowel. All withstand temperatures up to 1100°C and adhere to keyless surfaces.
		(No mixing or thinners required)
CERASTIL adhesive cements Industrial Science Ltd.	C3	Suitable for use up to 1000°C. Initial set begins at 32°C. Good chemical resistance is claimed and high electrical resistivity, but sodium silicate is mentioned as a constituent, so high temperature electrical insulating properties are questionable.
	C10	Suitable for use up to 1500°C. Curing needs 24 h at room temperature at 400°C.
	GC	Two-component graphite cement. Volume resistivity 10^5–10^6 ohm cm at 75°C. (Mix with water)

Table 2.15.30

TABLE 2.15.30 Heat resistant cements—*continued*

DEGUSSIT magnesia and zirconia cements Bush Beach Engineering Ltd.	4081–57030	Magnesia (analyses more than 97% MgO), maximum particle size ⅓ (0.33) mm. Bond with sufficient of a 5 wt % solution of magnesium chloride in distilled or deionised water to make a trowelling mix. High electrical resistivity at high temperature. Good thermal expansion match to stainless steel. ⎤ For high temperature furnace construction. ⎦
	4081–54020	Zirconia stabilised with 5% CaO; make into trowelling mix with a dilute solution of water soluble gum. This material becomes electrically CONDUCTING above 1000°C (resistivity less than 40 ohm cm) but is thermally insulating.
ECCOCERAM adhesive and potting cements Emerson & Cuming (UK) Ltd.	ECCOCERAM WL52	One part air setting ceramic adhesive and sealant. For use up to 600°C.
	ECCOCERAM SM25	Two part encapsulant for electronic components which will resist 1100°C. After curing at 150°C, can be used straight away at >1000°C.
FORTAFIX adhesive cements Fortafix Ltd.	Light	Milky consistency ⎤ For bonding metal parts.
	Medium	Easy flowing ⎦
	Heavy	Apply using spatula etc. For electrical applications up to 1200°C.
SECAR calcium aluminate cements Lafarge Aluminous Cement Co.		SECAR calcium aluminate cements are most usually used as bonds for refractory concretes but are hydraulic setting cements in their own right with temperature resistances up to 1700°C and quite good electrical properties.
TRIMOR alumina cements Morgan Refractories Ltd.	913	For nickel–chromium alloy furnace windings. Contains 93 wt % alumina and 6% silica. ⎤ Not adhesive. Little low temperature bonding. ⎦
	961	For Kanthal A1, platinum, molybdenum and tungsten windings. Contain 98.3 wt % alumina and 0.72% silica. (Mix with water)

Table 2.15.30—*continued*

TABLE 2.15.31 Cements compatible with specific furnace winding materials

Winding	Cement		
	ALCEM	SECAR	TRIMOR
Nichrome	CC60 and C40		913
Brightray	C60 and C40		913
Kanthal A1	CC60 and CC40	250 + alumina	961
Platinum	CC60 and CC40		961
Molybdenum	101		961

Table 2.15.31

2.15.10   Designing with ceramics

2.15.10.1   The statistical nature of brittle failure

Designing a component made of a ductile material to withstand a given loading is essentially a straightforward procedure. In many cases, the nominal maximum stress to which the loading gives rise is compared with the strength of the material, obtained from a simple mechanical test. A safety factor chosen on the basis of previous experience may be incorporated to allow for uncertainties in applied loads.

Design in brittle materials is more difficult because of the statistical nature of their strength.

Unlike ductile materials, ceramics have strengths which are variable in the sense that measurements on nominally identical specimens will show considerable scatter in strength results. The amount of scatter varies from ceramic to ceramic, but coefficients of variation (i.e. standard deviation/mean strength) of 20% or more are not uncommon.

This strength variability is not due to lack of quality control in ceramic manufacture or inadequate test procedures. It is a fundamental consequence of the nature of brittle failure.

All materials contain flaws which act as local stress raisers. In a brittle material, catastrophic failure usually occurs by the virtually instantaneous propagation of a crack as soon as the stress at the most severe flaw reaches a critical value. The strength of a ceramic therefore largely depends on the weakening caused by its most severe flaw, just as the strength of a chain depends on the strength of its weakest link. For most brittle materials, flaws on the surface of the component are more severe than internal flaws, and so determine the failure behaviour.

(In a ductile material, plastic deformation and work-hardening processes ensure that many flaws are involved in the fracture process so that the measured strength is an 'average' for many flaws.)

The 'weakest link' behaviour of ceramic materials has the following consequences:

1. Nominally identical specimens exhibit scatter in strength, since specimens will, in general, contain 'critical flaws' of varying severity.
2. Large components, on average, are weaker than small ones because there is greater chance of finding a flaw of given severity in a larger component.
3. A component subject to non-uniform stress may not fail at the point of maximum stress since a severe flaw, subject to a less than maximum stress, may go 'critical' before a less severe flaw experiencing the maximum stress.

These have the following implications for the designer:

Because of the size effect, no single measure of strength exists which is directly applicable both to the test specimen on which it was measured and the component under design.

Because of the variability in strength, no absolute guarantee that a particular component will not fail under its design load can be given. The aim of total safety must be relaxed and a probabilistic approach adopted in which the frequency of failure is kept to an acceptable figure; safety factors must be chosen to give a predetermined 'level of reliability' which will depend on the nature and economics of the application in question. For example, for minor components whose failure merely causes inconvenience, a failure probability of 1 in 100 might be acceptable; in general engineering where failures might be serious but not fatal, 10^{-4} might be selected as an allowable failure probability; while for manned space flight components might have to be reliable to 1 part in 10^6. These safety factors allow only for the uncertainty in material strength and a further factor will be needed to cope with any uncertainty in applied load.

Because the stress distribution throughout the component, rather than just the maximum stress present, determines the failure probability, a thorough stress analysis of the component will be required.

Because of the variability of strength, the relative merits of different ceramic materials can only be precisely compared if strength measurements are quoted together with appropriate measures of variability and some indication of the type of test and specimen size is given. Unfortunately it is not yet standard practice for manufacturers to give all this information so that comparisons between materials are sometimes difficult.

In view of this, strength comparisons must be treated with caution, and strengths quoted in tables must be taken as only rough guides to performance.

In Section 2.15.10.2 a method is given whereby the load-bearing ability of simple shapes can be calculated from suitable strength data. If the strength data are obtained for instantaneous failure of specimens tested under monotonically increasing loads, as is usual, the results of calculations can only be applied to this stress duration. Behaviour under long term loading or cyclic loading may be predicted if the methods described in Section 2.15.10.3 are used. The effect of combined stresses (Section 2.15.10.3) should be borne in mind when designing for a load-bearing situation.

2.15.10.2   A statistical theory for design in brittle materials

THE WEIBULL DISTRIBUTION

The strength variability of many ceramic materials can be conveniently described by a statistical distribution, known as the Weibull distribution. Justification for the use of this distribution is empirical rather than theoretical for all but the simplest stress states within a component.

The distribution states that the failure probability of a component is given by

$$F = 1 - \exp\left[-\left(\frac{\sigma_{nom}}{\bar{\sigma}_f}\right)^m \left(\frac{1}{m}!\right)^m \left(\frac{S_c}{S_s}\right)\right] \tag{1}$$

The parameters in this expression have the following significance: m is a material property known as the Weibull Modulus. m measures the variability in strength of the material under consideration, high values of m corresponding to consistent materials and low values representing materials with high strength variability. A value of $m = \infty$ corresponds to a material with no strength variability, i.e. a 'classical' material which is simply characterised by a single strength value and which exhibits no size effect. (The design procedure applied to ductile materials effectively assigns a value of $m = \infty$ to them.)

Values of m for ceramic materials vary between about $m = 5$ and $m = 20$.

σ_{nom} is a suitable nominal stress in the component subjected to the applied load. It is convenient to choose the maximum stress or an easily calculable stress for σ_{nom}. For example, in a bend test, the maximum fibre stress can be chosen as σ_{nom}. σ_{nom} defines the dependence of the failure probability on the applied load, to which it is , of course, directly related.

$\bar{\sigma}_f$ is the mean strength of the material measured in a test with effective surface S_s. S_c is the corresponding effective surface of the component.

The 'effective surface' of a component is a quantity with the dimensions of surface area which accounts for the dependence of failure probability on the size of the component and the nature of the stress distribution to which it is subjected. It is directly proportional to the actual surface area of the component and the constant of proportionality characterises the shape of the component and the type of loading while being independent of the magnitude of the load. The 'effective surface' admits of a simple physical interpretation; it is the surface area of a specimen subjected to a uniaxial tension of σ_{nom} which would have the same failure probability as the component. Methods of estimating m, $\bar{\sigma}_f$ and effective surfaces are given in the following sections. The examples below show how equation (1) can be used to design in ceramic materials once these parameters have been calculated.

EXAMPLE 1

A ceramic material has a mean strength of 200N/mm^2 when tested in a three-point bend test with an effective surface of 1.6 mm^2. The material is known to have a Weibull Modulus of 15. If the material is to be used to make a component and the load configuration gives an effective surface for this component of 9×10^4 mm^2, what is the appropriate working stress if only one failure per thousand components can be tolerated?

In equation (1) $\bar{\sigma}_f$ = 200 N/mm^2
$$m = 15$$
$$P_f = 10^{-3}$$
$$S_s = 1.6 \text{ mm}^2$$
$$S_c = 9 \times 10^4 \text{ mm}^2$$

$[(1/m)!]^m$ for $m = 15$ is (from Table 2.15.34) 0.59.

so $10^{-3} = 1 - \exp \left[- \left(\dfrac{\sigma_{nom}}{200} \right)^{15} \left(\dfrac{9 \times 10^4}{1.6} \right) 0.59 \right]$

and solving for σ_{nom} gives

σ_{nom} = 63 N/mm^2.

The appropriate design stress is therefore 63 N/mm^2.

EXAMPLE 2

Two ceramics are candidates for a certain application. Material A, when tested in a 1/4 point bend test had a mean strength of 170 N/mm^2 and a Weibull Modulus of 15 which gave an effective surface for this test of 13mm^2.

Material B was tested in a three-point bend test and had a mean strength of 440 N/mm^2, but a Weibull Modulus of only 7, which, for the three-point test on this specimen, corresponds to an effective surface of only 0.9mm^2. What are the relative merits of the two materials?

At first sight, it seems that Material B is much stronger. However, B was effectively tested on a much smaller specimen than A, and we need firstly to correct for the 'size effect' before valid comparisons can be made.

We begin, therefore, by adjusting the strength of B to correspond to the effective surface used in testing A.

The corrected strength = uncorrected strength $\times (S_B/S_A)$

$= 440 \times \left(\dfrac{0.9}{13} \right)^{1/7} = 300$ N/mm^2.

This means that, if B had been tested in the same way that A was, we should have expected a mean strength of 300 N/mm^2 for B. The real strength difference between the two materials is less marked than at first appears.

We can investigate the effect of the different Weibull moduli of A and B by calculating working stresses for various failure probabilities using equation (1) (see text). This has been done for both materials for an effective surface of 13mm^2 and the results are shown in:

Table 2.15.32   *Working stresses for failure probabilities of 10^{-1} and 10^{-4} in ceramic materials A and B*

Although material B has a higher mean strength, the working stress at 10^{-4} failure probability is higher for material A. This demonstrates that high mean strength may not outweigh a low *m*-value when high reliability is required. The choice of material for the application will depend on reliability level required.

ESTIMATION PARAMETERS

Mean strength $\bar{\sigma}_f$. Carefully conducted three-point or quarter-point bend tests ('Modulus of Rupture' tests) or thin ring-bursting tests are suitable. Tension tests are not recommended because of alignment difficulties.

50 tests on nominally identical specimens will be sufficient to estimate Weibull modulus and mean strength if the use of Weibull statistics can be accepted a priori.

Table 2.15.33 shows, for convenience, effective surfaces for the commonly used three-point and four-point bend tests for square-section and circular section bars.

Table 2.15.33 *Effective surfaces for common specimens*

WEIBULL MODULUS *m*

Computer programmes to estimate *m* from strength test data by least-squares or maximum likelihood techniques have been developed at Fulmer Research Institute.

An approximate value of *m* can be obtained from the mean strength and standard deviation of the test results by using the formula.

$$m = \frac{1.2 \times \text{mean strength}}{\text{standard deviation}} = \frac{1.2}{\text{coefficient of variation}}$$

This formula is especially useful for estimating *m* from strength results quoted in the literature.

The following graphical techniques may be employed.

1. Divide each strength result by the mean strength to give *n* 'reduced strengths' β.
2. Arrange the β – values in order of increasing β.
3. Calculate a P_s – value by assigning a value of

$$P_s = 1 - \frac{r - 0.3}{n + 0.4} \quad \text{to the rth β value.}$$

4. Plot corresponding values of P_s and β on the special chart of Fig 2.15.1 (values of P_s between 0.35 and 0.65 are difficult to plot accurately and may be omitted):
   Fig 2.15.1 *Graphical method for m*
5. Choose that value of *m* which divides the points equally. An example is given in:
   Fig 2.15.2 *Example of a Weibull fit to strength data for reaction-bonded silicon nitride*

Once *m* has been estimated the value of $(1/m!)^m$ can be obtained from tables of the γ-function or from :

Table 2.15.34 *Values of (1/m)! and [(1/m)]^m for m-values between 5 and 20*

CALCULATION OF THE EFFECTIVE SURFACE OF A COMPONENT

For components stressed only in uniaxial tension,

$$S_c = \int_A \left(\frac{\sigma}{\sigma_{\text{nom}}}\right)^m \cdot dA \tag{2}$$

integration being taken over the surface of the component. A completely general stress is described by three principal stresses σ_1, σ_2 and σ_3 which may be positive (i.e. tensile) of negative (compressive).

In order to extend equation (2) to cover this general case we need to know a 'failure criterion' for the ceramic which expresses the dependence of failure on the combined action of the stresses present like the Tresca or von Mises yield criterion for metals.

Experimental data on failure criteria for engineering ceramics is scarce but a commonly used criterion assumes that each principal stress acts independently.

$$ S_c = \int_A \left[\left(\frac{\sigma_1}{\sigma_{nom} H(\sigma_1)} \right)^m + \left(\frac{\sigma_2}{\sigma_{nom} H(\sigma_2)} \right)^m + \left(\frac{\sigma_2}{\sigma_{nom} H(\sigma_3)} \right)^m \right] dA $$

$H(\sigma)$ is a factor which takes account of the different effects compressive and tensile stresses have on failure behaviour.

$H(\sigma) = 1$ if σ is positive (tensile)

$H(\sigma) = - \dfrac{\text{mean compressive strength}}{\text{mean tensile strength}}$ if σ is negative (compressive)

Most ceramics are much stronger in compression than tension so that in many cases the effect of compressive stresses on the effective surface is negligible and may be ignored.

Calculation of effective surfaces requires a full stress analysis of the component. In all but the simplest cases, numerical methods are needed to evaluate the integral.

2.15.10.3   Notes on the properties of ceramics

DUCTILE AND NON-DUCTILE ENGINEERING MATERIALS

Ceramics and glasses are brittle but it is technologically preferable to use the term 'non-ductile' to describe them because, in many people's minds, brittleness has become associated with fragility or weakness. Though this assumption is not always justifiable, there are good reasons why it has arisen. For instance the brittleness of cast irons is associated with weakness when they are compared with the strong, tough, ductile steels but it is not always realised that thermally strengthened glass and fine-grained aluminium oxide ceramics have similar fracture stresses to that of steel. The difference lies in the ability of steel to deform plastically and to a very much greater extent than ceramics before failure.

'Brittle' ceramics and glasses are capable of permanent deformation if the temperature is high enough, although the mechanism involved differs from that which leads to ductile distortion of metals. Conversely a ductile metal may fail in a brittle manner when subjected to mechanical stress at a low temperature, especially when the stress is highly localised, as in one of the notched bar impact tests well known to engineers.

IMPACT STRENGTH

Impact testing of brittle materials is not of much value, except to demonstrate that their resistance to impact is low. When impact testing is done on these materials it is usually found that the materials which have the highest strengths, as measured in a slow monotonically increasing load test, also have the highest values of impact strength, but the values are about two orders of magnitude below those measured for steels. Although the figures are not justifiable in strictly scientific terms, comparative values are given below.

Mild steel	250 arbitrary units
Cast iron	25
Polystyrene	5
Polycrystalline alumina	2½
Annealed glass	1

The susceptibility of brittle materials to impact damage emphasises the need to avoid stress concentrating notches, grooves etc. (if at all possible) when designing for their use. Impact susceptibility is the antithesis of fracture toughness.

FRACTURE TOUGHNESS

The fracture toughness of a metal or alloy is usually quoted in terms of its critical stress intensity factor, i.e. the intensity of the stress at which fracture of the material becomes rapid. As an alternative way of comparing the resistance to fracture of a range of materials the work of fracture may be quoted. This is the work which has to be done per unit cross-sectional area to bring about rapid fracture proceeding from a pre-existing crack. Work of fracture values for six well known materials are shown in :
Fig 2.15.3   *Work of fracture for non-ductile and ductile materials*

The need to avoid the possibility of crack propagation in service when a design is developed for the use of a brittle material under load is vividly illustrated by the difference in work of fracture values between for instance, polycrystalline alumina and mild steel. It is not surprising that engineers have in the past been reluctant to make use of brittle materials. Even now, with the availability of new approaches based on fracture mechanics and failure prediction, caution is needed to ensure that premature fracture is avoided.

TENSILE AND BENDING STRENGTHS

From an engineering point of view, knowledge of the tensile strength of a material is of great value but the tensile strength of non-ductile materials is not always reported in manufacturers' data sheets because test pieces and satisfactory tensile strength measurements are more difficult to make than is the case with ductile metals. It is usual to determine and report the *bending* strength of glasses especially and of most ceramics since preparation of suitable specimens, e.g. by diamond wheel slitting of plate samples into bars is easily achieved. Bending strength determination might be described as an indirect method of measuring tensile strength since failure in bend is, in effect, tensile failure initiating on that face of the specimen which becomes convex as bending proceeds. The flexural strengths of ceramics are indicated in:
Fig. 2.15.4   *Flexural strength of ceramics*

COMPRESSIVE STRENGTH

The fact that the compressive strength of a brittle material is usually about five times as high as its tensile strength (determined by a bending test) is often quoted by manufacturers and others. *Pure* compressive stress is, however, rare in mechanical engineering. It is therefore difficult to take advantage of high compressive strength in an engineering design when, as is usual for non-ductile materials, it is allied with low tensile and shear strengths. Compressive strengths are not therefore normally quoted.

INFLUENCE OF COMBINED STRESSES ON STRENGTH

Load carrying components of engineering devices are likely to be subject to combined (or multiaxial stresses). For instance, a hollow cylinder might be subjected in use to an internal pressure simultaneously with a compressive end load applied to form a pressure seal. This biaxial stress system, with axial compressive stress normal to the tensile hoop stress, can be expected to increase the effective tensile strength of a brittle material. Conversely an axial tensile stress superimposed on internal pressure will decrease the effective strength.

Tensile strength may be raised by about 50% when a high compressive load is applied as suggested above, and reduced by about 50% when biaxial tension is applied.

Torsional strength, which is equivalent to failure under biaxial 1:1 tension–compression, may be about 30% higher than uniaxial tensile strength.

The above applies to initially stress free materials, but in the case of glass the strength can be increased by putting the surface layers where fracture initiates into a state of in-built compressive stress using a thermal or a chemical method.

'Thermally toughened' glass plate is commonly used in safety windshields, but the thermal toughening process cannot easily be used on components of complex shape, and once treated, toughened plate cannot be cut without catastrophic fracture.

Chemical toughening produces a much thinner compressive stress zone and smaller internal tensile stresses. Such glass can be cut and there are few restrictions on the complexity of shapes that can be treated. The process is, however, more expensive than thermal toughening, and applications may be restricted on economic grounds.

Schematic stress profiles for glass toughened by thermal and chemical means are given in:

Fig. 2.15.5 *Variation of stress through the glass thickness for*
(*a*) *thermally toughened glass*
(*b*) *glass chemically toughened by ion-exchange.*

Another range of strengthening techniques consists of coating the glass during component manufacture to protect the unflawed surface. The commonly used 'titanising' process deposits a thin film of titanium dioxide or stannic oxide on glass containers to increase abrasion resistance and impact strength.

Protective silicon coatings are also used on glass strengthened by acid-etching to remove surface flaws.

LOAD DURATION AND STRENGTH

Most materials tend to lose strength with the passage of time when they are under tensile loads. In the case of metals the effect is usually negligible at normal temperatures. Time dependent loss in strength occurs in glasses and most ceramics, without a change in the fracture inducing mechanism, whether the load is constant or whether it fluctuates. The mechanism which brings about loss in strength is generally (particularly for glasses and oxide ceramics) due to interaction between atmospheric moisture and the surface of the material at the tips of pre-existing flaws, which may be as small as the so-called Griffith cracks. Slow growth of cracks to a critical size takes place at these sites (which act as stress concentrators) due to a stress corrosion type of effect.

The study of ceramics and glasses under fatigue conditions is less well developed than is the study of metals, so that data are scarce and are not included in manufacturers' data sheets.

EFFECT OF MECHANICAL STRESS AT HIGH TEMPERATURES

In commenting on the mechanical properties of glasses and ceramics a combined approach has so far been used. At high temperatures, a clear division exists. Most glasses have to be excluded from a list of candidate materials for use at temperatures above about 500°C, whereas a ceramic material may be the only one which can be used at temperatures much above 1000°C.

The possibility of failure of a ceramic material, in a high-temperature application, due to thermal stress or thermal shock must however be borne in mind.

OTHER PROPERTIES

The properties of ceramics which have been discussed in the foregoing sections are structure sensitive and somewhat variable even within a single batch of material. Other properties—thermal, electrical, magnetic and optical (where appropriate)—may be found less so. However, as with the mechanical properties, the values given in the next section should be taken as indications only (even those quoted for proprietary materials) and full discussion with the manufacturers is essential.

TABLE 2.15.32 Working stresses for failure probabilities of 10⁻¹ and 10⁻⁴ in ceramic materials A and B

Mean strength in bend tests (N/mm²)	Material A 170 (13 mm² ¼ pt bend)	Material B 440 (0.9 mm² 3 pt bend)
Mean strength adjusted to equal effective surface (13 mm²)	170	300
Working stress for 10⁻¹ failure probability (13 mm²)	145	220
Working stress for 10⁻⁴ failure probability (13 mm²)	85	70

TABLE 2.15.33 Effective surfaces for common specimens

m	Effective surface/total prismatic surface			
	Square section bars		Circular section bars	
	3-point bend	¼ point bend	3-point bend	¼ point bend
5	.048611	.170139	.028294	.099030
6	.040816	.163265	.022321	.089286
7	.035156	.158203	.018189	.081851
8	.030864	.154321	.015191	.075955
9	.027500	.151250	.012935	.071140
10	.024793	.148760	.011186	.067116
11	.022569	.146701	.009799	.063693
12	.020710	.144970	.008676	.060735
13	.019133	.143495	.007753	.058147
14	.017778	.142222	.006982	.055859
15	.016602	.141113	.006332	.053818
16	.015571	.140138	.005776	.051983
17	.014660	.139275	.005297	.050322
18	.013850	.138504	.004881	.048808
19	.013125	.137812	.004516	.047422
20	.012472	.137188	.004195	.046147
21	.011880	.136622	.003910	.044968
22	.011342	.136106	.003656	.043875
23	.010851	.135634	.003429	.042857
24	.010400	.135200	.003224	.041907
25	.009985	.134800	.003038	.041017

Table 2.15.32 and Table 2.15.33

TABLE 2.15.34 Values of $(1/m)!$ and $[(1/m)!]^m$ for m-values between 5 and 20

m	$(1/m)!$	$[(1/m)!]^m$
5	0.918	0.65
6	0.930	0.65
7	0.935	0.62
8	0.942	0.62
9	0.947	0.61
10	0.951	0.61
12	0.957	0.59
15	0.965	0.59
20	0.974	0.59

Table 2.15.34

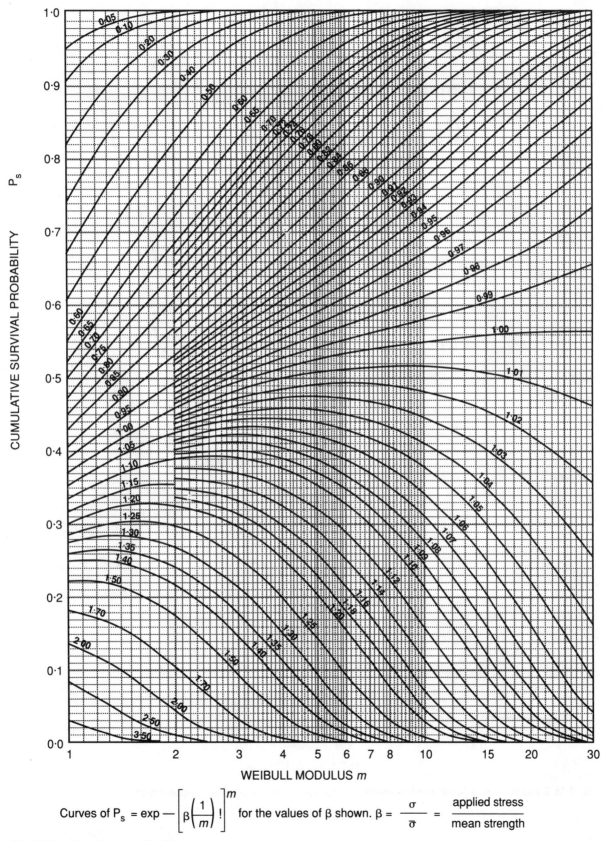

Curves of $P_s = \exp - \left[\beta\left(\dfrac{1}{m}\right)! \right]^m$ for the values of β shown. $\beta = \dfrac{\sigma}{\bar{\sigma}} = \dfrac{\text{applied stress}}{\text{mean strength}}$

Fig 2.15.1 Graphical method for *m*

Fig 2.15.1

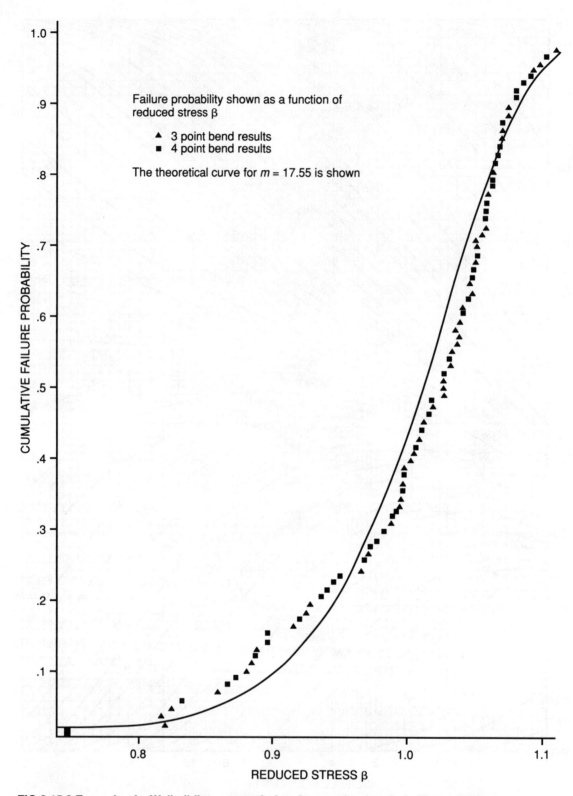

FIG 2.15.2 Example of a Weibull fit to strength data for reaction-bonded silicon nitride

Fig 2.15.2

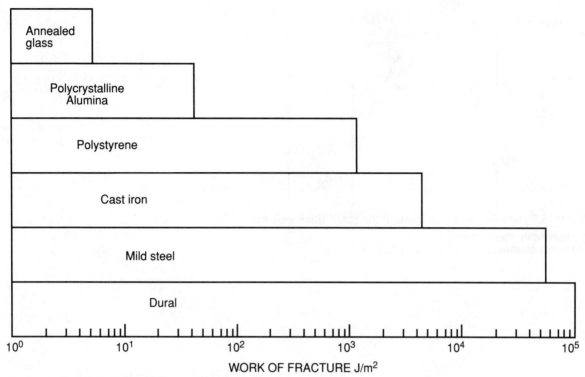

FIG 2.15.3 Work of fracture for non-ductile and ductile materials

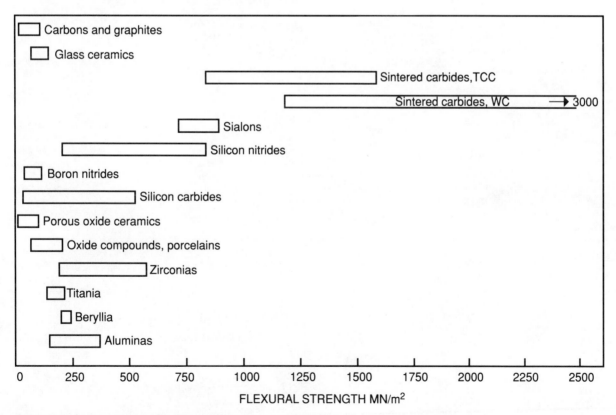

FIG 2.15.4 Flexural strength of ceramics

Fig 2.15.3 and Fig 2.15.4

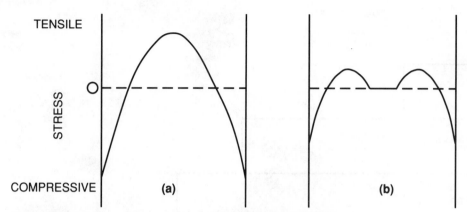

FIG 2.15.5 Variation of stress through the glass thickness for

(a) thermally toughened glass
(b) glass chemically toughened by ion-exchange

Fig 2.15.5

Index
Vols 1-3

Note: Tables are indicated by **bold page numbers**, and Figures by *italic page numbers*.
For longer Tables and Figures, the first page of the Table/Figure is given; please look through for the particular material/property

i24 **Index**

properties, **1336**, *1344*
Fuel permeability
thermoplastic elastomers, 2071, **2097**
Functional analysis, 5
examples, *9, 10*
Fungi
polymers affected by, **310**, **1536**, **1677**
Furan based polymers, 1967
adhesive bonding of
surface preparation for, **709**
advantages/limitations, **1968**
applications, 1967
asbestos filled grades
properties, **1968**
chemical/solvent resistance, **1536**, *1546*,
1968, **1969**
compressive strength, **1968**
environmental performance, **310**, **1536**,
1967, **1968**
flexural strength, *1508*, **1968**
glass filled/reinforced grades
characteristics, **2156**
properties, **1968**
hardness, **1968**
laminated plastics
characteristics, **2188**
processing techniques, 575, 1967, **1968**
service temperature range, **1968**
specific gravity, *1525*, **1968**
suppliers, 1967
tensile properties, *1505*, **1968**
trade names, 1967
water absorption, **1968**
Furane sand casting processes, **436**, **448**, **456**
Furnace brazing, **630**
Furnace hardware applications
carbons and graphites, **1443**
ceramics, **1436**
cobalt superalloys, **1336**, 1339, **1342**
stainless steels, *920*
Furnace soldering, **620**
Furnace windings
heat-resistant cements used, 1463, **1466**
platinum and alloys used, **1363**, **1366**
refractory metals used, **1371**
Fused salts *see* Molten salts
Fusible alloys, 1382
advantages, 1382
applications, **1384**
chemical composition, **1383**
dimensional stability, 1382
elongation data, **1384**
hardness, **1384**
mechanical properties, **1384**
melting range, **1383**
properties, 1382, **1383**, **1384**
tensile strength, **1384**
Fusion (surface coating) processes
advantages/limitations, **536**
applications, **544**
characteristics, **540**
compared with other coating processes,
539
materials suitable for deposition, **544**
selection guide to processes, **542**
Fusion welding
ceramics, 641
refractory metals, 1373
stainless steels, 897
titanium and alloys, 1253

G

Gallium–tin alloys, **1383**
Gallium–tin–zinc alloys, **1383**
Galvanic corrosion, 220
aluminium and alloys, **224**, **232**, 1170
cadmium and alloys, **224**, **236**
cobalt and alloys, **235**
copper and alloys, **224**, **230**
iron, **224**, **226**
lead and alloys, **224**, **238**
magnesium and alloys, **224**, **237**
nickel and alloys, **224**, **235**
precious metals, **224**, **240**
stainless steels, **224**, **228**
steels, **226**

susceptibility of various metals, **224**
tin and alloys, **224**, **238**
titanium and alloys, **224**, **234**, 1244
zinc and alloys, **224**, **236**
Galvanic series, 219, **223**
titanium compared with other metals, **1249**
Galvanised steel
adhesive bonding of, **695**
lifetime data, *296*
painting of, *297*
Galvanising, 255
applications, **287**
compared with other protection methods,
262
fastener finish/coating, **614**
process parameters, **287**
properties of coating, **287**, *296*
Gas brazing, **630**
Gas industry applications
ceramics, **1410**
stainless steels, *920*
Gas permeability
film/sheet plastics, **2134**
polyurethanes, **1982**, 2042
test methods for plastics, **1589**
units of, **790**
Gas shielded arc welding
surface coatings deposited using, **540**, **542**
Gas turbine applications
ceramics, **1436**
cobalt alloys, 1338, 1339
nickel alloys, **1263**, 1270
refractory metals, **1371**
stainless steels, *920*
titanium and alloys, **1222**
Gas welding
advantages/limitations, **644**
applications, **644**
dissimilar metals, **604**
metals suitable, **603**
surface coatings deposited using, **536**, **540**,
542
Gaskets
asbestos-based gaskets, 2207, **2214**
Gategorizing (superplastic forming) Process,
513
Gear materials, **361**
bending stress endurance limits, *378*
characteristics, **366**
combination selection, **365**
effect of mating material on surface load
capacity factor, *380*
surface fatigue resistance, *379*
Gears
wear in, **328**, 358
Geodetic structures, 2195
applications, 2195
Geometrical tolerances
ceramics, 586, **591**
Gerber's law, 751, *755*
GFRP *see* Glass fibre reinforced plastics
(GFRP)
Gilding metal, **1106**, **1108**, **1114**
Glass blowing, 711
Glass brazing, 582
Glass–ceramic seals, **717**
Glass–ceramics, **1396**, 1453
applications, 1453, **1457**
compressive strength, **1456**
density, *179*, **182**, **1454**, **1456**
dielectric strength, **1454**, **1456**
elastic modulus, *57*, **1456**
electrical resistivity, **1454**, **1456**
flexural strength, **1454**, *1479*
hardness, **1454**, **1456**
loss tangent (power factor), **1454**, **1456**
permittivity, **1454**
Poisson's ratio, **1454**
porosity, **1456**
production processes, *593*
properties, **1400**, **1454**
compared with other materials, **1456**
rupture modulus, **1456**
service temperature range, *157*, **1454**, **1456**
specific heat, **182**, **1454**
thermal conductivity, **182**, **184**, **1454**, **1456**
thermal expansion coefficient, **182**, *183*,
1454, **1456**

Young's modulus, **1454**, **1456**
Glass–epoxy laminates
in flexible printed circuits, **2226**
properties, **1915**
Glass fabric laminated plastics
characteristics, **2189**
properties, **2190**
Glass fibre based building board
properties, **2215**
Glass fibre reinforced plastics (GFRP), **1501**,
1617, 2153
adhesive bonding of
surface preparation for, **709**
elastic modulus, *57*, **376**
epoxies
characteristics, **2156**
properties, **1909**, **1915**, **2157**
erosive wear, *408, 410*
fatigue properties, 1503, *1519*
flammability, **1584**
flexure applications, **376**
furan based plastics
characteristics, **2156**
furan based polymers
properties, **1968**
honeycomb cored structure, 2179
load bearing capability, *2162*
stiffness data, *2163*
impact strength, *1512*
load-bearing capability for various
structures, *2162*
low-temperature applications, **173**
low-temperature contraction, **210**
mechanical properties, **376**, 2153
melamine–formaldehyde plastics
characteristics, **2156**
nylons
properties, **1617**, **1618**, *1632*, *1633*,
1635, *1636*, *1637*, **2157**
painting of, **314**
phenol based plastics
characteristics, **2156**
properties, **1894**
physical properties, **376**
polyesters
characteristics, **2156**
properties, **2157**, *2159*, *2160*, **2196**
polyethylenes
properties, **1754**
polyimides
characteristics, **2156**
processing techniques, 2154, **2158**
capital cost requirements, *2160*
labour requirements, *2160*
properties compared with other materials,
2157
PTFE
properties, **1710**, **1712**
reinforcement types used, 2153
resinous matrices for, **2156**
RRIM composites
epoxies
properties, **1919**
phenolics, 1891
polyurethanes, 2047, **2055**, *2060*
service temperature range, **2196**
silicones
characteristics, **2156**
properties, **1958**, **1959**
specific stiffness vs strength, *2177*
specific strength, *61*
stiffening techniques, 2154
corrugated panels, 2155, *2161*, *2162*
designs, *2161*
rib stiffeners used, 2155, *2161*
stiffness data for various structures, *2163*
tensile behaviour of, 736
tensile modulus
calculation of, 2153
tensile strength, *1505*
vinyl esters
properties, **2170**
Glass fibres
advantages/limitations, **2168**
in building boards, **2216**
cost comparison, *2224*
in friction materials, **2223**
in jointings/packings, **2217**

nylons, **1618**
polycarbonates, **1808**
polyesters, **1860**
polyethylenes, **1754**
polypropylenes, **1772**, **1776**
PVC, **1674**, **1676**, **1678**
styrene based plastics, **1730**, **1735**, **1736**

V

V groove welded joints, **660**
V process, **437**
Vacuum assisted resin injection moulding
  glass fibre reinforced plastics, **2158**
Vacuum bag moulding
  fibre reinforced plastics, **2158**, **2174**
Vacuum deposition
  precious metals, 1364
  zinc alloy die-castings, 1311
Vacuum evaporation process, **537**, **539**
Vacuum forming of polymers
  advantages/limitations, **560**
  polymer types suitable, **570**, **574**
  styrene based plastics, 1721
Vacuum furnace brazing, **630**
Vacuum induction melting
  cobalt superalloys, 1338
Vacuum melted steels
  chemical composition, **1030**
  mechanical properties, **1030**
Vacuum metallisation
  polypropylenes, 1769
Vacuum moulding
  metal castings, **437**, **456**
Vacuum sintering
  powder metallurgy, 474
Vacuum-tight ceramics, **1402**, **1412**
Valve steels, 956
  advantages, **815**
  applications, **962**
  creep properties, *970*
  environmental resistance, 958
  fatigue strength, **962**
  hardness
    variation with temperature, *1010*
  leaded-fuel resistance, **962**
  limitations, **815**
  mechanical properties, **815**, 957
  physical properties, 957, **962**
  selection criteria, 957
  standardisation, 957
  strength, **815**
  tensile properties
    variation with temperature, *1011*
  thermal properties, **962**
  weldability, 958, **962**
Vamac (acrylic) rubbers
  properties, **2012**
Vanadium
  steel properties affected by, **837**, **838**, **863**,
    864, 933
  weldability of steels affected by, **658**
Vanadium pentoxide
  ceramics attacked by, **301**
Vanadium steels
  weldability of, **654**
Vapour deposition processes
  advantages/limitations, **441**, 537
  choice of process dependent on quantity,
    456
  compared with other surface coating
    processes, **539**
Vapour-deposited aluminium alloys, 1208
  properties, **1214**
Varitherm (foamed plastics) process, **2113**
Vegetable fibres
  in jointings/packings, **2217**
Vegetable glues, **694**
Vegetable oils
  polymers affected by, *1546*
Velinvar (cobalt alloy), 1355, **1356**
Velocity
  units, **795**
Vermiculite
  insulation boards using, **2215**
  temperature limits, *2224*
Verticel paper honeycomb

characteristics, **2183**, *2185*
  properties, **2184**
Vibration-damping alloys
  copper–manganese alloys, **1148**
Vibratory trimming, **447**, **457**, **517**
Vicat softening temperature
  polybutylenes, **1788**
  polyesters, **2066**
  polymers (various), **768**, **1533**
  PVK, **1876**
  styrene based plastics, **1533**, **1736**
  thermoplastic elastomers, **2066**, **2101**
Vickers hardness values
  conversion to other scales, **799**, **800**, **802**,
    803
  see also main entry: Hardness
Vinyl based plastics
  abrasion wear resistance, **397**
  film/sheet
    properties, **2134**
    water permeability, **2134**, *2142*
  see also Chlorinated poly(vinyl chloride)
    (CPVC); Poly(vinyl chloride) (PVC);
    Poly(vinyl fluoride) (PVF);
    Poly(vinylidene chloride) (PVDC)
Vinyl chloride copolymers, 1662
  applications, 1662
  film/sheet
    environmental performance, *2143*
    properties, **2134**
    water permeability, **2134**, *2142*
  molecular structure, 1662
Vinyl esters, 1970
  advantages/limitations, 1970
  applications, 1970
  carbon fibre reinforced composites
    properties, **2170**
  environmental performance, 310, **1536**
  laminated plastics
    characteristics, **2188**
    properties, **2196**
  properties, 1970
  suppliers, 1971
  trade names, 1971
Vinyl paints
  properties, **265**, **266**
Viscosity
  units, **796**
Vistal alumina, **1402**
Vistalon thermoplastic elastomer, 2088,
    2092
Visual inspection of welds, **668**
Viton fluoroelastomers
  solvent effects, **2021**
Vitreous carbon, **1396**, 1442
  properties, **1444**
Vitreous enamelling
  advantages/limitations, **538**
  compared with other methods, **260**, **539**
  suitability for
    aluminium alloys, 249, **1181**, **1200**
    cast irons, 249
    copper, 250
    steels, 249
Vitreous enamels, 249
  acid resistance, 250
  alkali resistance, 250
  area maximum, **279**
  chemical properties, 250
  colour of, 250
  corrosion resistance, 250
  electrical resistance, 250
  mechanical properties, 251
  oxidation resistance, 250, *295*
  physical properties, 250
  steel protected by, *295*
  steel thicknesses required, **279**
  thermal properties, 251
Vitreous silica, **1396**
  applications, **1450**
  properties, **1400**, **1449**
  service temperature range, *157*
Volume
  units, **796**
Volume flow rate
  units, **796**
Volume resistivity *see* Electrical resistivity
Vulcanised rubber, 1986

W

W3 (silicide) coating, 1372
W152 (cobalt superalloy), **1341**
Wallex (cobalt) alloys, **1350**
Warm working, 496
Washers (on bolts), **608**
  materials used, **610**
Waspaloy (nickel) alloys, 516, 1270, **1276**
Watch spring materials, 1355
Water
  metals attacked by, **225**
  polymers affected by, 308, 310, 1536
Water absorption
  acetals, **1652**
  acrylics, **1696**, **1698**
  alkyds, **1935**
  allylics, **1942**
  amino based polymers, **1964**
  butadiene based rubbers, **1996**
  cellulosics, **1690**
  chlorinated polyether, **1874**
  epoxy based polymers, **1910**, **1915**, **1916**,
    1960
  fluoroplastics, **1712**
  foamed plastics, **2122**
  furan based polymers, **1968**
  high-temperature polymers, **1535**, 1703,
    1852
  laminated plastics, **1915**, **2190**
  nylons, **1612**, **1618**
    properties affected, **1624**
    thickness effects, **1624**
  phenol based polymers, **1894**, **1898**, **1903**,
    1960
  phenylene based polymers, **1535**, **1820**,
    1827
  polyamides, 2099, **2101**
  polyarylamide, **1883**
  polybutadiene, **1920**
  polybutylenes, **1788**
  polycarbonates, **1808**
  polyesters, **1535**, **1860**, **1870**, **1883**, **1930**,
    2066, **2096**
  polyethylenes, **1754**, **1757**, **1798**
  polyimides, **1535**, **1852**, **1948**
  polyolefins, **1798**, **1800**, **2066**
  polypropylenes, **1772**, **1776**
  polysulphones, **1535**, **1840**
  polyurethanes, **2078**, **2080**
  silicones, **1959**, *1960*
  styrene based plastics, **1734**, **1735**, **1736**,
    1738
  test methods for plastics, **1589**
  thermoplastic elastomers, **2066**, **2078**,
    2080, **2096**, 2099, **2101**
  TPX, **1880**
  vinyl based plastics, **1677**, **1876**
Water based paints, **264**, **265**
Water film lubricated bearings, *348*
Water vapour permeability
  film/sheet plastics, **2134**, *2142*
  test methods for plastics, **1589**
Waterline attack, 221
Wax-based adhesives, **694**
Waxes
  flammability of, **1584**
  polymers affected by, 310, **1536**
Weakest-link behaviour
  ceramic materials, 1467
Wear, 321
  conformal contact, **327**, 332
  counterformal contact, **327**, 358
  surface treatments to reduce, **420**
  titanium alloys, 1245
  types, 326, **327**
Wear mechanisms, **328**
Wear properties
  acetals, 1646, **1655**, *1954*
  fluoroplastics, 1704, *1954*
  lubricated bearings, *1954*
  nylons, 1604, **1626**, *1642*, *1954*
  polycarbonates, **1810**
  polyimides, **1836**, 1945, **1947**, **1948**, *1954*
  thermoplastic elastomers, **2074**, **2077**
Wear resistance
  ceramics and glasses, **1400**
  selection based on, 15, **420**